AF844118

NANOCOMPOSITES - NEW TRENDS AND DEVELOPMENTS

Edited by **Farzad Ebrahimi**

Nanocomposites - New Trends and Developments
http://dx.doi.org/10.5772/3389
Edited by Farzad Ebrahimi

Contributors

Priscila Anadão, Alexander Pogrebnjak, Jeong Hyun Yeum, Kuldeep Singh, Anil Ohlan, Sundeep Dhawan, Jow-Lay Huang, Pramoda Kumar Nayak, Xiaoli Cui, Bahman Nasiri-Tabrizi, Abbas Fahami, Reza Ebrahimi-Kahrizsangi, Farzad Ebrahimi, Masoud Mozafari, Dongfang Yang, Vladimir Pimenovich Dzyuba, Davide Micheli, Roberto Pastore, Giorgio Giannini, Ramon Bueno Morles, Mario Marchetti, Dmitri Muraviev, Julio Bastos Arrieta, Maria Muñoz Tapia, Amanda Alonso, Tito Trindade, Ricardo J.B. Pinto, Carlos Pascoal Neto, Márcia Neves, Jun Young Kim, Seunghun Lee, Do-Geun Kim, Jong-Kuk Kim, Elangovan Thangavel, Majda Sfiligoj-Smole, Manja Kurecic, Hema Bhandari, Anoop Kumar S, Nelcy Della Santina Mohallem, Rosendo Sanjines, Cosmin Sandu

Published by InTech

Janeza Trdine 9, 51000 Rijeka, Croatia

Copyright © 2012 InTech

All chapters are distributed under the Creative Commons Attribution 3.0 license, which allows users to download, copy and build upon published articles even for commercial purposes, as long as the author and publisher are properly credited, which ensures maximum dissemination and a wider impact of our publications. However, users who aim to disseminate and distribute copies of this book as a whole must not seek monetary compensation for such service (excluded InTech representatives and agreed collaborations). After this work has been published by InTech, authors have the right to republish it, in whole or part, in any publication of which they are the author, and to make other personal use of the work. Any republication, referencing or personal use of the work must explicitly identify the original source.

Notice

Statements and opinions expressed in the chapters are these of the individual contributors and not necessarily those of the editors or publisher. No responsibility is accepted for the accuracy of information contained in the published chapters. The publisher assumes no responsibility for any damage or injury to persons or property arising out of the use of any materials, instructions, methods or ideas contained in the book.

Publishing Process Manager Romina Skomersic
Technical Editor InTech DTP team
Cover InTech Design team

Additional hard copies can be obtained from orders@intechopen.com

Nanocomposites - New Trends and Developments, Edited by Farzad Ebrahimi
p. cm.
ISBN 978-953-51-0762-0

We are IntechOpen, the first native scientific publisher of Open Access books

3,400+
Open access books available

109,000+
International authors and editors

115M+
Downloads

151
Countries delivered to

Our authors are among the
Top 1%
most cited scientists

12.2%
Contributors from top 500 universities

BOOK CITATION INDEX
WEB OF SCIENCE™

Selection of our books indexed in the Book Citation Index in Web of Science™ Core Collection (BKCI)

Interested in publishing with us?
Contact book.department@intechopen.com

Numbers displayed above are based on latest data collected.
For more information visit www.intechopen.com

For EU product safety concerns:
IN TECH d.o.o., Prolaz Marije Krucifikse Kozulić 3, 51000 Rijeka, Croatia,
info@intechopen.com or visit our website at intechopen.com.

Meet the editor

Dr. Farzad Ebrahimi was born in Qazvin, Iran, in 1979. He graduated in mechanical engineering, from the University of Tehran, Iran, in 2002. He received his Msc and PhD in mechanical engineering, with a specialization in applied design from the University of Tehran, Iran, in 2009. Since 2002, he has been working at the "Smart Materials and Structures Lab" Research Center of the faculty of mechanical engineering at the University of Tehran, where he is a researcher of smart functionally graded materials and structures. He began his university carrier as an assistant professor in the department of mechanical engineering at Imam Khomeini International University, Qazvin. He is involved in several international journals as editor and reviewer. His research interests focus on areas of smart materials and structures, plate and shell theory, vibration analysis of continuous systems, composite materials and structures, functionally graded materials and structures, finite element analysis and fracture mechanics.

Contents

Preface

Nanoscience, nanotechnology and nanomaterials have become a central field of scientific and technical activity. Over the last years the interest in nanostructures and their applications in various electronic devices, effective optoelectronic devices, bio-sensors, photodetectors, solar cells, nanodevices, plasmonic structures has been increasing tremendously. This is caused by the unique properties of nanostructures and the outstanding performance of nanoscale devices. At the nanoscale, a material's property can change dramatically. With only a reduction in size and no change in the substance itself, materials can exhibit new properties such as electrical conductivity, insulating behavior, elasticity, greater strength, different color, and greater reactivity-characteristics that the very same substances do not exhibit at the micro- or macroscale. Additionally, as dimensions reach the nanometer level, interactions at interfaces of phases become largely improved, and this is important to enhance materials properties. Composite materials are multi-phased combinations of two or several components, which acquire new characteristic properties that the individual constituents, by themselves, cannot obtain. A composite material typically consists of a certain matrix containing one or more fillers which can be made up of particles, sheets or fibers. When at least one of these phases has dimensions less than 100 nm, the material is named a nanocomposite and offers in addition a higher surface to volume ratio. These are high performance materials that exhibit unusual property combinations and unique design possibilities and are thought of as the materials of the 21st century. Nowadays, nanocomposites offer new technology and business opportunities for all sectors of industry, in addition to being environmental- friendly. A glance through the pages of science and engineering literature shows that the use of nanocomposites for emerging technologies represents one of the most active areas of research and development throughout the fields of chemistry, physics, life sciences, and related technologies. In addition to being of technological importance, the subject of nanocomposites is a fascinating area of interdisciplinary research and a major source of inspiration and motivation in its own right for exploitation to help humanity. Based on the simple premise that by using a wide range of building blocks with dimensions in the nonosize region, it is possible to design and create new materials with unprecedented flexibility and improvements in their physical properties. Nanocomposites are attractive to researchers both from practical and theoretical point of view because of combination of special properties. Many efforts have been made in the last two decades using novel nanotechnology and nanoscience knowledge in order to get nanomaterials with determined functionality. This book reports on the state of the art research and development findings on this very broad matter through original and innovative research studies exhibiting various investigation directions.

The book **"Nanocomposites- New Trends and Developments"** meant to provide a small but valuable sample of contemporary research activities around the world in this field and it is expected to be useful to a large number of researchers. Through its 19 chapters the reader will have access to works related to the theory, preparation, and characterization of various types of nanocomposites such as composites of cellulose and metal nanoparticles, polymer/ clay, polymer/Carbon and polymer-graphene nanocomposites and several other exciting topics while it introduces the various applications of nanocomposites in water treatment, supercapacitors, green energy generation, anticorrosive and antistatic applications, hard coatings, antiballistic and electroconductive scaffolds. Besides it reviews multifunctional nanocomposites, photonics of dielectric nanostructures and electron scattering in nanocomposite materials.

The present book is a result of contributions of experts from international scientific community working in different aspects of nanocomposite science and applications. The introductions, data, and references in this book will help the readers know more about this topic and help them explore this exciting and fast-evolving field. The text is addressed not only to researchers, but also to professional engineers, students and other experts in a variety of disciplines, both academic and industrial seeking to gain a better understanding of what has been done in the field recently, and what kind of open problems are in this area.

I am pleased to have had the opportunity to have served as editor of this book which contains a wide variety of studies from authors from all around the world. I would like to thank all the authors for their efforts in sending their best papers to the attention of audiences including students, scientists and engineers throughout the world. The world will benefit from their studies and insights.

I also wish to acknowledge the help given by InTech Open Access Publisher, in particular Ms. Skomersic for her assistance, guidance, patience and support.

Dr. Farzad Ebrahimi
Faculty of Engineering,
Mechanical Engineering Department,
International University of Imam Khomeini
Qazvin, I.R.IRAN

Chapter 1

Polymer/ Clay Nanocomposites: Concepts, Researches, Applications and Trends for The Future

Priscila Anadão

Additional information is available at the end of the chapter

http://dx.doi.org/10.5772/50407

1. Introduction

On 29th December 1959, the physicist Richard Feynman delivered a lecture titled "There is plenty of room at the bottom" atthe American Physical Society. Such a lecture is a landmark of nanotechnology, asFeymann proposed the use of nanotechnology to store information as well as a series of new techniques to support this technology [1]. From then on, the technological and scientific mastership ofnanometric scale is becoming stronger due to the new research tools and theoretical and experimental developments. In this scenario, the worldwide nanotechnology market, in the next five years, is expected to be ofthe order of 1 trillion dollars [2].

Regarding polymer/ clay nanocomposite technology, the first mention in the literature was in 1949 and is credited to Bower that carried out the DNA absorption by the montmorillonite clay[3]. Moreover, other studies performed in the 1960s demonstrated that clay surface could act as a polymerization initiator [4,5] as well as monomers could be intercalated between clay mineral platelets [6]. It is also important to mention that, in 1963, Greeland prepared polyvinylalcohol/ montmorillonite nanocomposites in aqueous medium [7].

However, until the early1970s, the minerals were only used in polymers as fillers commercially aiming to reduce costs, since these fillers are typically heavier and cheaper than the added polymers. During the 1970s, there was a vertiginous and successive increase in thepetroleum price during and after the 1973 and 1979 crises [8]. These facts, coupled with polypropylene introduction in commercial scale, besides the development of compounds with mica, glass spheres and fibers, talc, calcium carbonate, led to an expansion of the ceramic raw materials as fillers and initiated the research as these fillers interacted with polymers.

INTECH
open science | open minds

© 2012 Anadão; licensee InTech. This is a paper distributed under the terms of the Creative Commons Attribution License (http://creativecommons.org/licenses/by/3.0), which permits unrestricted use, distribution, and reproduction in any medium, provided the original work is properly cited.

Nevertheless, only in the late 1980swas the great landmark in the polymer clay nanocomposite published by Toyota regarding the preparation and characterization of polyamide 6/ organophilic clay nanocomposite to be used as timing belts in cars [9-11]. This new material, that only had 4.2 wt.%, had an increase of 40% in the rupture tension, 68% in the Young modulus and 126% in the flexural modulus as well as an increase in the heat distortion temperature from 65°C to 152°C in comparison with pure polymer [12]. From then on, several companies introducedthermoplastic nanocomposites, such as polyamide and polypropylene,inautomotive applications [13]. Another highlightedapplication is as gas barrier, by using polyamides or polyesters [14].

2. Definitions

2.1. Polymer/ clay nanocomposites

Polymer/ clay nanocomposites are a new class of composites with polymer matrix in which the dispersed phase is the silicate constituted by particles that have at least one of its dimensions in the nanometer range (10^{-9} m).

2.2. Clays

The mineral particles most used in these nanocomposites are the smectitic clays, as, for example, montmorillonite, saponite and hectorire [15,16]. These clays belong to the philossilicate 2:1 family and they are formed by layers combined in a sucha waythat the octadedrical layers that have aluminum are between two tetrahedrical layers of silicon (Figure 1). The layers are continuous in the *a* and *b* directions and are stacked in the *c* direction.

The clay thickness is around 1 nm and the side dimensions can vary from 30 nm to various micrometers, depending on the clay. The layer stacking by Van der Waals and weak electrostatic forces originates the interlayer spaces or the galleries. In the layers, aluminum ions can be replaced by iron and magnesium ions, as well as magnesium ions can be replaced by lithium ions and the negative charge is neutralized by the alkaline and terrous- alkalinecations that are between the clay layers. Moreover, between these layers, water molecules and polar molecules can enter this space causing an expansion in the *c* direction. This resulting surface charge is known as cation exchange capacity (CEC) and is expressed as mequiv/ 100g. It should be highlighted that this charge varies according to the layer and is considered an average value in the whole crystal [17-20].

http://dx.doi.org/10.5772/50407

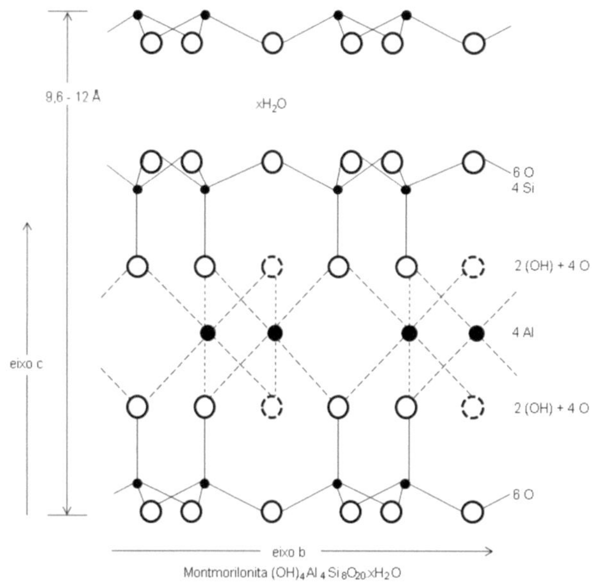

Figure 1. Schematic representation of montmorillonite.

2.3. Polymers

Polymers are constituted by largemolecules, called macromolecules, in which the atoms are linked between each other through covalent bonds. The great majority of the polymers are composed oflong and flexible chains whose rough sketch is generally made of carbon atoms (Figure 2). Such carbon atoms present two valence electrons notshared in the bonds between carbon atoms that can be part of the bonds between other atoms or radicals.

These chains are composed ofsmall repetitive units called *mero*. The origin of the word *mero*derives from the Greek word *meros,* which means part. Hence, one part is called by monomer and the word polymer means the presence of several *meros*.

When all the *meros* of the polymer are equal the polymer is a homopolymer. However, when the polymer is composed oftwo or more *meros*, the polymer is called copolymer.

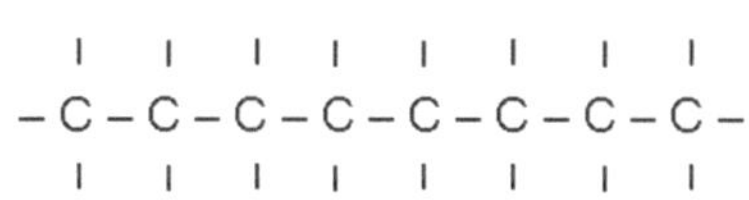

Figure 2. Representation of an organic polymer chain.

Regarding the polymer molecular structure, polymers are linear when the *meros* are united in a single chain. The ramified polymers present lateral ramifications connected to the main chain. Polymers with crossed bonds have united linear chain by covalent bonds. Network polymers have trifunctional*meros* that have three active covalent bonds, forming 3D networks (Figure 3)

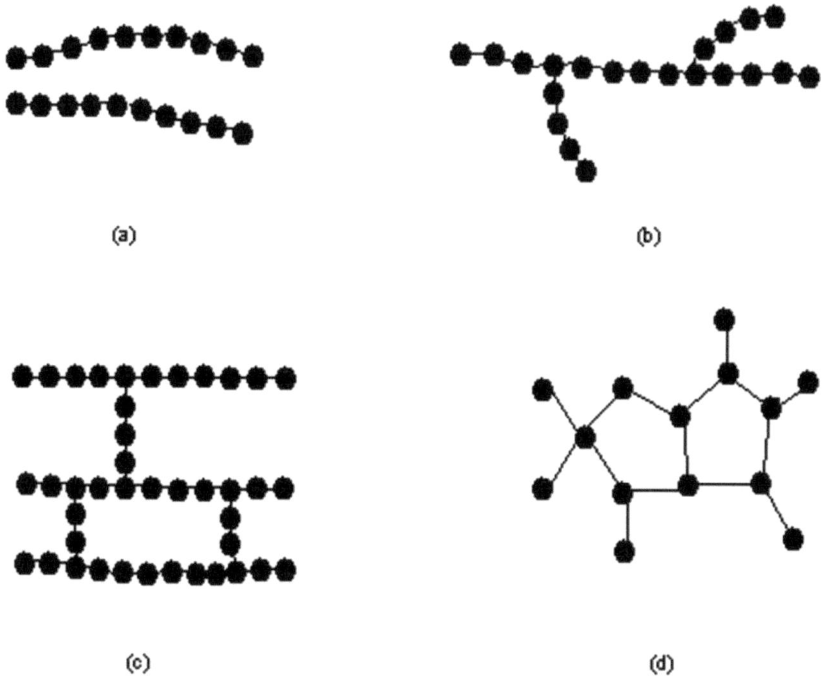

Figure 3. Schematic representation of: (a) linear, (b) ramified, (c) with crossed bonds and (d) network [21].

Polymers can be amorphous or semi-crystalline according to their structure. It is reasonable that the polymers that have a great number of radicals linked to the main chain are not able to have their molecules stacked as close as possible and, for this reason, the polymer chains are arranged in a disorganized manner, originating amorphous polymers. The polymers with linear chains and small groups are grouped in a more oriented form, forming crystals.

As a consequence of the polymer structure, there are two types of polymers: thermoplastic andthermofixes. Thermoplastic polymers can be conformed mechanically several times with reheating by the shear of the intermolecular bonds. Generally, linear and ramified polymers are thermoplastic and network polymers are thermofixes.

Thermofix polymers do not soften with temperature since there are crossed bonds in the 3D structure. Therefore, they cannot be recycled [21]

2.4. Polymer/ clay nanocomposite morphology

Depending on the interphase forces between polymer and clay, different morphologies are thermodynamically accepted (Figure4):

intercalated nanocomposite: the insertion of the polymer matrix in the silicate structure is crystalographicallyregular, alternating clay and polymer;

flocculated nanocomposites: it would be the same structure of the intercalated nanocomposite, except forthe formation of floccus due to the interaction between the hydroxile groups of the silicate;

exfoliated nanocomposites: individual clay layers are randomically separated in a continuous polymer matrix ata distance that depends on the clay charge [22,23]

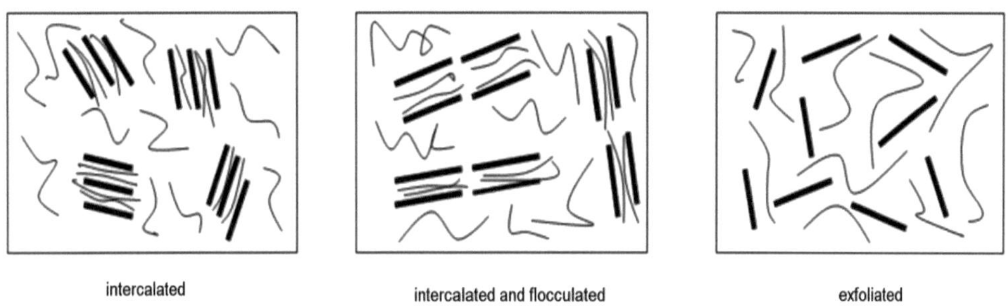

Figure 4. Polymer/ clay nanocomposites morphologies.

The formation and consequent morphology of the nanocomposites are related to entropic (ex.: molecular interactions) and enthalpic (changes in the configurations of the components) factors. Hence, efforts have been made to describe these systems. As an example, Vaia and Giannelis developed a model to predict the structure above according to the free energy variation of the polymer/ clay mixture in function of the clay layer separation.

The free energy variation, ΔH, associated to the clay layer separation and polymer incorporation is divided into two terms: the term related to the intern energy variation, ΔU, associated to the configuration changes of various components.

$$\Delta H = H(h) - H(h0) = \Delta U - T\Delta S \quad (1)$$

Where h and h_0 are the initial and final separation of the clay layers.Then, when ΔH<0, the intercalation process is favorable.

Such model presents as a limitation the separation of the configuration term, theintermolecular interactions and the separation of the entropy terms of various components.

Other mathematical models were also developed for studies of simulation of the thermodynamics of the polymer/ clay nanocomposites. These models consider the nanocomposite thermodynamics and architecture, the interaction between clay and polymer to the free energy and the polymer and clay conformation.

Specifically for polyamide 6 and 66/ clay nanocomposites, the study of the molecular dynamics was employed, which uses the bond energy between the various components that composes the nanocomposite.

The kinetics of polymer/ clay nanocomposite formation is also a very important issue to predict the resulting nanocomposite. Studies of the molecular dynamics were also employed to understand the system kinetics. Other mathematical models were also used to describe the system kinetics, but kinetics is less understood than thermodynamics.

There is still the needof developing models that are explored in individual time and length scales, besides the integration of concepts that permeate from smaller to larger scales, that is, in the quantum, molecular, mesoscopic and macroscopic dominium [24].

2.5. Preparation methods of polymer/ clay nanocomposite

Three methods are widely used in the polymer/ clay nanocomposite preparation. The first one is the *in situ*polymerization in which a monomer is used as a medium to the clay dispersion and favorable conditions are imposed to carry out the polymerization between the clay layers. As clay has high surface energy, it performs attraction by the monomer units to the inside of the galleries until equilibrium is reached and the polymerization reactions occur between the layers with lower polarity, dislocating the equilibrium and then, aiming at the diffusion of new polar species between the layers.

The second method is solution dispersion. Silicate is exfoliated in single layers by using a solvent in which the polymer or pre-polymer is soluble. Such silicate layers can be easily dispersed in a solvent through the entropy increase due to the disorganization of the layers that exceed the organizational entropy of the lamellae. Polymer is then sorved in the delaminated layers and when the solvent is evaporated, or the mixture is precipitated, layers are reunited, filled with the polymer.

Moreover, there is also the fusion intercalation, amethod developed by Vaia et al. in 1993 [25]. In this method, silicate is mixed with a thermoplastic polymer matrix in its melted state. Under these conditions, the polymer is dragged to the interlamellae space, forming a nanocomposite. The driving force in this process is the enthalpic contribution of the interactions between polymer and clay.

Besides these three techniques, the use of supercritical carbon dioxide fluids and sol-gel technology can also be mentioned [26].

3. Polymer and clay modifications to nanocomposite formation

As explained before, the great majority of polymers are composed of a carbon chain and organic groups linked to it, thus presentinga hydrophobic character. On the other hand, clays are generally hydrophilic, making them, at a first view, not chemically compatible. Aiming to perform clay dispersion and polymer chains insertion, it is necessary to modify these materials.

There are two possibilities to form nanocomposites: clay organomodification that will decrease clay hydrophilicity and the use of a compatibilizing agent in the polymer structure, by grafting, to increase polarity. The concepts that govern each of these modifications will be explored in this chapter.

3.1. Clay organomodification

This method consists in the interlamellae and surface cation exchange (generally sodium and calcium ions) by organic molecules that hold positive chains and that will neutralize the negative charges from the silicate layers, aiming to introduce hydrophobicity and then, producing an organophilic clay. With this exchange, the clay basal space is increased and the compatibility between the hydrophilic clay and hydrophobic polymer. Therefore, the organic cations decrease surface energy and improve the wettability by the polymer matrix.

The organomodification, also called as organophilization, can be reached through ion exchange reactions. Clay is swelled with water by using alkali cations. As these cations are not structural, they can be easily exchanged by other atoms or charged molecules, whichare called exchangeable cations.

The greaterdistance between the silicate galleries due to the size of the alquilammonium ions favor polymer and pre-polymer diffusion between the galleries. Moreover, the added cations can have functional groups in their structure that can react with the polymer or even begin the monomer polymerization. The longerthe ion chain is and the higher the charge density is, the greaterthe clay layer separation will be [4,11].

3.2. Use of a compatibilizing agent

Generally, a compatibilizing agent can be a polymer which offers a chemically compatible nature with the polymer and the clay. By a treatment, such as the graftization of a chemical element that has reactive groups, or copolymerization with another polymer which also has reactive groups, compatibility between the materials will form the nanocomposite. Amounts of the modified polymer are mixed with the polymer without modification and the clay to prepare the nanocomposites.

Parameters such as molecular mass, type and content of functional groups, compatibilizing agent/ clay proportion, processing method, among others, should be considered to have compatibility between polymer and clay. Maleic anidride is the organic substance most used to compatibilize polymer, especially with the polyethylene and polypropylene, since the polar character of maleic anidride results in favorable interactions, creating a special affinity with the silicate surfaces [27,28].

4. The most important polymers employed in polymer/ clay nanocomposites

In this item, examples of studies about the most important polymers that are currently employed in the polymer/ clay nanocomposite preparation will be presented. Fora better understanding, polymers are divided into general-purpose polymers, engineering plastics, conductive polymers and biodegradable polymers.

4.1. General-purposepolymers

General-purpose polymers, also called commodities, represent the majority of the total worldwide plastic production. These polymers are characterized by being used in low-cost applications due to theirprocessing ease and low level of mechanical requirement. The formation of nanocomposites is a way to addvalue to these commodities.

4.1.1. Polyethylene (PE)

PE is one of the polymers that most present scientific papers related to nanocomposite formation. Maleic anidride grafted PE/ Cloisite 20A nanocomposites were prepared by two techniques: fusion intercalation and solution dispersion. Only the nanocomposites produced by the first method produced exfoliated morphology. The LOI values, related to the material flammability, were lower in all composites and were highly reduced in the exfoliated nanocomposites due to the high clay dispersion [29].

Another work presented the choice of a catalyzer being supported on the clay layers that are able to promote *in situ* polymerization, besides exfoliation and good clay layer dispersion. The organophilic clays (Cloisite 20A, 20B, 30B and 93A) were used as a support to the Cp_2ZrCl_2 catalyzer. The higher polymerization rate was obtained with Cloisite 93A and the clay layers were dispersed and exfoliated in the PE matrix [30].

4.1.2. Polypropylene (PP)

Rosseau et al. prepared maleic anidride grafted PP/ Cloisite 30B nanocomposites by water assisted extrusion and by simple extrusion. The use of water improved clay delamination dispersion and, consequently, the rheological, thermal and mechanical properties [29].

The use of carbon dioxide in the extrusion of PP/ Cloisite 20A nanocomposites enabled a higher separation between the clay layers. The use of clay at lower contents in the foam formation also suppressed the cell coalescence, demonstrating that the nanocomposite is also favorable to produce foams [31].

4.1.3. PVC

The use of different clays (calcium, sodium and organomodified montmorillonite, aluminum magnesium silicate clay and magnesium lithium silicate clay) was studied in the preparation of rigid foam PVC nanocomposites. Although the specific flexure modulus and the density have been improved by the nanocomposite formation, the tensile strength and modulus have their values decreased in comparison with pure PVC [32].

4.2. Engineering plastics

Engineering plasticsare materials that can be used in engineering applications, as gear and structural parts, allowing the substitution of classic materials, especially metals, due to their superior mechanical and chemical properties in relation to the general-purpose polymers [33]. These polymers are also employed in nanocomposites aiming to explore their properties to the most.

4.2.1. Polyamide (PA)

Among all engineering plastics, this is the polymer that presents the highest number of researches involving the preparation of nanocomposites. PA/ organomodified hectorite nanocomposites were prepared by fusion intercalation. Advanced barriers properties were obtained by increasing clay content [34]. The flexure fatigue of these nanocomposites were studied in two environments: air and water. It was observed that the clay improved this property in both environments [35].

4.2.2. Polysulfone (PSf)

PSf/ montmorillonite clay nanocomposite membranes were prepared by using solution dispersion and also the method most employed in membrane technology, wet-phase inversion. A hybrid morphology (exfoliated/ intercalated) was obtained, and itsdispersion was efficient to increase a barrier to volatilization of the products generated by heat and, consequently, initial decomposition temperature. By the strong interactions between

polymers and silicate layers, the tensile strength was increased and elongation at break was improved by the rearrangement of the clay layers in the deformation direction. Furthermore, hydrophobicity was also increased,so that membranes couldbe used in water filtration operations, for example [36].

4.2.3. Polycarbonate (PC)

By in situ polycondensation, PC/ organophilic clay exfoliated nanocomposites were prepared. Although exfoliated nanocomposites were produced, transparency was not achieved [37].

4.3. Conductive polymers

Conductive polymers, also called synthetic metals, have electrical, magnetic and optical properties that can be compared to thoseof the semiconductors. They are also called conjugated polymers, since they have conjugated C=C bonds in their chains which allow the creation of an electron flux in specific conditions.

The conductive polymer conductivity is dependent on the polymer chains ordering that can be achieved by the nanocomposite formation.

4.3.1. Polyaniline (PANI)

PANI is the most studied polymer in the polymer/ clay nanocomposite technology. Exfoliated nanocomposites wereprepared with montmorillonite which contained transition by *in situ* polymerization. The thermal stability was improved in relation to the pure PANIduethe fact thatthe clay layers acted as a barrier towards PANI degradation [38].

4.3.2. Poly(ethylene oxide) (PEO)

PEO nanocomposites werepreparedwiththreetypes of organophilicclays (Cloisite 30B, Somasif JAD400 e Somasif JAD230) by fusion intercalation. The regularity and spherulites size of the PEO matrix were altered by only using Cloisite 30B. An improvement in the storage modulus of the other nanocomposites was not observed since the spherulites were similar in the other nanocomposites [39].

4.4. Biodegradable polymers

Biodegradable polymers are those that, under microbial activity, have their chains sheared. To have the polymer biodegradabilization, specific conditions, such as pH, humidity, oxygenation and the presence of some metals were respected. The biodegradable polymers can be made from natural resources, such as corn; cellulose can be produced by bacteria from molecules such as butyric, and valeric acid which produce polyhydrobutirate and polyhydroxivalerate or even can derive from petroleum; or fromthe biomass/ petroleum mixture, as the polylactides [40].

4.4.1. Polyhydroxibutirate (PHB)

The PHB disadvantages are stiffness, fragility and low thermal stability and because of this, improvements should be performed. One of the ways to improve these properties is by preparing nanocomposites.

PHB nanocomposites were prepared with the sodium montmorillonite and Cloisite 30B by fusion intercalation. A better compatibility between clay and polymer was established by using Cloisite 30 B and an exfoliated/ intercalated morphology was formed. Moreover, an increase in the crystallization temperature and a decrease in the spherulite size were also observed. The described morphology was responsible for increasing the Young modulus [41]. Besides that, thermal stability was increased in PHB/ organomodified montmorillonite in comparison with pure PHB [42].

5. Polymer/ Clay nanocomposite applications, market and future directons

Approximately 80% of the polymer/ clay nanocomposites is destined to the automotive, aeronautical and packaging industry.

The car part industry pioneered in the use of polymer/ clay nanocomposites, since these nanocomposites present stiffness and thermal and mechanical resistances able to replacemetals, and its use in car reduces powerconsumption. Moreover, its application is possible due to the possibility of being painted together with other car parts, as well as of undergoing the same treatments as metallic materials in vehicle fabrication.

General Motors was the first industry to use nanocomposites in car, reducing its mass byalmost one kilogram. In the past, car parts weremade of polypropylene and glass fillers, which hadthe disharmony with the other car partsas a disadvantage. By using lower filler content, as in the case of the nanocomposites, materials with a higher quality are obtained, as is the case of the nanoSeal™, which can be used in friezes, footboards, station wagon floors and dashboards. Basell, Blackhawk, Automotive Plastics, General Motors, Gitto Global produced polyolefines nanocomposites with, for example polyethylene and polypropylene, to be used in footboards of the Safari and Astro vehicles produced by General Motors.

Car parts, such as handles, rear view mirror, timing belt, components of the gas tank, engine cover, bumper, etc. also used nanocomposites, specially with nylon (polyamide), produced by the companies Bayer, Honeywell Polymer, RTP Company, Toyota Motors, UBE and Unitika.

In the packaging industry, the superior oxygen and dioxide carbon barrier properties of the nylon nanocomposites have been used to produce PET multilayer bottles and films for food and beverage packaging.

In Europe and USA, nanocomposites are used in soft drink and alcoholic beverage bottles and meat and cheese packaging since these materials present an increase in packaging flexibility and tear resistance as well as a humidity control.

Nanocor produced Imperm, a nylon MDXD6/ clay nanocomposite used as a barrier in beer and carbonated drink bottles, in meat and cheese packaging and in internal coating of juice and milk byproduct packaging. The addition of 5% of Imperm in PET bottles increase the shelf time bysix months and reduce the dioxide carbon lossto less than 10%.

Another commercial products can be cited, as for example the M9™, produced by the Mitsubish Gas Chemical Company, for application in juice and beer bottles and multilayer films; Durethan KU2-2601, a polyamide 6 nanocomposite produced by Bayer for coating juice bottles as barrier films and AEGIS™ NC which is polyamide 6/ polyamide nanocomposites, used as barrier in bottles and films, produced by Honeywell Polymer.

In the energy industry, the polymer nanocomposites positively affect the creation of forms of sustainable energy by offering new methods of energy extraction from benign and low-cost resources. One example is the fuel cell membranes; other applications include solar panels, nuclear reactors and capacitors.

In the biomedical industry, the flexibility of the nanocomposites is favorable, which allows their use in a wide range of biomedical applications as they fill several necessary premises for application in medical materials such as biocompatibility, biodegradability and mechanical properties. For this reason and forthe fact of being finely modulated by adding different clay contents, they can be applied in tissue engineering – the hydrogel form, in bone replacement and repair, in dental applications and in medicine control release.

Moreover, there is the starch/ PVA nanocomposite, produced by Novamont AS (Novara, Italy) that can replace the low density PE films to be used as water-soluble washing bags due to its good mechanical properties.

Other commercial applications include cables due to the slow burning and low released heat rate; the replacementof PE tubes withpolyamide 12 nanocomposites (commercial name SET™), produced by Foster Corporation and in furniture and domestic appliances withthe nanocomposite with brand name Forte™ produced by Noble Polymer.

Table 1 presents a summary of the application areas and products in which polymer/ clay nanocomposites are used.

The consumption of polymer/ clay nanocomposites was equal to 90 million dollars with a consumption of 11,300 ton in 2005. In 2011, a consumption of 71,200 ton was expected,corresponding to 393 million dollars.

The scenario that correspond to the areas in which polymer/ clay nanocomposite was used in 2005 is presented in Figure 5. By the end of 2011, the barrier applications were expected to exceed the percentage related to car parts.

In a near future, the PP nanocomposites produced by Bayer are expected to replace car parts that use pure PP and the PC nanocomposites produced by Exaltec are expected to be used in car glasses due to an improved abrasion resistance without loss of optical transparency.

Automotive	Packaging	Energy	Biomedical	Construction	Home furnishings
-footboards, -friezes, - station wagon floors, - dashboards, -timing belts, -handle, -gas tank components, -engine covers, - bumpers.	- beer and soft drink bottles, -meat and cheese packaging, -internal films of juice boxes,	-fuel cells, -lithium batteries, - solar panels - nuclear reactors, -capacitors.	-artificial tissues; -dental and bone prosthesis, -medicines.	-tubes, - cords.	-furniture, -home appliances.

Table 1. Application areas and products that use polymer/ clay nanocomposites.

The research about the application of these nanocomposites in car parts is still being developed since a reduction in the final car mass corresponds to benefits to the environment. The large use of nanocomposites would reduce 1.5 billion liters of gasoline a year and the CO_2 emission in more than 5 billion kilograms.

Another thriving field is the barrier applications, the use of which can increase food shelf life besides maintaining film transparency. As an example, by using Imperm nanocomposite in a Pet bottle, beer shelf life is increased to 28.5 weeks.

Great attention has been also paid to the biodegradable polymers which present a variety of applications. Moreover, another potential application is in nanopigment as an alternative to cadmium and palladium pigments which presenthigh toxicity.

The distant future of the applications of polymer/ clay nanocomposites is dependent on the results obtained from researches, commercial sectors, existing markets and the improvement level of the nanocomposite properties. Furthermore, the relevance of their application in large scale, the capital to be invested, production costs and the profits should be taken into account.

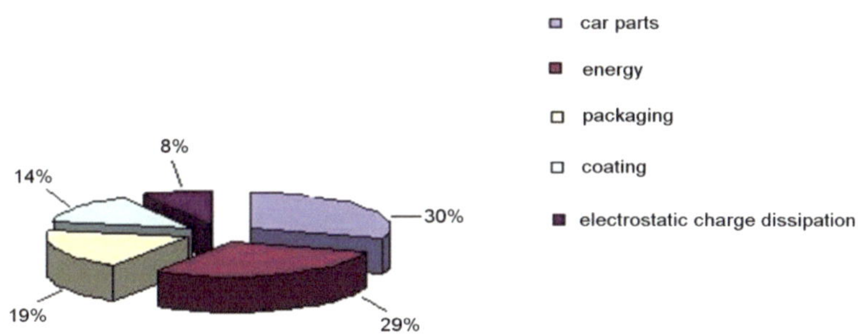

Figure 5. Applications of polymer/ clay nanocomposites in 2005.

Due to the aforementioned reasons, a considerable increase in investigations and the commercialization of nanocomposites in the packaging area, selective catalyzers, conductive polymers and filtration of toxic materials are expected. A light growth in the applications related to an increase of catalysis efficient and of material conductivity, new types of energy, storage information and improved membranes are also expected.

Although nanocomposites present a series of advanced properties, their production is still considered low in comparison with other materials due to the production costs. Once they become cheaper, polymer/ clay nanocomposites can be largely used in a series of applications [11, 43-45].

Author details

Priscila Anadão

Polytechnic School, University of São Paulo, Brazil

References

[1] Zyvez, R. (2012). Feynman's lecture. *There is plenty of room in the bottom.*, http://www.zyvex.com/nanotech/Feynman.html.

[2] *National Science Foundation. Investments in nanotechnology*, http://www.nsf.gov.

[3] Bower, C. A. (1949). Studies on the form and availability of organic soil phosphorous. IOWA Agriculture Experiment Station Research Bulletin; , 362-39.

[4] Ajayan, P. M. (2000). Nanocomposite. *Science and technology*, Weinheim, Wiley-VCH Verlag.

[5] Kollman, R. C. (1960). Clay complexes with conjugated unsaturated aliphatic compounds of four to five carbon atoms. National Lead Co.. US Patent 2951087.

[6] Gómez, Romero. P., & Sanchez, C. (2004). Functional Hybrid Materials. Weinheim: Wiley-VCH Verla.

[7] Greenland, D. J. (1963). Adsorption ofpoly(vinyl alcohols) by montmorillonite. Journal of Colloid Science ., 18-647.

[8] Rabello, M. S. (2000). *Aditivação de Polímeros*, São Paulo, Artliber.

[9] Okada, A., et al. (1988). Composite material and process for manufacturing same. *Kabushiki Kaisha Toyota Chou Kenkyusho.*, US Patent 4739007.

[10] Kawasumi, M., et al. (1989). Process for producing composite material. .Kabushiki Kaisha Toyota Chuo Kenkyusho. US Patent 4810734

[11] Pavlidou, S., & Papaspyrides, C. D. (2008). A review on polymer-layered silicate nanocomposite. Progress in Polymer Science ., 33-1119.

[12] Nguyen, Q. T., & Baird, D. G. (2006). Preparation of polymer-clay nanocomposites and their properties. Advances in Polymer Technology ., 25-270.

[13] Patterson, T. (2004). Nanocomposites- our revolutionary breakthrough. . In: 4th World Congress in Nanocomposites San Francisco, USA.

[14] Goldman, A. Y., & Copsey, C. J. (2004). Multilayer barrier liner material with nanocomposites for packaging applications. San Francisco, USA. 4th *World Congress in Nanocomposites,*.

[15] Alexandre, M., & Dubois, P. (2000). Polymer-layered silicate nanocomposites: preparation, properties and uses of a new class of materials. *Materials Science &Engineering R,*, 28-1.

[16] Esteves, A. C. C., Barros, Timmons. A., & Trindade, T. (2004). Nanocompósitos de matriz. polimérica: estratégias de síntese de matérias híbridos. *Química Nova*, 27-798.

[17] Brigatti, M. F., Galan, E., & Theng, B. K. G. (2006). Structures and mineralogy of clay minerals. *Handbook of Clay Science.*, Amsterdam, Elsevier, 19-86.

[18] Ke, Y. C., & Stroeve, P. (2005). Polymer-Layered Silicate and Silica. *Nanocomposites*, Amsterdam, Elsevier B.V.

[19] [19], Mittal. V. (1992). Polymer layered silicate nanocomposites. : A Review. Materials ., 2, 992-1057.

[20] Ray, S. S. (2005). Bousmina M. Biodegradable polymers and their layered silicate nanocomposites: In greening the 21 st century materials world. . Progress in Materials Science ., 50, 962-1079.

[21] Callister, W. D. (1994). Materials Science and Engineering: an introduction. New York John Wiley & Sons.

[22] Yeh, J. M., & Chang, K. C. (2008). Polymer/ layered silicate nanocomposite anticorrosive coatings. Journal of Industrial and Engineering Chemistry ., 14-275.

[23] Zaarei, D., Sarabi, A. A., Sharif, F., & Kassiriha, S. M. (2008). Structure, properties and corrosion resistivity of polymeric nanocomposite coatings based on layered silicates. . Journal of Coatings Technology and Research ., 5-241.

[24] Zeng, Q. H., Yu, A. B., & Lu, G. Q. (2008). Multiscale modeling and simulation of polymer nanocomposites. . Progress in Polymer Science ., 33191-269.

[25] Vaia, R. A., Ishii, H., & Giannelis, E. P. (1993). Synthesis and properties of two-dimensional nanostructures by direct intercalation of polymer melts in layered silicates. Chemistry of Materials ., 5-1694.

[26] Anadão, P., Wiebeck, H., & Díaz, F. R. V. (2011). Panorama da pesquisa acadêmica brasileira em nanocompósitos polímero/ argila e tendências para o futuro. Polímeros:. Ciência e Tecnologia ., 21-443.

[27] Mirzadeh, A., & Kokabi, M. (2007). The effect of composition and draw-down ratio on morphology and oxygen permeability of polypropylene nanocomposite blown films. .European Polymer Journal ., 43-3757.

[28] Lee, J. H., Jung, D., Hong, C. E., Rhee, K. Y., & Advani, S. G. (2005). Properties of polyethylene-layered silicate nanocomposites prepared by melt intercalation with a PP-g-MA compatibilizer. . Composites Science and Technology ., 65-1996.

[29] Minkova, L., & Filippi, S. (2011). Characterization of HDPE-g-MA/clay nanocomposites prepared by different preparation procedures: Effect of the filler dimension on crystallization, microhardness and flammability. Polymer Testing ., 30-1.

[30] Maneshi, A., Soares, J. B. P., & Simon, L. C. (2011). Polyethylene/clay nanocomposites made with metallocenes supported on different organoclays. Macromolecular Chemistry and Physics; ., 212-216.

[31] Zheng, W. G., Lee, Y. H., & Park, C. B. (2010). Use of nanoparticles for improving the foaming behaviors of linear PP. . Journal of Applied Polymer Science ., 117-2972.

[32] Alian, A. M., & Abu-Zahra, M. H. Mechanical properties of rigid foam PVC-clay nanocomposites. Polymer-Plastics. Technology and Engineering (20094). ., 48-1014.

[33] Mano, E. B. (2000). *Polímeros como materiais de engenharia.*

[34] Timmaraju, M. V., Gnanamoorthy, R., & Kannan, K. (2011). Influence of imbibed moisture and organoclay on tensile and indentation behavior of polyamide 66/ hectorite nanocomposites. . Composites Part B: Engineering ., 42-466.

[35] Timmaraju, M. V., Gnanamoorthy, R., & Kannan, K. (2011). Effect of environment on flexural fatigue behavior of polyamide 66/hectorite nanocomposites. International Journal of Fatigue ., 33-541.

[36] Anadão, P., Sato, L. F., Wiebeck, H., & Díaz, F. R. V. (2010). Montmorillonite as a component of polysulfone nanocomposite membranes. . Applied Clay Science , 48-127.

[37] Rama, [., Swaminathan, M. S., & , S. (2010). Polycarbonate/clay nanocomposites via in situ melt polycondensation. Industrial & Engineering Chemistry Research ., 49-2217.

[38] Narayanan, B. N., Koodathil, R., Gangadharan, T., Yaakob, Z., Saidu, F. K., & Chandralayam, S. (2010). Preparation and characterization of exfoliated polyaniline/montmorillonite nanocomposites. Materials Science and Engineering B ., 168-242.

[39] Abraham, T. N., Siengchin, S., Ratna, D., & Karger, Kocsis. J. (2010). Effect of modified layered silicates on the confined crystalline morphology and thermomechanical properties of poly(ethylene oxide) nanocomposites. . Journal of Applied Polymer Science ., 118-1297.

[40] Ray, S. S., & Bousmina, M. (2005). Biodegradable polymers and their layered silicate nanocomposites. : In Progress In Materials Science ., 50-962.

[41] Botana, A., Mollo, M., Eisenberg, P., & Zanchez, R. M. T. (2010). Effect of modified montmorillonite on biodegradable PHB nanocomposites. . Applied Clay Science ., 47-263.

[42] Achilias, D. S., Panayotidou, E., & Zuburtikudis, I. (2011). Thermal de gradation kinetics and isoconversional analysis of biodegradable poly (3 -hydroxybutyrate)/organomodified montmorillonite nanocomposites. Thermochimical Acta ., 514, 58-66.

[43] Leaversuch, R. (2001). Nanocomposites broaden roles in automotive, barrier packaging. . Plastics Technology ., 47-64.

[44] Buchholz, K. (2001). Nanocomposites debuts on GM vehicles. . Automotive Engineering International ., 109-56.

[45] Tang, X., & Alavi, S. (2011). Recent advances in starch, polyvynil alcohol based polymer blends, nanocomposites and their biodegradability. *Carbohydrate Polymers*, 85-7.

Chapter 2

Carbon Nanotube Embedded Multi-Functional Polymer Nanocomposites

Jeong Hyun Yeum, Sung Min Park, Il Jun Kwon,
Jong Won Kim, Young Hwa Kim,
Mohammad Mahbub Rabbani, Jae Min Hyun,
Ketack Kim and Weontae Oh

Additional information is available at the end of the chapter

http://dx.doi.org/10.5772/50485

1. Introduction

Polymer nanocomposites represent a new alternative to conventionally filled polymers which have significant commercial potential. Polymer nanocomposites are a class of materials in which nanometer scaled inorganic nanomaterials are dispersed in an organic polymer matrix in order to improve the structures and properties of the polymers effectively. An advanced morphologies and improved properties are expected from the polymer nanocomposite materials due to the synergetic effect of the comprising components which could not be obtained from the individual materials. The incorporation of a small amount of inorganic materials such as metal nanoparticles, carbon nanotubes (CNTs), clay into the polymer matrix significantly improve the performance of the polymer materials due to their extraordinary properties and hence polymer nanocomposites have a lot of applications depending upon the inorganic materials present in the polymers [34; 41; 58; 63].

There are many types of nanocomposites such as polymer/inorganic particle, polymer/polymer, metal/ceramic, and inorganic based nanocomposites which have attracted much interest to the scientists [59]. These types of polymer nanocomposites have diverse field of applications such as optics, electrical devices, and photoconductors, biosensors, biochips, biocompatible thin coatings, biodegradable scaffolds, drug delivery system and filter systems [81; 29; 30; 35; 46; 49; 51].

INTECH
open science | open minds

© 2012 Yeum et al.; licensee InTech. This is a paper distributed under the terms of the Creative Commons Attribution License (http://creativecommons.org/licenses/by/3.0), which permits unrestricted use, distribution, and reproduction in any medium, provided the original work is properly cited.

There are so many methods to produce polymer nanocomposites such as simple mixing of required inorganic materials with polymers [38], in-situ polymerization of monomers inside the galleries of the inorganic host [31], melt intercalation of polymers [53; 54] etc. On the other hand, to blend polymers directly with inorganic materials, microwaves, latex-colloid interaction, solvent evaporation, spray drying, spraying a polymer solution through a small orifice and Shirasu Porous Glass (SPG) membrane emulsification technique are employed [1; 7; 33; 36; 44; 59].

Electrospinning is one of the most important techniques for preparing polymer nanocomposites nanofibers that has attracted great interest among academic and industrial scientists. Electrospinning is a very simple, low cost, and effective technology to produce polymer nanocomposite nanofibers which have exhibited outstanding physicochemical properties such as high specific surface area, high porosity and resistance against microorganism. These nanofibers are widely used as separation filters, wound dressing materials, tissue engineering, scaffold engineering, drug delivery, sensors, protective clothing, catalysis reaction, etc. [3; 16; 19; 26; 28; 32; 38; 43; 47; 55; 56; 57; 59; 64]. Electrospraying is as the same as electrospinning and widely used to prepare polymer nanocomposite nanoparticles. The main distinguishable characteristics between electrospinning and electrospraying is the solution parameter that is low concentrated polymer solution is used during electrospraying.

Suspension polymerization is also another synthetic method to produce a whole range of polymer/inorganic nanocomposites. It is low cost, effective, and easy to manipulate and control particle size. In suspension polymerization technique there are some variables which have great effect on the polymerized microspheres. These variables include the type and amount of initiator and suspending agent, the polymerization temperature, the monomer to water ratio, and the agitation speed [14; 11; 17; 18; 25].

Fabrication of polymer nanocomposites with various morphologies by using different technique such as, electrospinning, electrospraying, and in-situ suspension polymerization has been discussed in this article. Inorganic nanomaterials such as, carbon nanotube (CNTs), gold (Au) and silver (Ag) nanoparticles, and inorganic clay, montmorillonite (MMT), were incorporated within the polymer, poly (vinyl alcohol) (PVA), matrix using the method mentioned above. These nanocomposites were characterized by field emission-type scanning electron microscope (FE-SEM), transmission electron microscopy (TEM), optical microscopy, and differential scanning calorimetry (DSC). The anti-bacterial performance of polymer nanofibers was also discussed.

2. Backgrownd

Inorganic nano-structured materials and their nano-composites have potential applications in microelectronics, optoelectronics, catalysis, information storage, textile, cosmetics and biomedicine. For instance, TiO_2, silver, gold, carbon nanotubes (CNTs), nano-clay and their nanocomposites are widely used in diverse fields for their anti-microbial, UV protecting, photo-catalyst, electrical conductive and flame retardant characteristics [4; 5; 6; 10; 15; 22; 39; 48; 52; 62].

Semi-crystalline structure, good chemical and thermal stability, high biocompatibility, non-toxicity, and high water permeability have made poly(vinyl alcohol) (PVA) the promising candidate for a whole range of applications especially in the medical, cosmetic, food, pharmaceutical and packaging industries [24; 27; 28; 42]. The outstanding physicochemical properties and unique structures of carbon nanotubes (CNTs) have made them attractive material for a whole range of promising applications such as supports for inorganic nanomaterials, central elements in electronic devices, building blocks for the fabrication of advanced nano devices and catalyst. They also have anti-microbial activity [39; 22].

Metal nanoparticles have potential application in diverse field of modern science [6]. Gold nanoparticles have novel biomedical applications for their anti-bacterial, anti-fungal, and electrical conductive characteristics. Antibacterial effectiveness against acne or scurf and no tolerance to the antibiotic have caused their commercial usage in soap and cosmetic industries [5; 15; 37; 60; 62]. Excellent structure depended physicochemical properties of silver nanoparticles have expanded their potential applications such as a photosensitive components, catalysts, chemical analysis, antibacterial and disinfectant agents. Silver nanoparticles have excellent resistance against microorganisms. Introducing Ag nanoparticles into polymer matrix improve the properties and expand the applications of polymer nanocomposites [6; 13; 38; 45; 59].

As an inorganic materials, MMT has been widely used in polymer nanocpomosites to improve their mechanical, thermal, flame-retardant, and barrier properties. A small amount of MMT is effective enough to promot preformance of polymer composites. It is regularly used for packaging and biomedical applications [9; 38; 50].

3. Experimental

3.1. Materials

PVA with P_n (number–average degree of polymerization) = 1,700 [fully hydrolyzed, degree of saponification = 99.9%] was collected from DC Chemical Co., Seoul, Korea. MMT was purchased from Kunimine Industries Co., Japan. Hydrogen tetrachloro aurate trihydrate ($HAuCl_4.3H_2O$), tetra-n-octylammonium bromide (TOAB), sodium borohydride ($NaBH_4$), 4-(dimethylamino)pyridine (DMAP), polyvinylpyrrolidone (PVP, Mw = 10,000) were purchased from Sigma–Aldrich, toluene from Junsei, MWNT (CM-95) from ILJIN Nanotech Co. Ltd., and aqueous silver nanoparticle dispersion (AGS-WP001; 10,000 ppm) with diameters ca.15–30 nm was purchased from Miji Tech., Korea. All of these chemicals were used as recieved. Gold (Au) nanoparticles were synthesized following the method described elsewhere by reducing gold salt between water/toluene interfaces and stabilized by TOAB in toluene. Finally to obtain highly polarized Au nanoparticles, an aqueous 0.1M DMAP solution was added to the as-made Au nanoparticles of the same volume [2; 12]. Doubly distilled water was used as a solvent to prepare all the solutions. Vinyl acetate (VAc) purchased from Aldrich was washed with aqueous $NaHSO_3$ solution and then water and dried with anhydrous $CaCl_2$, followed by distillation in nitrogen atmosphere under a reduced pressure. The initiator, 2,2′-azobis(2,4-dimethylvaleronitrile) (ADMVN) (Wako) was recrystallized twice in methanol before use [21] PVA with a number-average molecular weight of 127,000 and a degree of saponification of 88% (Aldrich) was used as a suspending agent.

3.2. Electrospinning nanocomposite nanofibers

The electrospinning was performed following our previous work [38]. Our group has optimized the best condition to make PVA blend nanofiber such as polymer concentration, electric voltage applied to create Taylor cone of polymer solutions, tip-collector distance (TCD), and solution flow rate etc. [20; 23; 26; 27; 38]. The polymer blend solution was contained in a syringe. During electrospinning, a high voltage power (CHUNGPA EMT Co., Korea) was applied to the polymer solution by an alligator clip attached to the syringe needle. The applied voltage was adjusted to 15 kV. The solution was delivered through the blunt needle tip by using syringe pump to control the solution flow rate. The fibers were collected on an electrically grounded aluminum foil placed at 15 cm vertical distance to the needle tip. The electrospinning process is shown schematically in Figure 1.

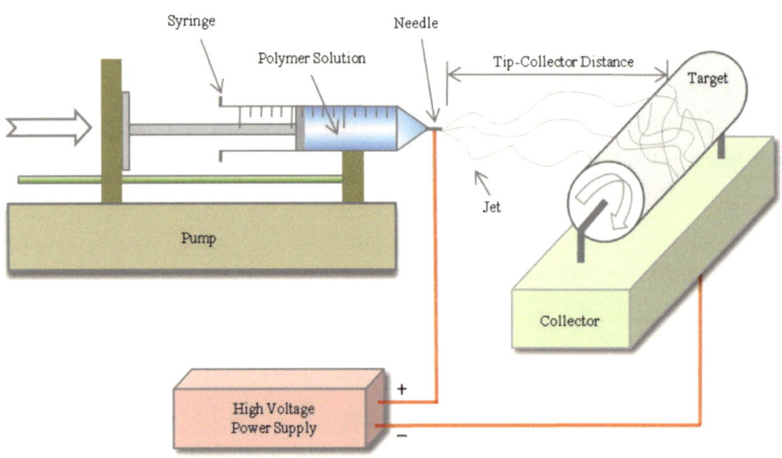

Figure 1. Schematic representation of electrospinning process

3.3. Electrospraying nanocomposite nanoparticles and nanosphere

The principle and apparatus setting of electrospraying and electrospinning techniques is the same. The most important variable distinguishing electrospraying and electrospinning is solution parameter such as polymer molecular weight, concentration and viscosity, etc. Our group has optimized the suitable conditions for electrospraying to prepare nanoparticles and nanosphere. During electrospraying 15-30 kV power was applied to the PVA solution to fabricate PVA/MWNT nanoparticles and PVA/MWNT/Ag nanospheres and the solution concentration was fixed at 5 wt% of PVA, 1 wt% of MWNTs and 1 wt % of Ag nanoparticles. The nanoparticles and nanospheres were collected on an electrically grounded aluminum foil placed at 15 cm vertical distance to the needle tip.

3.4. Suspension polymerization and saponification of nanocomposite microspheres

Vinyl acetate (VAc) was polymerized through suspension polymerization method to prepare PVAc/MWNT nanocomposite microspheres following the procedure described elsewhere [21]. Monomer and MWNTs were mixed together prior to suspension polymerization. Suspending agent, PVA, was dissolved in water under nitrogen atmosphere and ADMVN was used as an initiator. After 1 day cooling down of the reaction mixture, the collected PVAc/MWNTs nanocomposite microspheres were washed with warm water. To produce PVAc/PVA/MWNT core/shell microspheres, the saponification of PVAc/MWNT nanocomposite microspheres was conducted in an alkali solution containing 10 g of sodium hydroxide, 10 g of sodium sulfate, 10 g of methanol and 100 g of water following the method reported by [21]. PVAc/PVA/MWNT core/shell microspheres were washed several times with water and dried in a vacuum at 40 C for 1 day.

3.5. Anti-bacterial test

Resistance of PVA/MWNT-Au nanofibers against *Staphylococcus aureus* (ATCC6538) were performed following the conditions described in a report published by [38]. Samples were prepared by dispersing the nanofibers in a viscous aqueous solution containing 0.01 wt.% of neutralized polyacrylic acid (Carbopol 941, Noveon Inc.). A mixed culture of microorganism, *Staphylococcus aureus* (ATCC6538) was obtained on tryptone soya broth after 24 h incubation at 32 C. Then, 20 g of samples were inoculated with 0.2 g of the microorganism suspension to adjust the initial concentration of bacteria to 107 cfu/g. Then, the inoculant mixed homogeneously with the samples and was stored at 32 C.

3.6. Characterization

Field-emission scanning electron microscopic (FE-SEM) images were obtained using JEOL, JSM-6380 microscope after gold coating. The transmission electron microscopy (TEM) analysis was conducted on an H-7600 model machine (HITACHI, LTD) with an accelerating voltage of 100 kV. The thermal properties were studied with differential scanning calorimeter (DSC) (Q-10) techniques from TA instruments, USA under the nitrogen gas atmosphere. The core/shell structure of PVAc/PVA/MWNT nanocomposite microspheres was examined using an optical microscope (Leica DC 100). The degree of saponification (DS) of PVAc/PVA/MWNT nanocomposites microspheres was determined by the ratio of methyl and methylene proton peaks in the ^{1}H-NMR spectrometer (Varian, Sun Unity 300) [21]. The antibacterial performance was investigated to examine the biological function of PVA/MWNT/Au nanofibers by KSM 0146 (shake flask method) using ATCC 6538 (S. aureus) [38].

4. Results and discussion

4.1. PVA/MWNT-Au nanocomposite nanofibers

4.1.1. Morphology

Figure 2 shows the FE-SEM images of pure PVA and PVA/MWNT-Au nanocomposite nanofibers and they are compared each other. The high magnification images are shown in the insets of each respective image. It can be seen from Fig. 2 that the average diameter of PVA/MWNT-Au nanocomposite nanofiber is increased compared to pure PVA nanofiber due to the incorporation of MWNT-Au nanocomposites into PVA nanofiber. The average diameter of pure PVA nanofibers is estimated ca. 300 nm whereas that of the PVA/MWMT-Au composite nanofiber is ca. 400 nm. Moreover, the PVA/MWNT-Au nanofibers are found quite smooth and bead free as like as pure PVA nanofiber. This result indicates that MWNT-Au nanocomposites have expanded the morphology of PVA nanofiber and they have been embedded well within the PVA nanofiber.

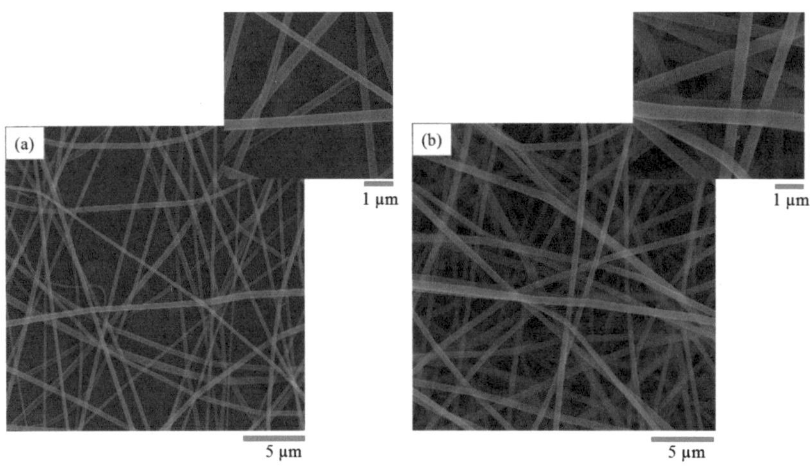

Figure 2. FE-SEM images of (a) pure PVA and (b) PVA/MWNT-Au nanocomposite nanofibers (PVA solution concentration = 10 wt%, TCD=15 cm, and applied voltage=15 kV; inset: high magnification morphologies of related images).

The detailed morphologies of the PVA/MWNT-Au nanocomposite nanofibers are investigated by transmission electron microscopy (TEM). Figure 3 demonstrates the TEM images of pure PVA and PVA/MWNT-Au composite nanofiber. Distributions of Au nano particles on the sidewalls of MWNTs and the structures of MWNT-Au composites are reported in our previous publication [40]. MWNT-Au nanocomposites are found unaltered into the polymer matrix comparing with our previous work [40]. A single isolated MWNT-Au nanocomposite is clearly seen in Figure 3 (b). This TEM image reveals that Au nanoparticles are remaining attached on the sidewalls of MWNTs and MWNT-Au nanocomposites are distributed along the PVA nanofiber which supports the smooth and uniform morphology of PVA/MWNT-Au composite nanofiber observed in the SEM images.

Moreover, this TEM image confirms that composites of MWNTs and Au nanoparticles were embedded well within the PVA nanofiber rather than cramming MWNTs and Au nanoparticles randomly. This might be a unique architecture of polymer nanofiber containing CNTs decorated with metal nanoparticles. However, some MWNT-Au composites were clustered together which is shown in Fig 3(c). This image indicates that in a polymer matrix MWNT-Au composites can be distributed randomly within the entire length of nanofiber.

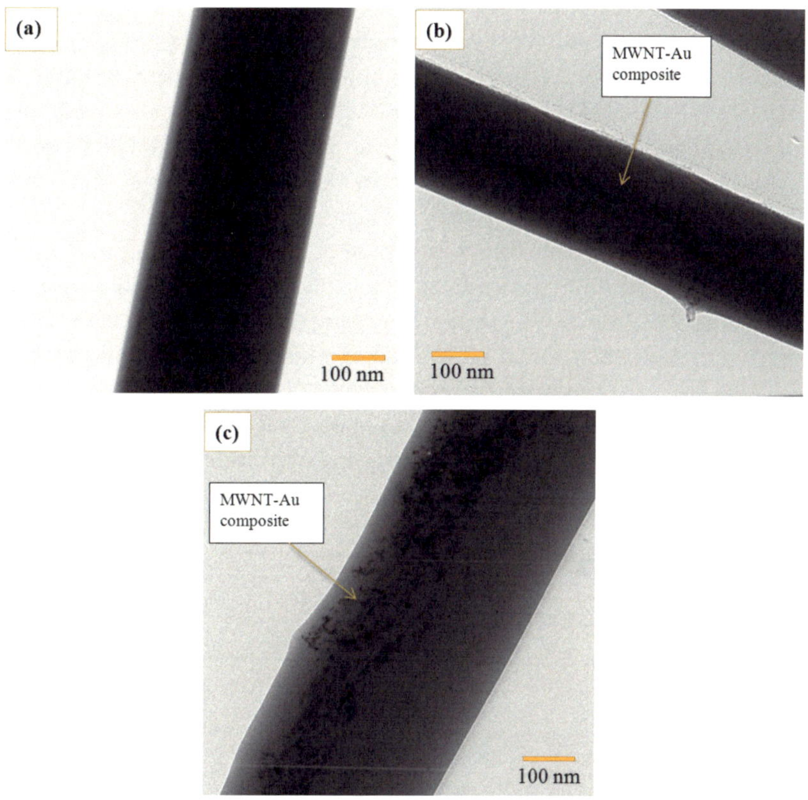

Figure 3. TEM images of (a) pure PVA nanofiber, and (b)-(c) PVA/MWNT-Au nanocomposite nanofibers. A single isolated (b) and an aggregated (c) MWNT-Au composites are clearly visible inside the fibers in which the Au nanoparticles are strongly attached to the surface of MWNTs. (PVA solution concentration= 10 wt%, TCD=15 cm, and applied voltage=15 kV.)

4.1.2. *Thermal properties*

Pyrolysis of PVA in nitrogen atmosphere undergoes dehydration and depolymerization at temperatures greater than 200 and 400 C, respectively. The actual depolymerization temperature depends on the structure, molecular weight, and conformation of the polymer [26]

Thermo gravimetric analysis (TGA) was conducted in nitrogen atmosphere to investigate the thermal stability of electrospun PVA/MWNT-Au nanocomposite nanofibers and the data were compared with pure PVA nanofibers. Figure 4 shows the TGA thermograms of pure PVA and PVA/MWNT-Au nanocomposite nanofiber at different decomposition temperature. Though the change is unclear but it can be assumed from the TGA thermograms that the thermal property of PVA/MWNT-Au nanocomposite nanofibers is different from pure PVA nanofiber [26].This result suggest that incorporating MWNT-Au nanocomposites can cause a change in thermal stability of PVA/ MWNT-Au nanocomposites nanofiber.

4.1.3. *Antibacterial efficacy*

CNTs and Au nanoparticles both have strong inhibitory and antibacterial effects as well as a broad spectrum of antimicrobial activities [5]. In this work, we have investigated the antibacterial efficacy of PVA/MWNT-Au nanocomposites nanofibers. The data obtained from the resistance of nanocomposite nanofiber against bacteria were compared with those of pure PVA nanofiber. The antibacterial test was performed in viscous aqueous test samples and shown in Fig. 5. The performance of nanofiber against bacteria was evaluated by counting the number of bacteria in the sample with the storage time at 32 °C. As shown in Fig. 5, pure PVA nanofibers are not effective enough to prevent the growth of bacteria and hence, a number of bacteria in the test samples remaining constant for a long time. On the other hand, PVA/MWNT-Au nanocomposites nanofibers exibit a remarkable inhibition of bacterial growth completely. This result indicates that only a small amount of MWNT-Au nanocomposites have improved anti-bacterial efficacy of PVA nanofibers and can make PVA nanofibers more efficient against bacteria. These featurs might have a potential medical applications.

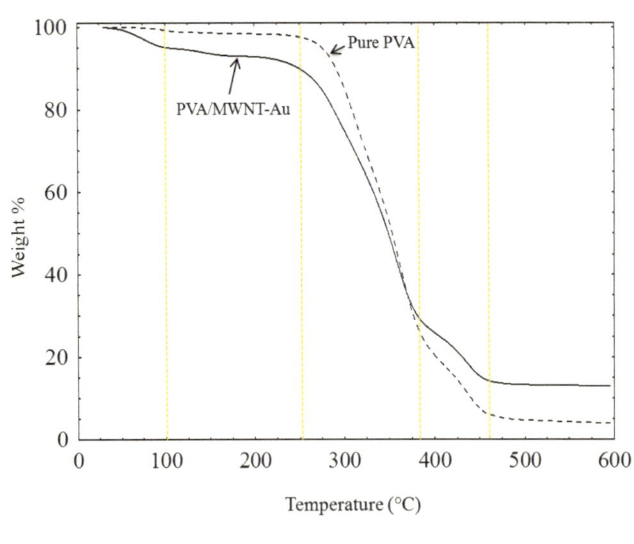

Figure 4. TGA thermographs of pure PVA and PVA/MWNT-Au composites nanofibers (PVA solution concentration = 10 wt%, TCD=15 cm, and applied voltage=15 kV)

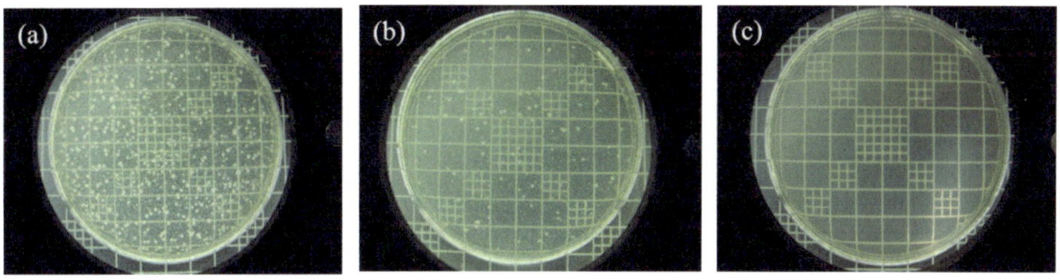

Figure 5. Anti-bacterial performance test of (a) blank, (b) pure PVA and (c) PVA/MWNT-Au nanocomposites nanofibers against the bacteria, *Staphylococcus aureus*. (PVA solution concentration = 10 wt%, TCD=15 cm, and applied voltage=15 kV)

4.2. PVA/MWNT/Ag nanocomposite nanoparticles and nanospheres

4.2.1. Morphology

Nanoparticles and nanospheres of PVA/MWNTs and PVA/MWNT/Ag nanocomposites were prepared by electrospraying technique following the methode describe in our previous report. Morphologies of these nanoparticles and nanospares are investigated by transmission electron microscopy and they were compared each other. Figure 6 shows the TEM images of PVA/MWNT nanocomposite nanoparticles. It can be seen from the TEM images that CNTs were crammed into PVA nanoarticles with a random manner and the CNTs were embeded within the particles rather than stiking on the surfaces of the nanoparticles. The incorporation of CNTs into the PVA nanoparticles expanded the morphologies of the nanocomposite nanoparticles. The shapes were lengthened and crinkled and the sizes were increased. This results suggest that CNTs have an effect on the morphologies of PVA nanoparticles.

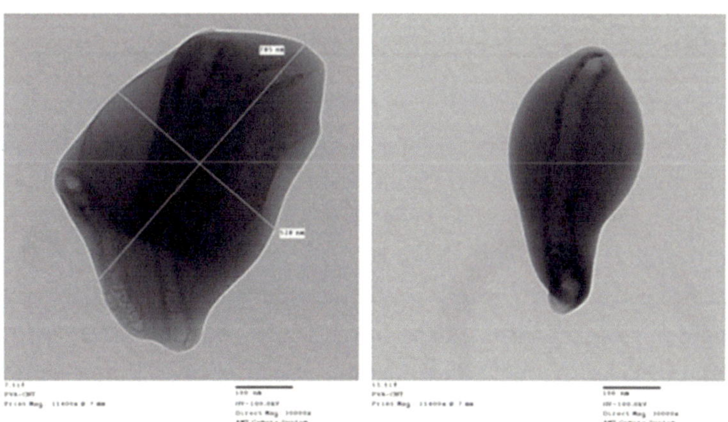

Figure 6. TEM images of the PVA/CNT nanoparticles using electrospraying (PVA solution concentration = 5 wt%, MWNTs concentration = 1 wt%, TCD=15 cm, and applied voltage=15 kV)

To prepare multifunctional nanocomposites, PVA/MWNT/Ag nanocomposites nanospheres were also prepared by electrospraying. TEM images in Figure 7 exhibit the morphologies of PVA/MWNT/Ag nanocomposites nanospheres.

A spherical morphology rather than particulates was obtained. Ag nanoparticles are distributed uniformly within the nanosphere together with CNTs but the Ag nanoparticles were not attached with the surfaces of CNTs. Moreover, Ag nanoparticles did not agglomate within the nanosphere.

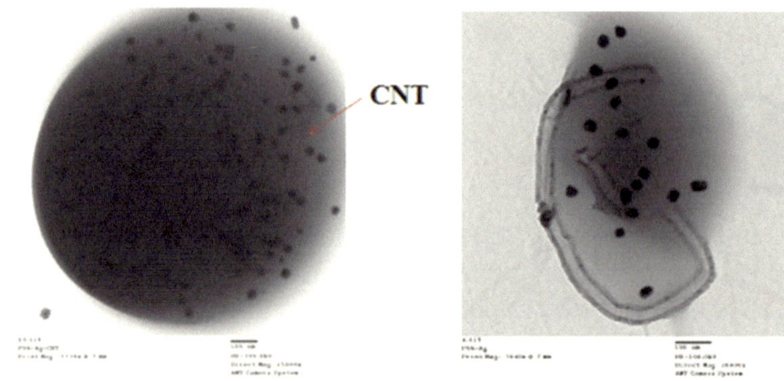

Figure 7. TEM images of the PVA/CNT/Ag nanosphere using electrospraying (PVA solution concentration = 5 wt%, MWNTs concentration = 1 wt%, Ag concentration = 1 wt.%, TCD= 15 cm, and applied voltage = 15 kV).

4.3. PVA/MWNT/Ag/MMT nanocomposite nanofibers

4.3.1. Morphology

Multifunctional nanocomposites nanofibers composed of PVA, MWNTs, Ag nanoparticles and clay, MMT, were also prepared in aqueous medium by electrospinning. Figure 8 represents the TEM images of PVA/MWNT/Ag/MMT multifunctional nanocomposites nanofibers electrospun from 5 wt% MMT solutions containing different amounts of carbon nanotubes (CNTs) (none, 0.1, and 0.5 wt%). PVA forms very smooth nanofibers but the addition of MMT clay and Ag nanoparticles into the polymer matrix increas the diameters of the nanofibers. The addition of MMT crinkled the fibers shape and may planes with many edges developed on surfaces of the nanofibers [38; 61]. It can be seen from Figure 8 (b) and (c) that CNTs were embeded along the fiber directions. Ag nanoparticles were unifromly distributed within the fibers and on the fiber cross-section [38]. It can be clearly seen that the increase of CNTs amount increased the diameter of the nanofibers and expand the morphology of the multifunctional nanocomposite nanofibers.

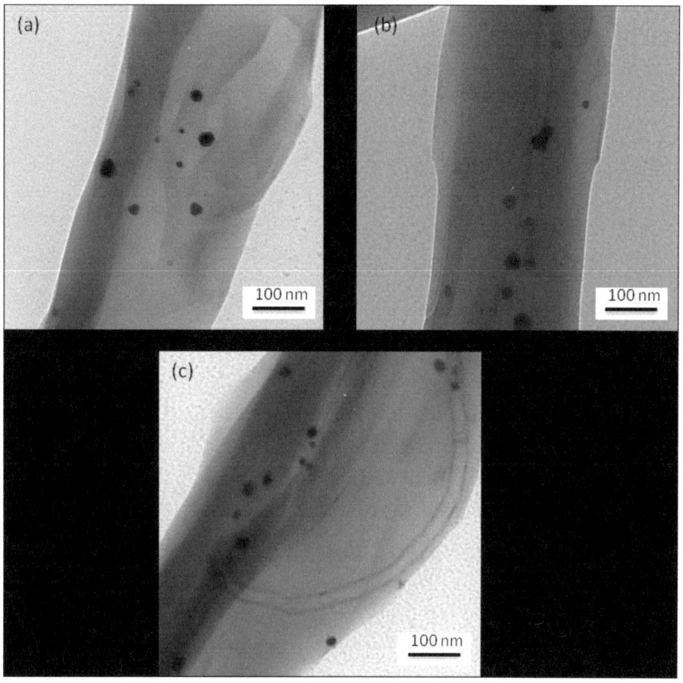

Figure 8. TEM images of electrospun PVA/MWNT/Ag/MMT multifunctional composite nanofibers with different CNT contents of 0 wt% (a), 0.1 wt% (b), and 0.5 wt% (c) (Polymer concentration = 10 wt%, MMT concentration= 5 wt%, Ag concentration = 1 wt%, TCD= 15 cm, and Applied voltage= 15 kV).

4.3.2. Thermal properties

Thermal properties of electrospun PVA/MWNT/Ag/MMT multifunctional composite nanofibers were measured using Differencial Scanning Calorometry (DSC) in nitrogen atmosphere. Figure 9 shows the DSC thermograms of electrospun PVA/MWNT/Ag/MMT multifunctional composite nanofibers containing different CNT contents (none, 0.1 and 0.5 wt%). A large endothermic peak was observed at 224 C in the DSC curve obtained from only PVA nanofibers (Figure 9a).

The peak of PVA/MMT/Ag was moved to higher temperature i.e 226.5 C while their was no CNTs (Figure 9b). This result indicates that Ag content increased the thermal stability [38]. With the addition and increase of CNTs content into the PVA/MMT/Ag nanocomposite nanofibers, the peaks of PVA/MWNT/Ag/MMT composite nanofibers in Figure 9 (c) and (d) shifted to 228 and 229 C, respectively. These results indicate that the addition of carbon nanotubes (CNTs) improves the thermal properties of PVA/MWNT/Ag/MMT composite nanofibers. Moreover, the increased amount of CNTs increase the thermal stability of PVA/MWNT/Ag/MMTcomposite nanofibers. These results suggest that the incorporation of CNTs into the multifunctional PVA composite nanofibers might increase their thermal stability significantly.

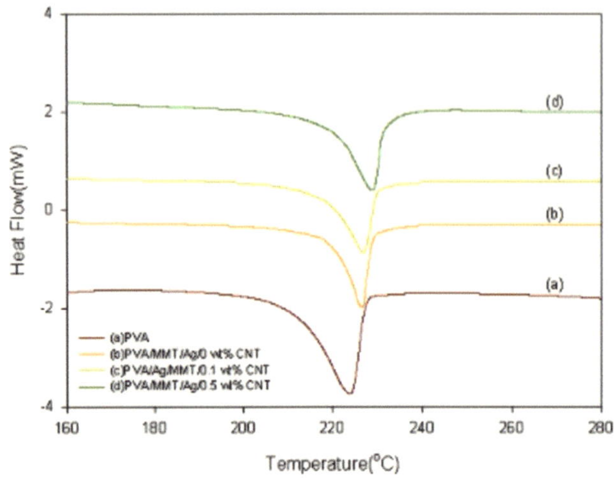

Figure 9. DSC data of electrospun PVA nanofibers (a), and PVA/MWNT/Ag/MMT multihybrid nanofibers with different CNT contents of 0 wt.% (b), 0.1 wt.% (c), and 0.5 wt.% (d) (Polymer concentration = 10 wt.%, MMT concentration= 5 wt.%, Ag concentration = 1 wt.%, TCD= 15 cm, and Applied voltage= 15 kV).

4.4. PVAc/PVA/MWNT microspheres

4.4.1. Morphology

Figure 10 represents the FE-SEM images of the PVAc/MWNT microspheres prepared by suspension polymerization [21]. It can be seen from Fig. 10 that sizes of the PVAc/MWNTs microspheres are not uniform. A single microsphere is enlarged and its rough surface is observed where as the surface of the PVAc microspheres is smooth [21]. The roughness of the surface was caused by the presence of MWNTs which is clearly seen in the highly magnified image in Figure 10. To understand the surface morphology of the PVAc/MWNT microspheres better, their fracture surface was investigated by SEM which is represented in Figure 11. The rough surface shown in the enlarged images cofirms that the MWNTs were evidently incorporated within the PVAc microspheres by suspension polymerization.

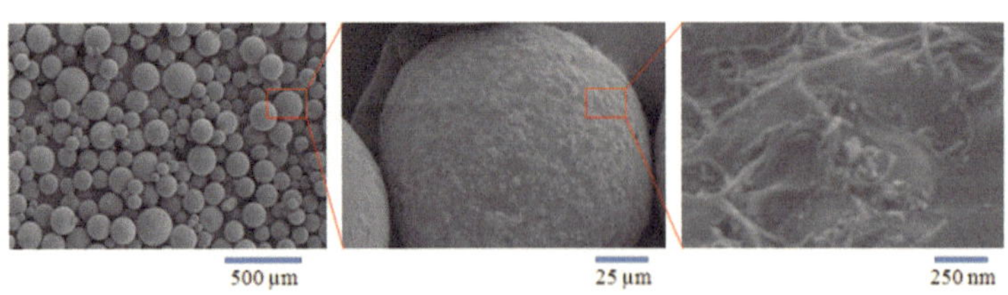

Figure 10. SEM images of the PVAc/MWNT microspheres prepared by suspension polymerization. A single PVAc/MWNT microsphere and its surfaces are enlarged with different magnifications

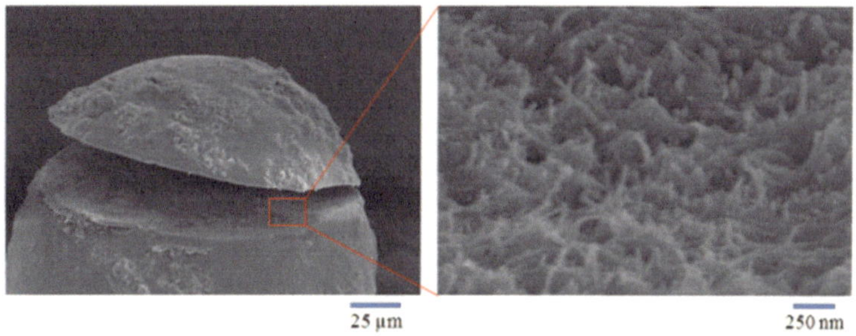

Figure 11. SEM images of the fracture part of a PVAc/MWNT microsphere prepared by suspension polymerization

4.4.2. Optical micrographs

PVA/MWNT nanocomposite microspheres were prepared by heterogeneous saponification following the method reported in our previous work [21]. The spherical shapes of PVAc/ MWNT nanocomposite particles were maintained during saponificaion process by dispersing PVAc/MWNT nanocomposite particles in aqueous alkali solution with very gentle agitation. The optical micrographs of PVAc/PVA/MWNT nanocomposite microspheres prepared by heterogeneous saponification are presented in Figure 12. It can be seen from the micrographs that composite microspheres with a PVAc core and PVA shell structure were obtained and MWNTs were distributed throughout the core/shell microshpere.

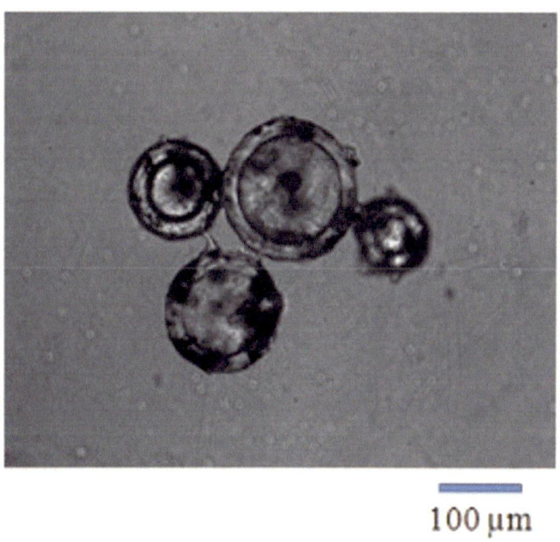

Figure 12. Optical micrograph of the PVAc/PVA/MWNT core/shell microspheres (The saponification times and DS value was 4 h and 18%.

5. Conclusions

Polymer nanocomposites of different types and structures have been successfully prepared and characterized by by FE-SEM, TEM, TGA, DSC, optical microscopy and antibacterial efficacy test. PVA/MWNT-Au, and PVA/MWNT/Ag/MMT nanocomposites nanofibers were prepared by electrospinning from aqueous solution. Electrospinning technique was employed to prepare PVA/MWNT/Ag nanoparticles and nanospheres. PVAc/PVA/MWNTs core/shell microsphere were prepared by saponication of PVAc/MWNTs microsphere prepared by suspension polymerization. Au nanoparticles were remaining attached with MWNTs within the PVA/MWNT-Au nanofibers. MWNT-Au nanocomposites expanded the morphologies and improved the properties of PVA/MWNT-Au nanofibers. MWNT-Au nanocomposites showed significatant performance against bacteria. MMT and MWNTs increased the diameters of the PVA/MWNT/Ag/MMT nanocomposites nanofibers. Silver nanoparticles were distibuted well within the PVA/MWNT/Ag nanocomposites nanoparticles. The results obtained in this study may help to fabricate polymer nanocomposite in order to improve their properties and expand their applications in the field of modern science.

Acknowledgements

This research was supported by Basic Science Research Program through the National Research Foundation of Korea (NRF) funded by the Ministry of Education, Science and Technology (2012-0003093 and 2012-0002689).

Author details

Jeong Hyun Yeum[1*], Sung Min Park[2], Il Jun Kwon[2], Jong Won Kim[2], Young Hwa Kim[1], Mohammad Mahbub Rabbani[1], Jae Min Hyun[1], Ketack Kim[3] and Weontae Oh[4]

*Address all correspondence to: jhyeum@knu.ac.kr

1 Department of Advanced Organic Materials Science & Engineering, Kyungpook National University, Korea

2 Korea Dyeing Technology Center, Korea

3 Department of Chemistry, Sangmyung University, Korea

4 Department of Materials and Components Engineering, Dong-eui University, Korea

References

[1] Berkland, C., Kim, C., & Pack, D. W. (2001). Fabrication of PLG microspheres with precisely controlled and monodisperse size distributions. *J. Control. Release*, 73(1), 59 -74 , 0168-3659.

[2] Brust, M., Schiffrin, D. J., Bethell, D., & Kiely, C. J. (1995). Novel gold-dithiol nano-networks with non-metallic electronic properties. *Adv. Mater.*, 7(9), 795 -797, 0935-9648.

[3] Choi, J. S., Lee, S. W., Jeong, L., Bae, S. H., Min, B. C., & Youk, J. H. (2004). Effect of organosoluble salts on the nanofibrous structure of electrospun poly(3 -hydroxybutyrate-co-3-hydroxyvalerate). *Int. J. Bio. Macromol.*, 34(4), 249-256, 0141-8130.

[4] Chu, H., Wei, L., Cui, R., Wang, J., & Li, Y. (2010). Carbon nanotubes combined with inorganic nanomaterials: Preparations and applications. *Coordin. Chem. Rev.*, 254(9-10), 1117 -1134 , 0010-8545.

[5] Dastjerdi, R., & Montazer, M. (2010). A review on the application of inorganic nano-structured materials in the modification of textiles: Focus on anti-microbial properties. *Colloid Surface B*, 79(1), 5 -18 , 0927-7765.

[6] Feldheim, D. L., & Foss, C. A. (2000). Metal Nanoparticles: Synthesis, Characterization and Applications,. Dekker, New York,, 978-0824706043.

[7] Fischer, H. R., Gielgens, L. H., & Koster, T. P. M. (1999). Nanocomposites from polymers and layered minerals. *Acta Polym.*, 50(4), 122-126, 0323-7648.

[8] Fleming, M. S., Mandal, T. K., & Walt, D. R. (2001). Nanosphere-microsphere assembly: methods for core-shell materials preparation. *Chem. Mater.*, 13(6), 2210 -2216 , 0897-4756.

[9] Gates, W. P., Komadel, P. K., Madefova, J., Bujdak, J., Stucki, J. W., & Kirkpatrick, R. J. (2000). Electronic and structural properties of reduced-charge maontmorillonites. *Appl. Clay Sci.*, 16(5-6), 257-271, 0169-1317.

[10] Georgakilas, V., Gournis, D., Tzitzios, V., Pasquato, L., Guldi, D. M., & Prato, M. (2007). Decorating carbon nanotubes with metal or semiconductor nanoparticles. *J. Mater. Chem.*, 17(26), 2679 -2694 , 0959-9428.

[11] Giannetti, E., & Mazzocchi, R. (1986). High conversion free-radical suspension polymerization: End groups in poly(methyl methacrylate) and their influence on the thermal stability. *J. Polym. Sci. Pol. Chem.*, 24(10), 2517-2551, 0087-624X.

[12] Gittins, D. I., & Caruso, F. (2001). Spontaneous Phase Transfer of Nanoparticulate Metals from Organic to Aqueous Media. *Angew. Chem. Int. Ed.*, 40(16), 3001 -3004 , 1433-7851.

[13] Glaus, S., Calzaferri, G., & Hoffmann, R. (2002). Electronic properties of the silver-silver chloride cluster interface. *Chem.-A Eur. J.*, 8(8), 1785 -94 , 0947-6539.

[14] Gotoh, Y., Igarashi, R., Ohkoshi, Y., Nagura, M., Akamatsu, K., & Deki, S. (2000). Preparation and structure of copper nanoparticle/poly(acrylic acid) composite films. *J. Mater.Chem.*, 10(11), 2548-2552, 0959-9428.

[15] Grace, A. N., & Pandian, K. (2007). Antibacterial efficacy of aminoglycosidic antibiotics protected gold nanoparticles: a brief study. *Colloids Surf. A*, 297(1-3), 63 -70 , 0927-7757.

[16] Han, X. J., Huang, Z. M., He, C. L., Liu, L., & Wu, Q. S. (2006). Coaxial electrospinning of PC(shell)/PU(core) composite nanofibers for textile application. *Polym. Composite.*, 27(4), 381-387, 0272-8397.

[17] Hatchett, D. W., Josowicz, M., Janata, J., & Baer, D. R. (1999). Electrochemical formation of au clusters in polyaniline. *Chem. Mater.*, 11(10), 2989 -2994 , 0897-4756.

[18] Huang, C. J., Yen, C. C., & Chang, T. C. (1991). Studies on the preparation and properties of conductive polymers. III. Metallized polymer films by retroplating out. *J. Appl. Polym. Sci.*, 42(8), 2237-2245, 0021-8995.

[19] Huang, Z. M., Zhang, Y. Z., Kotaki, M., & Ramakrishna, S. (2003). A review on polymer nanofibers by electrospinning and their applications in nanocomposites. *Compos. Sci. Technol.*, 63(15), 2223 -2253 , 0266-3538.

[20] Ji, H. M., Lee, H. W., Karim, M. R., Cheong, I. W., Bae, E. A., Kim, T. H., Islam, S., Ji, B. C., & Yeum, J. H. (2009). Electrospinning and characterization of medium-molecular-weight poly(vinyl alcohol)/high-molecular- weight poly(vinyl alcohol)/montmorillonite nanofibers. *Colloid and Polymer Science*, 287(7), 751-758, 0303-402X.

[21] Jung, H. M., Lee, E. M., Ji, B. C., Deng, Y., Yun, J. D., & Yeum, J. H. (2007). Poly(vinyl acetate)/poly(vinyl alcohol)/montmorillonite nanocomposite microspheres prepared by suspension polymerization and saponification. *Colloid Polym. Sci.*, 285(6), 705-710, 0303-402X.

[22] Kang, S., Herzberg, M., Rodrigues, D. F., & Elimelech, M. (2008). Antibacterial Effects of Carbon Nanotubes: Size Does Matter! *Langmuir*, 24(13), 6409-6413, 0743-7463.

[23] Karim, M. R., Lee, H. W., Kim, R., Ji, B. C., Cho, J. W., Son, T. W., Oh, W., & Yeum, J. H. (2009). Preparation and characterization of electrospun pullulan/montmorillonite nanofiber mats in aqueous solution. *Carbohydrate Polymers*, 78(2), 336 -342 , 0144-8617.

[24] Krumova, M., López, D., Benavente, R., Mijangos, C., & Pereña, J. M. (2000). Effect of crosslinking on the mechanical and thermal properties of poly(vinyl alcohol). *Polymer*, 41(26), 9265-9272, 0032-3861.

[25] Lee, J. E., Kim, J. W., Jun, J. B., Ryu, J. H., Kang, H. H., Oh, S. G., & Suh, K. D. (2004). Polymer/Ag composite microspheres produced by water-in-oil-in-water emulsion polymerization and their application for a preservative. *Colloid Polym. Sci.*, 282(3), 295 -299 , 0303-402X.

[26] Lee, H. W., Karim, M. R., Park, J. H., Ghim, H. D., Choi, J. H., Kim, K., Deng, Y., & Yeum, J. H. (2009a). Poly(vinyl alcohol)/Chitosan Oligosaccharide Blend Sub-micro-

meter Fibers Prepared from Aqueous Solutions by the Electrospinning Method. *J. Appl. Polym. Sci.*, 111(1), 132-140, 0021-8995.

[27] Lee, H. W., Karim, M. R., Ji, H. M., Choi, J. H., Ghim, H. D., Park, J. H., Oh, W., & Yeum, J. H. (2009b). Electrospinning Fabrication and Characterization of Poly(vinyl alcohol)/ Montmorillonite Nanofiber Mats. *J. Appl. Polym. Sci.*, 113(3), 1860-1867, 0021-8995.

[28] Li, D., & Xia, Y. (2004). Electrospinning of nanofibers: reinventing the wheel? Adv. Mater. 0935-9648 , 16(14), 1151-1170.

[29] Luna-Xavier, J. L., Bourgeat-Lami, E., & Guyot, A. (2001). The role of initiation in the synthesis of silica/poly(methylmethacrylate) nanocomposite latex particles through emulsion polymerization. *Colloid Polym. Sci.*, 279(10), 947-958, 0303-402X.

[30] Matsumoto, H., Mizukoshi, T., Nitta, K., Minagawa, M., Tanoika, A., & Yamagata, Y. (2005). Organic/inorganic hybrid nano-microstructured coatings on insulated substrates by electrospray deposition. *J. Colloid Interf. Sci.*, 286(1), 414 -6 , 0021-9797.

[31] Messersmith, P. B., & Giannelis, E. P. (1993). Polymer-layered silicate nanocomposites: In situ intercalative polymerization of ε-caprolactone in layered silicates. *Chem. Mater.*, 5(8), 1064-1066, 0897-4756.

[32] Min, B. M., Lee, G., Kim, S. H., Nam, Y. S., Lee, T. S., & Park, W. H. (2004). Electrospinning of silk fibroin nanofibers and its effect on the adhesion and spreading of normal human keratinocytes and fibroblasts in vitro. *Biomaterials*, 25(7-8), 1289 -1297 , 0142-9612.

[33] Mu, L., & Feng, S. S. (2001). Fabrication, characterization and in vitro release of paclitaxel (Taxol®) loaded poly(lactic co-glycolic acid) microspheres prepared by spray drying technique with lipid/cholesterol emulsifiers. *J. Control. Release*, 76(3), 239-254, 0168-3659.

[34] Okamoto, M., Morita, S., Taguchi, H., Kim, Y. H., Kotaka, T., & Tateyama, H. (2000). Synthesis and structure of smectic clay/poly(methyl methacrylate) and clay/polystyrene nanocomposites via in situ intercalative polymerization. *Polymer*, 41(10), 3887-3890, 0032-3861.

[35] Okuda, H., & Kelly, A. J. (1996). Electrostatic atomization-experiment, theory and industrial applications. Phys. Plasmas., , 5, 1070-664X, 3, 2191 -2196 .

[36] Oriakhi, C. O., & Lerner, M. M. (1995). Poly(pyrrole) and poly(thiophenb)/clay nanocomposites via latex-colloid interaction. *Mater. Res. Bull.*, 30(6), 723-729, 0025-5408.

[37] Park, S. H., Oh, S. G., Munb, J. Y., & Han, S. S. (2006). Loading of gold nanoparticles inside the DPPC bilayers of liposome and their effects on membrane fluidities,. *Colloid Surf. B*, 48(2), 112 -118 , 0927-7765.

[38] Park, J. H., Karim, M. R., Kim, I. K., Cheong, I. W., Kim, J. W., Bae, D. G., Cho, J. W., & Yeum, J. H. (2010). Electrospinning fabrication and characterization of poly(vinyl

alcohol)/montmorillonite/silver hybrid nanofibers for antibacterial applications. *Colloid Polym. Sci.*, 288(1), 115-121, 0303-402X.

[39] Popov, V. N. (2004). Carbon nanotubes: properties and application. *Mater. Sci. Eng. R*, 43(3), 61-102, 0927-796X.

[40] Rabbani, M. R., Ko, C. H., Bae, J. S., Yeum, J. H., Kim, I. S., & Oh, W. (2009). Comparison of some gold/carbon nanotube composites prepared by control of electrostatic interaction. *Colloid Surf. A*, 336(1-3), 183 -186 , 0927-7757.

[41] Ramos, J., Millan, A., & Palacio, F. (2000). Production of magnetic nanoparticles in a polyvinylpyridine matrix. *Polymer*, 41(24), 8461 -8464 , 0032-3861.

[42] Ren, G., Xu, X., Liu, Q., Cheng, J., Yuan, X., Wu, L., & Wan, Y. (2006). Electrospun poly(vinyl alcohol)/glucose oxidase biocomposite membranes for biosensor applications. *React. Funct. Polym.*, 66(12), 1559-1564, 1381-5148.

[43] Reneker, D. H., & Chun, I. (1996). Nanometre diameter fibres of polymer, produced by electrospinning. *Nanotechnology*, 7(3), 216 -223, 0957-4484.

[44] Rosca, I. D., Watari, F., & Uo, M. (2004). Microparticle formation and its mechanism in single and double emulsion solvent evaporation. J. Control. Release, 0168-3659 , 99(2), 271 -80 .

[45] Rujitanaroj, P. O., Pimpha, N., & Supaphol, P. (2008). Wound-dressing materials with antibacterial activity from electrospun gelatin fiber mats containing silver nanoparticles. *Polymer*, 49(21), 4723 -4732 , 0032-3861.

[46] Salata, O. V., Hull, P. J., & Dobson, P. J. (1997). Synthesis of nanometer-scale silver crystallites via a room-temperature electrostatic spraying process. *Adv. Mater*, 9(5), 413 -417 , 0935-9648.

[47] Sangamesh, G. K., Syam, P. N., Roshan, J., Hogan, M. V., & Laurencin, C. T. (2008). Recent Patents on Electrospun Biomedical Nanostructures: An Overview. *Recent Patents Biomed. Eng.*, 1(1), 68 -78 , 1874-7647.

[48] Schartel, B., Pötschke, P., Knoll, U., & Abdel-Goad, M. (2005). Fire behaviour of polyamide 6/multiwall carbon nanotube nanocomposites. *Eur. Polym. J.*, 41(5), 1061 -1070 , 0014-3057.

[49] Sinha, V. R., Bansal, K., Kaushik, R., & Trehan, A. (2004). Poly-ε-caprolactone microspheres and nanospheres: an overview. *Int. J. Pharmaceut*, 278(1), 1-23, 0378-5173.

[50] Svensson, P. D., & Staffan, Hansen. S. (2010). Freezing and thawing of montmorillonite-A time-resolved synchrotron X-ray diffraction study. *Appl. Clay Sci.*, 49(3), 127 -134 , 0169-1317.

[51] Tiarks, F., Landfester, K., & Antonietti, M. (2001). Silica nanoparticles as surfactants and fillers for latexes made by miniemulsion polymerization. *Langmuir*, 17(19), 5775 -5780 , 0743-7463.

[52] Unger, E., Duesberg, G. S., Liebau, M., Graham, A. P., Seidel, R., Kreupl, F., & Hoenlein, W. (2003). Decoration of multi-walled carbon nanotubes with noble- and transition-metal clusters and formation of CNT-CNT networks. *Appl. Phys. A*, 77(6), 735 -738, 0947-8396.

[53] Usuki, A., Kato, M., Olada, A., & Kurauchi, T. (1997). Synthesis of polypropylene-clay hybrid. *J. Appl. Polym. Sci*, 63(1), 137 -138, 0021-8995.

[54] Vaia, R., Vasudevan, S., Kraweic, W., Scanlon, L. G., & Giannelis, E. P. (1995). New polymer electrolyte nanocomposites: Melt intercalation of poly(ethylene oxide) in mica-type silicates. *Adv. Mater.*, 7(2), 154-156, 0935-9648.

[55] Vasita, R., & Katti, D. S. (2006). Nanofibers and their applications in tissue engineering. *Int. J. Nanomedicine*, 1(1), 15-30, 1176-9114.

[56] Wang, X., Drew, C., Lee, S. H., Senecal, K. J., Kumar, J., & Samuelson, L. A. (2002). Electrospun nanofibrous membranes for highly sensitive optical sensors. *Nano Lett.*, 2(11), 1273 -1275 , 1530-6984.

[57] Wu, L. L., Yuan, X. Y., & Sheng, J. (2005). Immobilization of cellulase in nanofibrous PVA membranes by electrospinning. J. Membr. Sci. - 0376-7388 , 250(1-2), 167 -173.

[58] Wu, W., He, T., & Chen, J. F. (2006). Study on in situ preparation of nano calcium carbonate/PMMA composite particles. *Mater. Lett.*, 60(19), 2410 -2415 , 0167-577X.

[59] Yeum, J. H., Park, J. H., Kim, I. K., & Cheong, I. W. (2011). Electrospinning Fabrication and Characterization of Water Soluble Polymer/Montmorillonite/Silver Nanocomposite Nanofibers out of Aqueous Solution. *Advances in Nanocomposites-Synthesis, Characterization and Industrial Applications,*, Intech Publishers,, Croatia,, 483-502, 978-9-53307-165-7.

[60] Yonezawa, T., & Kunitake, T. (1999). Practical preparation of anionic mercapto ligandstabilized gold nanoparticles and their immobilization,. *Colloids Surfaces A*, 149(1-3), 193-199, 0927-7757.

[61] Zhang, Z., & Han, M. (2003). One-step preparation of size-selected and well-dispersed silver nanocrystals in polyacrylonitrile by simultaneous reduction and polymerization. *J. Mater. Chem*, 13(4), 641-643, 0959-9428.

[62] Zhang, Y., Peng, H., Huanga, W., Zhou, Y., & Yan, D. (2008). Facile preparation and characterization of highly antimicrobial colloid Ag or Au nanoparticles. *J. Colloid Interface Sci.*, 325(2), 371 -376 , 0021-9797.

[63] Zhu, Z. K., Yin, J., Cao, F., Shang, X. Y., & Lu, Q. H. (2000). Photosensitive polyimide/silica hybrids. *Adv. Mater.*, 12(14), 1055-1057, 0935-9648.

[64] Zussmas, E., Theron, A., & Yarin, A. L. (2003). Formation of nanofiber crossbars in electrospinning. *Appl. Phys. Lett*, 82(6), 973-975, 0003-6951.

Chapter 3

Polymer-Graphene Nanocomposites: Preparation, Characterization, Properties, and Applications

Kuldeep Singh, Anil Ohlan and S.K. Dhawan

Additional information is available at the end of the chapter

http://dx.doi.org/10.5772/50408

1. Introduction

Carbon the 6th element in the periodic tables has always remains a fascinating material to the researcher and technologist. Diamond, graphite, fullerenes, carbon nanotubes and newly discovered graphene are the most studied allotropes of the carbon family. The significance of the these material can be understand as the discovery of fullerene and graphene has been awarded noble prizes in the years 1996 and 2010 to Curl, Kroto & Smalley and Geim & Novalec, respectively. After the flood of publications on graphite intercalated [1], fullerenes (1985) [2], and carbon nanotubes (1991) [3], graphene have been the subject of countless publications since 2004 [4,5]. Graphene is a flat monolayer of carbon atoms tightly packed into a two-dimensional (2D) honeycomb lattice, completely conjugated sp^2 hybridized planar structure and is a basic building block for graphitic materials of all other dimensionalities (Figure 1). It can be wrapped up into 0D fullerenes, rolled into 1D nanotube or stacked into 3D graphite.

In 2004, Geim and co-workers at Manchester University successfully identified single layers of graphene in a simple tabletop experiment and added a revolutionary discovery in the field of nano science and nanotechnology. Interest in graphene increased dramatically after Novoselov, Geim et al. reported on the unusual electronic properties of single layers of the graphite lattice. One of the most remarkable properties of graphene is that its charge carriers behave as massless relativistic particles or Dirac fermions, and under ambient conditions they can move with little scattering. This unique behavior has led to a number of exceptional phenomena in graphene [4]. First, graphene is a zero-band gap 2D semiconductor with a tiny overlap between valence and conduction bands. Second, it exhibits a strong ambipolar electric field effect so that the charge carrier concentrations of up to 1013 cm^{-2} and room-temperature mobility of ~10000 cm^{-2}s^{-1} are measured. Third, an unusual half-integer quantum Hall effect (QHE) for both electron and hole carriers in graphene has been observed by ad-

INTECH
open science | open minds

© 2012 Singh et al.; licensee InTech. This is a paper distributed under the terms of the Creative Commons Attribution License (http://creativecommons.org/licenses/by/3.0), which permits unrestricted use, distribution, and reproduction in any medium, provided the original work is properly cited.

justing the chemical potential using the electric field effect [5,6]. It has high thermal conductivity with a value of ~ 5000 WmK^{-1} for a single-layer sheet at room temperature. In addition, graphene is highly transparent, with absorption of ~ 2.3% towards visible light [7, 8]. Narrow ribbons of graphene with a thickness of 1-2 nm are, however, semiconductors with a distinct band gap, and these can be used to produce transistors [9-11].

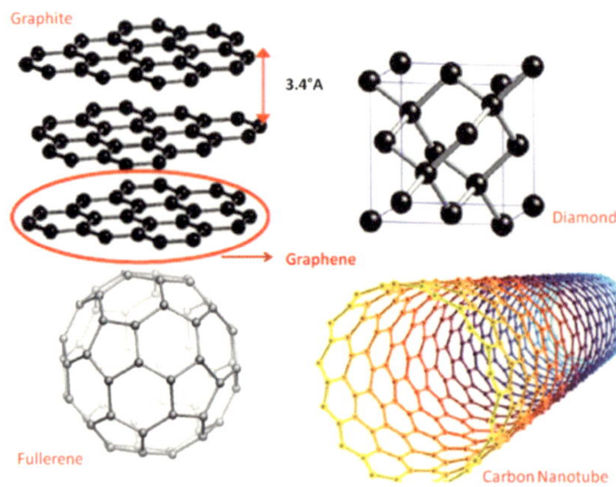

Figure 1. Different allotropes of carbon viz Graphite, Diamond, Fullerene, and Carbon nanotube

In last couples of years, graphene has been used as alternative carbon-based nanofiller in the preparation of polymer nanocomposites and have shown improved mechanical, thermal, and electrical properties [12-19]. The recent advances have shown that it can replace brittle and chemically unstable indium tin oxide in flexible displays and touch screens [20-21]. It is well established that the superior properties of graphene are associated with its single-layer. However, the fabrication of single-layer graphene is difficult at ambient temperature. If the sheets are not well separated from each other than graphene sheets with a high surface area tend to form irreversible agglomerates and restacks to form graphite through p–p stacking and Vander Waals interactions [22,23]. Aggregation can be reduced by the attachment of other small molecules or polymers to the graphene sheets. The presence of hydrophilic or hydrophobic groups prevents aggregation of graphene sheets by strong polar-polar interactions or by their bulky size [24]. The attachment of functional groups to graphene also aids in dispersion in a hydrophilic or hydrophobic media, as well as in the organic polymer. Therefore, an efficient approach to the production of surface-functionalized graphene sheets in large quantities has been a major focus of many researchers. The goal is to exploit the most frequently proposed applications of graphene in the areas of polymer nanocomposites, super-capacitor devices, drug delivery systems, solar cells, memory devices, transistor devices, biosensors and electromagnetic/ microwave absorption shields.

2. Methods of Graphene Synthesis

There have been continuous efforts to develop high quality graphene in large quantities for both research purposes and with a view to possible applications. The methods of preparation for graphene can be divided into two categories, top-down and bottom-up ones. The top-down methods include (1) mechanical exfoliation (2) chemical oxidation/exfoliation followed by reduction of graphene derivatives such as graphene oxide. While the bottom-up methods include (1) epitaxial growth on SiC and other substrates, (2) Chemical vapor deposition, and (3) arc discharging methods. Each of these methods has some advantages and limitations. Among them chemical synthesis of graphene using graphite, graphite oxide (GO) is a scalable process but it leads to more defent in the graphene layer.

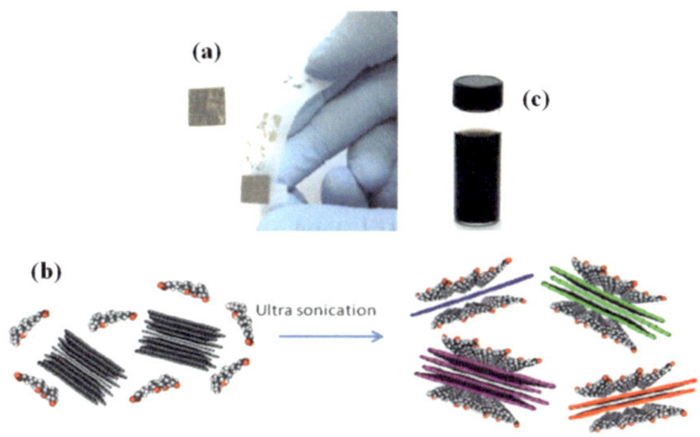

Figure 2. A) Mechanical exfoliation of graphene using scotch tape from HOPG (B) Schematic illustration of the graphene exfoliation process. Graphite flakes are combined with sodium cholate in aqueous solution. Horn-ultrasonication exfoliates few-layer graphene flakes that are encapsulated by sodium cholate micelles. & (C) Photograph of 90 µg/ml graphene dispersion in sodium cholate (Reprinted with permission from ref 25 Copyright.2009 American Chemical Society)

2.1. Mechanical exfoliation

The "perfect" graphene, necessary for the fundamental studies can be obtained by the mechanical exfoliation and epitaxial methods, but these methods have a limit for scale up. Mechanical exfoliation is a simple peeling process where a dried highly oriented pyrolytic graphite (HOPG) sheet are etched in oxygen plasma and then it is stuck onto a photo resist and peeled off layers by a scotch tape (Figure 2a). The thin flakes left on the photo resist were washed off in acetone and transferred to a silicon wafer. It was found that these thin flakes were composed of monolayer or a few layers of graphene [4].

2.2. Chemical exfoliation and intercalation of small molecules:

The first graphite intercalation compound (GIC), commonly known as expandable graphite was prepared by Schafhautl in 1841, while analyzing crystal flake of graphite in sulfuric acid solution. The intercalation of graphite by atoms or molecules such as alkali metals or mineral acids increases its interlayer spacing, weakening the interlayer interactions and facilitating the exfoliation of GIC by mechanical or thermal methods (Figure 2b& 2c) [25]. The intercalation of graphite by a mixture of sulfuric and nitric acid produces a higher-stage GIC that can be exfoliated by rapid heating or microwave treatment of the dried down powder, producing a material commonly referred to as expanded graphite [26]. It retained a layered structure but has slightly increased interlayer spacing relative to graphite and has been investigated as a composite filler [27-28]. However its effectiveness in enhancing the properties as compared to graphene oxide (GO) derived fliers is limited by its layered structure and relatively low specific surface area. To produce a higher surface area material, expanded graphite can be further exfoliated by various techniques to yield graphene nanoplates (GNPs) down to 5 nm thickness[29-30]. It has also been reported that sulfuric acid intercalated expanded graphite can be co-intercalated with tetrabutyl ammonium hydroxide. A monolayer like graphene can be obtained by sonicating the GIC in N,N-dimethylformamide (DMF) in the presence of a surfactant like poly(ethylene glycol)-modified phospholipid. Blake et al. and Hernandez et al.[31-32] have established a method for the preparation of defect free graphene by exfoliation of graphite in N-methyl-pyrrolidione. Such approach utilizes the similar surface energy of N-methyl-pyrrolidone and graphene that facilitates the exfoliation. However, the disadvantage of this process is the high cost of the solvent and the high boiling point of the solvent that makes the graphene deposition difficult. Lotya and coworkers have used a surfactant (sodium dodecyl benzene sulfonate, SDBS) to exfoliate graphite in water to produce graphene. The graphene monolayers are stabilized against aggregation by a relatively large potential barrier caused by the Coulomb repulsion between surfactant-coated sheets. The dispersions are reasonably stable with larger flakes precipitating out over more than 6 weeks [33].

2.3. Chemical vapor deposition & Epitaxial growth of graphene:

Chemical vapor deposition (CVD) is alternatives method to mechanical exfoliation and used to obtain high quality graphene for large-scale production of mono or few layer graphene films on metal substrate[34-40]. The CVD processes generally utilize transition metal surfaces for growth of Graphene nanosheets (GNS) using hydrocarbon gases as GNS precursors at the deposition temperature of about 1000 ºC. Ruoff et al. reported a CVD method for large-area synthesis of high-quality and uniform GNS films on copper foils using a mixture of methane and hydrogen as precursors. As obtained films are predominantly single-layer GNS with a small percentage (less than 5%) of the area having few layers, and continuous across copper surface steps and grain boundaries. Particularly, one of the major benefits of their process is that it could be used to grow GNS on 300 mm copper films on Si substrate and this GNS film could also be easily transferred to alternative substrates, such as SiO_2/Si or glass. Recently, Bae and coworkers reported a roll-to-roll production of 30 inch (Figure 3) graphene films using the CVD approach [41].

Another technique for the GNS synthesis is Epitaxial growth on silicon carbide (SiC). It is a very promising method for the synthesis of uniform, wafer-size graphene nano layers, in which single crystal SiC substrates are heated in vacuum to high temperatures in the range of 1200–1600 ºC. Since the sublimation rate of silicon is higher than that of carbon, excess carbon is left behind on the surface, which rearranges to form GNS[42-44] More recently Bao et al has reported an interesting route for the preparation of GNS that employed commercial polycrystalline SiC granules instead of single-crystal SiC [45] to formulate high-quality free-standing single-layer GNS.

2.4. Chemically converted Graphene

At present, the most viable method to afford graphene single sheets in considerable quantities is chemical conversion of graphite to graphene oxide followed by successive reduction [46-48]. Graphite oxide (GO) is usually synthesized through the oxidation of graphite using strong oxidants including concentrated sulfuric acid, nitric acid and potassium permanganate.

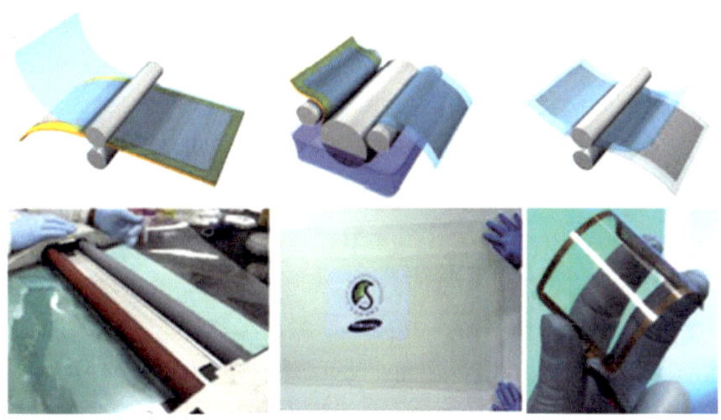

Figure 3. Schematic of the roll-based production of graphene films grown on a copper foil. A transparent ultra large-area graphene film transferred on a 35-in. PET sheet and an assembled graphene/PET touch panel showing outstanding flexibility. (Reprinted by permission from Macmillan Publishers Ltd: [Nature Nanotechnology] (ref 41: copyright (2010)

2.4.1. Synthesis of graphene oxide and its reduction

In 1859, Brodie was first to prepared graphite oxide by the oxidation of graphite with fuming nitric acid and potassium chlorate under cooling [49], In 1898, Staudenmaier improved this protocol by using concentrated sulfuric acid as well as fuming nitric acid and adding the chlorate in multiple aliquots over the course of the reaction. This small change in the procedure made the production of highly oxidized GO in a single reaction vessel [50]. In 1958, Hummers reported the method most commonly used today in which graphite is oxidized by treatment with $KMnO_4$ and $NaNO_3$ in concentrated H_2SO_4 [51]. These three methods comprise the primary routes for forming GO. Recently, an improved method was reported by Marcano et al. [52], they used $KMnO_4$ as the only oxidant and an acid mixture of concentrated H_2SO_4 and H_3PO_4 (9:1) as the acidic medium. This technique greatly increased the efficiency of oxidizing graphite to GO and also prevented the formation of toxic gases, such as NO_2 and N_2O_4. The graphene oxide prepared by this method is more oxidized than that prepared by Hummer's method and also possesses a more regular structure. Graphite can also be oxidized by benzoyl peroxide (BPO) at 110 C for 10 min in an opened system

(Caution! BPO is a strong oxidizer and may explode when heated in a closed container) to GO [53]. This technique provides a fast and efficient route to graphene oxide. The composition of anhydrous GO is approximately $C_8O_2\,(OH)_2$. Almost none of the carbon of the graphite used is lost during the formation of GO. Compared to pristine graphite, GO is heavily oxygenated bearing hydroxyl and epoxy groups on sp^3 hybridized carbon on the basal plane, in addition to carbonyl and carboxyl groups located the sheet edges on sp^2 hybridized carbon. Hence, GO is highly hydrophilic and readily exfoliated in water, yielding stable dispersion consisting mostly of single layered sheets (graphene oxide). It is important to note that although graphite oxide and graphene oxide share similar chemical properties (i.e. surface functional group), their structures are different. Graphene oxide is a monolayer material produced by the exfoliation of graphite oxide. Sufficiently dilute colloidal suspension of graphene oxide prepared by sonication are clear, homogeneous and stable indefinitely. AFM images of GO exfoliated by the ultrasonic treatment at concentrations of 1 mg/ml in water always revealed the presence of sheets with uniform thickness (1 nm). The pristine graphite sheet is atomically flat with the Vander Waals thickness of 0.34 nm, graphene oxide sheets are thicker due to the displacement of sp^3 hybridized carbon atoms slightly above and below the original graphene plane and presence of covalently bound oxygen atoms. A similar degree of exfoliation of GO was also attained for N,N-dimethylformamide (DMF), tetrahydrofuran (THF), N-methyl-2-pyrrolidone (NMP) and ethylene glycol [54]. Chung et al [55] has utilized the utilized a modified hummer method to produces a large sized highly functionalized graphene oxide, In a typical method a small amount of graphite was irradiated for 10 s in a microwave oven and expanded to about 150 times its original volume and carried out the further oxidation by modified Hummers method (Figure 4).

2.4.2. Reduction of Graphene oxide

As discussed the exfoliated sheets contain many hydrophilic functionality like –OH, —COOH, —C—O—C—, C=O which keep them highly dispersible and the layered sheets are named graphene oxide (GO). The most attractive property of GO is that it can be reduced to graphene-like sheets by removing the oxygen-containing groups with the recovery of a conjugated structure. The reduced GO (RGO) sheets are usually considered as one kind of chemically derived graphene (CCG). It is a very promising candidate for many applications such as electronic devices [56,57], polymer composites [58-61], energy conversion, storage materials [62,63], and sensors [64]. The most desirable goal of any reduction procedure is produce graphene-like materials similar to the pristine graphene. Though numerous efforts have been made, the final target is still a dream. Residual functional groups and defects dramatically alter the structure of the carbon plane and affect its conductivity which mainly depend on the long-range conjugated network of the graphitic lattice [65,66]. Functionalization breaks the conjugated structure and localizes p-electrons, which results in a decrease of both carrier mobility and carrier concentration therefore, it is not appropriate to refer to RGO/CCG, simply as graphene since the properties are considerably different [67-72]. Several reducing agents have been used to reduce graphene oxide, such as hydrazine [73], sodium borohydride [74], hydroiodic acid [75,76], sulfur-containing compounds [77], ascorbic acid [78], and vitamin C[79]

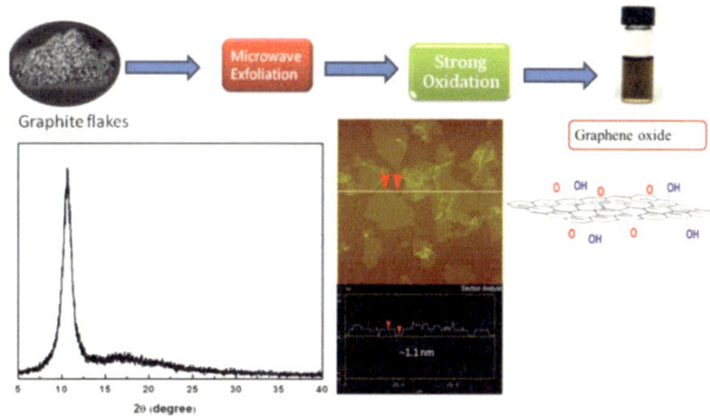

Figure 4. Scheme of synthesis of XRD and AFM image of GO.

Among them, hydrazine is widely used because it is an effective reducing agent and well suited to the reduction of graphene oxide in various media, including the aqueous phase, gas phase, and especially in organic solvents. The most obvious changes can be directly observed or measured to judge the reducing effect of different reduction processes. Since a reduction process can dramatically improve the electrical conductivity of GO, the increased charge carrier concentration and mobility will improve the reflection to incident light, and color changes to brown to black as shown in Figure 4. The variation of electrical conductivity of RGO can be a direct criterion to judge the effect of different reduction methods. Another important change is C/O ratio is usually obtained through elemental analysis measurements by combustion, and also by X-ray photo-electron spectrometry (XPS) analysis. Depending on the preparation method, GO with chemical compositions ranging from $C_8O_2H_3$ to $C_8O_4H_5$, corresponding to a C/O ratio of 4:1–2:1, is typically produced [80,81]. After reduction, the C/O ratio can be improved to approximately 12:1 in most cases, but values as large as 246:1 have been recently reported [82]. In addition to these other tools like Raman spectroscopy, solid-state FT-NMR spectroscopy, transmission electron microscopy (TEM), and atomic force microscopy (AFM), are most promising tools to show the structural changes of GO after reduction.

There are several routes to reduce the graphene oxide like thermal annealing, microwave and photo reduction, and chemical reduction (Chemical reagent reduction, Solvo-thermal reduction, Multi-step reduction, Electrochemical reduction, Photocatalyst reduction). Here we are only focusing on thermal annealing and solvo-thermal chemical reduction as these are most wide used method for the reduction.

2.4.3. Thermal annealing

GO can be reduced by thermal annealing and a temperature less than 2000 ºC was used in the initial stages of graphene research, to exfoliate graphite oxide to achieve graphene [83 84]]. The mechanism of exfoliation is mainly the sudden expansion of CO or CO_2 gases evolved into the spaces between graphene sheets during rapid heating of the graphite oxide. However, this technique is not so promising as it leads to the structural damage to graphene sheets caused by the release of carbon dioxide [85]. Approximately 30% of the mass of the graphite oxide is lost during the exfoliation process, leaving behind lattice defects throughout the sheet [83]. As a result, the electrical conductivity of the graphene sheets has a typical mean value of 10–23 S/cm that is much lower than that of perfect graphene, indicating a weak effect on reduction and restoration of the electronic structure of carbon plane.

An alternative way is to exfoliate graphite oxide in the liquid phase, which enables the exfoliation of graphene sheets with large lateral sizes [86]. The reduction is carried out after the formation of macroscopic materials, e.g. films or powders, by annealing in inert or reducing atmospheres. In this strategy, the heating temperature significantly affects reduction of GO. Schniepp et al [83] found that if the temperature was less than 500 ºC, the C/O ratio was not more than 7, while if the temperature reached 750 ºC, the C/O ratio could be higher than 13. The reduced GO film obtained at 500 ºC was only 50 S/cm, while for those at 700 and 1100 ºC it could be 100 and 550 S/cm respectively. In addition to annealing temperature, annealing atmosphere is important for the thermal annealing reduction of GO. Since the etching of oxygen will be dramatically increased at high temperatures, oxygen gas should be excluded during annealing. As a result, annealing reduction is usually carried out in vacuum [87], or an inert [88] or reducing atmosphere [89].

2.4.4. Chemical reduction

Chemical reduction has been evaluated as one of the most efficient methods for low-cost, large-scale production of Graphene. Another advantage of chemical reduction methods is that the produced GNS in the form of a monolayer can be conveniently deposited on any substrate with simple processing. The chemical reduction method involves graphite oxidation by a strong oxidant to create graphene oxide, which is subsequently reduced by reducing agents [90-94] thermal [94], solvo-thermal [95-98], or electrochemical [99] methods to produce chemically modified graphene. Among these reduction processes, hydrazine reduction and solvo-thermal reduction can create process able colloidal dispersions of reduced graphene oxide, which may be used in a wide range of applications. Chemical reduction using hydrazine is one of the most effective methods for converting graphene oxide to chemically converted graphene (CCG). Chung at al [100] has report a simple and effective method for reducing and functionalizing graphene oxide into chemically converted graphene by solvo-thermal reduction of a graphene oxide suspension in N-methyl-2-pyrrolidone (NMP). NMP is a powerful solvent for dispersing SWCNT and graphene and high boiling point (~202 ºC) of NMP facilitates the use of NMP as a solvent for solvo-thermal reduction in open systems. Dubin et al. [101] reported solvo-thermally reduced graphene oxide suspension in NMP for 24 h at 200 ºC under oxygen-free conditions. As obtained, graphene was well dispersed in various solvents such as dimethylsulfoxide, ethyl acetate, acetonitrile, ethanol, tetrahydrofuran (THF), DMF, chloroform, and acetone with minimum precipitation at 1 mg/ml after 6 weeks.

However, the electrical conductivity of free-standing paper of graphene prepared by filtration was very low i.e. 374 S/m, when dried in air and 1380 S/m, when dried at 250 ºC. Recently Chung et al. have report the superior disersibility of RGO in N,N-dimethylformamide

(DMF) by controlling the conditions of the hydrazine reduction. Instead of reducing the graphene oxide using hydrazine at high temperature (80–100 ºC) excess amounts of hydrazine were used. Reduction was carried out at ambient temperature to achieve extensive reduction with a C/O ratio of approximately 9.5, which is comparable to previous reports, while the RGO disersibility in NMP was as high as 0.71 mg/mL. The key to achieve highly dispersed RGO is performing the hydrazine reduction of graphene oxide at low temperature, which minimizes the formation of irreversible RGO aggregates [102]. Notably, the electrical conductivity of the hydrazine reduced graphene (HRGs) was sharply and inversely proportional to the dispersibility in DMF.

3. Conducting Polymer-graphene composite

Nanocomposites have been investigated since 1950, but industrial importance of the nanocomposites came nearly forty years later following a report from researchers at Toyota Motor Corporation that demonstrated large mechanical property enhancement using montmorillonite as filler in a Nylon-6 matrix and new applications of polymers. A nanocomposite is defined as a material with more than one solid phase, metal ceramic, or polymer, compositionally or structurally where at least one dimension falls in the nanometers range. Most of the composite materials are composed of just two phases; one is termed the matrix, which is continuous and surrounds the other phase, often called the dispersed phase and their properties are a function of properties of the constituent phases, their relative amounts, and the geometry of the dispersed phase. The combination of the nanomaterial with polymer is very attractive not only to reinforce polymer but also to introduce new electronic properties based on the morphological modification or electronic interaction between the two components. Depending on the nature of the components used and the method of preparation, significant differences in composite properties may be obtained. Nanocomposites of conducting polymers have been prepared by various methods such as colloidal dispersions, electrochemical encapsulation coating of inorganic polymers, and insitu polymerization with nanoparticles and have opened new avenues for material synthesis [103-105].

Conducting polymer composites with graphite, CNT, Metal/metal oxides are studied a lot because of their usual electrical and mechanical properties. For example, In case of electromagnetic interference shielding application, the combination of magnetic nanoparticles with conducting polymer leads to form a ferromagnetic conducting polymer composite possessing unique combination of both electrical and magnetic properties. This type of materials can effectively shield electromagnetic waves generated from an electric source. When conducting polymers are combined with carbons material like CNT graphite and graphene they show good thermal and electrical properties as electronic conduction occurs at long range. In last couples of years, a variety of processing routes have been reported for dispersing the graphene based and it derivative as fillers in the polymer matrices. Many of these procedures are similar to those used for other nanocomposite systems but some are different and unique and have enhanced the bonding interaction at the interface between the filler and matrix significantly. Most of the dispersion methods produce composites by non-covalent assemblies where the polymer matrix and the filler interact through relatively weak dispersive forces. However, there is a growing research focus on introducing covalent linkages between graphene-based filler and the supporting polymer to promote stronger interfacial bonding. It is well known that most of the π-conjugated conducting polymers (CPs) are quite different from classical insulating polymers. They have conjugated backbones, which provide them with unique electrical and optical properties. These polymers are conductive in their doped states while insulating in their neutral states. Furthermore they are usually brittle, weak in mechanical strengths and usually insoluble, intractable and decompose before melting, having poor processability [104]. Thus, CP/CCG composites were mostly prepared by in situ polymerizations using different approaches. The incorporation of CCG into conducting polymer is attractive for combining the properties of both components or improving the properties of resulting composites based on synergy effects. The major forerunner of conducting polymer family are polyaniline (PANI), polypyrrole (PPy), polythiophene and poly(3,4-ethylenedioxythiophene) PEDOT and most of the research work has been done on them, polyaniline [105,106-111], PPy [112], poly(3-hexylthiophene) (P3HT) [113], PEDOT [1114] have been hybridized with CCG to form composites.

3.3. *In-situ polymerization*

In situ polymerization combines a post-graphitization strategy and is most widely applied method for preparing CCG/CP composites. Most of the work is done on PANI as it has good environmental stability, reversible redox activity, and potential applications in sensing, energy conversion and storages and electromagnetic shielding application [109,110-111,115]. Graphene oxide/PANI composites can be prepared by polymerizing aniline in graphene oxide dispersion. After the reduction of GO with hydrazine, the corresponding composites with CCG can be obtained. The key point to note here is that, the polymerization must be carried out in an acidic medium (pH ~ 1) for producing high-quality PANI. However, over acidification of the solution will cause clogging of graphene oxide sheets. Thus, the pH value of the reaction system must be optimized carefully. A graphene oxide-polypyrrole composite was also prepared by in situ polymerization in water in the presence of a surfactant [116-120] although graphene oxide was not converted to CCG, the composite exhibited a higher electrical conductivity than pure PPy.

Graphene has a tendency of aggregation and shows poor solubility, is the dominant factor for limiting the application of this technique. In this case, special care should be taken to avoid the precipitation of CCG, especially when oxidant was added. In another work, Xu and Chen et al. polymerized 3,4-ethylenedioxythiophene (EDOT) in the dispersion of sulfonated graphene, giving a CCG/PEDOT composite [121]. They claimed that sulfonate groups could increase the solubility of CCG and acted as dopants of PEDOT.

3.4. Solution mixing

A very less number of research articles are available on the preparation of CCG/CP composites using solution mixing in comparison to insitu polymerization as most of the CPs are insoluble in common solvent. However, the solubility or dispersibility of CPs can be improved by chemical modifications or fabricating them into nanostructures. On the basis of this idea, CCG/sulfonated PANI (SPANI) [122] can be recognized as conjugated polyelectrolytes (CPE) according to their chain structures. The strong $\pi - \pi$ interaction between the CPE chains and the basal planes of CCG sheets enables the composites to form stable dispersions. Composite films can be fabricated by casting the blend solutions. Graphene oxide was also reduced in an organic solvent with the presence of P3HT, giving a CCG/P3HT composite. Transparent and conductive film of graphene –polymer composite can be spin-coated or evaporated to produce composite films and can be used as the counter electrode of a dye-sensitized solar cell [123-125].

3.5. Covalent Grafting of polyaniline on graphene sheet

Recently Kumar et al [126] has reported the covalent functionalization of amine-protected 4-aminophenol to acylated graphene oxide and simultaneously reduced and *in-situ* polymerized in the presence of aniline monomer and produces a highly conducting networks. In this the oxygen containing functional groups on the surface of graphene oxide make it easily dispersible in aqueous solution and act as nucleation sites for producing PANI on its surfaces. The fabrication of PANI-grafted RGO (PANI-g-RGO was carried out in three steps as shown in Figure 5.

First the GO was synthesized by modified Hummers method and acylated in the presence of excess $SOCl_2$ and then reacted with amine-protected 4-aminophenol. Futher, deprotection of N-(tert-butoxycarbonyl) groups by hydrolysis with trifluoroacetic acid. PANI-g-RGO was prepared from *in-situ* oxidative polymerization of aniline in the presence of an oxidant and amine-terminated RGO (RGO-NH_2) as an initiator. SEM image of the GO (Figure 6a) shows the layer-by-layer structure in stacking with a size of micrometers while PANI-g-RGO hybrid show typical fibrillar morphology (Figure 6 b and 6c), where in some areas the composites exhibit mainly an irregular morphology with multiple shapes including both fibrillar and a few rod-like structures. In some places embedded flakes of graphene in PANI matrix are seen, suggesting graphene interconnection with the polymer network and forming a highly conducting network. The electrical conductivity of these hybrid assemblies was observed as high as 8.66 S/cm. HR-TEM images of GO exhibit a transparent layered and wrinkled silk-like structure, representing a curled and corrugated morphology intrinsically associated with graphene. Interestingly, the TEM image (Figure 7d) showed a typical single layer GO sheet. After functionalization, tremendous changes in morphology have been observed which basically arises by the introduction of aminophenol (Figure 6e) and PANI (Figure 7f) on the RGO surfaces. As for the PANI-g-RGO composite (Figure 7c), the coating of PANI is clearly visible, and it clearly distinguishes itself from the highly crystalline graphitic support, which is attributed to the surrounding of PANI on the RGO host.

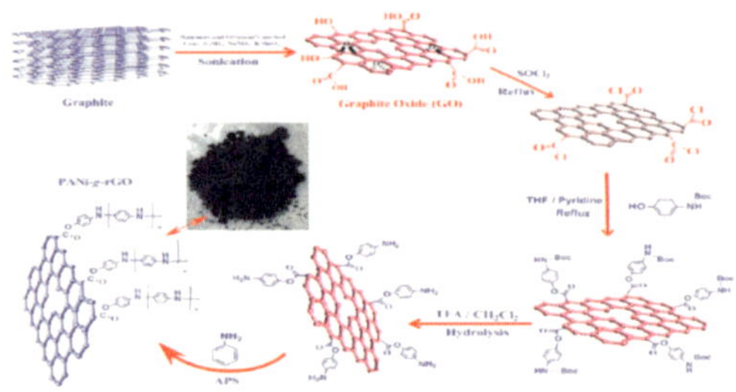

Figure 5. Scheme of direct grafting of polyaniline on the reduced graphene sheets (Reprinted with permission from ref 126 Copyright 2012 American Chemical Society)

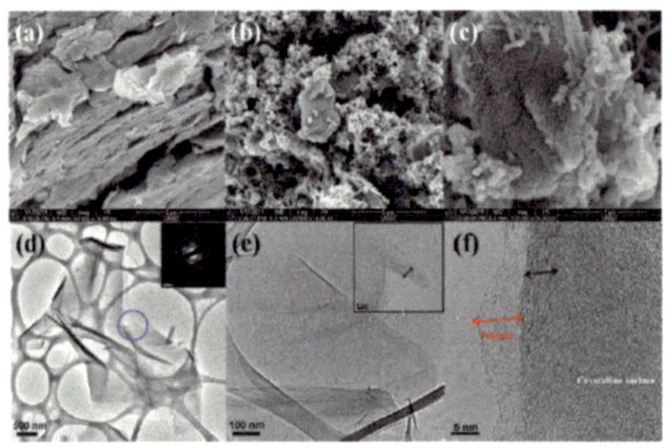

Figure 6. Typical FE-SEM images: (a) GO; (b and c) the surface of the PANI-g-RGO hybrid. HR-TEM images: (d) GO. Inset image is of a selected-area electron diffraction (SAED) pattern; (e) RGO-NH_2. Inset image is at higher magnification; (f) PANI-g-RGO. (Reprinted with permission from ref 126 Copyright 2012 American Chemical Society)

4. Application of conducting polymer graphene composites in EMI shielding

Graphene being a two-dimensional (2D) structure of carbon atoms own exceptional chemical, thermal, mechanical, and electrical properties and mechanical properties. Extensive research has shown the potential of graphene or graphene-based sheets to impact a wide range of technologies. In this section, graphene based conducting polymer composites are discussed focusing their use an Electromagnetic interference shielding material [127-130].

The development made in the Nano sciences & nanotechnology had flourished the electronic industries. Electronic systems have compact with increased the density of electrical components within an instrument. The operating frequencies of signals in these systems are also increasing and have created a new kind of problem called electromagnetic interference (EMI). Unwanted EMI effects occur when sensitive devices receive electromagnetic radiation that is being emitted whether intended or not, by other electric or electronic devices such as microwaves, wireless computers, radios and mobile phones. As a result, the affected receiving devices may malfunction or fail. The effects of electromagnetic interference are becoming more and more pronounced, caused by the demand for high-speed electronic devices operating at higher frequencies, more intensive use of electronics in computers, communication equipment and the miniaturisation of these electronics. For example, mobile phones and smartphones are typically operating at 2-3 GHz for data transmission through Universal Mobile Telecommunications Systems (UMTS). Compact, densely packed electronic components produce more electronic noise. Due to the increase in use of high operating frequency and band width in electronic systems, especially in X-band and broad band frequencies, there are concerns and more chances of deterioration of the radio wave environment. These trends indicate the need to protect components against electromagnetic interference (EMI) in order to decrease the chances of these components adversely affecting each other or the outer world. The effects of electromagnetic interference can be reduced or diminished by positioning a shielding material between the source of the electromagnetic field and the sensitive component. Shielding can be specified in the terms of reduction in magnetic (and electric) field or plane-wave strength caused by shielding. The effectiveness of a shield and its resulting EMI attenuation are based on the frequency, the distance of the shield from the source, the thickness of the shield and the shield material. Shielding effectiveness (SE) is normally expressed in decibels (dB) as a function of the logarithm of the ratio of the incident and exit electric (E), magnetic (H), or plane-wave field intensities (F): SE (dB) = 20 log (E_o/E_1), SE (dB) = 20 log (H_o/H_1), or SE (dB) = 20 log (F_o/F_1), respectively. With any kind of electromagnetic interference, there are three mechanisms contributing to the effectiveness of a shield. Part of the incident radiation is reflected from the front surface of the shield, part is absorbed within the shield material and part is reflected from the shield rear surface to the front where it can aid or hinder the effectiveness of the shield depending on its phase relationship with the incident wave, as shown in Figure 7

Therefore, the total shielding effectiveness of a shielding material (SE) equals the sum of the absorption factor (SE_A), the reflection factor (SE_R) and the correction factor to account for multiple reflections (SE_M) in thin shields

$$SE = SE_A + SE_R + SE_M \tag{1}$$

All the terms in the equation are expressed in dB. The multiple reflection factor SE_M, can be neglected if the absorption loss SE_A is greater than 10 dB. In practical calculation, SE_M can also be neglected for electric fields and plane waves.

4.3. Absorption Loss

Absorption loss SE_A, is a function of the physical characteristics of the shield and is independent of the type of source field. Therefore, the absorption term SE_A is the same for all three waves. As shown in Figure 8, when an electromagnetic wave passes through a medium its amplitude decreases exponentially. This decay or absorption loss occurs because currents induced in the medium produce ohmic losses and heating of the material, where E_1 and H_1 can be expressed as $E_1 = E_o e^{-t/\delta}$ and $H_1 = H_o e^{-t/\delta}$. The distance required by the wave to be attenuated to 1/e or 37% is defined as the skin depth. Therefore, the absorption term SE_A in decibel is given by the expression:

$$SE_A = 20(t / \delta)\log e = 8.69(t / \delta) = 131.t\sqrt{f\mu\sigma} \quad (2)$$

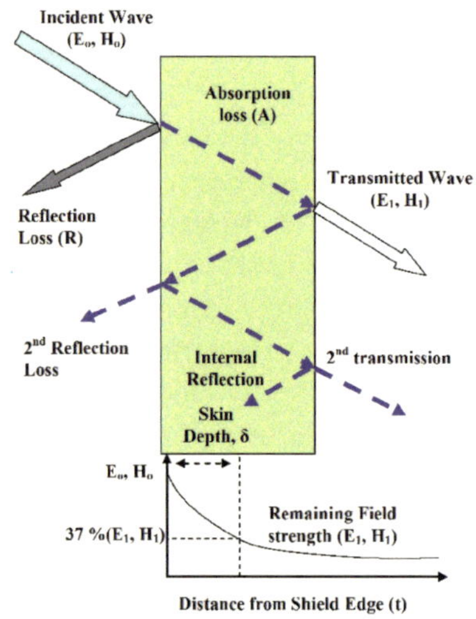

Figure 7. Graphical representation of EMI shielding

where, t is the thickness of the shield in mm; f is frequency in MHz; μ is relative permeability (1 for copper); σ is conductivity relative to copper. The skin depth δ can be expressed as:

The absorption loss of one skin depth in a shield is approximately 9 dB. Skin effect is especially important at low frequencies, where the fields experienced are more likely to be predominantly magnetic with lower wave impedance than 377 Ω. From the absorption loss point of view, a good material for a shield will have high conductivity and high permeability along with a sufficient thickness to achieve the required number of skin depths at the lowest frequency of concern.

$$\delta = \frac{1}{\sqrt{\pi f\mu\sigma}} \quad (3)$$

4.4. Reflection Loss

The reflection loss is related to the relative mismatch between the incident wave and the surface impedance of the shield. The computation of refection losses can be greatly simplified by considering shielding effectiveness for incident electric fields as a separate problem from that of electric, magnetic or plane waves. The equations for the three principle fields are given by the expressions

$$R_E = K_1 10\log\left(\frac{\sigma}{f^3 r^2 \mu}\right) \quad (4)$$

$$R_H = K_2 10\log\left(\frac{f r^2 \sigma}{\mu}\right) \quad (5)$$

$$R_P = K_3 10\log\left(\frac{f\mu}{\sigma}\right) \quad (6)$$

where, R_E, R_H, and R_P are the reflection losses for the electric, magnetic and plane wave fields, respectively, expressed in dB; σ is the relative conductivity relative to copper; f is the frequency in Hz; μ is the relative permeability relative to free space; r is the distance from the source to the shielding in meter.

4.5. Multiple Reflections

The factor SE_M can be mathematically positive or negative (in practice, it is always negative) and becomes insignificant when the absorption loss SE_A> 6 dB. It is usually only important when metals are thin and at low frequencies (i.e., below approximately 20 kHz). The formulation of factor SE_M can be expressed as

$$SE_M = -20\log(1 - e^{-2t/\delta}) \quad (7)$$

Due to their high electrical conductivity, metals are particularly suitable as shielding material against electromagnetic fields. This can be a self-supporting full metal shielding, but also a sprayed, painted or electro-less applied conducting coating (e.g. nickel) on a supporting material such as plastic. Another option is the incorporation of metal (stainless steel) powder or fibres as conducting filler in a plastic matrix.

However, there are a certain draw backs to use metal as a shielding material. The weight of the 'heavy' metal can be an issue in the case of full metal shielding, processing and corrosion are other draw back to prohibit their use. In order to produce metal coatings, at least two processing techniques have to be applied one for the support and one for the coating, which can be costly. It will also be difficult to apply these coatings onto complicated shaped objects. In addition, the long-term adhesion of the coating to the support has to be reliable.

To solve the EMI problems, spinel-type ferrites, metallic magnetic materials, and carbon nanotube (CNT) composites [131-138] have been extensively studied. To achieve higher SE and to overcome the drawbacks of the metal-based art, polymer material with appropriate conductive fillers can be shaped into an EMI shielding substrate, which exhibit improved EMI shielding and absorption properties. The conductive composites in the form of coatings, strips or molded materials have been prepared by the addition of highly conductive fillers or powders to non-conductive polymer substrates. Conductive polymer composites give a significantly better balance of mechanical and electrical properties than some of the current generation of commercially available EMI-shielding material. It is observed that the high conductivity and dielectric constant of the materials contribute to high EMI shielding efficiency (SE). The combination of conducting polymer with nanostructured ferrite along with graphene offers potentials to fight with EM pollution. Recently Dhawan et al have reported that if magnetic particles of barium ferrite or Fe_2O_3 are incorporated in the polymer matrix they improve the magnetic and dielectric properties of host materials [139-141]. Therefore, conjugated polymers combined with magnetic nanoparticles to form ferromagnetic nanocomposites provide an exciting system to investigate the possibility of exhibiting novel functionality. The unique properties of nanostructured ferrite offer excellent prospects for designing a new kind of shielding materials. The absorption loss in the material is caused by the heat loss under the action between electric dipole and/or magnetic dipole in the shielding material and the electromagnetic field so that the absorption loss is the function of conductivity and the magnetic permeability of the material. The designing of ferrite based conducting polymer nanocomposites increases the shielding effectiveness. Conducting and magnetic properties of conducing polymer-ferrite nanocomposites can be tuned by suitable selection of polymerization conditions and controlled addition of ferrite nanoparticles. The contribution to the absorption value comes mainly due the magnetic losses (μ'') and dielectric losses (ε''). The dependence of SE_A on magnetic permeability and conductivity demonstrates that better absorption value has been obtained for material with higher conductivity and magnetization. Therefore, it has been concluded that the incorporation of magnetic and dielectric fillers in the polymer matrix lead to better absorbing material which make them futuristic radar absorbing material.

5. Preparation of conducting polyaniline- graphene/ ferrite Composites

There are many methods for the preparation of conducting polyaniline (PANI) like chemical or electrochemical oxidation of a monomer where the polymerization reaction is stoichiometric in electrons. However, number of methods such as photochemical polymerization, pyrolysis, metal-catalyzed polymerization, solid-state polymerization, plasma polymerization, ring-forming condensation, step-growth polymerization, and soluble precursor polymer preparation, have been reported in literature for synthesis of conjugated polymers. However, as discussed earlier good quality of polymer graphene composite can synthesized *in-situ* polymerization technique [140].

5.3. Synthesis of nanocomposites

Prior to the synthesis of polyaniline graphene composite, graphene oxide was synthesis using modified Hummers method followed by hydrazine reduction at 80 ºC to get CCG/RGO and *in-situ* polymerized can carried out. The Oxidative polymerization of aniline in aqueous acidic media using ammonium persulfate as an oxidant is the most common and widely

http://dx.doi.org/10.5772/50408

used method [141]. However by taking cationic or anionic surfactant one can easily controlled the morphology of the polymer. Therefore, emulsion polymerization is an appropriate method as the polymerization reaction takes place in a large number of loci dispersed in a continuous external phase. In a typical synthesis process, functional protonic acid such as dodecyl benzene sulfonic acid (DBSA) is used which being a bulky molecule, can act both as a surfactant and as dopant. The polymerization of aniline monomer in the presence DBSA (dodecyl benzene sulfonic acid) leads to the formation of emeraldine salt form of polyaniline. When the graphene nanosheets are dispersed and homogenized with DBSA in aqueous solution, micelles are formed over the graphene sheets. Anilinium cations sit between the individual DBSA molecules near the shell of the micelle complexed with sulfonate ion. When polymerization proceeds, anilinium cations are polymerized within the micelle with DBSA & over the graphene sheets resulting in the formation of polyaniline graphene composite. Pictorial representation for the formation of polyaniline-graphene composite is shown in figure 8. The same methodology can be used to prepare ferromagnetic conducting polymer graphene composite.

Here key to synthesized good quality of polymer composite is the weight ratio of ferrite and graphene to monomer. In this process, water is the continuous phase and DBSA is a surfactant that acts as discontinuous phase. Monomer aniline is emulsified to form the micro micelles of oil in water type. The shape of a micelle is a function of the molecular geometry of its surfactant molecules and solution conditions such as surfactant concentration, temperature, pH and ionic strength. Addition of the APS to the aniline monomer leads to the formation of cation radicals which combine with another monomer moiety to form a dimer, which on further oxidation and combination with another cation radical forms a termer and ultimately to a long chain of polymer.

Recently our group has synthesized the graphene oxide coated Fe_2O_3 nanoparticles and prepared polyaniline GO- Fe_2O_3 (PGF) nanocomposite by the same procedure as depicted in scheme (Figure 9)and reports the SE and dielectric measurement. Here we have varied the weight ratio of monomer to γ-Fe_2O_3 as An: GO: γ-Fe_2O_3: 1:1:1 (PGF11), 1:1:2 (PGF12) and compared results with pristine polyaniline doped with DBSA (PD13) without ferrite particles and GO/polyaniline composite having weight ratio of aniline: GO in 2:1 (PG21) has also been synthesized in similar manner.

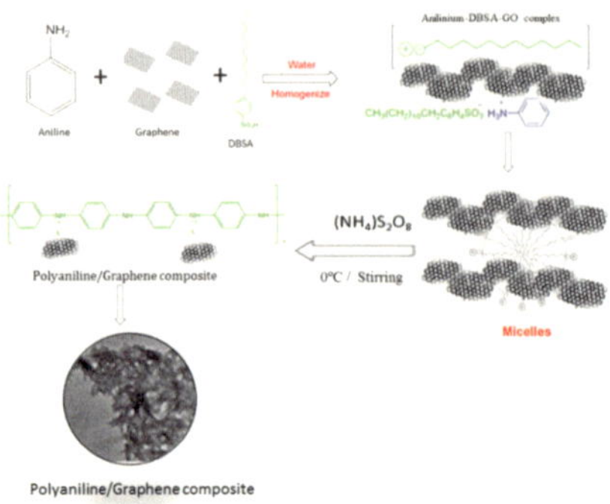

Figure 8. Schematic representation of the polymerization of graphene polyaniline composite

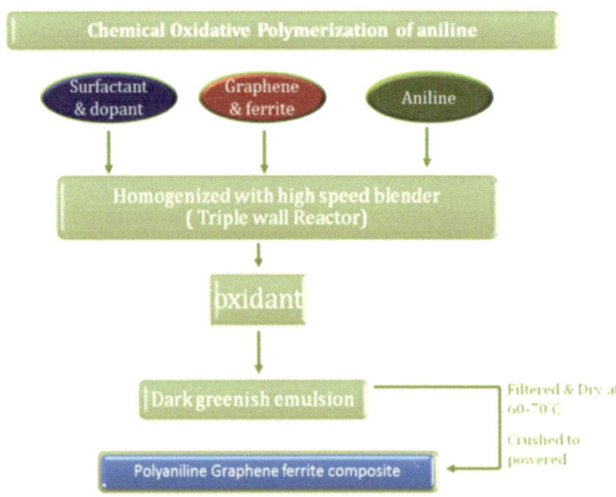

Figure 9. Pictorial representation for the formation of polyaniline nanocomposite by chemical oxidative polymerization

5.4. Shielding Measurements

Figure 10 shows the variation of the SE_A and SE_R with frequency for single layer of PG21, PGF11 and PGF12 composites in 12.4-18 GHz frequency range having thickness of ~2mm. It has been observed that conducting composites of polyaniline with nanosize γ-Fe_2O_3 and GO have SE mainly attributed by absorption. The maximum shielding effectiveness due to absorption $SE_{A(max)}$ has been ca. 41.6 dB at 16.1 GHz for PGF12 sample whereas for PG21 and PGF11 samples the $SE_{A(max)}$ has been ca. 20 dB at 18 GHz and 24.8 dB at 13.8 GHz, respectively. For the reflection part, the $SE_{R(max)}$ has been ca. 7.7 dB at 12.4 GHz for PG21 sample whereas for PGF11 and PGF12 samples the $SE_{R(max)}$ has been ca. 1.3 and 2 dB at 18 GHz, respectively. The higher values of SE_A strongly suggest that the microwave absorption in the PGF nanocomposites results mainly from the absorption loss rather than the reflection loss. In addition, it is observed that SE increases with the concentration of γ-Fe_2O_3 in the polymer matrix. The increase in the absorption part is mainly attributed to be due to the presence of GO and a magnetic γ-Fe_2O_3 nanomaterial which increase more scattering which in turn results in more attenuation of the electromagnetic radiations. Moreover, with the change in the frequency in 12–18 GHz, the variation in the SE_A value is very small, showing high bandwidth, which is commercially important for wide band absorbers. Clearly, compared to the other carbon coated magnetic nanoparticle as reported by Zhang et al. [142] (R_{max} is ca. –32 dB) and Tang et al. [143] (R_{max} is ca. –36 dB) these PGF composites demonstrate superior absorption properties.

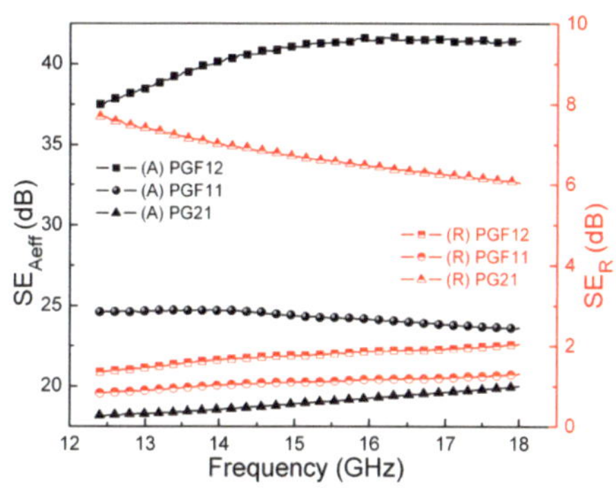

Figure 10. Dependence of shielding effectiveness (SE_A& SE_R) of polyaniline composites PG21, PGF11 and PGF12 on frequency in 12.4-18 GHz

The total shielding effectiveness ($SE_T = SE_R + SE_A$) of the respective samples has been calculated and it is observed that the PGF12 composite show maximum SE_T value of 43.5 dB whereas total SE for PG21 and PGF11 composites is of same order i.e. ~ 26 dB. In PG21 composite, incorporation of GO in the polymer matrix increase the total SE to 26 dB in which ~18 dB is due to absorption and ~ 8 dB is due to the reflection. With the addition of γ-Fe_2O_3 nanoparticles the absorption part increases to ~ 24.5 dB while reflection part decreases to ~ 1.5 dB and further by doubling the concentration of γ-Fe_2O_3 nanoparticles the absorption value enhanced to 41.6 dB. This increase in the absorption of microwave is due to the fact that in PG21 only dielectric losses contributes to the SE_A whereas in PGF11 both dielectric and magnetic losses contributes to the absorption of microwaves. The dependence of SE on complex permittivity and permeability can be expressed as [144]

$$SE_A(dB) = 20\frac{d}{\delta}\log e = 20d\sqrt{\frac{\mu_r \omega \sigma_{AC}}{2}}.\log e \quad (8)$$

$$SE_R(dB) = 10\log\left(\frac{\sigma_{AC}}{16\omega\mu_r\varepsilon_0}\right) \tag{9}$$

where, d is the thickness of the shield, μ_r is the magnetic permeability, δ is the skin depth, $\sigma_{AC} = \omega\varepsilon_0\varepsilon''$ is the frequency dependent conductivity [145], ε″ is imaginary part of permittivity (dielectric loss factor), ω is the angular frequency (ω = 2πf) and ε_0 is the permittivity of the free space. From equations 8& 9, it is observed that with the increase in frequency, the SE_A values increases while the contribution of the reflection decreases. Dependence of SE_A and SE_R on conductivity and permeability revel that the material having higher conductivity and magnetic permeability can achieve better absorption properties.

5.5. Complex permittivity and permeability

To investigate the possible mechanism and effects giving rise to improve microwave absorption, complex permittivity (ε_r = ε′–jε″) and permeability (μ_r = μ′–jμ″) of the samples have been calculated using scattering parameters (S_{11}& S_{21}) based on the theoretical calculations given in Nicholson, Ross and Weir method [146,147]. The dielectric performance of the material depends on ionic, electronic, orientational and space charge polarization. The contribution to the space charge polarization appears due to the heterogeneity of the material. The real (ε′) and imaginary (ε″) part of complex permittivity vs. frequency has been shown in Fig. 11 (a& b). The real part (ε′) is mainly associated with the amount of polarization occurring in the material while the imaginary part (ε″) is related with the dissipation of energy. In polyaniline, strong polarization occurs due to the presence of polaron/bipolaron and other bound charges, which leads to high value of ε′ & ε″. With the increase in frequency, the dipoles present in the system cannot reorient themselves along with the applied electric field as a result of this dielectric constant decreases.

The main characteristic feature of GO is that it has high dielectric constant (ε′~32) with dominant dipolar polarization and the associated relaxation phenomenon constitutes the loss mechanism. With the addition of GO and γ-Fe_2O_3 in polyaniline matrix, significant increase in the imaginary part of complex permittivity has been observed. The higher values of the dielectric loss is attributed to the more interfacial polarization due to the presence of GO and γ-Fe_2O_3 particles which consequently leads to more shielding effectiveness due to absorption. Fig. 12 (a& b) shows the variation of real part and imaginary part of magnetic permeability with frequency. The magnetic permeability of all the samples decreases with the increase in frequency whereas, higher magnetic loss has been observed for higher percentage of γ-Fe_2O_3 in the polymer matrix.

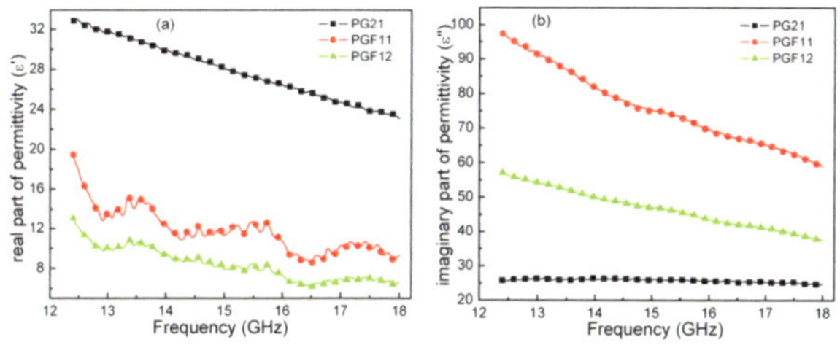

Figure 11. Behavior of (a) real and (b) imaginary part of permittivity of PG21, PGF11 and PGF12 composites as a function of frequency

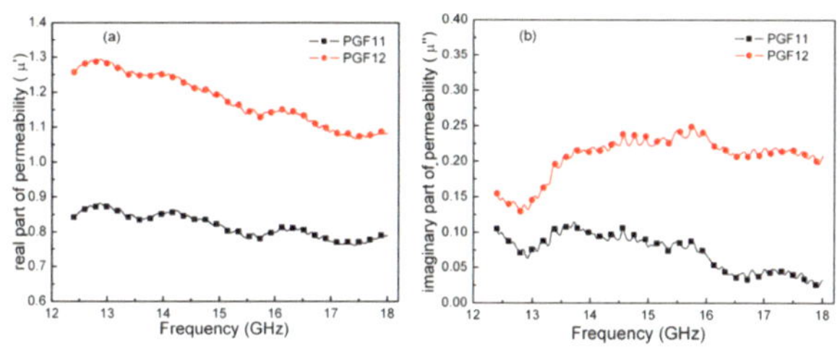

Figure 12. Variation of real and imaginary part of magnetic permeability of PGF11 and PGF12 composites as a function of frequency

The magnetic loss caused by the time lag of magnetization vector (M) behind the magnetic field vector. The change in magnetization vector generally brought about by the rotation of magnetization and the domain wall displacement. These motions lag behind the change of the magnetic field and contribute to the magnetic loss (μ″). The rotation of domain of magnetic nanoparticles might become difficult due to the effective anisotropy (magneto-crystalline anisotropy and shape anisotropy). The surface area, number of atoms with dangling bonds and unsaturated coordination on the surface of polymer matrix are all enhanced [148-150]. These variations lead to the interface polarization and multiple scattering, which is useful for the absorption of large number of microwaves. Therefore we can conclude that, incorporation of graphene along with ferrite nanoparticles in the polyaniline matrix by *in-situ* emulsion polymerization.leads to increase the absorption of the electromagnetic wave to a large extent. The high value of shielding effectiveness due to absorption (41.6 dB that demonstrates >99.99% attenuation of microwave) has been obtained because of the interfacial dipolar polarization and higher anisotropic energy due to the nano-size of the GO and γ-Fe_2O_3. The dependence of SE_A on magnetic permeability and ac conductivity shows that better absorption value can be obtained for a material having higher conductivity and magnetization.

In another article, Basavaraja et al [151] has synthesized polyaniline-gold-GO nanocomposite by an in situ polymerization and reports the microwave absorption property in the 2 –12 GHz frequency range. They found electromagnetic interference shielding effectiveness of polyaniline gold nanocomposite (PANI-GNP) has been enhanced due to the inclusion of

25% by weight GO in the polyaniline matrix. In Figure 13a, FT-IR spectra of GO, PANI-GNP, and PANI-GNP-GO has shown which clearly shows that some small deviations from the characteristic band of polyaniline that may be attributed some molecular interaction between GO with polyaniline ring has taken place this can be supported by UV–Vis spectra as shown in figure 14b. The spectrum for PANI-GNP shows three sharp absorption bands at around 320, 415, and 550 nm attributed to the π–π* transition of the benzenoid rings, and the polaron/bipolaron transition. The presence of GNPs is shown by the absorption peak at 520–530 nm. The peak at 550 nm indicates the presence of GNPs in PANI-GNP and their conjugation with PANI. The spectrum for PANI-GNP-GO shows all three absorption bands with slightly larger area as compared to that of PANI-GNP and red shift has taken place. However the GO peak in PANI-GNP-GO appeared to merged with the π–π* transition of the benzenoid rings. Figure 14c shows the SEM images Here the lump- and fiber-like structures of PANI-GNP disappeared after incorporation of GO into the matrix while the Figure 13d shows the TEM images for PANI-GNP and PANI-GNP-GO. In PANI-GNP, spherical GNPs covered by PANI polymers formed nano-capsules. These particles had a diameter between 25 and 45 nm. After the incorporation of GO in PANI-GNP, the surface morphology of PANI-GNP- GO changed. The spherical PANI-GNP particles disappeared and new pellet/flake-like structures were formed.

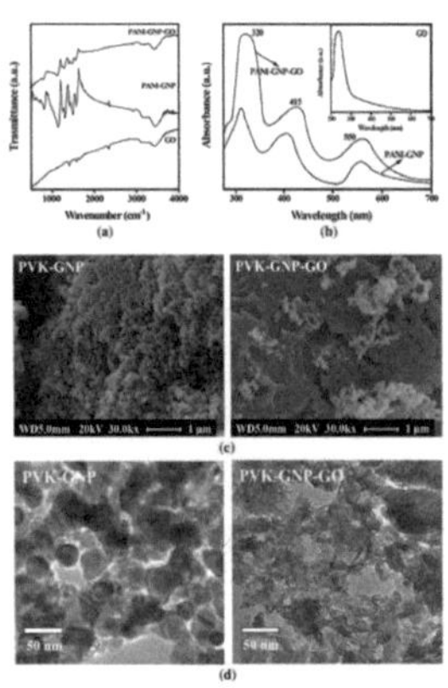

Figure 13. a) FT-IR spectra of GO, PANI-GNP, and PANI-GNP-GO, (b) UV–vis spectra of GO, PANI-GNP, and PANI-GNP-GO, (c) SEM images of PANI-GNP and PANI-GNP-GO, (d) TEM images of PANI-GNP and PANI-GNP-GO (Reprinted from ref 151 Copyright (2011), with permission from Elsevier)

The variation of the electromagnetic interference (EMI) shielding effectiveness (SE) as a function of frequency measured in the 2.0–12.0 GHz for GO, PANI-GNP, and PANI-GNP-GO films are shown in figure 14a Here GO exhibited lower values of SE, The SE values observed for GO and PANI-GNP in this frequency range were 20–33 and 45–69 dB, respectively. The higher values in the PANI-GNP are mainly attributed to the presence of GNPs. The highest values of SE have been observed in the PANIGNP- GO composite. The observed SE values for PANI-GNP-GO were within 90–120 dB. This range of values is very high compared with other carbon-based materials [152]. The EMI-SE data suggest that the electrochemical responses of PANI-GNP have been enhanced due to the inclusion of GO. Figure 14b shows the SE values variation with the thickness at 9.0 GHz. The SE values increase with increasing thickness of the sheets. This probably would overcome the poor cycling life, processability and solubility of the homo-polymer.

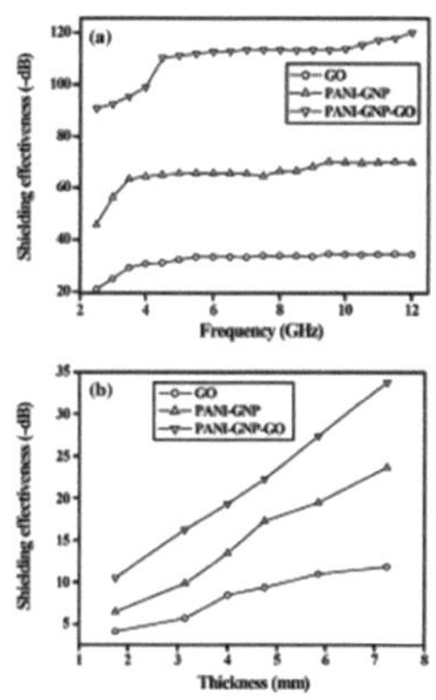

Figure 14. a) EMI-SE values as a function of frequency measured at 2.0–12.0 GHz. (b) EMI-SE values as a function of sheet thickness at 9.0 GHz for GO, PANI-GNP, and PANI-GNP-GO. (Reprinted from ref 151 Copyright (2011), with permission from Elsevier)

Conclusion

Although most of the research progress has been made in understanding the structure, processing, and properties of GO/RGO-based compound, there is significantly more to be explored and exploited given the highly versatile properties of the material. GO provides an exciting platform to study engineering, physics, chemistry, and materials science of unique 2D systems as well as offers a route towards realizing conducting polymer graphene composite. Continued involvement of researchers from all disciplines should further uncover the potential of GO/RGO polymer to processible and highly user friendly end product The enhancement in the microwave shielding and absorption properties of the polyaniline nanocomposite has been achieved by the incorporation of GO & RGO along with the magnetic filler in the polyaniline matrix. Now there is a need to form Graphene polymer composite paint that can be easily coat over the electronic encloser. Therefore, from the present studies, it can be concluded that the incorporation of magnetic and dielectric fillers in the polymer matrix lead to better absorbing material which make them futuristic radar absorbing material.

In spite of these interesting developments, a lot remains to be done with regard to both fundamental understanding and the much needed improvement of the method of the designing of electromagnetic shielding materials to operate at higher frequencies for their application.

Author details

Kuldeep Singh[1], Anil Ohlan[2] and S.K. Dhawan[1*]

*Address all correspondence to: skdhawan@mail.nplindia.ernet.in

1 Polymeric & Soft Material Section, National Physical Laboratory (CSIR), New Delhi –110 012, India

2 Department of Physics, Maharshi Dayanand University Rohtak – 124001, India

References

[1] Lincoln Vogel, F. (1977). The electrical conductivity of graphite intercalated with superacid fluorides: experiments with antimony pentafluoride. *Journal of Materials Science*, 12(5), 982-986.

[2] Kroto, H. W., Heath, J. R., O'Brien, S. C., Curl, R. F., & Smalley, R. E. (1985). C60: Buckminsterfullerene. *Nature*, 318(6042), 162-163.

[3] Iijima, S. (1991). Helical microtubules of graphitic carbon. *Nature*, 354(6348), 56-58.

[4] Novoselov, K.S., Geim, A.K., Morozov, S.V., Jiang, D., Zhang, Y., Dubonos, S.V., Grigorieva, I.V., & Firsov, A.A. (2004). Electric Field Effect in Atomically Thin Carbon Films. *Science*, 306(5696), 666-669.

[5] Novoselov, A. K., Geim, S. V., Morozov, D., Jiang, M. I., Katsnelson, I.V., Grigorieva, S. V., Dubonos, S. V., & Firsov, A. A. (2005). Two-dimensional gas of massless Dirac fermions in graphene. *Nature*, 438(7065), 197-200.

[6] Zhang, Y., Tan, Y.W., Stormer, H.L., & Kim, P. (2005). Experimental observation of the quantum Hall effect and Berry's phase in graphene. *Nature*, 438(7065), 201-204.

[7] Balandin, A. A., Ghosh, S., Bao, W., Calizo, I., Teweldebrhan, D., Miao, F., & Lau, C. N. (2008). Superior Thermal Conductivity of Single-Layer Graphene. *Nano Letters*, 8(3), 902-907.

[8] Nair, R. R., Blake, P., Grigorenko, A. N., Novoselov, K. S., Booth, T. J., Stauber, T., Peres, N. M. R., & Geim, A. K. (2008). Fine Structure Constant Defines Visual Transparency of Graphene. *Science*, 320(5881), 1308.

[9] Li, X., Wang, X., Zhang, L., Lee, S., & Dai, H. (2008). Chemically Derived, Ultrasmooth Graphene Nanoribbon Semiconductors. *Science*, 319(5867), 1229-1232.

[10] Ritter, K. A., & Lyding, J. W. (2009). The influence of edge structure on the electronic properties of graphene quantum dots and nanoribbons. *Nat Mater*, 8(3), 235-242.

[11] Cai, J., Ruffieux, P., Jaafar, R., Bieri, M., Braun, T., Blankenburg, S., Muoth, M., Seitsonen, A. P., Saleh, M., Feng, X., Mullen, K., & Fasel, R. (2010). Atomically precise bottom-up fabrication of graphene nanoribbons. *Nature*, 466(7305), 470-473.

[12] Ansari, S., & Giannelis, E. P. (2009). Functionalized graphene sheet-Poly(vinylidene fluoride) conductive nanocomposites. *Journal of Polymer Science Part B: Polymer Physics*, 47(9), 888-897.

[13] Ramanathan, T., Abdala, A. A., Stankovich, S., Dikin, D. A., Herrera, Alonso. M., Piner, R. D., Adamson, D. H., Schniepp, H. C., Chen, X., Ruoff, R. S., Nguyen, S. T., Aksay, I. A., Prud'Homme, R. K., & Brinson, L. C. (2008). Functionalized graphene sheets for polymer nanocomposites. *Nat Nano*, 3(6), 327-331.

[14] Stankovich, S., Dikin, D. A., Dommett, G. H. B., Kohlhaas, K. M., Zimney, E. J., Stach, E. A., Piner, R. D., Nguyen, S. T., & Ruoff, R. S. (2006). Graphene-based composite materials. *Nature*, 442(7100), 282-286.

[15] Fan, H., Wang, L., Zhao, K., Li, N., Shi, Z., Ge, Z., & Jin, Z. (2010). Fabrication, Mechanical Properties, and Biocompatibility of Graphene-Reinforced Chitosan Composites. *Biomacromolecules*, 11(9), 2345-2351.

[16] Zhang, K., Zhang, L. L., Zhao, X. S., & Wu, J. (2010). Graphene/Polyaniline Nanofiber Composites as Supercapacitor Electrodes. Chemistry of Materials. ; , 22(4), 1392-1401.

[17] Zhao, X., Zhang, Q., Chen, D., & Lu, P. (2010). Enhanced Mechanical Properties of Graphene-Based Poly(vinyl alcohol) Composites. *Macromolecules*, 43(5), 2357-2363.

[18] Kuila, T., Bose, S., Hong, C. E., Uddin, M. E., Khanra, P., Kim, N. H., & Lee, J. H. (2011). Preparation of functionalized graphene/linear low density polyethylene composites by a solution mixing method. *Carbon*, 49(3), 1033-1037.

[19] Wu, J., Becerril, H. A., Bao, Z., Liu, Z., Chen, Y., & Peumans, P. (2008). Organic solar cells with solution-processed graphene transparent electrodes. *Applied Physics Letters*, 92(26), 263-302.

[20] Huang, J., Wang, X., de Mello, A. J., de Mello, J. C., & Bradley, D. D. C. (2007). Efficient flexible polymer light emitting diodes with conducting polymer anodes. *Journal of Materials Chemistry*, 17(33), 3551-3554.

[21] Park, H., Rowehl, J. A., Kim, K. K., Bulovic, V., & Kong, J. (2010). Doped graphene electrodes for organic solar cells. *Nanotechnology*, 21(505204).

[22] Li, D., Muller, M. B., Gilje, S., Kaner, R. B., & Wallace, G. G. (2008). Processable aqueous dispersions of graphene nanosheets. *Nat Nano*, 3(2), 101-105.

[23] Shan, C., Yang, H., Han, D., Zhang, Q., Ivaska, A., & Niu, L. (2009). Water-Soluble Graphene Covalently Functionalized by Biocompatible Poly-l-lysine. *Langmuir*, 25(20), 12030-12033.

[24] Si, Y., & Samulski, E. T. (2008). Synthesis of Water Soluble Graphene. *Nano Letters*, 8(6), 1679-1682.

[25] Green, A. A., & Hersam, M. C. (2009). Solution Phase Production of Graphene with Controlled Thickness via Density Differentiation. *Nano Letters*, 9(12), 4031-4036.

[26] Chen, G., Wu, D., Weng, W., & Wu, C. (2003). Exfoliation of graphite flake and its nanocomposites. *Carbon*, 41(3), 619-62.

[27] Pötschke, P., Abdel, Goad. M., Pegel, S., Jehnichen, D., Mark, J. E., Zhou, D., & Heinrich, G. (2009). Comparisons Among Electrical and Rheological Properties of Melt-Mixed Composites Containing Various Carbon Nanostructures. *Journal of Macromolecular Science*, 47(1), 12-19.

[28] Yasmin, A., Luo, J. J., & Daniel, I. M. (2006). Processing of expanded graphite reinforced polymer nanocomposites. *Composites Science and Technology*, 66(9), 1182-1189.

[29] Jang, B., & Zhamu, A. (2008). Processing of nanographene platelets (NGPs) and NGP nanocomposites: a review. *Journal of Materials Science*, 43(15), 5092-5101.

[30] Li, X., Zhang, G., Bai, X., Sun, X., Wang, X., Wang, E., & Dai, H. (2008). Highly conducting graphene sheets and Langmuir-Blodgett films. *Nat Nano*, 3(9), 538-542.

[31] Blake, P., Brimicombe, P. D., Nair, R. R., Booth, T. J., Jiang, D., Schedin, F., Ponomarenko, L. A., Morozov, S. V., Gleeson, H. F., Hill, E. W., Geim, A. K., & Novoselov, K. S. (2008). Graphene-Based Liquid Crystal Device. *Nano Letters*, 8(6), 1704-1708.

[32] Hernandez, Y., Nicolosi, V., Lotya, M., Blighe, F. M., Sun, Z., De , S., Mc Govern, I. T., Holland, B., Byrne, M., Gun, Ko. Y. K., Boland, J. J., Niraj, P., Duesberg, G., Krishnamurthy, S., Goodhue, R., Hutchison, J., Scardaci, V., Ferrari, A. C., & Coleman, J. N. (2008). High-yield production of graphene by liquid-phase exfoliation of graphite. *Nat Nano*, 3(9), 563-568.

[33] Lotya, M., Hernandez, Y., King, P. J., Smith, R. J., Nicolosi, V., Karlsson, L. S., Blighe, F. M., De , S., Wang, Z., Mc Govern, I. T., Duesberg, G. S., (2009, , & Coleman, J. N. Liquid Phase Production of Graphene by Exfoliation of Graphite in Surfactant/Water Solutions. *Journal of the American Chemical Society*, 131(10), 3611-3620.

[34] Li, X., Cai, W., An, J., Kim, S., Nah, J., Yang, D., Piner, R., Velamakanni, A., Jung, I., Tutuc, E., Banerjee, S. K., Colombo, L., & Ruoff, R. S. (2009). Large-Area Synthesis of High-Quality and Uniform Graphene Films on Copper Foils. *Science*, 324(5932), 1312-1314.

[35] Malesevic, A., Vitchev, R., Schouteden, K., Volodin, A., Zhang, L., Tendeloo, G. V., Vanhulsel, A., & Haesendonck, C. V. (2008). Synthesis of few-layer graphene via microwave plasma-enhanced chemical vapour deposition. *Nanotechnology*, 19(305604).

[36] Dervishi, E., Li, Z., Watanabe, F., Biswas, A., Xu, Y., Biris, A. R., Saini, V., & Biris, A. S. (2009). Large-scale graphene production by RF-cCVD method. *Chemical Communications*, 27-4061.

[37] Reina, A., Jia, X., Ho, J., Nezich, D., Son, H., Bulovic, V., Dresselhaus, M. S., & Kong, J. (2008). Large Area, Few-Layer Graphene Films on Arbitrary Substrates by Chemical Vapor Deposition. *Nano Letters*, 9(1), 30-35.

[38] Sutter, P. W., Flege, J. I., & Sutter, E. A. (2008). Epitaxial graphene on ruthenium. *Nat Mater*, 7(5), 406-411.

[39] Srivastava, A., Galande, C., Ci, L., Song, L., Rai, C., Jariwala, D., Kelly, K. F., & Ajayan, P. M. (2010). Novel Liquid Precursor-Based Facile Synthesis of Large-Area Continuous, Single, and Few-Layer Graphene Films. *Chemistry of Materials*, 22(11), 3457-3461.

[40] Nandamuri, G., Roumimov, S., & Solanki, R. (2010). Chemical vapor deposition of graphene films. *Nanotechnology*, 21-145604.

[41] Bae, S., Kim, H., Lee, Y., Xu, X., Park, J. S., Zheng, Y., Balakrishnan, J., Lei, T., Ri, Kim. H., Song, Y. I., Kim, Y. J., Kim, K. S., Ozyilmaz, B., Ahn, J. H., Hong, B. H., & Iijima, S. (2010). Roll-to-roll production of 30-inch graphene films for transparent electrodes. *Nat Nano*, 5(8), 574-578.

[42] Shivaraman, S., Barton, R. A., Yu, X., Alden, J., Herman, L., Chandrashekhar, M. V. S., Park, J., Mc Euen, P. L., Parpia, J. M., Craighead, H. G., & Spencer, M. G. (2009). Free-Standing Epitaxial Graphene. *Nano Letters*, 9(9), 3100-3105.

[43] Aristov, V. Y., Urbanik, G., Kummer, K., Vyalikh, D. V., Molodtsova, O. V., Preobrajenski, A. B., Zakharov, A. A., Hess, C., Ha◎nke, T., Bu◎chner, B., Vobornik, I., Fujii, J., Panaccione, G., Ossipyan, Y. A., & Knupfer, M. (2010). Graphene Synthesis on Cubic SiC/Si Wafers. Perspectives for Mass Production of Graphene-Based Electronic Devices. *Nano Letters*, 10(3), 992-995.

[44] Emtsev, K. V., Bostwick, A., Horn, K., Jobst, J., Kellogg, G. L., Ley, L., Mc Chesney, J. L., Ohta, T., Reshanov, S. A., Rohrl, J., Rotenberg, E., Schmid, A. K., Waldmann, D., Weber, H. B., & Seyller, T. (2009). Towards wafer-size graphene layers by atmospheric pressure graphitization of silicon carbide. *Nat Mater*, 8(3), 203-207.

[45] Deng, D., Pan, X., Zhang, H., Fu, Q., Tan, D., & Bao, X. (2010). Freestanding Graphene by Thermal Splitting of Silicon Carbide Granules. *Advanced Materials*, 22(19), 2168-2171.

[46] Stankovich, S., Dikin, D. A., Dommett, G. H. B., Kohlhaas, K. M., Zimney, E. J., Stach, E. A., Piner, R. D., Nguyen, S. T., & Ruoff, R. S. (2006). Graphene-based composite materials. *Nature*, 442(7100), 282-286.

[47] Verdejo, R., Bernal, M. M., Romasanta, L. J., & Lopez, Manchado. M. A. (2011). Graphene filled polymer nanocomposites. *Journal of Materials Chemistry*, 21(10), 3301-3310.

[48] Gilje, S., Han, S., Wang, M., Wang, K. L., & Kaner, R. B. (2007). A Chemical Route to Graphene for Device Applications. *Nano Letters*, 7(11), 3394-3398.

[49] Brodie, B. C. (1859). On the Atomic Weight of Graphite. *Philosophical Transactions of the Royal Society of London*.

[50] Staudenmaier, L. (1898). Verfahren zur Darstellung der Graphitsäure. *Berichte der deutschen chemischen Gesellschaft*, 31(2), 1481-1487.

[51] Hummers, W. S., & Offeman, R. E. (1958). Preparation of Graphitic Oxide. *Journal of the American Chemical Society*, 80(6), 1339.

[52] Marcano, D. C., Kosynkin, D. V., Berlin, J. M., Sinitskii, A., Sun, Z., Slesarev, A., Alemany, L. B., Lu, W., & Tour, J. M. (2010). Improved Synthesis of Graphene Oxide. *ACS Nano*, 4(8), 4806-4814.

[53] Shen, J., Hu, Y., Shi, M., Lu, X., Qin, C., Li, C., & Ye, M. (2009). Fast and Facile Preparation of Graphene Oxide and Reduced Graphene Oxide Nanoplatelets. *Chemistry of Materials*, 21(15), 3514-3520.

[54] Paredes, J. I., Villar, Rodil. S., Martínez, Alonso. A., & Tascón, J. M. D. (2008). Graphene Oxide Dispersions in Organic Solvents. *Langmuir*, 24(19), 10560-10564.

[55] Pham, V. H., Cuong, T. V., Dang, T. T., Hur, S. H., Kong, B. S., Kim, E. J., Shin, E. W., & Chung, J. S. (2011). Superior conductive polystyrene chemically converted graphene nanocomposite. *Journal of Materials Chemistry*, 21(30), 11312-11316.

[56] Gilje, S., Han, S., Wang, M., Wang, K. L., & Kaner, R. B. (2007). A Chemical Route to Graphene for Device Applications. *Nano Letters*, 7(11), 3394-3398.

[57] Eda, G., Fanchini, G., & Chhowalla, M. (2008). Large-area ultrathin films of reduced graphene oxide as a transparent and flexible electronic material. *Nat Nano*, 3(5), 270-274.

[58] Eda, G., & Chhowalla, M. (2010). *Adv. Mater.*, 22(2392).

[59] Stankovich, S., Dikin, D. A., Dommett, G. H. B., Kohlhaas, K. M., Zimney, E. J., Stach, E. A., Piner, R. D., Nguyen, S. T., & Ruoff, R. S. (2006). Graphene-based composite materials. *Nature*, 442(7100), 282-286.

[60] Ramanathan, T., Abdala, A. A., Stankovich, S., Dikin, D. A., Herrera, Alonso. M., Piner, R. D., Adamson, D. H., Schniepp, H. C., Chen, X., Ruoff, R. S., Nguyen, S. T., Aksay, I. A., Prud, Homme. R. K., & Brinson, L. C. (2008). Functionalized graphene sheets for polymer nanocomposites. *Nat Nano*, 3(6), 327-331.

[61] Kim, H., Abdala, A. A., & Macosko, C. W. (2010). Graphene/Polymer. *Nanocomposites. Macromolecules*, 43(16), 6515-6530.

[62] Pham, V. H., Cuong, T. V., Dang, T. T., Hur, S. H., Kong, B. S., Kim, E. J., Shin, E. W., & Chung, J. S. (2011). Superior conductive polystyrene chemically converted graphene nanocomposite. *Journal of Materials Chemistry*, 21(30), 11312-11316.

[63] Stoller, M. D., Park, S., Zhu, Y., An, J., & Ruoff, R. S. (2008). Graphene-Based Ultracapacitors. *Nano Letters*, 8(10), 3498-3502.

[64] Kamat, P. V. (2011). Graphene-Based Nanoassemblies for Energy Conversion. The Journal of Physical Chemistry. *Letters*, 2(3), 242-251.

[65] Shao, Y., Wang, J., Wu, H., Liu, J., Aksay, I. A., & Lin, Y. (2010). Graphene Based Electrochemical Sensors and Biosensors:. *A Review. Electroanalysis*, 22(10), 1027-1036.

[66] Kaiser, A. B. (2001). Electronic transport properties of conducting polymers and carbon nanotubes. *Rep Prog Phys.*, 64(1), 1-49.

[67] Kopelevich, Y., & Esquin, P. (2007). Graphene physics in graphite. *Adv Mater.*, 19(24), 455-9.

[68] Stankovich, S., Dikin, D. A., Piner, R. D., Kohlhaas, K. A., Kleinhammes, A., Jia, Y., Wu, Y., Nguyen, S. T., & Ruoff, R. S. (2007). Synthesis of graphene-based nanosheets via chemical reduction of exfoliated graphite oxide. *Carbon*, 45(7), 1558-1565.

[69] Li, D., Muller, M. B., Gilje, S., Kaner, R. B., Wallace, G., & , G. (2008). Processable aqueous dispersions of graphene nanosheets. *Nat Nano*, 3(2), 101-105.

[70] Pham, V. H., Cuong, T. V., Nguyen, Phan. T. D., Pham, H. D., Kim, E. J., Hur, S. H., Shin, E. W., Kim, S., & Chung, J. S. (2010). One-step synthesis of superior dispersion of chemically converted graphene in organic solvents. Chemical Communications , 46(24), 4375-4377.

[71] Eda, G., Fanchini, G., & Chhowalla, M. (2008). Large-area ultrathin films of reduced graphene oxide as a transparent and flexible electronic material. *Nat Nano*, 3(5), 270-274.

[72] Park, S., An, J., Jung, I., Piner, R.D., An, S.J., Li, X., Velamakanni, A., & Ruoff, R.S. ((2009).).Colloidal Suspensions of Highly Reduced Graphene Oxide in a Wide Variety of Organic Solvents. Nano Letters , 9(4), 1593-1597.

[73] Villar, P., Rodil, S., Paredes, J. I., Martinez, Alonso. A., & Tascon, J. M. D. (2009). Preparation of graphene dispersions and graphene-polymer composites in organic media. *Journal of Materials Chemistry*, 19(22), 3591-3593.

[74] Shin, H. J., Kim, K. K., Benayad, A., Yoon, S. M., Park, H. K., , I., Jung, S., Jin, M. H., Jeong, H. K., Kim, J. M., Choi, J. Y., & Lee, Y. H. (2009). *Adv. Funct. Mater.*, 19(1987).

[75] Moon, I. K., Lee, J., Ruoff, R. S., & Lee, H. (2010). Reduced graphene oxide by chemical graphitization. *Nat Commun*, 1(73).

[76] Pham, H. D., Pham, V. H., Cuong, T. V., Nguyen, Phan. T. D., Chung, J. S., Shin, E. W., & Kim, S. (2011). Synthesis of the chemically converted graphene xerogel with superior electrical conductivity. *Chemical Communications*, 47(34), 9672-9674.

[77] Chen, W., Yan, L., & Bangal, P. R. (2010). Chemical Reduction of Graphene Oxide to Graphene by Sulfur-Containing Compounds. *The Journal of Physical Chemistry*, 114(47), 19885-19890.

[78] Zhang, J., Yang, H., Shen, G., Cheng, P., Zhang, J., & Guo, S. (2010). Reduction of graphene oxide vial-ascorbic acid. *Chemical Communications*, 46(7), 1112-1114.

[79] Gao, J., Liu, F., Liu, Y., Ma, N., Wang, Z., & Zhang, X. (2010). Environment-Friendly Method To Produce Graphene That Employs Vitamin C and Amino Acid. *Chemistry of Materials*, 22(7), 2213-2218.

[80] Hontoria,]., Lucas, C., López Peinado, A. J., López, González. J. D., Rojas, Cervantes. M. L., & Martín-Aranda, R. M. (1995). Study of oxygen-containing groups in a series of graphite oxides: Physical and chemical characterization. *Carbon*, 33(11), 1585-1592.

[81] Jeong, H. K., Lee, Y. P., Lahaye, R. J. W. E., Park, M. H., An, K. H., Kim, I. J., et al. (2008). Evidence of graphitic ab stacking order of graphite oxides. *J Am Chem Soc.*, 130(4), 136-2.

[82] Gao, W., Alema, L.B., Ci, L., & Ajayan, P. M. (2009). New insights into the structure and reduction of graphite oxide. *Nat Chem.*, 1(5), 403-408.

[83] Schniepp, H. C., Li, J. L., Mc Allister, M. J., Sai, H., Herrera, Alonso. M., Adamson, D. H., Prud'homme, R. K., Car, R., Saville, D. A., & Aksay, I. A. (2006). Functionalized Single Graphene Sheets Derived from Splitting Graphite Oxide. *The Journal of Physical Chemistry B*, 110(17), 8535-8539.

[84] Wu, Z. S., Ren, W., Gao, L., Liu, B., Jia, C., & Cheng, H. M. (2009). Synthesis of high-quality graphene with a pre-determined number of layers. *Carbon*, 47(2), 493-499.

[85] Kudin, K. N., Ozbas, B., Schniepp, H. C., Prudhomme, R. K., Aksay, I. A., & Car, R. (2007). Raman Spectra of Graphite Oxide and Functionalized Graphene Sheets. *Nano Letters*, 8(1), 36-41.

[86] Zhao, J., Pei, S., Ren, W., Gao, L., & Cheng, H. M. (2010). Efficient Preparation of Large-Area Graphene Oxide Sheets for Transparent Conductive Films. *ACS Nano*, 4(9), 5245-5252.

[87] Becerril, H. A., Mao, J., Liu, Z., Stolten, berg. R. M., Bao, Z., & Chen, Y. (2008). Evaluation of Solution-Processed Reduced Graphene Oxide Films as Transparent Conductors. *ACS Nano*, 2(3), 463-470.

[88] Wang, X., Zhi, L., & Mullen, K. (2007). Transparent, Conductive Graphene Electrodes for Dye-Sensitized Solar Cells. *Nano Letters*, 8(1), 323-327.

[89] Li, X., Wang, H., Robinson, J. T., Sanchez, H., Diankov, G., & Dai, H. (2009). Simultaneous nitrogen doping and reduction of graphene oxide. *J Am Chem Soc.*, 131(43), 15939-15944.

[90] Stankovich, S., Dikin, D. A., Piner, R. D., & Kohlhaas, K. A. (2007). Kleinhammes A, Jia Y, Wu Y, Nguyen ST, Ruoff RS. Synthesis of graphene-based nanosheets via chemical reduction of exfoliated graphite oxide. *Carbon*, 45(7), 1558-1565.

[91] Li, D., Muller, M. B., Gilje, S., & Kaner, R. B. (2008). Wallace GG. Processable aqueous dispersions of graphene nanosheets. *Nat Nano*, 3(2), 101-105.

[92] Tung, V. C., Allen, M. J., Yang, Y., & Kaner, R. B. (2009). High-throughput solution processing of large-scale graphene. Nat. *Nano*, 4(1), 25-29.

[93] Park, S., An, J., Jung, I., Piner, R. D., An, S. J., Li, X., Velamakanni, A., & Ruoff, R. S. (2009). Colloidal Suspensions of Highly Reduced Graphene Oxide in a Wide Variety of Organic Solvents. *Nano Letters*, 9(4), 1593-1597.

[94] Mc Allister, M. J., Li, J. L., Adamson, D. H., Schniepp, H. C., Abdala, A. A., Liu, J., Herrera-Alonso, M., Milius, D. L., Car, R., & Prud'homme, R. K. (2007). Aksay IA.Single Sheet Functionalized Graphene by Oxidation and Thermal Expansion of Graphite. Chemistry of Materials .., 19(18), 4396-4404.

[95] Nethravathi, C., & Rajamathi, M. (2008). Chemically modified graphene sheets produced by the solvothermal reduction of colloidal dispersions of graphite oxide. *Carbon*, 46(14), 1994-1998.

[96] Wang, H., Robinson, J. T., Li, X., & Dai, H. (2009). Solvothermal Reduction of Chemically Exfoliated Graphene Sheets. *Journal of the American Chemical Society*, 131(29), 9910-9911.

[97] Zhou, Y., Bao, Q., Tang, L., Zhong, A. L., , Y., & Loh, K. P. (2009). Hydrothermal Dehydration for the "Green" Reduction of Exfoliated Graphene Oxide to Graphene and Demonstration of Tunable Optical Limiting Properties. *Chemistry of Materials*, 21(13), 2950-2956.

[98] Zhu, Y., Stoller, M. D., Cai, W., Velamakanni, A., Piner, R. D., Chen, D., & Ruoff, R. S. (2010). Exfoliation of Graphite Oxide in Propylene Carbonate and Thermal Reduction of the Resulting Graphene Oxide Platelets. *ACS Nano*, 4(2), 1227-1233.

[99] Ramesha, G. K., & Sampath, S. (2009). Electrochemical Reduction of Oriented Graphene Oxide Films: An in Situ Raman Spectroelectrochemical Study. *The Journal of Physical Chemistry*, 113(19), 7985-7989.

[100] Pham, V. H., Cuong, T. V., Hur, S. H., Oh, E., Kim, E. J., Shin, E. W., & Chung, J. S. (2011). Chemical functionalization of graphene sheets by solvothermal reduction of a graphene oxide suspension in N-methyl-2-pyrrolidone. *Journal of Materials Chemistry;*, 21(10), 3371-3377.

[101] Dubin, S., Gilje, S., Wang, K., Tung, V. C., Cha, K., Hall, A. S., Farrar, J., Varshneya, R., Yang, Y., & Kaner, R. B. (2010). A one-step, solvothermal reduction method for producing reduced graphene oxide dispersions in organic solvents. *ACS Nano*, 4(7), 3845-3852.

[102] Dang, T. T., Pham, V. H., Hur, S. H., Kim, E. J., Kong, B. S., & Chung, J. S. (2012). Superior dispersion of highly reduced graphene oxide in N,N-dimethylformamide. *Journal of Colloid and Interface Science*, 376(1), 91-96.

[103] Shimizu, F., Mizoguchi, K., Masubuchi, S., & Kume, K. (1995). Metallic temperature dependence of resistivity in heavily doped polyacetylene by NMR. *Synthetic Metals*, 69(1-3), 43-44.

[104] Roncali, J. (1992). Conjugated poly(thiophenes): synthesis, functionalization, and applications. *Chemical Reviews*, 92(4), 711-738.

[105] Higashika, S., Kimura, K., Matsuo, Y., & Sugie, Y. (1999). Synthesis of polyaniline-intercalated graphite oxide. *Carbon*, 37(2), 354-356.

[106] Wang, P., Li, D. W., Zhao, J., Ren, W., Chen, Z. G., Tan, J., Wu, Z. S., Gentle, I., Lu, G. Q., & Cheng, H. M. (2009). Fabrication of Graphene/Polyaniline Composite Paper via In Situ Anodic Electropolymerization for High-Performance Flexible Electrode. *ACS Nano*, 3(7), 1745-1752.

[107] Wang, H., Hao, Q., Yang, X., Lu, L., & Wang, X. (2009). Graphene oxide doped polyaniline for supercapacitors. *Electrochemistry Communications*, 11(6), 1158-1161.

[108] Wang, H., Hao, Q., Yang, X., Lu, L., & Wang, X. (2010). Effect of Graphene Oxide on the Properties of Its Composite with Polyaniline. *ACS Applied Materials & Interfaces*, 2(3), 821-828.

[109] Yan, J., Wei, T., Shao, B., Fan, Z., Qian, W., Zhang, M., & Wei, F. (2010). Preparation of a graphene nanosheet/polyaniline composite with high specific capacitance. *Carbon*, 48(2), 487-493.

[110] Yang, N., Zhai, J., Wan, M., Wang, D., & Jiang, L. (2010). Layered nanostructures of polyaniline with graphene oxide as the dopant and template. *Synthetic Metals*, 160(15-16)), 1617, 1622.

[111] Zhang, K., Zhang, L. L., Zhao, X. S., & Wu, J. (2010). Graphene/Polyaniline Nanofiber Composites as Supercapacitor Electrodes. *Chemistry of Materials*, 22(4), 1392-1401.

[112] Bissessur, R., Liu, P. K. Y., & Scully, S. F. (2006). Intercalation of polypyrrole into graphite oxide. *Synthetic Metals*, 156(16-17), 1023-102.

[113] Liu, Q., Liu, Z., Zhang, X., Yang, L., Zhang, N., Pan, G., Yin, S., Chen, Y., & Wei, J. (2009). Polymer Photovoltaic Cells Based on Solution-Processable Graphene and 3HT. *Advanced functional material*, 19, 894-904.

[114] Choi, K. S., Liu, F., Choi, J. S., & Seo, T. S. (2010). Fabrication of Free-Standing Multilayered Graphene and Poly(3,4-ethylenedioxythiophene) Composite Films with Enhanced Conductive and Mechanical Properties. *Langmuir*, 26(15), 12902-12908.

[115] Murugan, A. V., Muraliganth, T., & Manthiram, A. (2009). Rapid, Facile Microwave-Solvothermal Synthesis of Graphene Nanosheets and Their Polyaniline Nanocomposites for Energy Strorage. *Chemistry of Materials*, 21(21), 5004-5006.

[116] Yan, J., Wei, T., Shao, B., Fan, Z., Qian, W., Zhang, M., & Wei, F. (2010). Preparation of a graphene nanosheet/polyaniline composite with high specific capacitance. *Carbon*, 48(2), 487-493.

[117] Zhang, K., Zhang, L. L., Zhao, X. S., & Wu, J. (2010). Graphene/Polyaniline Nanofiber Composites as Supercapacitor Electrodes. *Chemistry of Materials*, 22(4), 1392-1401.

[118] Wu, Q., Xu, Y., Yao, Z., Liu, A., & Shi, G. (2010). Supercapacitors Based on Flexible Graphene/Polyaniline. *Nanofiber Composite Films, ACS Nano*, 4(4), 1963-1970.

[119] Xu, Y., Wang, Y., Liang, J., Huang, Y., Ma, Y., Wan, X., & Chen, Y. (2009). A hybrid material of graphene and poly (3,4-ethyldioxythiophene) with high conductivity, flexibility, and transparency. *Nano Research*, 2(4), 343-348.

[120] Bai, H., Xu, Y., Zhao, L., Li, C., & Shi, G. (2009). Non-covalent functionalization of graphene sheets by sulfonated polyaniline. *Chemical Communications*, 13-1667.

[121] Yang, H., Zhang, Q., Shan, C., Li, F., Han, D., & Niu, L. (2010). Stable, Conductive Supramolecular Composite of Graphene Sheets with Conjugated Polyelectrolyte. *Langmuir*, 26(9), 6708-6712.

[122] Yang, H., Zhang, Q., Shan, C., Li, F., Han, D., & Niu, L. (2010). Stable, Conductive Supramolecular Composite of Graphene Sheets with Conjugated Polyelectrolyte. *Langmuir*, 26(9), 6708-6712.

[123] Chunder,]., Liu, J., & Zhai, L. (2010). Reduced Graphene Oxide/Poly(3-hexylthiophene) Supramolecular Composites. *Macromolecular Rapid Communications*, 31(4), 380-384.

[124] Liu, Q., Liu, Z., Zhang, X., Zhang, N., Yang, L., Yin, S., & Chen, Y. (2008). Organic photovoltaic cells based on an acceptor of soluble graphene. *Applied Physics Letters*, 92(22), 223303.

[125] Liu, Q., Liu, Z., Zhang, X., Yang, L., Zhang, N., Pan, G., Yin, S., Chen, Y., & Wei, J. (2009). Polymer photovoltaic cell based on a solution processable graphene and 3HT. *Advanced functional material*, 19, 894-904.

[126] Kumar, N. A., Choi, H. J., Shin, Y. R., Chang, D. W., Dai, L., & Baek, J. B. (2012). Polyaniline-Grafted Reduced Graphene Oxide for Efficient Electrochemical Supercapacitors. *ACS Nano*, 6(2), 1715-1723.

[127] Wang, X., Zhi, L., & Mullen, K. (2007). Transparent, Conductive Graphene Electrodes for Dye-Sensitized Solar Cells. *Nano Letters*, 8(1), 323-327.

[128] Yoo, E., Kim, J., Hosono, E., Zhou, H., Kudo, T., & Honma, I. (2008). Large Reversible Li Storage of Graphene Nanosheet Families for Use in Rechargeable Lithium Ion Batteries. *Nano Letters*, 8(8), 2277-2282.

[129] Tang, Z., Shen, S., Zhuang, J., & Wang, X. (2010). Noble-Metal-Promoted Three-Dimensional Macroassembly of Single-Layered Graphene Oxide. Angewandte. *Chemie International Edition*, 49(27), 4603-4607.

[130] Varrla, E., Venkataraman, S., & Sundara, R. (2011). Functionalized Graphene-PVDF Foam Composites for EMI Shielding. *Macromol. Mater. Eng.*, 296-894.

[131] Ghasemi, A., Saatchi, A., Salehi, M., Hossienpour, A., Morisako, A., & Liu, X. (2006). Influence of matching thickness on the absorption properties of doped barium ferrites at microwave frequencies. *Phys Status Solidi A*, 203, 358-65.

[132] Xie, J. L., Han, M., Chen, L., Kuang, R., & Deng, L. (2007). Microwave-absorbing properties of NiCoZn spinel ferrites. *J Magn. Magn. Mater*, 314-37.

[133] Lu, B., Dong, X. L., Huang, H., Zhang, X. F., Zhu, X. G., & Lei, J. P. (2008). Microwave absorption properties of the core/shell-type iron and nickel nanoparticles. *J. Magn. Magn. Mater*, 320-1106.

[134] Liu, X. G., Geng, D. Y., Meng, H., Shang, P. J., & Zhang, Z. D. (2008). Microwave-absorption properties of ZnO-coated iron nanocapsules. *Appl Phys Lett*, 92(173117).

[135] Li, B. W., Shen, Y., Yue, Z. X., & Nan, C. W. (2006). Enhanced microwave absorption in nickel/hexagonal-ferrite/polymer composites. *Appl Phys Lett*, 89(132504).

[136] Che, R. C., Zhi, C. Y., Liang, C. Y., & Zhou, X. G. (2006). Fabrication and microwave absorption of carbon nanotubes/CoFe2O4 spinel nanocomposite. *Appl Phys Lett*, 88(033105).

[137] Watts, P. C. P., Hsu, W. K., Barnes, A., & Chambers, B. (2003). High permittivity from defective multiwalled carbon nanotubes in the X-band. *Adv Mater*, 15-600.

[138] Xu, H., Anlage, S. M., Hu, L., & Grunera, G. (2007). Microwave shielding of transparent and conducting single-walled carbon nanotube films. *Appl Phys Lett*, 90, 183119.

[139] Singh, K., Ohlan, A., Bakhshi, A. K., & Dhawan, S. K. (2010). Synthesis of conducting ferromagnetic nanocomposite with improved microwave absorption properties. *Mater. Chem. Phys.*, 119, 201-7.

[140] Ohlan, A., Singh, K., Chandra, A., & Dhawan, S. K. (2010). Microwave absorption behavior of core–shell structured poly (3,4-ethylenedioxy thiophene)–barium ferrite nanocomposites. *ACS Appl. Mater. Interfaces*, 2, 927-33.

[141] Singh, K., Ohlan, A., Saini, P., & Dhawan, S. K. (2008). Poly (3,4-ethylenedioxythiophene) γ-Fe2O3 polymer composite-super paramagnetic behavior and variable range hopping 1D conduction mechanism-synthesis and characterization. *Polym. Adv. Technol*, 19, 229-36.

[142] Zhang, X. F., Dong, X. L., Huang, H., Liu, Y. Y., Wang, W. N., & Zhu, X. G. (2006). Microwave absorption properties of the carbon-coated nickel nanocapsules. *Appl Phys Lett.*, 89, 053115.

[143] Tang, N., Zhong, W., Au, C., Yang, Y., Han, M., & Lin, K. (1992). Synthesis, microwave electromagnetic, and microwave absorption properties of twin carbon nanocoils. *J Phys Chem.*, C112, 19316-23.

[144] Colaneri, N. F., & Shacklette, L. W. (1992). EMI shielding measurements of conductive polymer blends. *IEEE Trans Instru Meas*, 41, 291-7.

[145] Singh, R., Kumar, J., Singh, R. K., Rastogi, R. C., & Kumar, V. (2007). Low frequency ac conduction and dielectric relaxation in pristine poly(3 -octylthiophene) films. *New J Phys*, 9-40.

[146] Nicolson, A. M., & Ross, G. F. (1970). Measurement of the intrinsic properties of materials by time-domain techniques. *IEEE Trans Instrum Meas*, 19-377.

[147] Weir,]., & , W. B. (1974). Automatic measurement of complex dielectric constant and permeability at microwave frequencies. *Proceedings of the IEEE*, 62-33.

[148] Ishino, K., & Narumiya, Y. (1987). Development of magnetic ferrites: control and application of losses. *Ceram Bull*, 66-1469.

[149] Dimitrov, D. A., & Wysin, G. M. (1995). Magnetic properties of spherical fcc clusters with radial surface anisotropy. *Phys Rev B*, 51(11947).

[150] Shilov, V. P., Bacri, J. C., Gazeau, F., Gendron, F., Perzynski, R., & Raikher, Y. L. (1999). Ferromagnetic resonance in ferrite nanoparticles with uniaxial surface anisotropy. *J Appl Phys.*, 85(6642).

[151] Basavaraja, C., Kim, W. J., Kim, Y. D., & Huh, D. S. (2011). Synthesis of polyaniline-gold/graphene oxide composite and microwave absorption characteristics of the composite films. *Materials Letters*, 65(19-20), 3120-3123.

[152] Liu, D. Y., & Reynolds, J. R. (2010). Dioxythiophene-Based Polymer Electrodes for Supercapacitor Modules. *ACS Applied Materials & Interfaces*, 2(12), 3586-3593.

Chapter 4

Composites of Cellulose and Metal Nanoparticles

Ricardo J. B. Pinto, Márcia C. Neves,
Carlos Pascoal Neto and Tito Trindade

Additional information is available at the end of the chapter

http://dx.doi.org/10.5772/50553

1. Introduction

Research on inorganic/organic nanocomposite materials is a fast growing interdisciplinary area in materials science and engineering. In particular, extensive work has been undertaken in the development of sustainable and environmentally friendly resources and methods. A key idea has been the production of nanocomposites comprising biopolymers that in specific contexts can replace conventional materials such as synthetic polymers. It is well known that the properties of nanocomposite materials depend not only on the properties of their individual components but also on morphological and interfacial characteristics arising from the combination of distinct materials [1]. Therefore the use of polymers such as cellulose, starch, alginate, dextran, carrageenan, and chitosan among others, gain great relevance not only due to their renewable nature and biodegradability, but also because a variety of formulations can be exploited depending on the envisaged functionality [2, 3].

This chapter has focus on the use of cellulose as the matrix in the production of nanocomposites. Cellulose has critical importance namely because is the most abundant and widespread biopolymer on Earth. Owing to its abundance and specific properties, it is important noted for the development of environmental friendly, biocompatible, and functional composites, quite apart from its traditional and massive use in papermaking and cotton textiles [4]. Additionally different types of cellulose are available for the preparation of nanocomposites, namely vegetable cellulose (VC), bacterial cellulose (BC) and nanofibrillated cellulose (NFC). Although sharing similar chemistry and molecular structure, the different kinds of cellulose show important differences in terms of morphology and mechanical behavior. For example, BC and NFC are composed of fibers with nanosized dimensions as compared to VC, which might impart new properties, and in some cases improvements to the ensuing nanocomposite materials [5].

INTECH
open science | open minds

© 2012 Pinto et al.; licensee InTech. This is a paper distributed under the terms of the Creative Commons Attribution License (http://creativecommons.org/licenses/by/3.0), which permits unrestricted use, distribution, and reproduction in any medium, provided the original work is properly cited.

The association of cellulose with different fillers can bring benefits like improvement of properties (optical, mechanical, ...) and delivering unique functions by their use [6]. Cellulose has been used as a soft matrix to accommodate inorganic fillers to produce composites that bring together the intrinsic functionalities of the fillers and the biointerfaces offered by cellulose fibers [2]. Among the wide range of available inorganic fillers, in this review metal nanoparticles (Au, Ag, and Cu, among others) will be considered. Metal NPs exhibit properties that differ from the bulk analogues due to size and surface effects, thus the properties of the final materials can be adjusted as a function of the size, shape, particle size distribution of the nanofillers as well as by interactions occurring with the cellulose fibers' surfaces. Preparative strategies play a determinant role in the performance and properties of the nanocomposites, hence chemical approaches for the synthesis of these materials are reviewed namely for *in situ* and *ex situ* methods. Examples will be given for applications of cellulose nanocomposites by taking in consideration the type of nanoparticles used. As a concluding note, the development of new multifunctional cellulose nanocomposites will be put in perspective.

2. General aspects on the chemistry and properties of cellulose

The last years have seen great interest in research and application of cellulose nanocomposites namely due to the technological interest in renewable materials and environmentally friendly and sustainable resources [7]. In fact, within the polymers obtained from renewable sources, cellulose is the most abundant natural polymer in Nature as well as the most important component of the plants'"skeleton". This biopolymer formed by repeated connection of glucose building blocks is the structural basis of cell walls of virtually all plants and is usually considered an almost inexhaustible source of raw materials [8, 9]. Cellulose has particular significance owing its unique structure and distinct tendency to form intra- and intermolecular bonding. These characteristics influence the cellulose supramolecul ararrangement that together with other practical aspects such as the product origin and processing treatment, have important consequences on the final properties of cellulose. This polymer is the main constituent of softwood and hardwood, representing about 40-45% of dry wood, with wood pulp remaining the most important source for cellulose processing namely in paper fabrication [8, 10, 11]. Wood pulp is also the main industrial feedstock for the production of cellulose regenerate fibers and films. This biopolymer is also used in the synthesis of different cellulose derivatives such as esters and ethers. These derivatives are well-kwon active components in applications which include coatings, pharmaceutics and cosmetics, among others [11] and also used in numerous hybrids containing metal and metal oxides NPs.

Besides extraction from plants, cellulose can be produced by alternative methods, namely by using different types of microorganisms (certain bacteria, algae or fungi). Among the cellulose-forming bacteria, *Acetobacter* strains have been widely used because they are not pathogenic. In fact, these Gram-negative bacteria are usually found in fruits and can be used in laboratorial conditions in order to obtain significant amounts of cellulose [8, 11-13]. Nowa-

http://dx.doi.org/10.5772/50553

days, it is observed a growing interest in the use of BC, not only within applications in nanocomposites but also in other fields including food industry (e.g. calorie–free dessert) and medical field (e.g. wound dressing). Apart their three-dimensional (3D) network of nanofibers, BC has high purity (do not have lignin, hemicelluloses, pectin and other compounds associated to VC), high degree of polymerization (DP up to 8000), high crystallinity (60 to 90%), high water content and high elasticity and mechanical stability (particularly in wet form) [8, 11-15].

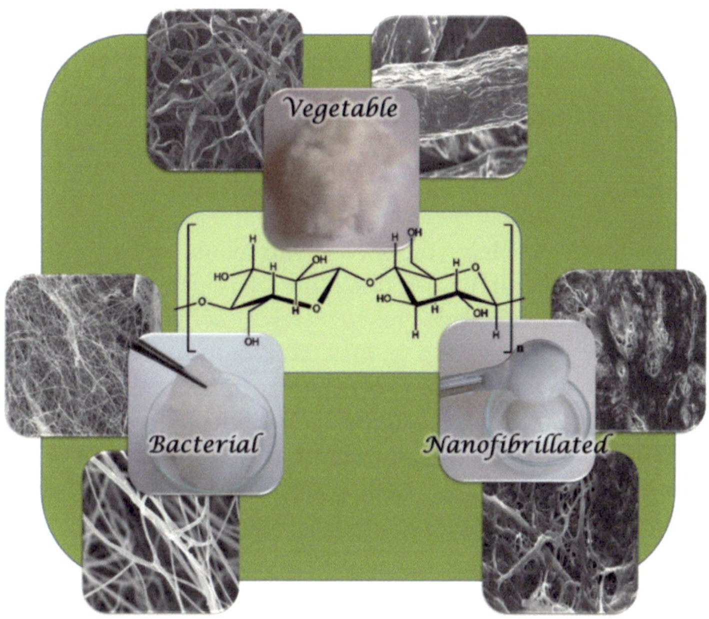

Figure 1. Chemical structure and morphological characteristicsof different forms of cellulose.

Due to the complex and expensive process to produce BC, there has been interest to find other ways to obtain cellulose fibers of nanometric dimensions, namely at an industrial scale [11]. NFC can be obtained from VC fibers by distinct methods including chemical treatment and mechanical disintegration procedures, in the form of aqueous suspensions of nanoscale fibers, leading to high aspect ratio materials (5-30 nm diameter and lengths in micrometer range) with remarkable strength and flexibility [16, 17]. The mechanical properties of NFC make this polymer a good candidate for reinforcement materials in nanocomposites. However, besides their interesting mechanical properties, NFC shows other properties of practical interest. For example, NFC has a large surface area which makes it a promising candidate for filtering membranes. Appropriate chemical modifications performed on NFC result in a versatile additive for paints, lacquers or latex. Due to its biocompatibility, NFC might also be used in food and medical applications [11, 17]. As will be clear in the next section, the properties of the cellulose nanocomposites depend not only on the NPs employed as fillers but also on the type of cellulose matrix used.

From all the cellulose derivatives commonly use on the chemistry market, esters and ethers are the predominant. Although produced since the middle of the eighteen century the actually research is related with their manufacture technological improvement. The mail goal is related with the development of greener processes being investigated the use of ionic liquids (IL), microwaves irradiation and solvent-free systems in the synthesis of this cellulose derivatives. This strategy has been followed for the cellulose derivatives used in specifics applications, such as biomedical and optoelectronic and produced in small amounts. The industries responsible for the production of high amounts of these derivatives (as cellulose acetate) have neglected this mandatory necessity [18].

3. Cellulose/metal nanocomposites

3.1. Preparative strategies of cellulose/metal nanocomposites

A key aspect to consider in combining metal NPs with cellulose fibers is the methodology to be employed namely by taking in consideration the envisaged applications. In order to exploit the properties of nanocomposites, the NPs should be well dispersed over the matrix without the formation of large aggregates that may compromise the final properties and should as much as possible exhibit a small narrow size distribution. There is critical need to find effective techniques that allow the large-scale production that at the same time maintain control of the NPs dispersion in the cellulose matrix. A number of approaches have been developed to attach metal NPs onto cellulose fibers. Table 1 gives examples of methods employed in the preparation of cellulose/metal nanocomposites.

3.1.1. Blending of components

The blending of inorganic NPs and polymers by promoting their homogeneous mixture to form nanocomposites materials has been widely employed [19]. Although this method offers the advantage of simplicity, the use of cellulose as matrix commonly lead to NPs aggregates that decrease the benefits associated to the presence of nanosized fillers. This process often leads to poor laundering durability of the materials and, for example when Ag NPs have been used, the antibacterial efficiency are lower than expected and discontinuous in time [20]. The direct deposition of Ag and Au NPs, by dropwise addition of the respective colloids onto filter papers, has been reported [21, 22]. Usually, this methodology does not lead to an homogeneous distribution of NPs on the paper substrates and the formation of aggregates at the edge of the droplets during the drying process is common [23].

3.1.2. In situ reduction of metal salts

The preparation of cellulose/metal nanocomposites by the *in situ* reduction of metal salts in cellulose aqueous suspensions has been extensively investigated. Typically this involves the use of a soluble metal salt as precursor, a reducing agent and a co-stabilizer to avoid agglomeration. However, the *in situ* method can be employed without addition of an external reducing agent, because adsorption of metal ions on the cellulose surfaces may bet subsequently reduced to metal NPs by organic moieties such as terminal aldehyde or carboxylic groups, whose presence depend on pulp bleaching. In this case, the unique structure and the presence of ether and hydroxyl groups in cellulose fibers constitute an effective nanoreactor for *in situ* synthesis of the NPs. The ether and hydroxyl functions not only anchor the metal ions tightly onto the fibers via ion–dipole interactions, but also after reduction stabilizes the as prepared NPs via surface interactions [27, 57]. This process presents some advantages compared to the simple mixture of the composite components. The template role of the host macromolecular chains for the synthesis of NPs helps to improve their distribution inside the cellulose matrix and also prevents formation of aggregates. At the same time the polymer chains play an important role leading to a narrow size distribution and well defined shape for the metal NPs [69].

Cellulose matrix	**Preparative method**		**Metal NPs**
Vegetable	Blending of components		Ag [20, 21], Au [22]
	In situ reduction	Cellulose reducing groups	Ag [24-29], Au [29, 30], Cu [31, 32], Pt [33, 34]
		External reducing agents	Ag [20, 27, 28, 35-38], Au [27, 36, 39, 40], Cu [31, 41, 42], Pt [27, 36, 43, 44], Co [45], Pd [27, 36]
		UV reduction	Ag [28, 46, 47]
	Electrostatic assembly		Ag [28, 36], Au [36, 39], Cu [48], Pt [36], Pd [36]
	Microwave-assisted preparation		Ag [49-51]
	Surface pre-modification		Ag [52], Au [52-54]
Bacterial	*In situ*	Cellulose reducing groups	Ag [55]
		Reducing agents	Ag [19, 28, 37, 43, 56-60], Au [39, 43, 61-63], Cu [64], Pt [43, 65], Co [66]
	Electrostatic assembly		Au [39]
	UV reduction		Ag [28]
	Surface pre-modification		Ag [67]
	Diffusion		Ag [28]
Nanofibrillated	Blending of components		Ag [68]
	Electrostatic assembly		Ag [16]

Table 1. Preparative methods of cellulose/metal nanocomposites.

3.1.2.1. Addition of reducing agents

The most commonly used *in situ* approach to prepare a dispersion of NPs in cellulose matrices involves the entrapment of metal cations in the fibers followed by their reduction with an external reducing agent. In this procedure the reducing agent also act as a co-stabilizer (together with the cellulose fibers) for the metal NPs. Sodium borohydride has been extensively used to reduce metal ions in cellulose matrices. The particle size distribution is adjusted by varying the $NaBH_4$: metal salt molar ratio. The use of tri-sodium citrate has also been reported as reducing and stabilizing agent. Some reports have described the loading ofAg NPs into grafted filter paper [35], in BC and VC matrices [37].

The use of hydrazine, hydroxylamine and ascorbic acid together with gelatin or polyvinylpyrrolidone (PVP) as colloidal stabilizers has been investigated [58]. Ascorbic acid acted as an efficient reductant for Ag^+ and gelatin a good colloidal stabilizer toavoid NPs coalescence and to control particle size. *In situ* Ag ions reduction by the chelating-reducing agent triethanolamine (TEA) has been reported to produce small spherical particles with 8.5 nm average size, appearing well dispersed in the BC bulk ultrafine reticulated structure [59].

Reduction of gold salts by flowing H_2 over the cellulose matrix has been reported [40]. This methodology allows the preparation of NPs about 2 nm mean diameter. A facile one-step method, in aqueous medium, makes use of poly (ethyleneimine) (PEI) as reducing and macromolecular linker [61]. In this case the thickness of the Au coating surrounding the cellulose fibers can be adjusted by adding different halides (Fig. 2).

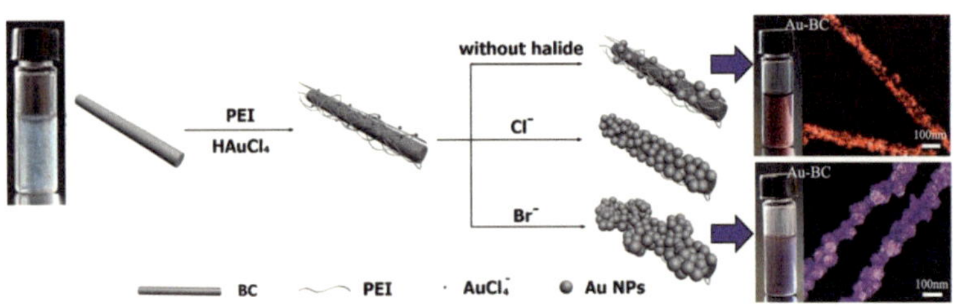

Figure 2. Scheme illustrating the formation of Au–BC nanocomposites using the polyelectrolyte PEI (Adaptedby permission from ref [61] (Copyright 2010 WILEY-VCH Verlag GmbH & Co. KGaA, Weinheim).

3.1.2.2. Reduction of metals salts by cellulose reducing groups

An alternative route for the *in situ* preparation of cellulose based nanocomposites involves the reducing groups of cellulose that simultaneously can entrap NPs within the fibers net. This process shows the advantage that external chemicals are not added to the reacting mixture, thus avoiding adventitious contaminations that may interfere in some applications such as catalysis [36].

This methodology constitutes a green approach to the synthesis of a variety of metal NPs in cellulose matrices in which no additional reducing agents or colloidal stabilizers are used. Kunitake et al. [27] have reported pioneer research using VC fibers with following work reporting the use of BC fibers for the production of silver and gold nanostructures [28, 39]. This strategy has been reported for other types of biomaterials, hence Ag NPs have been prepared by using the *in situ* reduction of a silver(I) salt in the presence of cotton fibers. The washing durability of these nanocomposites and the small amounts of silver NPs required, make this an alternative path to produce cellulose based functional textiles. BC and porous cellulose have also been used as reducing and stabilizer for several metal NPs using a hydrothermal method [43, 55].

In the context of composite science, ionic liquids have attracted substantial interest because of their ability to dissolve biopolymers like cellulose. This has been illustrated in the formation of cellulose/Au nanocomposites [30]. The combined use of cellulose and IL allowed the NPs morphology controlin a process in which the IL was retrieved after metal reduction.The use of unbleached kraft fibers have the advantage of limiting metal leaching due chemical affinity between the NPs and the substrate. In this case NPs are formed directly on the VC fiber surfaces by a redox reaction with the associated lignin. This has not been observed for fibers that do not contain lignin [29].

3.1.2.3. Photo-induced metal growth using irradiation

The *in situ* reduction using UV irradiation is a simple method to produce metal NPs on the surface of cellulose fibers. The preparation of the nanocomposites is based on the photo-activation of cellulose surface by photons, followed by chemical reduction of metal salts. A possible mechanism is based on the number of reducing sites at the surface of cellulose fibers that are activated by UV photons [46]. The active role of reducing ends of cellulose chains in this mechanism has been demonstrated by employing cellulose fibers (VC and BC) in which such groups had been removed to show that metal NPs are not formed [28]. For cellulose/Ag nanocomposites [28, 46, 47] it was demonstrated the relevance of UV light intensity and time of irradiation as important parameters to control the amount of silver and their dispersion in the final composites. The metal NPs formed by this method tend to coat the cellulose fibers, with tendency to aggregate over prolonged times of UV irradiation, eventually leading to NPs with variable morphologies.

3.1.3. Electrostatic assembly

The electrostatic assembly of NPs is based on the sequential adsorption of oppositely charged species on a solid substrate which very often is mediated by ionizable polymers [70]. This assembly technique offers some advantages over other methodologies due to the possibility of a better control of inorganic content in the final nanocomposites, full control of NPs size and morphology, and normally leads to less agglomeration of previously prepared NPs.

Cellulose fibers dispersed in water are negatively charged over a wide pH range (2-10), due to the presence of ionizable moieties such as carboxyl and hydroxyl groups, resulting from chemical processing or from minor polysaccharides such as glucuronoxylans [71]. The deposition of Au NPs [39, 72] onto cellulosic fibers was achieved by previous treatment of fibers using multi-layers of poly (diallyldimethylammonium chloride) (polyDADMAC), poly (sodium 4-styrenesulfonate) (PSS), and again polyDADMAC. The use of a positively charged polyelectrolyte as the outer layer favored electrostatic interactions of the fibers with negatively surface charged Au NPs. This methodology has been also applied to the fabrication of Ag/NFC composites using distinct polyelectrolytes as macromolecular linkers [16]. Another example of an electrostatic assembly procedure was based on the chemical modification of cellulose with (2,3-epoxypropyl) trimethylammonium chloride (EPTAC) [36, 73]. This methodology allowed the grafting of the cellulose substrates with positive ammonium ions which is particularly useful for attachment of metal NPs with surface negatives charge.

3.1.4. Other methodologies

Chemical modification of cellulose can be performed to produce distinct types of cellulose/metal nanocomposites. In this context, common cellulose derivatives such as carboximetilcellulose, cellulose acetate and hydroxypropil cellulose have been used [74-76]. 2,2,6,6-tetramethylpyperidine-1-oxy radical (TEMPO) has been used to oxidize selectively the C6 primary hydroxyl groups of cellulose resulting in the corresponding polyuronic acids [67]. In this context, BC acts as an efficient template with the surface carboxylate groups used to quantitatively anchor metal ions via an ion-exchange reaction. The subsequent reduction of the cationsat the nanofibers' surfaces originated metal NPs with a narrow size distribution. Chemical surface modification of hydroxyl groups into aminic groups, which act as selective coordination sites [52] and the use of thiol labeled cellulose through spontaneous chemisorption [53] has been demonstrated. In the latter, chemical attachment of the NPs onto the fibers' surface limits particle desorption, hence extending the lifetime of the resulting hybrid materials.

The fabrication of size-controlled metal nanowires using cellulose nanocrystals as biomolecular templates has been reported [54]. This method allowed designing Au nanowires of variable sizes that exhibit unique optical properties by controlling the thicknesses of gold shells. In another approach microwave irradiation was used as an efficient method to prepare cellulose/metal nanocomposites [49]. In this study cellulose was treated in a lithium chloride (LiCl)/ N,N-dimethylacetamide (DMAc) and ascorbic acid mixture to produce a homogeneous distribution of Ag NPs within the cellulose matrix. More recently the same group has reported the use of ethylene glycol as solvent, reducing reagent, and microwave absorber, thus excluding an additional reducing agent [50]. This one-step simultaneous formation of Ag NPs and precipitation of the cellulose is a suitable method due to its characteristics of rapid volumetric heating, high reaction rate, short reaction time, enhanced reaction selectivity, and energy saving [49, 77]. A similar methodology was applied in a one-pot process to produce Ag–cellulose nanocomposites, however in this case the cellulose matrix was used as the reducing and stabilizer agent in water suspensions [51].

3.2. Metal nanoparticles as cellulosic composite fillers and their applications

There are a variety of metal NPs that can be used as dispersed phases in bionanocomposites with cellulose. In the last decades there has been great progress in the colloidal synthesis of inorganic NPs. Colloidal metal NPs have received great attention due to their unique optical, electronic, magnetic, antimicrobial properties. Their small size, large specific surface area and tunable physico-chemical properties that differ significantly from the bulk analogues led to intense research on their use in composite materials [78]. This section gives examples of research on metal NPs used as fillers in cellulose nanocomposites. The applications of these materials are related with the type of NPs present though new properties arise due to the combined use of the metal NPs and cellulose. Table 2 summarizes important applications of cellulose/metal nanocomposites and a brief description will follow in this section.

3.2.1. Silver

Nowadays a renewed interest in Ag antimicrobial materials has reappeared mainly due to the increase of multi-drug resistance of microbial strains to conventional antibiotics. The design of protective medical clothing or antibacterial packaging materials are examples of this current trend [35]. Ag NPs are well known by their strong cytotoxicity towards a broad range of microorganisms, such as bacteria and fungi [79].

Similarly to other applications, well dispersed Ag NPs in the cellulose matrix are required otherwise the antimicrobial effect decreases. However, important parameters such as particle size distribution, metal content, cationic silver release and interaction with the surface of cellulose are also relevant parameters that influence the antimicrobial activity of these nanocomposites [23, 69]. Due to the high water holding capacity and biocompatibility, BC wound dressing materials with improved antimicrobial activity have been prepared using Ag [57, 60]. Other examples include the development of antibacterial food-packaging materials [35, 80], bactericidal paper for water treatment [20] and the study of laundering properties of nanocomposites [24, 81].

The cellulose fibers can be chemically functionalized creating reactive sites in order to control the *in situ* synthesis of Ag NPs. Few examples are known of composites of NFC and metal NPs [16, 68]. Thus NFC functionalization with fluorescent Ag nanoclusters has been performed by dipping nanocellulose films into a colloid of Ag protected with poly(methacrylic acid) (PMAA) [64]. The electrostatic assembly of commercial Ag NPs onto NFC mediated by polyelectrolyte linkers have been described as a possible route to scale up the preparation of Ag/NFC composites [16].

Nanostructured metals such Ag and Au are well known substrates for surface enhanced Raman scattering (SERS). Strong enhancement of the Raman signals is observed for certain molecules chemisorbed to the surface of these metals. Therefore the combined use of these metal NPs and cellulose is of great interest to develop molecular detection and biosensing platforms [37]. In this context, the use of cellulose nanocomposites might bring several advantages such as the fabrication of handy and low cost substrates in the form of paper products. A study on the use of distinct cellulosic matrices containing Ag NPs has shown that the BC/Ag nanocomposites were more sensitive as compared to the vegetable analogues, namely in biodetection of amino acids [37]. The use of filter paper with Ag NPs or Au NPs demonstrate the potential of these materials as SERS platforms to study diverse analytes such as *p*-hydroxybenzoic [21], single-walled carbon nanotubes [82] and binary mixtures of 9-aminoacridine-acridine and acridine-quinacrine separated by paper chromatography [83].

A simple and low-cost approach to the fabrication of fuel cells has been described based on a nanostructured Ag electrocatalyst and cellulose. Heat removal of the template and combination with graphite improved oxygen reduction in basic medium [38].

Metal NPs	**Application**		**Metal NPs**	**Application**
Ag	Anti-counterfeiting		Au	Biosensors
Ag	Antimicrobial	Artificial skin	Au	Catalytic
Ag	Antimicrobial	Food-packaging	Au	Conducting
Ag	Antimicrobial	Water treatment	Au	Electronic devices
Ag	Antimicrobial	Wound dressing	Au	Medical (Drug and protein delivery)
Ag	Biosensors		Au	SERS
Ag	Catalytic		Au	Smart papers and textiles
Ag	Paper industry		Cu	Antimicrobial
Ag	SERS		Cu	Catalytic
Ag	Textiles		Pd	Catalytic
Pt	Catalytic	Electrocatalytic	Co	Electronic actuators
Pt	Catalytic	Photocatalytic	Co	Magnetic
Pt	Fuel cells		Co	Microfluidics devices

Table 2. Common applications of cellulose/metal nanocomposites.

3.2.2. Gold

The noble metal gold has long been a cornerstone precious metal occupying a premier position in the world economy, representing wealth and high value. Traditionally it has been used in its yellow lustrous bulk metallic form for monetary and jewelry applications and over recent decades as an electrical conductor and chemically inert contact material in the electronics industry [29]. Au NPs are among the most studied particles in modern materials science namely due to the number of available methods to produce colloids with uniform-particle sizes and well-defined morphologies. Stable Au NPs colloids have been prepared whose particles surfaces are efficiently stabilized by citrate anions in hydrosols or by alkanethiols when organic solvents are used [84].

Cellulose/Au nanocomposites have been used as catalysts in glucose oxidation [40]. It has been reported that good dispersion of Au NPs in cellulose allowed effective contact with reactants making these materials good catalysts for the reduction of 4-nitrophenol. [85] Furthermore, cellulose can be used in several solvents having potential applicability in a variety of reactions. Another interesting possibility is the transformation of renewable biomass resources into valuable chemicals. Selective conversion of cellulose or cellobiose into gluconic acid catalyzed by polyoxometalate- [86] or CNT-supported by Au NPs [87] has been demon-

http://dx.doi.org/10.5772/50553

strated. Agglomeration of the Au NPs in the cellulose nanocomposites has been described as a major limitation decreasing the catalytic activity of the composite materials.

Cellulose based sensors have great interest in several applications including in fields as diverse as medical diagnosis, environmental control and food safety. It is important to develop materials that show good electron transfer capability, biocompatibility, stability and easy accessibility towards the analyte. Additionally large surface area for immobilization of the analytes, rapid response, high sensitivity, good reproducibility and anti-interference are also required characteristics. As expected, it is a great challenge to develop a single material that include all these important characteristics [63].

Figure 3. SEM images and micrographs (inset) of bacterial cellulose (BC) and derived composites: a) BC; b) BC/Ag; BC/Au and d) BC/Cu (bar: 1.5 µm).

Cellulose based sensors are inexpensive, disposable, and environmentally friendly. These materials transport liquids via capillary action with no need of external power [88]. BC/Au nanocomposites have been reported to exhibit good sensitivity, low detection limit and fast response toward hydrogen peroxide makingthese materials suitable matrices for enzyme immobilization [61-63]. The practical application of these nanocomposites for glucose biosensors in human blood samples has been reported. The values obtained showed good agreement with corresponding values obtained in hospital trials. The entrapment of Au NPs and enzymes in a paper coating material of sol–gel derived silica has been reported as a versatile material to be used as an entrapment medium and hydrophobic agent. This characteristic allowed more reproducible enzyme loading on rough and non-uniform paper surfaces [88].

Conductive or semi-conductive nanocomposites containing Au NPs are very attractive for electronic applications. Although uniform NPs dispersions are required for many applications, for some cases controlled aggregation isused as an advantage. Electrical conductive cellulose films containing Au NPs have been prepared by self-assembly showing electrically conducting above a gold loading of 20 wt % [84]. The mechanism of electronic conduction in Au NPs-cellulose films is strongly dependent on the resistivity of the film [89].

3.2.3. *Copper*

Copper NPs were found to be good candidates as efficient catalysts in hydrogen production [90]. An important use for Cu NPs include the fabrication of low electrical resistance materials due to their remarkable conductive properties [91]. In addition Cu NPs and their oxides show broad spectrum biocide effects and the antimicrobial activity has been reported in studies of growth inhibition of bacteria, fungus, and algae [48].

Antimicrobial nanocoatings of Cu NPs on cellulose have been fabricated via electrostatic assembly [48]. In this process, a chemical pre-treatment step was performed in order to impart surface charge on the cotton substrate that promoted binding of cupric ions, followed by chemical reduction to yield metal nanostructured coatings. The resulting composites showed high effectiveness killing to multi-drug resistant pathogen *A. baumannii*. Compared to the Ag analogous there was no particle leaching for the copper nanocomposites.

The use of microcrystalline cellulose as a porous natural material supporting copper ions has been demonstrated [31, 42]. Reducing agent and respective amount have critical importance to achieve Cu NPs (or the metal oxides) with controlled particle size. Conversion of CuO into Cu using cellulose as a reducing agent under alkaline hydrothermal conditions was described as a green process for the production of Cu at power low cost [32]. This process gives rise to the conversion of cellulose into value-added chemicals, such as lactic acid and acetic acid. The possibility of modifying the surface of cellulosic fibers and using chitosan has also been reported [41]. In this case, chitosan-attached cellulose fibers were used in the immobilization of Cu ions followed by a reduction step in the presence of borohydride to obtain Cu NPs. Unlike Au e Ag, non-coated Cu NPs oxidize extensively under ambient atmosphere. Although this detrimental effect is limited by incorporating the Cu NPs in bacterial cellulose, stable Cu/cellulose composites have been prepared by using Cu nanowires as inorganic fillers [64]. These nanocomposites are attractive for the emerging technologies based on electronic paper.

3.2.4. *Platinum*

Platinum is an useful material for numerous industrial catalytic applications and several reports have described the synthesis of Pt NPs using a variety of methods [33]. This metal is also considered the best electrocatalyst for the four-electron reduction of oxygen to water in acidic environments as it provides the lowest overpotential and the highest stability [65]. The preparation of Pt/cellulose nanocomposites generally involves the reduction of ionic Pt by addition of a reducing agent ($NaBH_4$, HCHO, ...) in the presence of cellulose, which might act as a structural-directing agent [43].

Nanocomposites of Pt and amorphous carbon films were obtained by the catalyzed carbonization of cellulose fibers [44]. This type of NPs has been synthesized using $NaBH_4$ as reducing agent in hydrothermal conditions in the presence of nanoporous cellulose [43]. The Pt NPs were well dispersed and stabilized in the cellulose network thus avoiding particle aggregation. Cationic cellulose bearing ammonium ions at the surface has also been used to produce this type of nanocomposites [36]. In this method, the attachment of negatively charged Pt NPs onto cationic cellulose substrates was promoted via electrostatic interactions, which result into high surface coverage of the fibers.

Thermally stable proton-conducting membranes have been prepared by the *in situ* deposition of Pt NPs on BC membranes, via liquid phase chemical deoxidization method in the presence of the reducing agents $NaBH_4$ or HCHO [65]. The obtained black nanocomposite have been reported to display high electrocatalytic activity, with good prospects to be explored as membranes in fuel cell field [63]. However in case of Pt/cellulose nanocomposites, the reducing groups of cellulose are less effective in the reduction of metal precursors. A supercritical CO_2/ water system for reducing H_2PtCl_6 precursor to PtNPs using suspended crystalline cellulose nanofibrils of cotton has been described [33]. In this methodology VC was employedin a direct reduction route to form cellulose/Pt nanocomposites using a renewable reducing agent. The same authors have reported the use of cellulose nanocrystals (large surface area per unit weight in relation to normal cellulose fibers) for the same purpose. In this alternative the reaction temperature can be lowered to achieve Pt NPs with an average diameter of approximately 2 nm and with narrow particle size distribution [34].

The incorporation of irregularly shaped Pt NPs dispersed in IL and cellulose acetate lead to a nanocomposite that exhibits synergistic effects in the activity and durability enhancement of the catalyst [92]. The authors have suggested that the presence of IL caused higher separation of the cellulose macromolecules which result in a higher flexible and lower viscous material. The ensuing nanocomposites displayed higher catalytic activity and stability when compared to the Pt NPs dispersed in the IL. Potential uses of cellulose/Pt nanocomposites in catalysis comprise the hydrogenation of cyclohexene [92] and hydrogen production [93].

3.2.5. *Cobalt*

Co NPs in cellulose matrices has been a topic of interest due to the potential application as magnetic nanocomposites. However, due to easy oxidation their use has been associated to the formation of metal alloys such as FeCo, as will be described in section 3.2.6. The properties of magnetic Co NPs are determined in large extent by surface atoms. In addition, crystallinity, size distribution, particles shape and neighboring particles, affect the response of the material when submitted to a magnetic field. Therefore the matrix in which the NPs are embedded, in this case cellulose, has strong influence on the magnetic properties of the NPs as well as the distance between them [45].

The structure and morphology of Co NPs synthesized in cellulose matrix and resulting magnetic properties have been reported [45]. The authors have used two distinct chemical routes to investigate the effect on the structural properties of the NPs. In the borohydride reduction amorphous Co–B or Co oxide composites were obtained in which a detrimental effect on the magnetic properties was observed as compared to bulk Co. In contrast, using a NaH_2PO_2 reduction method, well-ordered ferromagnetic cobalt nanocrystals were obtained in which the magnetic properties of the samples resemble those of bulk cobalt.

3.2.6. Metal alloys

The properties of metallic systems can be significantly varied by blending the metal components into intermetallic compounds and alloys. The diversity of compositions, structures, and properties of metallic alloys not only can originate new properties but might also improve certain properties of the metal components due to synergistic effects [94]. The association of metal alloys (typically bimetallic) to cellulose yields interesting materials with well-defined, controllable properties and structures on the nanometer scale coupled with easier processing capability of the matrix. A tubular FeCo bimetallic nanostructure was obtained by using a cellulose/cobalt hexacyanoferrate (Fe–CN–Co) composite material as precursor [95]. The metal was then deposited onto a cellulose template via H_2 gas-phase reduction that converted the precursor in FeCo bimetallic NPs. The FeCo NPs formed hollow tubular structures that mimic the original precursor composite morphology via a template-direct assisted method.

Lightweight porous magnetic aerogels made of nanofibrils of VC and BC have been compressed into a stiff magnetic nanopaper [66]. The thick cellulose fibrils act as templates for the growth of discrete ferromagnetic cobalt ferrite NPs forming a dry, lightweight, porous and flexible magnetic aerogel with potential application in microfluidics devices and as electronic actuators. PdCu/BC nanocomposites showing high catalytic activity have been obtained, in which the PdCu NPs were homogeneously and densely precipitated at the surface [96]. Although the cost of these materials need to be considered, these compositesare of potential interest in water remediation processes because the Pd Cu alloy is considered the best catalyst for the denitrification of polluted water.

3.3. Multifunctional cellulose/metal nanocomposites

The combination of cellulose with distinct metal NPs to design multifunctional nanocomposites is an interesting approach to extend the scope of these materials to several areas of applications. A fluorescent nanocomposite exhibiting antibacterial activity has been achieved through the functionalization of NFC with luminescent silver metal nanoclusters [68]. A novel type of supramolecular native cellulose nanofiber/nanocluster adduct was obtained by using poly(methacrylic acid) (PMMA) as the mediator between Ag nanocluster and cellulose. The PMMA not only stabilized the Ag nanoclusters but also allowed hydrogen bonding between the particles and cellulose. Another example reports Au and Ag NPs as colorfast colorants in cellulose materials for textiles with antimicrobial and catalytic properties [29].

Also the combination of metal NPs with metal oxides is an emerging strategy to produce a range of multifunctional cellulose nanocomposites. The linkage of Ag NPs on magnetite containing BC substrates has been reported to produce magnetic and antimicrobial composites [97]. The possibility of bringing together diverse types of inorganics NPs and distinct cellulose matrices opens a new field for future applications, where the design of natural based multifunctional materials will be privileged.

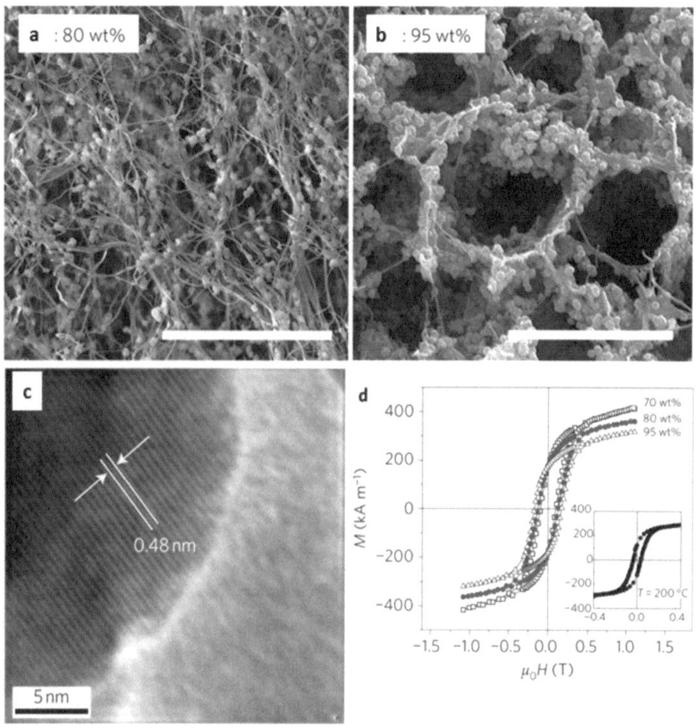

Figure 4. Magnetic aerogels at different loadings of cobalt ferrite nanoparticles. SEM images of sample (a) 80 wt% of particles and sample (b) 95 wt% of particles (Scale bars, 4 μm). (c) HRTEM image of a single particle from sample (b) showing the lattice fringescorresponding to the {111} reflections of the spinel structure, and the corresponding distance. The image was fast Fourier transform (FFT) filtered for clarity.(d) Magnetic hysteresis loops of cobalt-ferrite-based aerogels. Inset: hysteresis loop of cobalt ferrite-based of sample (a) at $T = 200$ °C. (Adapted by permission from Macmillan Publishers Ltd: Nature Nanotechnology([66]) (Copyright 2010).

4. Concluding remarks and future trends

The combination of metal nanoparticles and distinct types of cellulose matrices takes benefit of the properties of both components and simultaneously might result in properties due to synergistic effects. Besides the nature of the components in the final nanocomposite, the performance of the final material depends on the preparative methodologies employed in their production. This review has shown the relevance on the nanocomposite performance not only of the type of metal NPs used as fillers but also the origin of the cellulose matrix. In this context, methods that allow the chemical modification of both components, metal NPs and cellulose matrices, appear a very promising field of research to develop new functional materials. The combination of diverse metal NPs in cellulosic matrices is an important but less exploited strategy to prepare multifunctional composites. Fundamental studies concerning physico-chemical interactions that occur between the composite components have been scarce despite their obvious relevance in the optimization of the materials properties. Finally, the impact of these nanocomposites on health and environment is an issue in the agenda of the scientific community but whose importance will increase due to the commercialization of products based on these materials.

5. List of abbreviations

BC	Bacterial cellulose
CNT	Carbon nanotubes
DMAc	N,N-dimethylacetamide
DP	Degree of polymerization
EPTAC	(2,3-epoxypropyl)trimethylammonium chloride
FFT	Fast Fourier transform
HRTEM	High resolution transmission electron microscopy
IL	Ionic liquids
NFC	Nanofibrillated cellulose
NPs	Nanoparticles
PEI	poly(ethyleneimine)
PMMA	poly(methacrylic acid)
polyDADMAC	poly(diallyldimethylammonium chloride)
PSS	poly(sodium 4-styrenesulfonate)
PVP	polyvinylpyrrolidone
SEM	Scanning electron microscopy
SERS	Surface enhanced Raman scattering
TEA	Triethanolamine
TEMPO	2,2,6,6-tetramethylpyperidine-1-oxy radical
UV	Ultraviolet
VC	Vegetable cellulose

Acknowledgements

R.J.B. Pinto and M.C. Neves thank Fundação para a Ciência e Tecnologia (FCT) for the grants SFRH/BD/45364/2008 and SFRH/BPD/35046/2007, respectively. The authors acknowledge FCT (Pest-C/CTM/LA0011/2011), FSE and POPH for funding.

Author details

Ricardo J. B. Pinto*, Márcia C. Neves, Carlos Pascoal Neto and Tito Trindade

*Address all correspondence to: r.pinto@ua.pt

Department of Chemistry and CICECO, University of Aveiro,, Portugal

References

[1] Sanchez, C., Rozes, L., Ribot, F., Laberty-Robert, C., Grosso, D., Sassoye, C., Boissiere, C., & Nicole, L. (2010). Chimie douce: A land of opportunities for the designed construction of functional inorganic and hybrid organic-inorganic nanomaterials. *C. R. Chim.*, 1, 3 -39 .

[2] Manorama, S. V., Basak, P., & Singh, S. (2011). Anti-Microbial Polymer Nanocomposites. In Trindade T. & Daniel-da-Silva A. L. (Eds.) Nanocomposite Particles for Bio-Applications. Singapura Pan Stanford Publishing , 249-264.

[3] Belgacem, M. N., & Gandini, A. (2008). Monomers Polymers and Composites from Renewable Resources. Amesterdam:Elsevier

[4] Goncalves, G., Marques, P. A. A. P., Pinto, R. J. B., Trindade, T., & Pascoal, Neto. C. (2009). Surface modification of cellulosic fibres for multi-purpose TiO(2) based nanocomposites. *Compos. Sci. Technol.*, 69(7-8), 1051 -1056 .

[5] Martins, N. C. T., Freire, C. S. R., Pinto, R. J. B., Fernandes, S. C. M., Neto, C. P., Silvestre, A. J. D., Causio, J., Baldi, G., Sadocco, P., & Trindade, T. (2012). Electrostatic assembly of Ag nanoparticles onto nanofibrillated cellulose for antibacterial paper products. Submitted.

[6] Shen, J., Song, Z., Qian, X., & Ni, Y. (2011). A Review on Use of Fillers in Cellulosic Paper for Functional Applications. *Ind. Eng. Chem. Res*, 50(2), 661-666.

[7] Kim, J., Yun, S., & Ounaies, Z. (2006). Discovery of cellulose as a smart material. *Macromolecules*, 39(12), 4202-4206.

[8] Klemm, D., Heublein, B., Fink, H. P., & Bohn, A. (2005). Cellulose: Fascinating biopolymer and sustainable raw material. *Angew. Chem.-Int. Edit.*, 44(22), 3358-3393.

[9] Roberts, J. C. (1996). Paper Chemistry. Glasgow: Blackieeditor.

[10] Sjöström, E. (1993). *WOOD CHEMISTRY Fundamentals and Applications.*, London, Academic Press.

[11] Klemm, D., Schumann, D., Kramer, F., Hessler, N., Hornung, M., Schmauder, H. P., & Marsch, S. (2006). Nanocelluloses as innovative polymers in research and application. In Klemm D. (ed.) Polysaccharides II , 205, 49-96.

[12] Czaja, W., Krystynowicz, A., Bielecki, S., & Brown, R. M. (2006). Microbial cellulose-the natural power to heal wounds. *Biomaterials*, 27(2), 145-151.

[13] Trovatti, E., Serafim, L. S., Freire, C. S. R., Silvestre, A. J. D., & Neto, C. P. (2011). Gluconacetobacter sacchari: An efficient bacterial cellulose cell-factory. *Carbohydr. Polym*, 86(3), 1417-1420.

[14] Iguchi, M., Yamanaka, S., & Budhiono, A. (2000). Bacterial cellulose- a masterpiece of nature's arts. *J. Mater. Sci.*, 35261-270.

[15] Klemm, D., Schumann, D., Udhardt, U., & Marsch, S. (2001). Bacterial synthesized cellulose- artificial blood vessels for microsurgery. *Prog. Polym. Sci.*, 26(9), 1561-1603.

[16] Martins, N. C. T., Freire, C., Pinto, R. J. B., Fernandes, S., Pascoal, Neto. C., Silvestre, A., Causio, J., Baldi, G., Sadocco, P., & Trindade, T. (2012). Electrostatic assembly of silver nanoparticles onto nanofibrillated cellulose for the development of antibacterial paper products. *Cellulose*, DOI: 10.1007/s10570-10012-19713-10575.

[17] Klemm, D., Kramer, F., Moritz, S., Lindstrom, T., Ankerfors, M., Gray, D., & Dorris, A. (2011). Nanocelluloses: A New Family of Nature-Based Materials. *Angew. Chem.-Int. Edit.*, 50(24), 5438-5466.

[18] Belgacem, M. N., & Gandini, A. (2011). Production, Chemistry and Properties of Cellulose-Based Materials. In Plackett D. (ed.) Biopolymers- New Materials for Sustainable Films and Coatings John Wiley & Sons, Ltd , 151-178.

[19] Balazs, A. C., Emrick, T., & Russell, T. P. (2006). Nanoparticle polymer composites: Where two small worlds meet. *Science*, 314(5802), 1107-1110.

[20] Dankovich, T. A., & Gray, D. G. (2011). Bactericidal Paper Impregnated with Silver Nanoparticles for Point-of-Use Water Treatment. *Environ. Sci. Technol.*, 45(5), 1992-1998.

[21] Wu, D., & Fang, Y. (2003). The adsorption behavior of p-hydroxybenzoic acid on a silver-coated filter paper by surface enhanced Raman scattering. *J. Colloid Interface Sci.*, 265(2), 234-238.

[22] Ma, W., & Fang, Y. (2006). Experimental (SERS) and theoretical (DFT) studies on the adsorption of p-, m-, and o-nitroaniline on gold nanoparticles. *J. Colloid Interface Sci.*, 303(1), 1-8.

[23] Ngo, Y. H., Li, D., Simon, G. P., & Gamier, G. (2011). Paper surfaces functionalized by nanoparticles. *Adv. Colloid Interface Sci.*, 163(1), 23-38.

[24] Jiang, T., Liu, L., & Yao, J. (2011). In situ Deposition of Silver Nanoparticles on the Cotton Fabrics. *Fiber. Polym.*, 12(5), 620-625.

[25] El -Shishtawy, R. M., Asiri, A. M., Abdelwahed, N. A. M., & Al-Otaibi, M. M. (2011). In situ production of silver nanoparticle on cotton fabric and its antimicrobial evaluation. *Cellulose*, 18(1), 75-82.

[26] Montazer, M., Alimohammadi, F., Shamei, A., & Rahimi, M. K. (2012). In situ synthesis of nano silver on cotton using Tollens' reagent. *Carbohydr. Polym.*, 87(2), 1706-1712.

[27] He, J. H., Kunitake, T., & Nakao, A. (2003). Facile in situ synthesis of noble metal nanoparticles in porous cellulose fibers. *Chem. Mater.*, 15(23), 4401-4406.

[28] Pinto, R. J. B., Marques, P. A. A. P., Pascoal, Neto. C., Trindade, T., Daina, S., & Sadocco, P. (2009). Antibacterial activity of nanocomposites of silver and bacterial or vegetable cellulosic fibers. *Acta Biomater.*, 5(6), 2279-2289.

[29] Johnston, J. H., & Nilsson, T. (2012). Nanogold and nanosilver composites with lignin-containing cellulose fibres. *J. Mater. Sci.*, 47(3), 1103-1112.

[30] Li, Z., Friedrich, A., & Taubert, A. (2008). Gold microcrystal synthesis via reduction of HAuCl4 by cellulose in the ionic liquid 1-butyl-3-methyl imidazolium chloride. *J. Mater. Chem.*, 18(9), 1008-1014.

[31] Vainio, U., Pirkkalainen, K., Kisko, K., Goerigk, G., Kotelnikova, N. E., & Serimaa, R. (2007). Copper and copper oxide nanoparticles in a cellulose support studied using anomalous small-angle X-ray scattering. *Eur. Phys. J. D*, 42(1), 93-101.

[32] Li, Q., Yao, G., Zeng, X., Jing, Z., Huo, Z., & Jin, F. (2012). Facile and Green Production of Cu from CuO Using Cellulose under Hydrothermal Conditions. *Ind. Eng. Chem. Res.*, 51(7), 3129-3136.

[33] Benaissi, K., Johnson, L., Walsh, D. A., & Thielemans, W. (2010). Synthesis of platinum nanoparticles using cellulosic reducing agents. *Green Chem.*, 12(2), 220-222.

[34] Johnson, L., Thielemans, W., & Walsh, D. A. (2011). Synthesis of carbon-supported Pt nanoparticle electrocatalysts using nanocrystalline cellulose as reducing agent. *Green Chem.*, 13(7), 1686-1693.

[35] Tankhiwale, R., & Bajpai, S. K. (2009). Graft copolymerization onto cellulose-based filter paper and its further development as silver nanoparticles loaded antibacterial food-packaging material. *Colloid Surf. B-Biointerfaces*, 69(2), 164-168.

[36] Dong, H., & Hinestroza, J. P. (2009). Metal Nanoparticles on Natural Cellulose Fibers: Electrostatic Assembly and In Situ Synthesis. *ACS Appl. Mater. Interfaces*, 1(4), 797-803.

[37] Marques, P. A. A. P., Nogueira, H. I. S., Pinto, R. J. B., Neto, C. P., & Trindade, T. (2008). Silver-bacterial cellulosic sponges as active SERS substrates. *J. Raman Spectrosc.*, 39(4), 439-443.

[38] Sharifi, N., Tajabadi, F., & Taghavinia, N. (2010). Nanostructured silver fibers: Facile synthesis based on natural cellulose and application to graphite composite electrode for oxygen reduction. *Int. J. Hydrogen Energy*, 35(8), 3258-3262.

[39] Pinto, R. J. B., Marques, P. A. A. P., Martins, M. A., Neto, C. P., & Trindade, T. (2007). Electrostatic assembly and growth of gold nanoparticles in cellulosic fibres. *J. Colloid Interface Sci.*, 312(2), 506-512.

[40] Ishida, T., Watanabe, H., Bebeko, T., Akita, T., & Haruta, M. (2010). Aerobic oxidation of glucose over gold nanoparticles deposited on cellulose. *Appl. Catal. A-Gen.*, 377(1-2), 42 -46 .

[41] Mary, G., Bajpai, S. K., & Chand, N. (2009). Copper (II) Ions and Copper Nanoparticles-Loaded Chemically Modified Cotton Cellulose Fibers with Fair Antibacterial Properties. *J. Appl. Polym. Sci.*, 113(2), 757-766.

[42] Kotelnikova, N., Vainio, U., Pirkkatainen, K., & Serimaa, R. (2007). Novel approaches to metallization of cellulose by reduction of cellulose-incorporated copper and nickel ions. *Macromol. Symp.*, 25474-79.

[43] Cai, J., Kimura, S., Wada, M., & Kuga, S. (2009). Nanoporous Cellulose as Metal Nanoparticles Support. *Biomacromolecules*, 10(1), 87-94.

[44] He, J. H., Kunitake, T., & Nakao, A. (2004). Facile fabrication of composites of platinum nanoparticles and amorphous carbon films by catalyzed carbonization of cellulose fibers. *Chem. Commun* [4], 410 -411 .

[45] Pirkkalainen, K., Leppanen, K., Vainio, U., Webb, M. A., Elbra, T., Kohout, T., Nykanen, A., Ruokolainen, J., Kotelnikova, N., & Serimaa, R. (2008). Nanocomposites of magnetic cobalt nanoparticles and cellulose. *Eur. Phys. J. D*, 49(3), 333-342.

[46] Omrani, A. A., & Taghavinia, N. (2012). Photo-induced growth of silver nanoparticles using UV sensitivity of cellulose fibers. *Appl. Surf. Sci.*, 258(7), 2373-2377.

[47] Son, W. K., Youk, J. H., & Park, W. H. (2006). Antimicrobial cellulose acetate nanofibers containing silver nanoparticles. *Carbohydr. Polym.*, 65(4), 430-434.

[48] Cady, N. C., Behnke, J. L., & Strickland, A. D. (2011). Copper-Based Nanostructured Coatings on Natural Cellulose: Nanocomposites Exhibiting Rapid and Efficient Inhibition of a Multi-Drug Resistant Wound Pathogen, A. baumannii, and Mammalian Cell Biocompatibility In Vitro. *Adv. Funct. Mater.*, 21(13), 2506-2514.

[49] Li, S. M., Jia, N., Zhu, J. F., Ma, M. G., Xu, F., Wang, B., & Sun, R. C. (2011). Rapid microwave-assisted preparation and characterization of cellulose-silver nanocomposites. *Carbohydr. Polym.*, 83(2), 422-429.

[50] Li, S. M., Jia, N., Ma, M. G., Zhang, Z., Liu, Q. H., & Sun, R. C. (2011). Cellulose-silver nanocomposites: Microwave-assisted synthesis, characterization, their thermal stability, and antimicrobial property. *Carbohydr. Polym.*, 86(2), 441-447.

[51] Silva, A. R., & Unali, G. (2011). Controlled silver delivery by silver-cellulose nanocomposites prepared by a one-pot green synthesis assisted by microwaves. *Nanotechnology*, 22(31), 315605.

[52] Boufi, S., Ferraria, A. M., Botelho, doRego. A. M., Battaglini, N., Herbst, F., & Vilar, M. R. (2011). Surface functionalisation of cellulose with noble metals nanoparticles through a selective nucleation. *Carbohydr. Polym.*, 86(4), 1586-1594.

[53] Yokota, S., Kitaoka, T., Opietnik, M., Rosenau, T., & Wariishi, H. (2008). Synthesis of Gold Nanoparticles for In Situ Conjugation with Structural Carbohydrates. *Angew. Chem.-Int. Edit.*, 47(51), 9866-9869.

[54] Gruber, S., Taylor, R. N. K., Scheel, H., Greil, P., & Zollfrank, C. (2011). Cellulose-biotemplated silica nanowires coated with a dense gold nanoparticle layer. *Mater. Chem. Phys.*, 129(1-2), 19 -22 .

[55] Yang, G., Xie, J., Deng, Y., Bian, Y., & Hong, F. (2012). Hydrothermal synthesis of bacterial cellulose/AgNPs composite: A "green" route for antibacterial application. *Carbohydr. Polym.*, 87(4), 2482-2487.

[56] Yang, G., Xie, J., Hong, F., Cao, Z., & Yang, X. (2012). Antimicrobial activity of silver nanoparticle impregnated bacterial cellulose membrane: Effect of fermentation carbon sources of bacterial cellulose. *Carbohydr. Polym.*, 87(1), 839-845.

[57] Maneerung, T., Tokura, S., & Rujiravanit, R. (2008). Impregnation of silver nanoparticles into bacterial cellulose for antimicrobial wound dressing. *Carbohydr. Polym.*, 72(1), 43-51.

[58] de Santa, Maria. L. C., Santos, A. L. C., Oliveira, P. C., Barud, H. S., Messaddeq, Y., & Ribeiro, S. J. L. (2009). Synthesis and characterization of silver nanoparticles impregnated into bacterial cellulose. *Mater. Lett.*, 63(9-10), 797-799.

[59] Barud, H. S., Barrios, C., Regiani, T., Marques, R. F. C., Verelst, M., Dexpert-Ghys, J., Messaddeq, Y., & Ribeiro, S. J. L. (2008). Self-supported silver nanoparticles containing bacterial cellulose membranes. *Mater. Sci. Eng. C-Mater. Biol. Appl.*, 28(4), 515-518.

[60] Jung, R., Kim, Y., Kim, H. S., & Jin, H. J. (2009). Antimicrobial Properties of Hydrated Cellulose Membranes With Silver Nanoparticles. *J. Biomater. Sci.-Polym. Ed.*, 20(3), 311-324.

[61] Zhang, T., Wang, W., Zhang, D., Zhang, X., Ma, Y., Zhou, Y., & Qi, L. (2010). Biotemplated Synthesis of Gold Nanoparticle-Bacteria Cellulose Nanofiber Nanocomposites and Their Application in Biosensing. *Adv. Funct. Mater.*, 20(7), 1152-1160.

[62] Wang, W., Zhang, T. J., Zhang, D. W., Li, H. Y., Ma, Y. R., Qi, L. M., Zhou, Y. L., & Zhang, X. X. (2011). Amperometric hydrogen peroxide biosensor based on the immobilization of heme proteins on gold nanoparticles-bacteria cellulose nanofibers nanocomposite. *Talanta*, 84(1), 71-77.

[63] Wang, W., Li, H. Y., Zhang, D. W., Jiang, J., Cui, Y. R., Qiu, S., Zhou, Y. L., & Zhang, X. X. (2010). Fabrication of Bienzymatic Glucose Biosensor Based on Novel Gold Nanoparticles-Bacteria Cellulose Nanofibers Nanocomposite. *Electroanalysis*, 22(21), 2543-2550.

[64] Pinto, R. J. B., Neves, M. C., Pascoal, Neto. C., & Trindade, T. (2012). Growth and chemical stability of copper nanostructures on cellulosic fibers ACS . Submitted.

[65] Yang, J., Sun, D., Li, J., Yang, X., Yu, J., Hao, Q., Liu, W., Liu, J., Zou, Z., & Gu, J. (2009). In situ deposition of platinum nanoparticles on bacterial cellulose membranes and evaluation of PEM fuel cell performance. *Electrochim. Acta*, 54(26), 6300-6305.

[66] Olsson, R. T., Samir, M. A. S. A., Salazar-Alvarez, G., Belova, L., Strom, V., Berglund, L. A., Ikkala, O., Nogues, J., & Gedde, U. W. (2010). Making flexible magnetic aerogels and stiff magnetic nanopaper using cellulose nanofibrils as templates. *Nat. Nanotechnol.*, 5(8), 584-588.

[67] Ifuku, S., Tsuji, M., Morimoto, M., Saimoto, H., & Yano, H. (2009). Synthesis of Silver Nanoparticles Templated by TEMPO-Mediated Oxidized Bacterial Cellulose Nanofibers. *Biomacromolecules*, 10(9), 2714-2717.

[68] Diez, I., Eronen, P., Osterberg, M., Linder, M. B., Ikkala, O., & Ras, R. H. A. (2011). Functionalization of Nanofibrillated Cellulose with Silver Nanoclusters: Fluorescence and Antibacterial Activity. *Macromol. Biosci.*, 11(9), 1185-1191.

[69] Dallas, P., Sharma, V. K., & Zboril, R. (2011). Silver polymeric nanocomposites as advanced antimicrobial agents: Classification, synthetic paths, applications, and perspectives. *Adv. Colloid Interface Sci.*, 166(1-2), 119 -135 .

[70] Hubsch, E., Ball, V., Senger, B., Decher, G., Voegel, J. C., & Schaaf, P. (2004). Controlling the growth regime of polyelectrolyte multilayer films: Changing from exponential to linear growth by adjusting the composition of polyelectrolyte mixtures. *Langmuir*, 20(5), 1980-1985.

[71] Roberts, J. C. (1996). Paper Chemistry. Glasgow, Blackie.

[72] Ribitsch, V., Stana-Kleinschek, K., & Jeler, S. (1996). The influence of classical and enzymatic treatment on the surface charge of cellulose fibres. *Colloid Polym. Sci.*, 274(4), 388-394.

[73] Hyde, K., Dong, H., & Hinestroza, J. P. (2007). Effect of surface cationization on the conformal deposition of polyelectrolytes over cotton fibers. *Cellulose*, 14(6), 615-623.

[74] Song, J., Birbach, N. L., & Hinestroza, J. P. (2012). Deposition of silver nanoparticles on cellulosic fibers via stabilization of carboxymethyl groups. *Cellulose*, 19(2), 411-424.

[75] Chou, W. L., Yu, D. G., & Yang, M. C. (2005). The preparation and characterization of silver-loading cellulose acetate hollow fiber membrane for water treatment. *Polym. Adv. Technol.*, 16(8), 600-607.

[76] Abdel-Halim, E. S., & Al-Deyab, S. S. (2011). Utilization of hydroxypropyl cellulose for green and efficient synthesis of silver nanoparticles. *Carbohydr. Polym.*, 86(4), 1615-1622.

[77] Nadagouda, M. N., Speth, T. F., & Varma, R. S. (2011). Microwave-Assisted Green Synthesis of Silver Nanostructures. *Acc. Chem. Res.*, 44(7), 469-478.

[78] Thomas, V., Namdeo, M., Mohan, Y. M., Bajpai, S. K., & Bajpai, M. (2008). Review on polymer, hydrogel and microgel metal nanocomposites: A facile nanotechnological approach. *J. Macromol. Sci. Part A-Pure Appl. Chem.*, 45(1), 107-119.

[79] Carlson, C., Hussain, S. M., Schrand, A. M., Braydich-Stolle, L. K., Hess, K. L., Jones, R. L., & Schlager, J. J. (2008). Unique Cellular Interaction of Silver Nanoparticles: Size-Dependent Generation of Reactive Oxygen Species. *J. Phys. Chem. B*, 112(43), 13608-13619.

[80] Fernandez, A., Picouet, P., & Lloret, E. (2010). Cellulose-silver nanoparticle hybrid materials to control spoilage-related microflora in absorbent pads located in trays of fresh-cut melon. *Int. J. Food Microbiol.*, 142(1-2), 222 -228 .

[81] Kim, S. S., Park, J. E., & Lee, J. (2011). Properties and Antimicrobial Efficacy of Cellulose Fiber Coated with Silver Nanoparticles and 3-Mercaptopropyltrimethoxysilane (3-MPTMS). *J. Appl. Polym. Sci.*, 119(4), 2261-2267.

[82] Niu, Z., & Fang, Y. (2006). Surface-enhanced Raman scattering of single-walled carbon nanotubes on silver-coated and gold-coated filter paper. *J. Colloid Interface Sci.*, 303(1), 224-228.

[83] Cabalin, L. M., & Laserna, J. J. (1995). Fast spatially resolved surface-enhanced Raman spectrometry on a silver coated filter paper using charge-coupled device detection. *Anal. Chim. Acta*, 310(2), 337-345.

[84] Liu, Z., Li, M., Turyanska, L., Makarovsky, O., Patane, A., Wu, W., & Mann, S. (2010). Self-Assembly of Electrically Conducting Biopolymer Thin Films by Cellulose Regeneration in Gold Nanoparticle Aqueous Dispersions. *Chem. Mater.*, 22(8), 2675-2680.

[85] Koga, H., Tokunaga, E., Hidaka, M., Umemura, Y., Saito, T., Isogai, A., & Kitaoka, T. (2010). Topochemical synthesis and catalysis of metal nanoparticles exposed on crystalline cellulose nanofibers. *Chem. Commun.*, 46(45), 8567-8569.

[86] An, D., Ye, A., Deng, W., Zhang, Q., & Wang, Y. (2012). Selective Conversion of Cellobiose and Cellulose into Gluconic Acid in Water in the Presence of Oxygen, Catalyzed by Polyoxometalate-Supported Gold Nanoparticles. *Chemistry-A European Journal*, 18(10), 2938-2947.

[87] Tan, X., Deng, W., Liu, M., Zhang, Q., & Wang, Y. (2009). Carbon nanotube-supported gold nanoparticles as efficient catalysts for selective oxidation of cellobiose into gluconic acid in aqueous medium. *Chem. Commun.* [46], 7179 -81 .

[88] Luckham, R. E., & Brennan, J. D. (2010). Bioactive paper dipstick sensors for acetylcholinesterase inhibitors based on sol-gel/enzyme/gold nanoparticle composites. *Analyst*, 135(8), 2028-2035.

[89] Turyanska, L., Makarovsky, O., Patane, A., Kozlova, N. V., Liu, Z., Li, M., & Mann, S. (2012). High magnetic field quantum transport in Au nanoparticle-cellulose films. *Nanotechnology*, 23(4), 045702 .

[90] Gamarra, D., Munuera, G., Hungría, A. B., Fernández-García, M., Conesa, J. C., Midgley, P. A., Wang, X. Q., Hanson, J. C., Rodríguez, J. A., & Martínez-Arias, A. (2007). Structure–Activity Relationship in Nanostructured Copper–Ceria-Based Preferential CO Oxidation Catalysts. *J. Phys. Chem. C*, 111(29), 11026-11038.

[91] Guillot, S., Chemelli, A., Bhattacharyya, S., Warmont, F., & Glatter, O. (2009). Ordered Structures in Carboxymethylcellulose-Cationic Surfactants-Copper Ions Precipitated Phases: in Situ Formation of Copper Nanoparticles. *J. Phys. Chem. B*, 113(1), 15-23.

[92] Gelesky, M. A., Scheeren, C. W., Foppa, L., Pavan, F. A., Dias, S. L. P., & Dupont, J. (2009). Metal Nanoparticle/Ionic Liquid/Cellulose: New Catalytically Active Membrane Materials for Hydrogenation Reactions. *Biomacromolecules*, 10(7), 1888-1893.

[93] Himeshima, N., & Amao, Y. (2003). Photoinduced hydrogen production from cellulose derivative with chlorophyll-a and platinum nanoparticles system. *Energy Fuels*, 17(6), 1641-1644.

[94] Ferrando, R., Jellinek, J., & Johnston, R. L. (2008). Nanoalloys: From theory to applications of alloy clusters and nanoparticles. *Chem. Rev.*, 108(3), 845-910.

[95] Yamada, M., Tsuji, T., Miyake, M., & Miyazawa, T. (2009). Fabrication of a tubular FeCo bimetallic nanostructure using a cellulose-cobalt hexacyanoferrate composite as a precursor. *Chem. Commun.* [12], 1538 -1540 .

[96] Sun, D., Yang, J., Li, J., Yu, J., Xu, X., & Yang, X. (2010). Novel Pd-Cu/bacterial cellulose nanofibers: Preparation and excellent performance in catalytic denitrification. *Appl. Surf. Sci.*, 256(7), 2241-2244.

[97] Sureshkumar, M., Siswanto, D. Y., & Lee, C. K. (2010). Magnetic antimicrobial nanocomposite based on bacterial cellulose and silver nanoparticles. *J. Mater. Chem.*, 20(33), 6948-6955.

Chapter 5

High Performance PET/Carbon Nanotube Nanocomposites: Preparation, Characterization, Properties and Applications

Jun Young Kim and Seong Hun Kim

Additional information is available at the end of the chapter

http://dx.doi.org/10.5772/50413

1. Introduction

Poly(ethylene terephthalate) (PET) is one of aromatic polyesters, widely used polyester resin in conventional industry because of its good mechanical properties, low cost, high transparency, high processability, and moderate recyclability. Thus, PET holds a potential for industrial application, including industrial fibers, films, bottles, and engineering plastics [1–3]. In this regard, much research has been extensively performed to develop commercial application of aromatic polyesters or its composites, such a high performance polymer [4–11]. Although promising, however, insufficient mechanical properties and thermal stability of PET have hindered its practical application in a broad range of industry. From both an economic and industrial perspective, the major challenges for high performance polymer nanocomposites are to fabricate the polymer nanocomposites with low costs and to facilitate large scale-up for commercial applications.

Carbon nanotubes (CNTs), which were discovered by Iijima [12], have attracted a great deal of scientific interest as advanced materials for next generation. The CNT consisting of concentric cylinder of graphite layers is a new form of carbon and can be classified into three types [12-14]: single-walled CNT (SWCNT), double-walled CNT (DWCNT), and multi-walled CNT (MWCNT). SWCNT consists of a single layer of carbon atoms through the thickness of the cylindrical wall with the diameters of 1.0~1.4 nm, two such concentric cylinders forms DWCNT, and MWCNT consists of several layers of coaxial carbon tubes, the diameters of which range from 10 to 50 nm with the length of more than 10 μm [12-14]. The graphite nature of the nanotube lattice results in a fiber with high strength, stiffness, and conductivity, and higher aspect ratio represented by very small diameter and long length

INTECH
open science | open minds

© 2012 Young Kim and Hun Kim; licensee InTech. This is a paper distributed under the terms of the Creative Commons Attribution License (http://creativecommons.org/licenses/by/3.0), which permits unrestricted use, distribution, and reproduction in any medium, provided the original work is properly cited.

makes it possible for CNTs to be ideal nanoreinforcing fillers in advanced polymer nanocomposites [15]. Both theoretical and experimental approaches suggest the exceptional mechanical properties of CNTs ~100 times higher than the strongest steel at a fraction of the weight [16-19]: The Young's modulus, strength, and toughness of SWCNT shows 0.32~1.47 TPa of Young's modulus, 10~52 GPa of strength, and ~770 J/g of toughness, respectively [18]. For MWCNT, the values of strength, Young's modulus, and toughness were found to be 11-63 GPa, 0.27-0.95 TPa, and ~1240 J/g, respectively [19]. In addition, CNTs exhibit excellent electrical properties and electric current carrying capacity ~1000 times higher than copper wires [20]. In general, MWCNTs show inferior mechanical performance as compared to SWCNTs. However, MWCNTs have a cost advantage, in that they can be produced in much larger quantities at lower cost compared with the SWNT. In addition, MWCNTs are usually individual, longer, and more rigid than SWCNTs. Because of their remarkable physical properties such as high aspect ratio and excellent mechanical strength, MWCNTs are regarded as prospective reinforcing fillers in high performance polymer nanocomposites. For these reasons, extensive research and development have been directed towards the potential applications of CNTs for novel composite materials in a wide range of industrial fields. The fundamental research progressed to date on applications of CNTs also suggests that CNTs can be utilized as promising reinforcements in new kinds of polymer nanocomposites with remarkable physical/chemical characteristics [14].

During the rapid advancement in the materials science and technology, much research has extensively undertaken on high-performance polymer composites for targeted applications in numerous industrial fields. Furthermore, a great number of efforts have been made to develop high-performance polymer nanocomposites with the benefit of nanotechnology [21-25]. Polymer nanocomposites, which is a new class of polymeric materials based on the reinforcement of polymers using nanofillers, have attracted a great deal of interest in fields ranging from basic science to the industrial applications because it is possible to remarkably improve the physical properties of composite materials at lower filler loading [21-25]. These attempts include studies of the polymer composites with the introduction of nanoreinforcing fillers such as CNT, carbon nanofibers, inorganic nanoparticles, and polymer nanoparticles into the polymer matrix [21-25, 35-48]. In particular, excellent mechanical strength, thermal conductivity, and electrical properties of CNT have created a high level of activity in materials research and development for potential applications such as fuel cell, hydrogen storage, field emission display, chemical or biological sensor, and advanced polymer nanocomposites [26-34]. This feature has motivated a number of attempts to fabricate CNT/polymer nanocomposites in the development of high-performance composite materials. In this regard, much research and development have been performed to date for achieving the practical realization of excellent properties of CNT for advanced polymer nanocomposites in a broad range of industrial applications. However, because of their high cost and limited availability, only a few practical applications in industrial field such as electronic and electric appliances have been realized to date. In addition, potential applications as nanofillers have not been fully realized, despite extensive studies on CNT-filled polymer nanocomposites. Therefore, the fabrication of the polymer nanocomposites reinforced with various nanofillers is believed to a key technology on advanced composites for next generation.

http://dx.doi.org/10.5772/50413

For the fabrication of the CNT/polymer nanocomposites, major goal realize the potential applications of CNT as effective nanoreinforcements, leading to high performance polymer nanocomposites, are uniform dispersion of CNT in the polymer matrix and good interfacial adhesion between CNT and polymer matrix [49]. The functionalization of CNT, which can be considered as an effective method to achieve the homogeneous dispersion of CNT in the polymer matrix and its compatibility with a polymer, can lead to the enhancement of interfacial adhesion between CNT and polymer matrix, thereby improving the overall properties of the CNT/polymer nanocomposites [50-53]. Currently, four processing techniques are in common use to fabricate the CNT/polymer nanocomposites in situ polymerization, direct mixing, solution method, and melt compounding. Of these processing techniques, melt compounding has been accepted as the simplest and the most effective method from an industrial perspective because this process makes it possible to fabricate high performance polymer nanocomposites at low process cost, and facilitates commercial scale-up. Furthermore, the combination of a very small quantity of expensive CNT with conventional cheap thermoplastic polymers may provide attractive possibilities for improving the physical properties of polymer nanocomposites using a simple and cost-effective method [35-48].

This chapter focused on the fabrication and characterization of PET/CNT nanocomposites. The PET nanocomposites reinforced with a very small quantity of the modified CNT were prepared by simple melt compounding using a twin-screw extruder to fabricate high-performance polymer nanocomposites at low cost, and the resulting nanocomposites were characterized by means of Fourier transform infrared (FT-IR) spectroscopy, thermogravimetric analysis (TGA), rheological measurement, transmission electron microscopy (TEM), scanning electron microscopy (SEM), tensile testing, and differential scanning calorimetry (DSC) to clarify the effects of modified CNT on the physical properties and non-isothermal crystallization behavior of PET/CNT nanocomposites. This study demonstrates that the mechanical and rheological properties, thermal stability, and the non-isothermal crystallization behavior of PET/CNT nanocomposites were strongly dependent on the dispersion of the modified CNT in the PET matrix and the interfacial interactions between the modified CNT and the PET matrix. This chapter also suggests a simple and cost-effective method that can facilitate the industrial realization of CNT-reinforced PET nanocomposites with enhanced physical properties.

2. Fabrication of PET/CNT Nanocomposites

2.1. General features

PET nanocomposites reinforced with a very small quantity of modified CNT were prepared by melt compounding using a twin-screw extruder to create high performance polymer nanocomposites at low manufacturing cost for practically possible application in a broad range of industry. The introduction of carboxylic acid groups on the surfaces of the nanotube leads to the enhanced interactions between the nanotube and the polymer matrix through hydrogen bonding formation. The thermal stability, mechanical, and rheological properties of the PET nanocomposites are strongly dependent on the interfacial interactions between the PET and the modified CNT as well as the dispersion of the modified CNT in the PET. The introduction of the nanotube can significantly influence the non-isothermal crystallization behavior of the PET nanocomposites. This study demonstrates that a very small quantity of the modified CNT can substantially improve the thermal stability and mechanical properties of the PET nanocomposites, depending on the dispersion of the modified CNT and the interfacial interactions between the polymer matrix and the modified CNT. The key to improve the overall properties of PET nanocomposites depend on the optimization of the unique geometry and dispersion state of CNT in PET nanocomposites. This study also suggests a simple and cost-effective method that facilitates the industrial realization of PET/CNT nanocomposites with enhanced physical properties.

2.2. PET nanocomposites containing modified CNT

Conventional thermoplastic polymer used was the PET with an intrinsic viscosity of 1.07 dl/g, supplied by Hyo Sung Corp., Korea. The nanotubes used are multiwalled CNT (degree of purity > 95%) synthesized by a thermal chemical vapor deposition process, purchased from Iljin Nanotech, Korea. According to the supplier, their length and diameter were 10–30 nm and 10–50 μm, respectively, indicating that their aspect ratio reaches 1000.

The pristine CNT was added to the mixture of concentrated HNO_3 and H_2SO_4 with a volumetric ratio of 1:3 and this mixture was sonicated at 80 °C for 4 h to create the carboxylic acid groups on the nanotube surfaces [50]. After this chemical modification, the carboxylic acid groups-induced CNT (c-CNT) is expected to enhance the chemical affinity of the nanotube with the PET as well as the dispersion of the nanotube in the PET matrix. All materials were dried at 120 °C *in vacuo* for at least 24 h, before use to minimize the effect of moisture. The PET nanocomposites were prepared by a melt compounding in a Haake rheometer (Haake Technik GmbH, Germany) equipped with a twin-screw. The temperature of the heating zone, from the hopper to the die, was set to 270, 280, 285, and 275 °C, and the screw speed was fixed at 20 rpm for the fabrication of the PET nanocomposites, PET was melt blended with the addition of CNT content, specified as 0.1, 0.5, and 1.0 wt% in the polymer matrix, respectively. Upon completion of melt blending, the extruded strands were allowed to cool in the water-bath, and then cut into pellets with constant diameter and length using a rate-controlled PP1 pelletizer (Haake Technik GmbH, Germany)

Chemical structures of CNT and the PET nanocomposites were characterized by means of FT-IR measurement using a Magma-IR 550 spectrometer (Nicolet) in the range of 400–4000 cm^{-1}. TGA of the PET nanocomposites was performed with a TA Instrument SDF-2960 TGA over a temperature range of 30~800 °C at a heating rate of 10 °C/min under N_2. Rheological properties of the PET nanocomposites were performed on an ARES (Advanced Rheometer Expanded System) rheometer (Rheometric Scientific) in oscillation mode with the parallel-plate geometry using the plate diameter of 25 mm and the plate gap setting of 1 mm at 270, 280, and 290 °C, covering the temperature processing windows of the PET nanocomposites. The frequency ranges were varied between 0.05 and 450 rad/s, and the strain amplitude was applied to be within the linear viscoelastic ranges. Morphologies of CNT and the PET nano-

http://dx.doi.org/10.5772/50413

composites were observed using a JEOL 2000FX TEM and a JEOL JSM-6300F SEM. Mechanical properties of the PET nanocomposites were measured with an Instron 4465 testing machine, according to the procedures in the ASTM D 638 standard. The gauge length and crosshead speed were set to 20 mm and 10 mm/min, respectively. Thermal behavior of the PET nanocomposites was measured with a TA Instrument 2010 DSC over a temperature range of 30~295 °C at a scan rate of 10 °C/min under N_2. Samples were heated to 295 °C at a heating rate of 10 °C/min, held at 295 °C for 10 min to eliminate any previous thermal history and then cooled to room temperature at a cooling rate of 10 °C/min. The non-isothermal crystallization kinetics was investigated by cooling samples from 295 to 30 °C at constant cooling rates of 2.5, 5, 10, 15, and 20 °C/min, respectively. The relative degree of crystallinity, $X(T)$, of the PET nanocomposites at various cooling rates can be calculated from the ratio of the area of the exothermic peak up to temperature (T) divided by that of the total exotherms of the crystallization

3. Influence of Modified CNT on PET Nanocomposites

3.1. CNT modification

The FT-IR spectra of CNT and the PET nanocomposites are shown in Figure 1. The characteristic peaks observed at ~1580 cm^{-1} was attributed to the IR-phonon mode of multi-walled CNT [54]. The characteristic peaks observed at 1080, 1190, and 1720 cm^{-1}, respectively, for the c-CNT were attributed to the stretching vibrations of the carboxylic acid groups [55]. This result demonstrates that carboxylic acid groups on the surface of the c-CNT were effectively induced via chemical modification. After chemical modification, the c-CNT exhibits less entangled structures as compared to pristine CNT showing some agglomerated structures, indicating that the dispersion of the c-CNT in the PET matrix will be more effective than that of pristine CNT. Thus, it is expected that the functional groups effectively induced on the surface of the nanotube via chemical modification are helpful for enhancing the interactions between the polymer matrix and the nanotube.

As shown in Figure 1B, the PET nanocomposites exhibited similar absorption bands of pure PET, which were observed at 1715 (C=O), 1454 (CH_2), 1407 (aromatic ring), 1247 (C-O), 1101 (O=CH_2), 1018 (aromatic ring), and 723 cm^{-1} (CH), respectively [56]. However, the stretching vibration peaks for the PET nanocomposites shifted from 1715, 1247, and 1101 to 1708, 1232, and 1085 cm^{-1}, respectively, as compared to pure PET. This result indicated the existence of some interactions between the c-CNT and the PET matrix through hydrogen bonding formation, as shown in Figure 1C. Thus, it is expected that the enhanced interactions between the c-CNT and the PET matrix can lead to the good interfacial adhesion between them, resulting in the improvement in the overall mechanical properties of the PET nanocomposites due to the nanoreinforcing effect of the c-CNT.

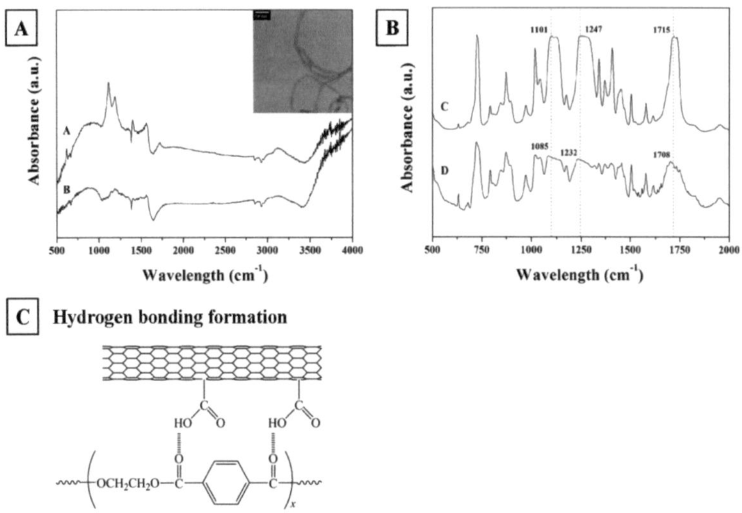

Figure 1. FT-IR spectra of (A) CNT and (B) the PET nanocomposites (a: c-CNT; b: pristine CNT; c: PET; and d: the PET nanocomposite containing 1.0 wt% of the c-CNT). The inset shows TEM images of the c- CNT after chemical modification. (C) Schematic showing possible interactions between the c-CNT and the PET matrix through hydrogen bonding formation. Reproduced with the permission from Ref. [46]. © 2010 Wiley Periodicals, Inc.

3.2. Thermal stability and thermal decomposition kinetics

TGA thermograms of the PET nanocomposites are shown in Figure 2, and their results are summarized in Table 1. The curve patterns of the PET nanocomposites are similar to that of pure PET, indicating that the features of the weigh-loss for thermal decomposition of the PET nanocomposites may mostly stem from PET matrix. As shown in Table 1, the thermal decomposition temperatures, thermal stability factors, and residual yields of the PET nanocomposites increased with increasing the c-CNT content. The presence of the c-CNT can lead to the stabilization of the PET matrix, and good interfacial adhesion between the c-CNT and the PET may restrict the thermal motion of the PET molecules [57], resulting in the increased thermal stability of the PET nanocomposites. Shaffer and Windle [58] suggested that the thermal decomposition of CNT-filled polymer nanocomposites was retarded by high protecting effect of CNT against the thermal decomposition. In the PET nanocomposites, the effective function of the c-CNT as physical barriers to prevent the transport of volatile decomposed products in the polymer nanocomposites during thermal decomposition resulted in the enhanced thermal stability of the PET nanocomposites. Similar observation has been also reported that thermal stability of poly(ethylene 2,6-naphthalte) (PEN)/CNT nanocomposites was improved by physical barrier effects of CNT layers acting as effective thermal insulators in the PEN nanocomposites [40].

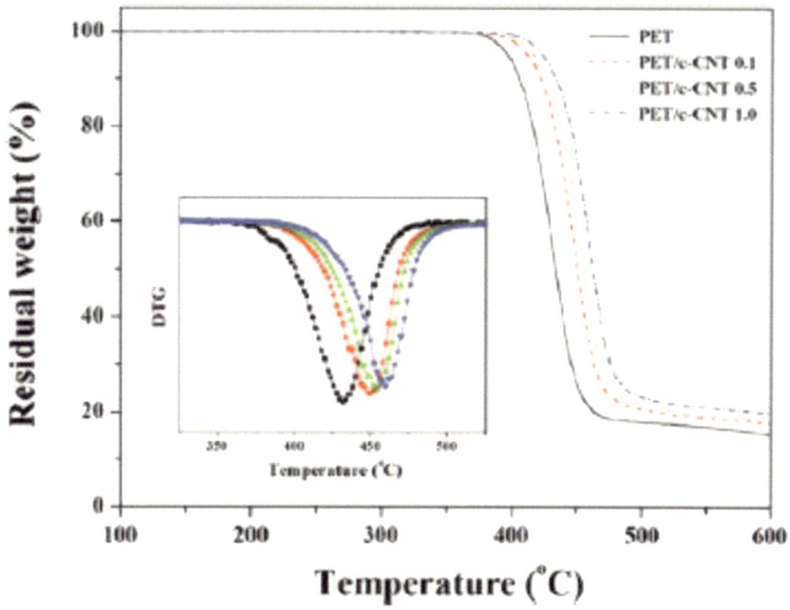

Figure 2. TGA thermograms of the PET nanocomposites. The inset shows the first derivative curves corresponding to TGA thermograms of the PET nanocomposite. Reproduced with the permission from Ref. [46]. © 2010 Wiley Periodicals, Inc.

Materials	$T_{di}{}^{a}$ (°C)	$T_{dm}{}^{b}$ (°C)	*A*	*K*	*IPDT*^{c} (°C)	$W_R{}^{d}$ (%)
PET	387.8	432.1	0.7558	1.2627	573.8	15.7
PET/c-CNT 0.1	398.8	449.5	0.7803	1.2934	605.3	17.7
PET/c-CNT 0.5	402.7	452.9	0.7887	1.3025	615.3	18.3
PET/c-CNT 1.0	411.4	460.2	0.8017	1.3274	636.9	19.8

Table 1. Table 1. Effect of the c-CNT on the thermal stability of the PET nanocomposites. [[a] Initial decomposition temperatures at 2% of the weight-loss; [b] Decomposition temperature at the maximum rate of the weight-loss; [c] Integral procedure decomposition temperatures, $IPDT = A\,K(T_f - T_i) + T_i$, where *A* is the area ratio of total experimental curve divided by total TGA curves; *K* is the coefficient of *A*; T_i is the initial experimental temperature, and T_f is the final experimental temperature [59]; [d] Residual yield in TGA thermograms at 600 °C under N_2].

3.3. Rheological properties

As shown in Figure 3A, the $|\eta^*|$ of the PET nanocomposites decreased with increasing frequency, indicating that the PET nanocomposites exhibited a non-Newtonian behavior over the whole frequency range measured. The shear thinning behavior of the PET nanocomposites resulted from the random orientation of entangled molecular chains in the polymer nanocomposites during the applied shear deformation. The $|\eta^*|$ of the PET nanocomposites increased with the c-CNT content, and this effect was more pronounced at low frequency than at high frequency, indicating the formation of the interconnected or network-like structures in the PET nanocomposites as a result of nanotube-nanotube and nanotube-polymer interactions. In addition, the PET nanocomposites exhibited higher $|\eta^*|$ values and more distinct shear thinning behavior as compared to pure PET, suggesting that better dispersion of the c-CNT or stronger interactions between the nanotubes and the polymer matrix [60]. The increase in the $|\eta^*|$ of the PET nanocomposites with the introduction of the c-CNT was closely related to the increase in the storage modulus of the PET nanocomposites, which will be described in the following section. As shown in Figure 3B, the $|\eta^*|$ of the PET nanocomposites decreased with increasing temperature. The temperature had little effect on the $|\eta^*|$ of the PET nanocomposites at lower frequency, while at higher frequency, the rheological properties of the PET nanocomposites were affected by the temperature, and the $|\eta^*|$ values of the PET nanocomposites decreased significantly with increasing temperature. This result indicated the enhancement in the flow behavior of the PET nanocomposite melts with increasing temperature.

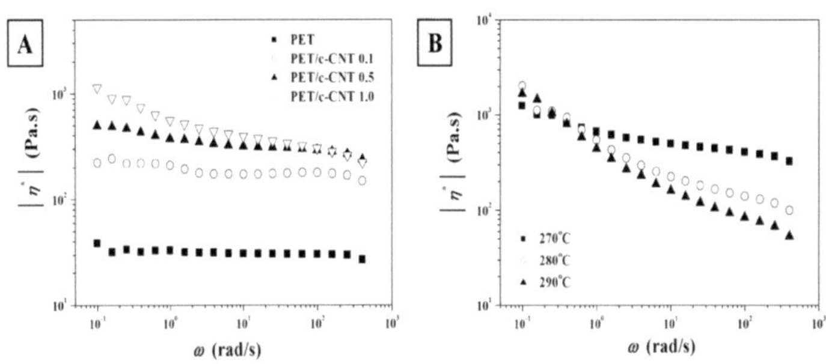

Figure 3. Complex viscosity of (A) the PET nanocomposites with the c-CNT content, and (B) the PET nanocomposites containing 1.0 wt% of the c-CNT at different temperatures as a function of frequency. Reproduced with the permission from Ref. [46]. © 2010 Wiley Periodicals, Inc.

The storage modulus (G') and loss modulus (G'') of the PET nanocomposites as a function of frequency are shown in Figure 4. The storage and loss moduli of the PET nanocomposites increased with increasing frequency and c-CNT content and this effect was more pronounced at low frequency than at high frequency. This feature may be explained by the fact that the interactions between the nanotubes-nanotube and nanotube-polymer with the introduction of the c-CNT can lead to the formation of interconnected or network-like structures in the polymer nanocomposites [36, 53]. Furthermore, the values of G' and G'' of the PET nanocomposites were higher than those of pure PET over the whole frequency range measured, and this enhancing effect was more pronounced at low frequency than at high frequency. The non-terminal behavior observed in the PET nanocomposites at low frequency, which was similar to the relaxation behavior of typical filled-polymer composite system [61], was related to the variation of the terminal slope of the flow curves based on the power-law equations: $|\eta^*| \approx \omega^n$ and $G' \approx \omega^m$ (where ω is the frequency, n is the shear-thinning exponent, and m is the relaxation exponent) [62]. The variations of the shear-thinning exponent and relaxation exponent of the PET nanocomposites are shown in Figure 5. As compared to pure PET, the lower shear thinning exponent of the PET nanocomposites indicated the significant dependence of shear-thinning behavior of the PET nanocomposites on the presence of the c-CNT, resulting from the increased interfacial interactions between the c-CNT and the PET as well as good dispersion of the c-CNT in the PET matrix. In addition, the decrease in the relaxation exponent of the PET nanocomposite with increasing the c-CNT content may be attributed to the formation of interconnected or network-like structures in the PET nanocomposites, resulting in the pseudo solid-like behavior of the PET nanocomposites. Similar non-terminal low frequency rheological behavior has been observed in ordered block copolymers, smectic liquid-crystalline small molecules, polymer/silicate nanocomposites, and CNT/polycarbonate composites [63–66]

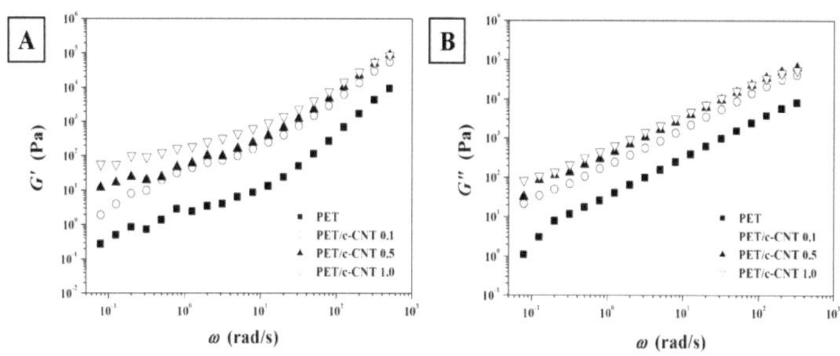

Figure 4. A) Storage modulus (*G′*) and (B) loss modulus (*G″*) of the PET nanocomposites as a function of frequency. Reproduced with the permission from Ref. [46[. © 2010 Wiley Periodicals, Inc.

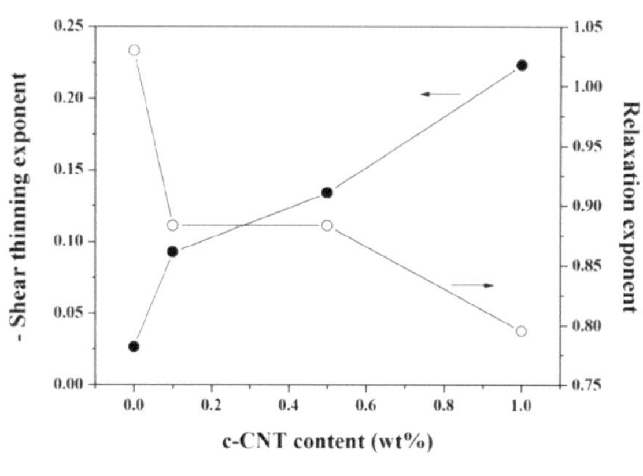

Figure 5. Variations of the shear thinning exponent and relaxation exponent of the PET nanocomposites with the c-CNT content. Reproduced with the permission from Ref. [46]. © 2010 Wiley Periodicals, Inc.

3.4. Morphology and mechanical properties

Pristine CNT typically tends to bundle together and to form some agglomeration due to the intrinsic van der Waals attractions between the individual nanotubes in combination with high aspect ratio and large surface area, making it difficult for CNT to disperse in the polymer matrix. In this study, the chemical modification was performed to achieve the enhanced adhesion between CNT and polymer matrix as well as good dispersion state of CNT. TEM image of the PET nanocomposites containing 0.1 wt% of the c-CNT is shown in Figure 6A. The c-CNT exhibited less entangled structures due to the functional groups formed on the nanotube surfaces via chemical modification as compared to pristine CNT (refer Figure 1), and the c-CNT was well dispersed in the PET matrix. SEM image of the fracture surfaces of the PET nanocomposites containing 1.0 wt% of the c-CNT is shown in Figure 6B. It can be observed that some nanotubes were broken with their two ends still embedded in the PET matrix and other nanotubes were bridging the local microcracks, which may delay the failure of the polymer nanocomposites [67]. This result indicates good wetting and adhesion of the c-CNT with the PET matrix. Similar observation has been reported that the presence of fractured tubes, along with the matrix still adhered to the fractured tubes matrix in terms of a crack interacting with the nanotube reinforcement could increased the elastic modulus and ultimate strength of CNT/polystyrene composites [68]. In the PET nanocomposites, the c-CNT stabilized their dispersion by good interactions with the PET matrix, resulting from the interfacial interactions of the -COOH groups at the c-CNT and the C=O groups in the PET macromolecular chains through hydrogen bonding formation due to the functional groups onto the nanotube surfaces induced effectively via chemical modification as illustrated in Figure 1C. This enhanced interfacial adhesion between the c-CNT and the PET matrix may be considered as the evidence for efficient load transfer from the polymer matrix to the nanotubes, thus leading to the nanoreinforcing effects of the c-CNT on the improvement in the mechanical properties of the PET nanocomposites

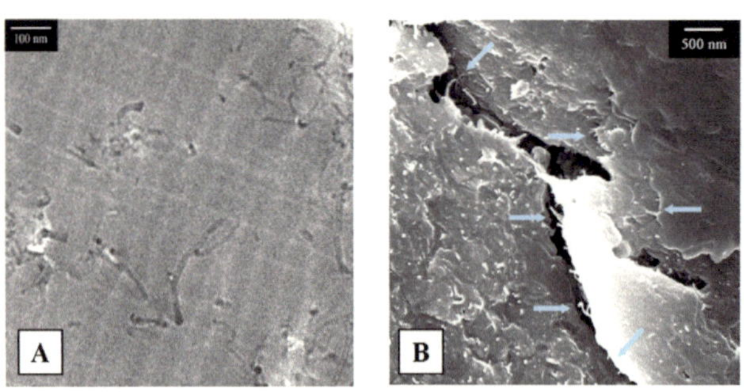

Figure 6. TEM image of (a) the PET nanocomposites containing 0.1 wt% of the c-CNT and (b) SEM image of the fracture surfaces of the PET nanocomposites containing 1.0 wt% of the c-CNT. The arrows indicates that the nanotubes were to be broken with their ends still embedded in the PET matrix or they were bridging the local micro-cracks in the nanocomposites, suggesting good wetting with the matrix and enhanced adhesion between the nanotubes and the matrix, thus being favorable to efficient load transfer form the polymer matrix to the nanotubes. Reproduced with the permission from Ref. [46]. © 2010 Wiley Periodicals, Inc.

The mechanical properties of the PET nanocomposites with the c-CNT are shown in Figure 7A. The tensile strength (σ) and tensile modulus (E) of the PET nanocomposites increased significantly with increasing the c-CNT content due to the nanoreinforcing effect of the c-CNT with high aspect ratio. As illustrated in Figure 1C, possible interactions between the carboxylic acid groups of the c-CNT and the ester groups in PET macromolecular chains through hydrogen bonding formation results in the enhanced interfacial adhesion between them as well as good dispersion of the c-CNT in the PET matrix, thus being more favorable to more effective load transfer from the polymer matrix to the nanotubes, and lead to the substantial improvement in the mechanical properties of the PET nanocomposites. The elongation at break (E_b) of the PET nanocomposites decreased with the introduction of CNT (Table 2), which may be attributed to the increase in the stiffness of the PET nanocomposites by the c-CNT and the micro-voids formed around the nanotube during the tensile testing [40, 41]. However, the PET nanocomposites containing the c-CNT exhibited higher E_b than in the case of pristine CNT, resulting from the enhanced interfacial interactions between the c-CNT and the PET as well as the good dispersion of the c-CNT in the PET matrix. Meng et al. [69] reported that modified CNT/polyamide (PA) nanocomposites showed higher tensile strength, tensile modulus, and elongation at break than those of pristine CNT/PA nanocomposites because of uniform dispersion and good interfacial adhesion in the modified CNT/PA nanocomposites.

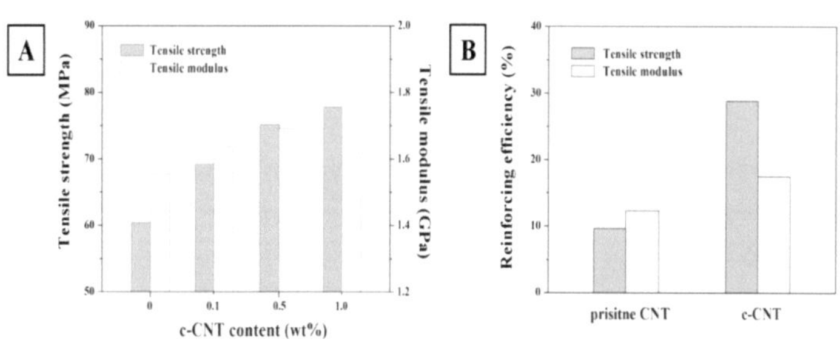

Figure 7. A) Mechanical properties of the PET nanocomposites and (B) the reinforcing efficiency of pristine CNT and the c-CNT on the mechanical properties of the PET nanocomposites containing 1.0 wt% of the CNT. Reinforcing efficiency (%) = $[(M_c - M_m)/M_m] \times 100$, where M_c and M_m represent the mechanical properties, such as tensile strength and tensile modulus, of the PET nanocomposites and pure PET, respectively. Reproduced with the permission from Ref. [46]. © 2010 Wiley Periodicals, Inc.

Materials	σ^a (MPa)	E (GPa)	E_b (%)
PET	60.4 ± 1.8	1.55 ± 0.072	174 ± 19
PET/c-CNT 1.0	77.8 ± 3.2	1.82 ± 0.028	96 ± 12
PET/p-CNT 1.0 [a]	66.2 ± 1.2	1.74 ± 0.031	81 ± 15

Table 2. Mechanical properties of the PET nanocomposites. [[a] The p-CNT represents pristine CNT without chemical modification].

For characterizing the effect of the c-CNT on the mechanical properties of the PET nanocomposites, it is also very instructive to compare the reinforcing efficiency of the c-CNT for a given content in the PET nanocomposites. The reinforcing efficiency is defined as the normalized mechanical properties of the PET nanocomposites with respect to those of pure PET as follows:

Reinforcing efficiency (%)

$$= \frac{M_c - M_m}{M_m} \times 100$$

where M_c and M_m represent the mechanical properties, including tensile strength and tensile modulus of PET nanocomposites and pure PET, respectively. As shown in Figure 7B, the enhancing effect of the mechanical properties for the PET nanocomposites was more significant in the PET nanocomposites containing the c-CNT than in the case of pristine CNT. This result indicated that the introduction of the c-CNT into the PET matrix was more effective in improving the mechanical properties of the PET nanocomposites as compared to pristine CNT. The incorporation of the c-CNT into the PET matrix resulted in the increased interfacial adhesion between the c-CNT and the PET matrix, thus being favorable to more efficient load transfer form the polymer matrix to the nanotubes. Thus, the enhanced interfacial adhesion between the c-CNT and the PET as well as good dispersion of the c-CNT result in the improvement in the overall mechanical properties of the PET nanocomposites.

3.5. Non-isothermal crystallization behavior

The incorporation of the c-CNT had little effect on the melting temperatures of the PET nanocomposites, whereas the glass transition temperature of the PET nanocomposites increased with the introduction of the c-CNT, resulting from the hindrance of the segmental motions of the PET macromolecular chains by the c-CNT. As shown in Figure 8A, the crystallization temperatures of the PET nanocomposites significantly increased by incorporating the c-CNT and this enhancing effect was more pronounced at lower content. This result indicated the efficiency of the c-CNT as strong nucleating agents for the PET crystallization. As the c-CNT content increased, the decrease in the ΔT for crystallization as well as the increase in the T_{mc} of the PET nanocomposites (Table 3) suggested that a very small quantity of the c-CNT acted as effective nucleating agents in PET, enhancing the crystallization of the PET nanocomposites with the presence of the c-CNT. The non-isothermal crystallization curves of the PET nanocomposites at various cooling rates are shown in Figure 8B. As the cooling rate increased, the crystallization peak temperature range becomes broader and shifts to lower temperatures, indicating that the lower the cooling rate, the earlier crystallization occurs. The PET nanocomposites exhibited higher peak temperature and lower overall crystallization time at a given cooling rate, as compared to pure PET. Homogeneous nucleation started spontaneously below the melting temperature and required longer times, whereas heterogeneous nuclei formed as soon as samples reached the crystallization temperature [70]. As the crystallization of polymer nanocomposites proceeds through heterogeneous nucleation, the introduction of the c-CNT increased the PET crystallization because of high nucleation induced by the c-CNT. Similar observations have been reported that the crystallization of CNT/polymer nanocomposites was accelerated by the presence of CNT through heterogeneous nucleation [35, 38, 40-43].

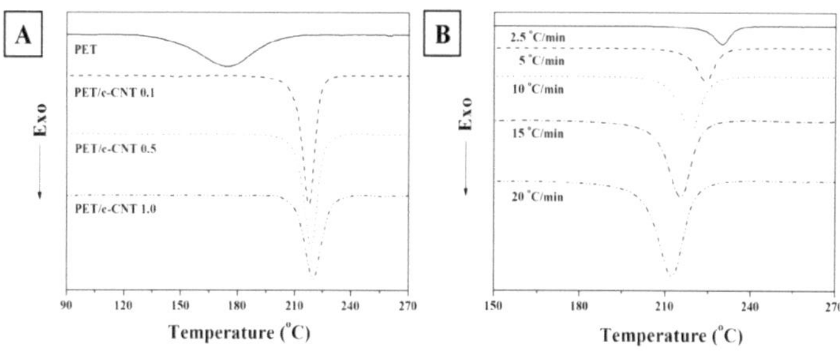

Figure 8. A) DSC cooling traces of the PET nanocomposites at a cooling rate of 10 °C/min and (B) the non-isothermal crystallization curves of the PET nanocomposites containing 0.5 wt% of the c-CNT at various cooling rates. Reproduced with the permission from Ref. [46]. © 2010 Wiley Periodicals, Inc.

Materials	T_g (°C)	T_m (°C)	T_c (°C)	ΔT [a] (°C)
PET	84.4	255.4	193.7	61.7
PET/c-CNT 0.1	90.4	255.3	218.6	36.7
PET/c-CNT 0.5	91.3	255.2	219.0	36.2
PET/c-CNT 1.0	92.0	255.9	219.6	36.3

Table 3. Table 3. Thermal behaviour of PET nanocomposites [[a] The values obtained from the DSC heating traces at 10 °C/min; [b] The crystallization temperatures measured from the DSC cooling traces at 10 °C/min; [c] The degree of supercooling, $\Delta T = T_m - T_{mc}$]

During the non-isothermal crystallization, the relative degree of crystallinity, $X(T)$, with temperature and time for the PET nanocomposites at various cooling rates are shown in Figure 9. The crystallization of the PET nanocomposites occurred at higher temperature and over a longer time with decreasing cooling rate, suggesting that crystallization may be controlled by nucleation [71]. As the cooling rate increased, the time for completing crystallization was decreased and the X(T) values of the PET nanocomposites were higher than that of pure PET. The variations of the crystallization peak temperature (T_p) and the crystallization half-time ($t_{0.5}$) of the PET nanocomposites are shown in Figure 10. The T_p of the PET nanocomposites were higher than that of pure PET at a given cooling rate, while the $t_{0.5}$ were lower than that of pure PET. This result suggested that the introduction of the c-CNT increased the crystallization rate of the PET nanocomposites by its effective function as strong nucleating agents for enhancing the PET crystallization. This enhancing effect of the crystallization rate induced by the c-CNT may be attributed to the interactions between the carboxylic acid groups on the surface of the c-CNT and the PET macromolecular chains as well as the physical adsorption of PET molecules onto the surface of the c-CNT, resulting in the enhancement of the crystallization rate of the PET nanocomposites. Similar observation has been reported that the introduction of carboxylated CNT increased more efficiently the crystallization rate of polyamide [72]

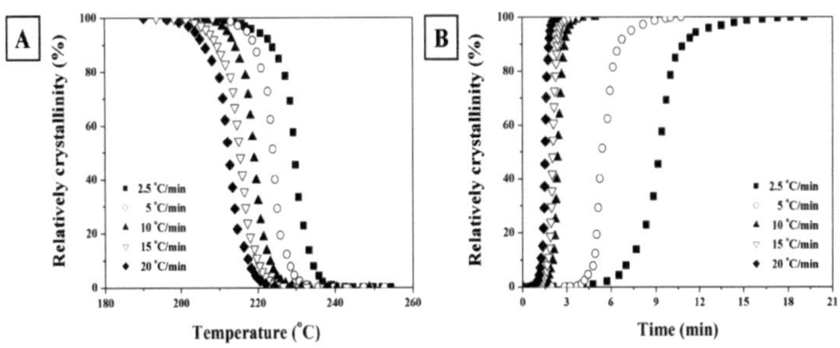

Figure 9. Relative degree of crystallinity of the PET nanocomposites containing 0.5 wt% of the c-CNT with (A) the temperature and (B) the time at various cooling rates. Reproduced with the permission from Ref. [46]. © 2010 Wiley Periodicals, Inc.

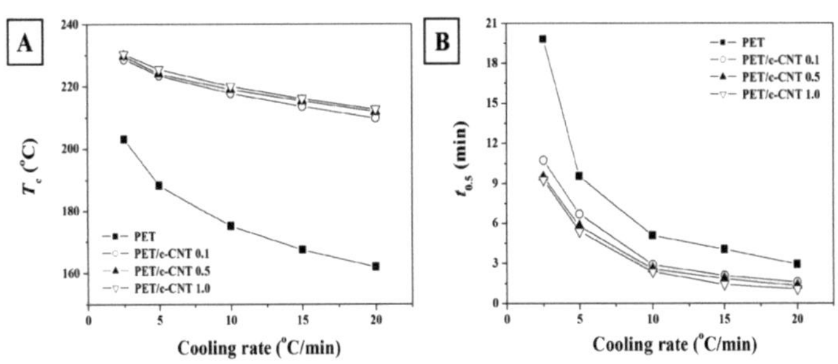

Figure 10. A) Crystallization temperatures (T_c) and (B) crystallization half-time ($t_{0.5}$) of the PET nanocomposites as a function of cooling rate during the non-isothermal crystallization. The crystallization half-time ($t_{0.5}$) can be defined as the time taken to complete half of the non-isothermal crystallization process, i.e., the time required to attain a relative degree of crystallinity of 50%. Reproduced with the permission from Ref. [46]. © 2010 Wiley Periodicals, Inc.

On the basis of the T_p values obtained from the non-isothermal crystallization curves, the fastest crystallization time (t_{max}) at various cooling rates (a) can be determined and their results are summarized in Table 4. The tmax represents the time from the onset temperature (T_0) to the peak temperature (T_p) of the crystallization and can be expressed by the relationship of $t_{max} = (T_0 - T_c)/a$ [73]. At a given cooling rate, the tmax values of the PET nanocomposites decreased with the introduction of the c-CNT. This result indicated that the nucleation effect by the c-CNT was significant even with a very small quantity of the c-CNT, providing possible evidence of the enhancement of the crystallization rate for the PET nanocomposites due to high nucleation induced by the c-CNT. In addition, the effect of the c-

CNT on the non-isothermal crystallization rate of the PET nanocomposites was characterized by means of the crystallization rate constant (CRC) suggested by Khanna [74]. The CRC can be calculated from the slope of the plots of the cooling rate versus the crystallization temperature plot, i.e., CRC = $\Delta a/\Delta T_c$, meaning that the larger the CRC, the faster the crystallization rate. The CRC value of pure PET in this study was similar to those reported by Khanna [74], who found the CRC values of PET in the range of 30~60/h, depending on the molecular weight and the processing method. As shown in Figure 11A, the CRC values of the PET nanocomposites significantly increased with the introduction of the c-CNT, indicating the higher crystallization rate of the PET nanocomposites as compared to pure PET because of high nucleation effect induced by the c-CNT. In addition, the CRC values of the PET nanocomposites significantly increased with the addition of 0.1 wt% of the c-CNT, and increased slightly with further addition of the c-CNT. This result revealed that the acceleration of the non-isothermal crystallization for the PET nanocomposites was not directly proportional to the increase in the c-CNT content and that the enhancement of the crystallization rate of the PET nanocomposites induced by the c-CNT could not be described by means of simple linear regression method, implying the complex mechanism of the non-isothermal crystallization process in the presence of the c-CNT. The enhanced interfacial interactions between the functional groups induced on the surface of the c-CNT and the PET as well as the strong nucleation effect of the c-CNT could increase the crystallization rate of the PET nanocomposites

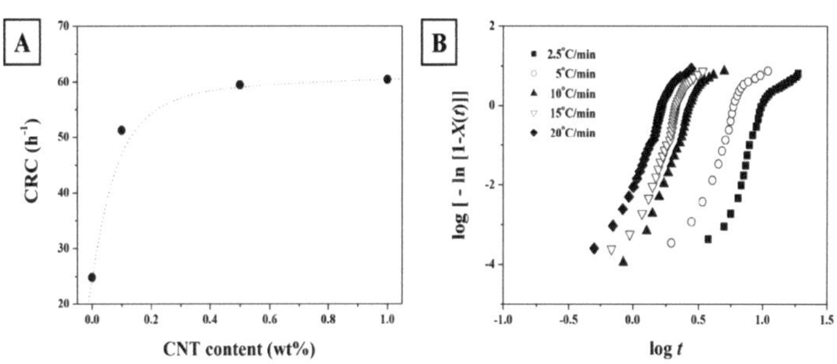

Figure 11. A) Crystallization rate constant (CRC) for the PET nanocomposites with the c-CNT content and (B) the modified Avrami plots of the PET nanocomposites containing 0.1 wt% of the c-CNT during the non-isothermal crystallization. Reproduced with the permission from Ref. [46]. © 2010 Wiley Periodicals, Inc.

The modified Avrami equation [75, 76] is in common use to characterize the non-isothermal crystallization kinetics, and can be expressed as:

$$1 - X(t) = \exp(-Z_t t^n)$$

where $X(t)$ is the relative degree of crystallinity; Z_t is the crystallization rate parameter involving the nucleation and growth rate parameters; t is the crystallization time, and n is the Avrami constant depending on the type of nucleation and growth process. The kinetic parameters such as Z_t and n explicit physical meanings for the isothermal crystallization, while in the non-isothermal crystallization their physical meaning does not have the same significance due to constant changes in the temperature, influencing the nucleation and crystal growth. On the basis of the non-isothermal character of the process suggested by Jeziorny [77], the rate parameter (Z_t) should be corrected by assuming the cooling rate to be constant or approximately constant, according to the relationship of $\log Z_c = \log Z_t / a$ (where a is the cooling rate). The plots of $\log[-\ln\{1 - X(t)\}]$ versus $\log t$ for the PET nanocomposites are shown in Figure 11B. The kinetic parameters such as n and Z_t can be determined from the slope and intercept of the plot of $\log[-\ln\{1 - X(t)\}]$ versus $\log t$. The kinetic parameters for the non-isothermal crystallization of the PET nanocomposites estimated from the kinetic data selected in the linear region are summarized in Table 4. The n values were in the range of 3.21~3.56 for pure PET, whereas 4.14~6.29 for the PET nanocomposites, depending on the cooling rate and the c-CNT content. The PET nanocomposites exhibited values of n higher than 4, suggesting that the mechanism of the non-isothermal crystallization of the PET nanocomposites was very complicated and the c-CNT significantly influenced the non-isothermal crystallization behavior, leading to the fact that the incorporation of the c-CNT into the PET matrix could change the non-isothermal crystallization of the PET nanocomposites. As the cooling rate increased, the $t_{0.5}$ values decreased and the Z_c values increased for the PET nanocomposites in comparison with pure PET. In addition, the PET nanocomposites exhibited higher Z_c and lower $t_{0.5}$ values than those of pure PET at a given cooling rate. This result revealed that the c-CNT dispersed in the PET matrix could induce heterogeneous nucleation and enhance the rate of the non-isothermal crystallization of the PET nanocomposites. The introduction of the c-CNT into the PET matrix can lead to faster crystallization kinetics of the PET nanocomposites, and significantly influence the non-isothermal crystallization process involving the nucleation and the crystal growth.

3.6. Nucleation activity and crystallization activation energy

The nucleation activity is a factor by which the work of three dimensional nucleation decreases with the addition of a foreign substrate [78], the nucleation activity of different substrates can be estimated from the relationship of $\log a = A - B/2.3\Delta T_p^{\ 2}$ (where a is the cooling rate; A is a constant; ΔT_p is the degree of supercooling, and B is parameter related to three dimensional nucleation) [35, 40, 78]. The B values were obtained from the slope of the plot of log a versus $1/T_p^{\ 2}$ as shown in Figure 12A, and then the nucleation activity ((ϕ)) can be calculated from the relationship of $\phi = B^*/B^0$ (where B^0 and B^* are the values of B for homogeneous and heterogeneous nucleation, respectively). The value of 0 implied strong nucleation activity and that of 1 implied inert nucleation activity. The calculated values in the PET nanocomposites were found to be 0.227, 0.231, and 0.211, respectively. This result demonstrates that a very small quantity of the c-CNT can act as excellent nucleating agents for the PET nanocomposites during the non-isothermal crystallization, which was corresponded well with the results for the non-isothermal crystallization kinetics of the PET nanocomposites. The incorporated c-CNT in the PET matrix exhibits much higher nucleation activity than any other nanoreinforcing filler reported to date, with even a very small quantity of the c-CNT.

Materials	Cooling rate (°C/min)	Z_c	N	t_{max} *(min)*
	2.5	4.54 x 10-6	3.55	12.46
	5	2.27 x 10-3	3.33	7.87
PET	10	5.70 x 10-2	3.56	4.66
	15	1.74 x 10-1	3.37	3.50
	20	2.99 x 10-1	3.21	2.73
	2.5	2.01 x 10-2	5.09	10.02
	5	2.27 x 10-1	4.93	5.76
PET/c-CNT 0.1	10	4.89 x 10-1	5.56	3.00
	15	7.46 x 10-1	5.83	1.91
	20	8.31 x 10-1	5.36	1.43
	2.5	1.95 x 10-2	5.02	9.86
	5	2.14 x 10-1	4.14	5.70
PET/c-CNT 0.5	10	5.57 x 10-1	5.71	2.99
	15	7.02 x 10-1	6.29	1.85
	20	8.52 x 10-1	6.17	1.28
	2.5	1.94 x 10-2	4.87	9.72
	5	2.93 x 10-1	4.78	5.51
PET/c-CNT 1.0	10	6.01 x 10-1	5.89	2.96
	15	8.05 x 10-1	4.59	1.87
	20	9.09 x 10-1	5.13	1.30

Table 4. Kinetic parameters of PET nanocomposites during the non-isothermal crystallization

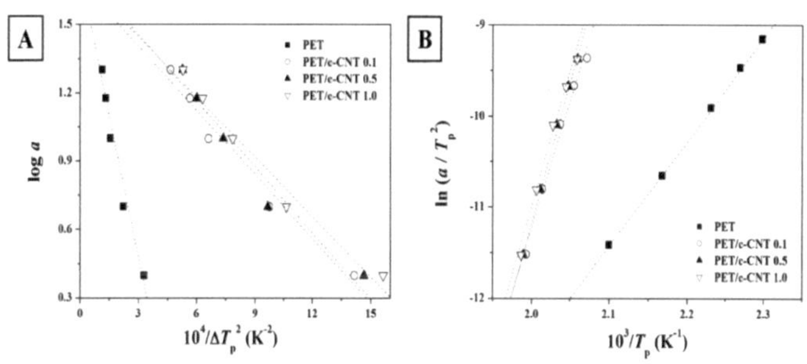

Figure 12. Plots of log *a* versus of $1/\Delta T_{p}^{2}$, for the PET nanocomposites. Using the slope of the plot of log *a* versus of $1/\Delta T_{p}^{2}$, the nucleation activity () can be calculated from the relationship of = B^{*}/B^{0} and (b) Plots of ln(a/T_{p}^{2},) versus $1/T_{p}$, for the PET nanocomposites. The slopes of the plots of ln(a/T_{p}^{2},) versus $1/T_{p}$, provide an estimate of the activation energy for non-isothermal crystallization of the PET nanocomposites. Reproduced with the permission from Ref. [46]. © 2010 Wiley Periodicals, Inc.

The activation energy for the non-isothermal crystallization can be derived from the combination of the cooling rate and the crystallization peak temperature, suggested by Kissinger [79]. The ΔE_{a} values of the PET nanocomposites were obtained from the slope of the plot of

$\ln(a/T_p{}^2)$ versus $1/T_p$ as shown in Figure 12B. The calculated ΔE_a values of the PET nanocomposites were found to be 231.1, 255.9, and 248.9 kJ/mol, respectively, and they were higher than that of pure PET (ΔE_a = 5.3 kJ/mol). This result indicated that the introduction of the c-CNT probably reduced the transportation ability of polymer chains in the PET nanocomposites during the non-isothermal crystallization [80], leading to the increase in the ΔE_a values. However, the addition of 1 wt% of the c-CNT induced more heterogeneous nucleation, and could lead to the slight decrease in the ΔE_a value of the PET nanocomposite during the non-isothermal crystallization. Kim et al. [35, 40] studied the unique nucleation of multiwalled CNT and PEN nanocomposites during the non-isothermal crystallization and they suggested that the introduced CNT could perform two functions in the PEN nanocomposites: CNT acted as good nucleating agents, thus accelerating the non-isothermal crystallization of the PEN nanocomposites, and CNT also adsorbed the PEN molecular segments and restricted the movement of chain segments, thereby making crystallization difficult. Consequently, the PEN molecular segments require more energy to rearrange, leading to the increase in the activation energy for the non-isothermal crystallization

4. Summary and Outlook

This chapter describes the fabrication and characterization of poly(ethylene terephthalate) (PET) nanocomposites containing modified carbon nanotube (CNT). The PET nanocomposites reinforced with a very small quantity of the c-CNT were prepared by simple melt blending in a twin-screw extruder to create high performance polymer nanocomposites for practical applications in a broad range of industries. The carboxylic acid groups effectively induced on the surfaces of the c-CNT via chemical modification can significantly influence the mechanical and rheological properties, thermal stability, and non-isothermal crystallization behavior of the PET nanocomposites. Morphological observation revealed that the c-CNT was well dispersed in the PET matrix and enhanced the interfacial adhesion between the nanotubes and the PET matrix. The enhancement of thermal stability of the PET nanocomposites resulted from the physical barrier effect of the c-CNT against the thermal decomposition. The incorporation of the c-CNT into the PET matrix increased the shear thinning nature of the PET nanocomposites, and the non-terminal behavior observed in the PET nanocomposites was attributable to the nanotube-nanotube and nanotube-polymer interactions. The improvement in the mechanical properties of PET nanocomposites with the introduction of the c-CNT resulted from the enhanced interfacial interactions between the c-CNT and the PET as well as good dispersion of the c-CNT in the PET matrix. The variations of the nucleation activity and the crystallization activation energy of the PET nanocomposites reflected the enhancement of crystallization of the PET nanocomposites effectively induced by a very small quantity of the c-CNT. The incorporation of the c-CNT into the PET matrix has a significant effect on the non-isothermal crystallization kinetics of the PET nanocomposites in that the c-CNT dispersed in the PET matrix can effectively act as strong nucleating agents and lead to the enhanced crystallization of the PET nanocomposites through heterogeneous nucleation. The uniform dispersion of modified CNT and strong interfacial

http://dx.doi.org/10.5772/50413

adhesion or intimate contact between the nanotubes and the polymer matrix can lead to more effect load transfer from the polymers to the nanotubes, resulting in the substantial enhancement of mechanical properties of PET/CNT nanocomposites even with a very small quantity of modified CNT. Future development of PET/CNT nanocomposites for targeted applications in a broad range of industry will be performed by balancing high performance against multiple functionalities and manufacturing cost.

Acknowledgements

Authors thank Mr. H. J. Choi, Mr. C. S. Kang, and Mr. D. K. Kim for their assistance in part with the experiment and characterization of PET nanocomposites in this study.

Author details

Jun Young Kim[1*] and Seong Hun Kim[2]

*Address all correspondence to: junykim74@hanmail.net

1 Corporate Research & Development Center, Samsung SDI Co. Ltd., Republic of Korea

2 Department of Organic & Nano Engineering, Hanyang University,, Republic of Korea

References

[1] Imai, Y., Nishimura, S., Abe, E., Tateyama, H., Abiko, A., Yamaguchi, A., & Taguchi, H. (2002). High-Modulus Poly(ethylene terephthalate)/Expandable Fluorine Mica Nanocomposites with a Novel Reactive Compatibilizer. *Chem. Mater.*, 14, 477-479.

[2] Jeol, S., Fenouillot, F., Rousseau, A., & Varlot, K. M. (2007). Drastic Modification of the Dispersion State of Submicron Silica During Biaxial Deformation of Poly(ethylene terephthalate. *Macromolecules*, 40, 3229-3237.

[3] Giannelis, E.P. (1996). Polymer Layered Silicate Nanocomposites. *Adv. Mater.*, 8, 29-35.

[4] Kim, J. Y., Seo, E. S., Kim, S. H., & Kikutani, T. (2003). Effects of Annealing on Structure and Properties of TLCP/PEN/PET Ternary Blend Fibers. *Macromol. Res.*, 11, 62-68.

[5] Kim, J. Y., Kim, O. S., Kim, S. H., & Jeon, H. Y. (2004). Effects of Electron Beam Irradiation on Poly(ethylene 2 6 -naphthalate)/Poly(ethylene terephthalate) Blends. Polym. Eng. Sci. , 44, 395-405.

[6] Kim, J. Y., Kim, S. H., & Kikutani, T. (2004). Fiber Property and Structure Development of Polyester Blend Fibers Reinforced with a Thermotropic Liquid-Crystal Polymer. *J. Polym. Sci. Part B: Polym. Phys.*, 42, 395-403.

[7] Kim, J. Y., & Kim, S. H. (2005). Situ Fibril Formation of Thermotropic Liquid Crystal Polymer. *Polyesters Blends. J. Polym. Sci. Part B: Polym. Phys.*, 43, 3600-3610.

[8] Kim, J. Y., Kang, S. W., Kim, S. H., Kim, B. C., & Lee, J. G. (2005). Deformation Behavior and Nucleation Activity of a Thermotropic Liquid-crystalline Polymer. *Poly(butylene terephthalate)-Based Composites. Macromol. Res.*, 13, 19-29.

[9] Kim, J. Y., & Kim, S. H. (2006). Structure and Property Relationship of Thermotropic Liquid Crystal Polymer and Polyester Composite Fibers. *J. Appl. Polym. Sci.*, 99, 2211-2219.

[10] Kim, J. Y., & Kim, S. H. (2006). Influence of viscosity ratio on processing and morphology of thermotropic liquid crystal polymer-reinforced poly(ethylene 2 6 -naphthalate) blends. *Polym. Int.*, 55, 449-455.

[11] Kim, J. Y., Kim, D. K., & Kim, S. H. (2009). Thermal Decomposition Behavior of Poly(ethylene 2 6 -naphthalate)/Silica Nanocomposites. *Polym. Compos*, 30, 1779-1787.

[12] Iijima, S. (1991). Helical Microtubles of Graphitic Carbon. *Nature*, 354, 56-58.

[13] Dresselhaus, M. S., Dresselhaus, G., & Avouris, P. H. (2001). Carbon Nanotubes. *Synthesis, Structure, Properties, and Applications*, Berlin, Springer.

[14] Shonaike, G. O., & Advani, S. G. (2003). *Advanced Polymeric Materials*, New York, CRC Press.

[15] Thostenson, E. T., Ren, Z., & Chou, T. W. (2001). Advanced in the Science and Technology of Carbon Nanotubes and their Composites : A Review. *Compos. Sci. Technol.*, 61, 1899-1912.

[16] Goze, C., Bernier, P., Henrard, L., Vaccarini, L., Hernandez, E., & Rubio, A. (1999). Elastic and Mechanical Properties of Carbon Nanotubes. *Synth. Metals*, 103, 2500-2501.

[17] Yao, Z., Zhu, C. C., Cheng, M., & Liu, J. (2001). Mechanical Properties of Carbon Nanotube by Molecular Dynamics Simulation. *Comput. Mater. Sci.*, 22, 180-184.

[18] Yu, M. F., Files, B. S., Arepalli, S., & Ruoff, R. S. (2000). Tensile Loading of Ropes of Single Wall Carbon Nanotubes and their Mechanical Properties. *Phys. Rev. Lett.*, 84, 5552-5555.

[19] Yu, M. F., Lourie, O., Dyer, M. J., Moloni, K., Kelly, T. F., & Ruoff, R. S. (2000). Strength and Breaking Mechanism of Multiwalled Carbon Nanotubes Under Tensile Load. *Science*, 287, 637-640.

[20] Frank, S., Poncharal, P., Wang, Z. L., & De Heer, W. A. (1998). Carbon Nanotube Quantum Resistors. *Science*, 280, 1744-1746.

[21] Ebbesen, T. (1997). Carbon Nanotubes:. *Preparation and Properties*, New York, CRC Press.

[22] Schadler, L. S., Giannaris, S. C., & Ajayan, P. M. (1998). Load Transfer. *Carbon Nanotube Epoxy Composites. Appl. Phys. Lett.*, 73, 3842-3844.

[23] AjayanP.M., . (1999). Nanotubes from Carbon. *Chem. Rev.*, 99, 1787-1800.

[24] Bokobza, L. (2007). Multiwall Carbon Nanotube Elastomeric Composites. *Polymer*, 48, 4907-4920.

[25] Paul, D. R., & Robesson, L. M. (2008). Polymer Nanotechnology:. *Nanocomposites. Polymer*, 49, 3187-3204.

[26] De Heer, W. A. (1995). Chatelain A., Ugarte D. Carbon Nanotube Field-emission Electron Source. *Science*, 270, 1179-1180.

[27] Wong, E. W., Sheehan, P. E., & Lieber, C. M. (1997). Nanobeam Mechanics: Elasticity, Strength, and Toughness of Nanorods and Nanotubes. *Science*, 277, 1971-1975.

[28] Fan, S., Chapline, M. G., Franklin, N. R., Tombler, T. W., Casell, A. M., & Dai, H. (1999). Self-oriented Regular Arrays of Carbon Nanotubes and Their Field Emission Properties. *Science*, 283, 512-514.

[29] Kim, P., & Lieber, C. M. (1999). Nanotube Nanotweezers. *Science*, 286, 2148-2150.

[30] Liu, C., Fan, Y. Y., Liu, M., Kong, H. T., & Cheng, H. M. (1999). Dresselhaus M.S. Hydrogen Storage in Single-walled Carbon Nanotubes at Room Temperature. *Science*, 286, 1127-1129.

[31] Kong, J., Franklin, N., Zhou, C., Peng, S., Cho, J. J., & Dai, H. (2000). Nanotube Molecular Wires as Chemical Sensor. *Science*, 287, 622-625.

[32] Ishihara, T., Kawahara, A., Nishiguchi, H., Yoshio, M., & Takita, Y. (2001). Effects of Synthesis Condition of Graphitic Nanocarbon Tube on Anodic Property of Li-ion Rechargeable Battery. *J. Power Sources*, 97(98), 129-132.

[33] Alan, B., Dalton, A. B., Collins, S., Munoz, E., Razal, J. M., Ebron, V. H., Ferraris, J. P., Coleman, J. N., Kim, B. G., & Baughman, R. H. (2003). Super-tough Carbon Nanotube Fibres. *Nature*, 423, 703.

[34] Wu, M., & Shaw, L. A. (2005). Novel Concept of Carbon-filled Polymer Blends for Applications in PEM Fuel Cell Bipolar Plates. *Int. J. Hydrogen Energy*, 30, 373-380.

[35] Kim, J. Y., Park, H. S., & Kim, S. H. (2006). Unique Nucleation of Multiwalled Carbon Nanotube and Poly(ethylene 2 6 -naphthalate) Nanocomposites During Non-isothermal Crystallization. Polymer , 47, 1379-1389.

[36] Kim, J. Y., & Kim, S. H. (2006). Influence of Multiwall Carbon Nanotube on Physical Properties of Poly(ethylene 2 6 -naphthalate) Nanocomposites. *J. Polym. Sci. Part B: Polym. Phys.*, 44, 1062-1071.

[37] Kim, J. Y., Park, H. S., & Kim, S. H. (2007). Multiwall Carbon Nanotube-Reinforced Poly(ethylene terephthalate) Nanocomposites by Melt Compounding. *J. Appl. Polym. Sci.*, 103, 1450-1457.

[38] Kim, J. Y., Han, S. I., & Kim, S. H. (2007). Crystallization Behavior and Mechanical Properties of Poly(ethylene 2 6 -naphthalate)/Multiwall Carbon Nanotube Nanocomposites. *Polym. Eng. Sci.*, 47, 1715-1723.

[39] Kim, J. Y., & Kim, S. H. (2008). Multiwall Carbon Nanotube Reinforced Polyester Nanocomposites. *Polyesters: Properties, Preparation and Applications*, New York, Nova Science Publishers, 33-107.

[40] Kim, J. Y., Han, S. I., & Hong, S. (2008). Effect of Modified Carbon Nanotube on the Properties of Aromatic Polyester Nanocomposites. *Polymer*, 49, 3335-3345.

[41] Kim, J.Y. (2009). The Effect of Carbon Nanotube on the Physical Properties of Poly(butylene terephthalate) Nanocomposite by Simple Melt Blending. *J. Appl. Polym. Sci.*, 112, 2589-2600.

[42] Kim, J. Y., Kim, D. K., & Kim, S. H. (2009). Effect of Modified Carbon Nanotube on Physical Properties of Thermotropic Liquid Crystal Polyester Nanocomposites. *Eur. Polym. J.*, 45, 316-324.

[43] Kim, J. Y., Han, S. I., Kim, D. K., & Kim, S. H. (2009). Mechanical Reinforcement and Crystallization Behavior of Poly(ethylene 2 6 -naphthalate) Nanocomposites Induced by Modified Carbon Nanotube. *Composites: Part A*, 40, 45-53.

[44] Kim, J. Y., Park, H. S., & Kim, S. H. (2009). Thermal Decomposition Behavior of Carbon Nanotube-Reinforced Poly(ethylene 2 6 -naphthalate. *Nanocomposites. J. Appl. Polym. Sci.*, 113, 2008-2017.

[45] Kim, J.Y. (2009). Carbon Nanotube-reinforced thermotropic Liquid Crystal Polymer Nanocomposites. *Materials*, 2, 1955-1974.

[46] Kim, J. Y., Choi, H. J., Kang, C. S., & Kim, S. H. (2010). Influence of Modified Carbon Nanotube on Physical Properties and Crystallization Behavior of Poly(ethylene terephthalate) Nanocomposites. *Polym. Compos*, 31, 858-869.

[47] Kim, J.Y. (2011). Aromatic Polyester Nanocomposites Containing Modified Carbon Nanotube. *Thermoplastic and Thermosetting Polymers and Composites.*, New York, Nova Science Publishers, 1-36.

[48] Kim, J.Y. (2011). Poly(butylene terephthalante) Nanocomposites Containing Carbon Nanotube. *Advances in Nanocomposites: Synthesis, Characterization and Industrial Applications*, Rijeka, InTech, 707-726.

[49] Wang, C., Guo, Z. X., Fu, S., Wu, W., & Zhu, D. (2004). Polymers Containing Fullerene of Carbon Nanotube Structures. *Prog. Polym. Sci.* , 29, 1079-1141.

[50] Liu, J., Rinzler, A. G., Dai, H., Hafner, J. H., Bradly, R. K., Boul, P. J., Lu, A., Iverson, T., Shelimov, K., Huffman, C. B., Rodriguez-Macias, F., Shon, Y. S., Lee, T. R., Colbert, D. T., & Smally, R. E. (1998). Fullerene Pipes. *Science*, 280, 1253-1256.

[51] Hirsch, A. (2002). Functionalization of Single-Walled Carbon Nanotubes. *Angew. Chem. Int. Ed.*, 41, 1853-1859.

[52] Bellayer, S., Gilman, J. W., Eidelman, N., Bourbigot, S., Flambard, X., Fox, D. M., De Long, H. C., & Trulove, P. C. (2005). Preparation of Homogeneously Dispersed Multiwalled Carbon Nanotube/Polystyrene Nanocomposites via Melt Extrusion Using Trialkyl Imidazolium Compatibilizer. *Adv. Funct. Mater.*, 15, 910-916.

[53] Pötschke, P., Fornes, T. D., & Paul, D. R. (2002). Rheological Behavior of Multiwalled Carbon Nanotube/Polycarbonate Composites. *Polymer*, 43, 3247-3255.

[54] Liu, L., Qin, Y., Guo, Z. X., & Zhu, D. (2003). Reduction of Solubilized Multi-walled Carbon Nanotubes. *Carbon*, 41, 331-335.

[55] Wu, T. M., & Chen, E. C. (2006). Crystallization Behavior of Poly(-caprolactone)/Multiwalled Carbon Nanotube Composites. *J. Polym. Sci. Part B: Polym. Phys.*, 44, 598-606.

[56] Kazarian, S. G., Brantley, N. H., & Eckert, C. A. (1999). Applications of Vibrational Spectroscopy to Characterize Poly(ethylene terephthalate) Processed with Supercritical CO2. *Vibrational Spectrosc.*, 19, 277-283.

[57] Wu, C. S., & Liao, H. T. (2007). Study on the Preparation and Characterization of Biodegradable Polylactide/Multi-walled Carbon Nanotubes Nanocomposites. *Polymer*, 48, 4449-4458.

[58] Shaffer, M. S. P., & Windle, A. H. (1999). Fabrication and Characterization of Carbon Nanotube/Poly(vinyl alcohol) Composites. *Adv. Mater.*, 11, 937-941.

[59] Park, S. J., & Cho, M. S. (2000). Thermal Stability of Carbon-MoSi2-Carbon Composites by Thermogravimetric Analysis. *J. Mater. Sci.*, 35, 3525-3527.

[60] Abdalla, M., Derrick, D., Adibempe, D., Nyairo, E., Robinson, P., & Thompson, G. (2007). The Effect of Interfacial Chemistry on Molecular Mobility and Morphology of Multiwalled Carbon Nanotubes Epoxy Nanocomposite. *Polymer*, 48, 5662-5670.

[61] Krishnamoorti, R., Vaia, R. A., & Giannelis, E. P. (1996). Structure and Dynamics of Polymer-Layered Silicate Nanocomposites. *Chem. Mater.*, 8, 1728-1734.

[62] Abdel-Goad, M., & Pötscke, P. (2005). Rheological Characterization of Melt Processed Polycarbonate-Multiwalled Carbon Nanotube Composites. *J. Non-Newtonian Fluid Mech.*, 128, 2-6.

[63] Rosedalev, J. H., & Bates, F. S. (1990). Rheology of Ordered and Disordered Symmetric Poly(ethylenepropylene)-poly(ethyl ethylene) Diblock Copolymer. *Macromolecules*, 23, 2329-2338.

[64] Larson, R. G., Winey, K. I., Patel, S. S., Watanabe, H., & Bruinsma, R. (1993). The Rheology of Layered Liquids: Lamellar Block Copolymers and Smectic Liquid Crystals. *Rheol. Acta*, 32, 245-253.

[65] Krishnamoorti, R., & Giannelis, E. P. (1997). Rheology of End-tethered Polymer layered silicate Nanocomposites. Macromolecules. 30, 4097-4102.

[66] Pötschke, P., Abdel-Goad, M., Alig, I., Dudkin, S., & Lellinger, D. (2004). Rheological and Dielectrical Characterization of Melt Mixed Polycarbonate-Multiwalled Carbon Nanotube Composites. *Polymer*, 45, 8863-8870.

[67] Cho, J., & Daniel, I. M. (2008). Reinforcement of Carbon/Epoxy Composites with Multi-wall Carbon Nanotubes and Dispersion Enhancing Block Copolymers. *Scripta Mater.*, 58, 533-536.

[68] Thostenson, E. T., & Chou, T. W. (2002). Aligned Multi-walled Carbon Nanotube-Reinforced Composites: Processing and Mechanical Charactrerization. *J. Phys. D: Appl. Phys.*, 35, L77-L80.

[69] Meng, H., Sui, G. X., Fang, P. F., & Yang, R. (2008). Effects of Acid- and Diamine-modified MWNTs on the Mechanical Properties and Crystallization Behavior of Polyamide 6. *Polymer*, 49, 610-620.

[70] Cheng, S. Z. D., & Wunderlich, B. (199821). Glass Transition and Melting Behavior of Poly(ethylene 2 6 -naphthalenedicarboxylate. *Macromolecules*, 21, 789-797.

[71] Lopez, L.C., & Wilkes, GL. (1989). Non-isothermal Crystallization Kinetics of Poly(p-phenylene sulphide. *Polymer*, 30, 882-887.

[72] Wang, B., Sun, G., Liu, J., He, X., & Li, J. (2006). Crystallization Behavior of Carbon Nanotubes-filled Polyamide 1010. *J. Appl. Polym. Sci.*, 100, 3794-3800.

[73] Kong, X., Yang, X., Li, G., Zhao, X., Zhou, E., & Ma, D. (2001). Non-isothermal Crystallization Kinetics: Poly(ethylene terephathalate -Poly(ethylene oxide) Segmented Copolymer and Poly(ethylene oxide). *Homopolymer. Eur. Polym. J.*, 37, 1855-1862.

[74] Khanna, Y.P. (1990). A Barometer of Crystallization Rates of Polymeric Materials. *Polym. Eng. Sci.*, 30, 1615-1619.

[75] Avrami, M. (1940). Kinetics of Phase Change. II. Tranformation-Time Relations for Randonm Distribution of Nuclei. *J. Chem. Phys.*, 8(212).

[76] Ozawa, T. (1971). Kinetics of Non-isothermal Crystallization. *Polymer*, 12, 150-158.

[77] Jeziorny, A. (1978). Parameters Characterizing the Kinetics of the Non-isothermal Crystallization of Poly(ethylene terephthalate) Determined by DSC. *Polymer*, 19, 1142-1144.

[78] Dobreva, A., & Gutzow, I. J. (1993). Activity of Substrates in the Catalyzed Nucleation of Glass-forming Melts. *I. Theory. J. Non-Cryst. Solids*, 162, 1-12.

[79] Kissinger, H.E. (1956). The Crystallization Kinetics with Heating Rate. *Differential Thermal Analysis, J. Res. Natl. Stand.*, 57(217).

[80] Chen, E. C., & Wu, T. M. (2008). Isothermal and Nonisothermal Crystallization Kinetics of Nylon 6/Functionalized Multi-walled Carbon Nanotube Composites. *J. Polym. Sci. Part B: Polym. Phys.*, 46, 158-169.

Chapter 6

Hard Nanocomposite Coatings, Their Structure and Properties

A. D. Pogrebnjak and V. M. Beresnev

Additional information is available at the end of the chapter

http://dx.doi.org/10.5772/50567

1. Introduction

The development of the new nanostructured coating with high hardness (40 GPa) and thermal stability (> 1200°C) is one of the most important problem of the modern material science. According to the previous experimental results it can be considered that not only grains size has strong influence on properties of the solid but also structural states of interfaces (grains boundary) [1-7]. As the quantity of atoms at grains boundary reaches about 30-50%, properties of the material are strongly depend on condition of the grains boundary: gap of the border band (in this band lattice parameter deviate from standard value), disorientation of the grains and interfaces, concentration of the defects at boundary and value of the free volume.

So, nanocrystalline materials, that contain nanosized crystallite along with rather extensive and partially disordered boundaries structure, present new properties by comparison with the large-grained materials [8-15].

These stable nanocrystalline materials can be created on base of multi-component compound, since such materials have the heterogeneous structure that include practically non-interacting phases with average linear dimension about 7-35 nm. In this case nanocrystalline materials demonstrate high thermal stability and long-term stable properties Recently, there are many papers related to the research of the structure and properties of the multi-component hard nanostructures (nanocomposite coating based on Zr-Ti-Si-N, Zr-Ti-N and Mo-Si-N etc.) were already published However, the development of the new type of the coating is

INTECH
open science | open minds

© 2012 Pogrebnjak and Beresnev; licensee InTech. This is a paper distributed under the terms of the Creative Commons Attribution License (http://creativecommons.org/licenses/by/3.0), which permits unrestricted use, distribution, and reproduction in any medium, provided the original work is properly cited.

still continuing. It is well known that superhard coating can be formed on base of nc - TiN or nc-(Zr, TiN) covered with a-Si_3N_4, or BN amorphous or quasiamorphous phase, Hardness of such coating can reach 80 GPa and higher. In addition, the deposition of the coating at temperature about 550 - 600°C allows to finalize spinodal segregation along grain boundaries and hence improve properties of the coatings New superhard coatings based on Ti-Hf-Si-N featuring high physical and mechanical properties were fabricated. We employed a vacuum-arc source with HF stimulation and a cathode sintered from Ti-Hf-Si. Nitrides were fabricated using atomic nitrogen (N) or a mixture of Ar/N, which were leaked-in a chamber at various pressures and applied to a substrate potentials. RBS, SIMS, GT-MS, SEM with EDXS, XRD, and nanoindentation were employed as analyzing methods of chemical and phase composition of thin films. We also tested tribological and corrosion properties. The resulting coating was a two-phase, nanostructured nc-(Ti, Hf)N and a-Si_3N_4. Sizes of substitution solid solution nanograms changed from 3.8 to 6.5 nm, and an interface thickness surrounding a-$Si3N_4$ varied from 1.2 to 1.8 nm. Coatings hardness, which was measured by nanoindentation was from 42.7 GPa to 48.6 GPa, and an elastic modulus was E = (450 to 515) GPa. [14-18].

The films stoichiometry was defined for various deposition conditions. It was found that in samples with superhard coatings of 42.7 to 48.6GPa hardness and lower roughness in comparison with other series of samples, friction coefficient was equal to 0.2, and its value did not change over all depth (thickness) of coatings. A film adhesion to a substrate was essentially high and reached 25MPa.

Zr-Ti -Si-N coating had high thermal stability of phase composition and remained structure state under thermal annealing temperatures reached 1180°C in vacuum and 830°C in air. Effect of isochronous annealing on phase composition, structure, and stress state of Zr-Ti-Si-N-ion-plasma deposited coatings (nanocomposite coatings) was reported. Below 1000°C annealing temperature in vacuum, changing of phase composition is determined by appearing of siliconitride crystallites (β-S_3N_4) with hexagonal crystalline lattice and by formation of $Zr0_2$ oxide crystallites. Formation of the latter did not result in decay of solid solution (Zr,Ti)N but increased in it a specific content of Ti-component.

Vacuum annealing increased sizes of solid solution nanocrystallites from (12 to 15) in as-deposited coatings to 25nm after annealing temperature reached 1180°C. One could also find macro- and microrelaxations, which were accompanied by formation of deformation defects, which values reached 15.5 vol.%.

Under 530°C annealing in vacuum or in air, nanocomposite coating hardness increased, demonstrating, however, high spread in values from 29 to 54GPa (first series of samples). When Ti and Si concentration increased (second series) and three phases nc-ZrN, (Zr, Ti)N-nc, and α-Si_3N_4 were formed, average hardness increased to 40,8 ± 4GPa (second series of samples). Annealing to 500°C increased hardness and demonstrated lower spread in values H = 48 ± 6GPa and E = (456 ± 78)GPa.

http://dx.doi.org/10.5772/50567

2. Enhanced hardness of nanocomposite coatings

The enhanced hardness of the nanocomposite coating Hn can be more than two times greater than that of its harder component. Main mechanisms, which are responsible for the hardness enhancement, are: (1) the dislocation-induced plastic deformation, (2) the nanostructure of materials, and (3) cohesive forces between atoms. The dislocation-induced plastic deformation dominates in the materials composed of large grains with size d < 10 nm. On the contrary, the nanostructure is dominant in materials composed of small grains with size d ≤ 10 nm. It means that the hardness enhancement of coating strongly depends on the grain size d, see Figure 1. From this figure it is seen that there is a critical value of the grain size d_c≈10 nm at which a maximum value of hardness Hmax of the coating is achieved. The regions of d around H_{max}, achieved at $d=d_c$, corresponds to a continuous transition from the activity of the intragranular processes at $d > d_c$, dominated by the dislocations and described by the Hall-Petch law ($H\sim d^{-1/2}$), to the activity of the intergranular processes at $d < d_c$ dominated by the interactions between atoms of neighbouring grains and/or by the small-scale sliding in grain boundaries.

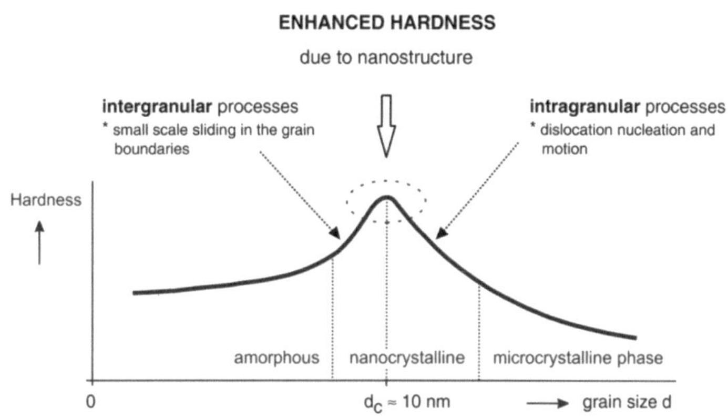

Figure 1. Schematic illustration of coating hardness as a function of the size d of grains. Adapted after reference.

In materials with the grain size $d \leq d_c$ (1) dislocations are not generated (grain size d is smaller than the length of dislocation) and (2) processes in grain boundary regions play a dominant role over those inside grains. Therefore, besides chemical and electronic bonding between atoms the nanostructure of material plays a dominant role when $d \leq d_c$. It was found that there are at least four types of nanostructures that result in the enhanced hardness of nanocomposite coatings: (1) bilayers with nanosize period λ, (2) the columnar nanostructure, (3) nanograins surrounded by very thin (~1 to 2 ML) tissue phase and (4) the mixture of nanograins with different crystallographic orientations and/or different phases, see Figure2; here $\lambda = h_1+h_2$, h_1 and h_2 is the thickness of first and second layer of the bilayer, respectively, and ML denotes the monolayer. [1,2]

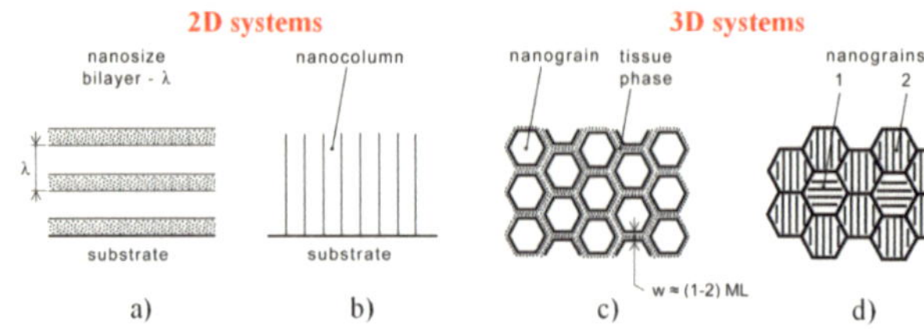

Figure 2. Schematic illustration of four nanostructures of the nanocomposite coating with enhanced hardness: (a) nanosize bilayers, (b) columnar nanostructure, (c) nanograins surrounded by a tissue phase and (d) mixture of nanograins with different crystallographic orientation.

Individual nanostructures are formed under different conditions using eiher a sequential deposition of individual layers in the nanosize bilayers or in transition regions where the coating structure changes from crystalline through nanocrystalline to amorphous. There are three transition regions: (1) the transition from the crystalline to the X-ray amorphous material, (2) the transition between two crystalline phases of different materials and (3) the transition between two crystallographic orientations of grains of the same material. More details are given in the references.

3. Phase composition and thermal properties (stability)

3.1. Thermal stability of the properties

However, the nano- structure constitutes a metastable phase: if the temperature at which a film forms exceeds a certain threshold value T_c its material undergoes crystallization, leading to the destruction of the nanostructure and the appearance of new crystalline phases that account for the loss of unique properties by nanocomposite films for $T > T_c$. In other words, temperature T_c at which the nanostructure turns to the crystalline phase determines the thermal stability of a given nanocompo- site. However, these materials not infrequently have to be employed at temperatures above 1000°C, hence the necessity to develop new ones with maximum thermal stability in excess of 1000°C. [3-5]

3.2. Resistance to high-temperature oxidation

Oxidation resistance is a most attractive property of hard nanocomposite coatings. Oxidation resistance of hard films stronglydepends on their elemental composition. Figure 3 illustrates the increase in the film weight D_m as a function of annealing temperature T. The temperature at which D_m sharply increases is described as maximum temperature T_{max} at which film oxidation can be avoided. The higher T_{max}, the greater the oxidation resistance. All the films represented in Figure3 and characterized by a sharp growth in D_m with increasing temperature are crystalline or nanocrystalline. All of them possess oxidation resistance T_{max} below 1000°C. This is not surprising, since they are composed of grains that are constantly in contact with the air through grain boundaries at the film/substrate interface. This phenomenon sharply decreases oxidation resistance in the bulk of the film and is thereby responsible for the impaired efficiency of the barrier formed by the upper layer of an oxide film. For all that, an improvement is feasible by the utilization of the intergranular vitreous phase.

Thus far, only one efficacious method for increasing oxidation resistance in hard coatings is known, namely, interruption of the continuous path along grain boundaries from the coating surface to the underlying substrate across the bulk. It is possible to realize in solid amorphous films such as those formed by a new family of composites a-Si_3 N_4 =MeNx with a high content (> 50 vol.%) of the amorphous phase a-Si_3N_4. This possibility is illustrated by Figure 3b, c showing a polished section of nanocomposite Ta-Si-N and Mo-Si-N films. Change in mass D_m remains practically unaltered after annealing the Ta-Si-N film at temperatures up to 1300°C (Figure 3a).

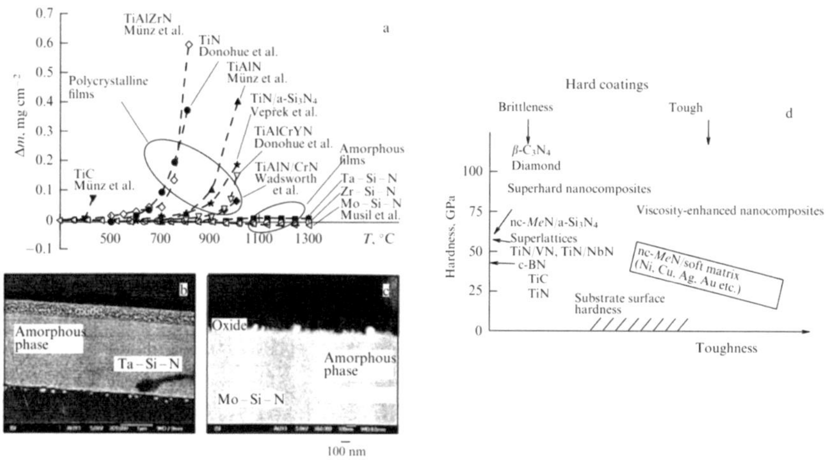

Figure 3. a) Oxidation resistance of selected hard coatings, characterized by the dependence of D_m on annealing temperature T. Polished section of Ta-Si-N (b) and Mo-Si-N (c) films on an Si (100) substrate after high-temperature annealing in flowing air atT=1300°C. (d) Classification of nanocomposites by hardness and viscosity.

3.3. Amorphous nanocomposites resistant to high-temperature oxidation

Nanocomposites containing >50 vol. % of silicon nitride are amorphous (Figure 3b, c). It can be seen that the bulk of the Ta-Si-N film possesses an amorphous structure and only the surface of the film underwent oxidation; the oxide surface layer of Ta_2O_5 is about 400 nm thick. This film exhibits the highest oxidation resistance (Figure 3a); its hardness H varies from 20 to 40 GPa. Such characteristics of Ta-Si-N films account for the wide range of their applications, e.g., as protective coatings for cutting tools.

However, a high content of silicon nitride phase alone is not sufficient to ensure resistance to high-temperature oxidation. Certain elements, like Mo, W and some others, tend to form volatile oxides released from a nanocomposite upon oxidation. This results not only in the formation of a porous structure of the oxide layer surface (Figure 3b, c) but also in impaired oxidation resistance. The pores appear because newly formed volatile oxides MoO_x diffuse to the outside from the surface layer at T= 800– 1000°C. But the main cause of impaired oxidation resistance is disintegration of the metal nitride (MeN_x) phase in the nanocomposite; hence the importance of choosing films with proper elemental composition.

Oxidation resistance at maximum annealing temperatures can be achieved by ensuring high thermal stability of both phases in a given nanocomposite, i.e., of amorphous siliconnitride against crystallization and of metal nitride against degradation ($MeNx \rightarrow Me + N_2$). In this context, such nanocomposites as Zr-Si-N, Ta-Si-N, and Ti-Si-N with a high (>50 vol.%) silicon nitride phase content, as well as silicon oxide- and oxynitride-based nanocomposites, appear especially promising. [4]

4. Effect of thermal annealing in vacuum and in air on nanograin sizes in hard and superhard coatings Zr-Ti-Si-N

Analyzing phase composition of Zr-Ti-Si-N films, we found that a basic crystalline component of as-deposition on state was solid solution (Zr, Ti)N based on cubic lattice of structured NaCl.

Crystallites of solid (Zr, Ti)N solution underwent compressing elastic macro stresses occurring in a "film-substrate" system. Compressing stresses, which were present in a plane of growing film, indicated development of compressing deformation in a crystal lattice, which was identified by a shift of diffraction lines in the process of angular surveys ("sin2ψ – method") and reached – 2.93 % value. With E ≈ 400 GPa characteristic elastic modulus and 0.28 Poisson coefficient, deformation value corresponded to that occurring under action of compressing stresses $\sigma_c \approx 8.5$ GPa.

Figure 4 shows morphology of surface on base (Zr-Ti-Si)N formed with U = –150 V, P = 0,8 Pa. Investigated, that a change in direction of increasing the potential applied to the substrate, the roughness decreases. [5]

Figure 4. Surface morphology of coatings on base (Zr-Ti-Si)N with U= –150V, P=0,8 Pa.

The resulting coatings have the following hardness: TiN (H = 28 GPa and E = 312 GPa); Ti-Si-N (H = 38-39 GPa, E = 356 GPa); Ti-Zr-Si-N hardness values H = 38... 41 GPa, E = 478 GPa. Tables 1 and 2 show the results of tribological tests.

Coatings	Temperature tests, °C	Wear coating factor, mm^3/nm	Wear factor of the sample, mm^3/nm	f_{mp}
Ti-Zr-Si-N	30	$7{,}59\times10^{-5}$	$1{,}93\times10^{-5}$	0,80
	300	$2{,}22\times10^{-5}$	$3{,}14\times10^{-5}$	0,71
	500	$1{,}49\times10^{-5}$	$2{,}81\times10^{-5}$	0,58
Ti-Zr-Si-N (a)	30	$7{,}559\times10^{-5}$	$3{,}214\times10^{-5}$	0,805
Ti=22,73	300	$1{,}84\times10^{-5}$	$4{,}726\times10^{-5}$	0,836
Zr=2,12 Si=3,05	500	$1{,}47\times10^{-5}$	$3{,}047\times10^{-5}$	0,582
Ti-Zr-Si-N (b)	30	$6{,}75\times10^{-5}$	$3{,}304\times10^{-5}$	0,793
Ti=28,32	300	$3{,}62\times10^{-5}$	$3{,}83\times10^{-5}$	0,813
Zr=2,67 Si=3,64	500	$1{,}985\times10^{-5}$	$2{,}749\times10^{-5}$	0,585
Ti-Zr-Si-N (c)	30	$7{,}697\times10^{-5}$	$3{,}279\times10^{-5}$	0,877
Ti=27,46	300	$2{,}635\times10^{-5}$	$3{,}486\times10^{-5}$	0,825
Zr=2,51 Si=3,76	500	$1{,}955\times10^{-5}$	$2{,}749\times10^{-5}$	0,632
Ti-Si-N	30	$7{,}69\times10^{-5}$	$3{,}28\times10^{-5}$	0,88
	300	$2{,}63\times10^{-5}$	$3{,}49\times10^{-5}$	0,82
	500	$1{,}95\times10^{-5}$	$2{,}75\times10^{-5}$	0,69
TiN	30	$6{,}75\times10^{-5}$	$3{,}30\times10^{-5}$	0,81
	300	$3{,}62\times10^{-5}$	$3{,}51\times10^{-5}$	0,87
	500	$5{,}16\times10^{-5}$	$3{,}83\times10^{-5}$	0,91

Table 1. The results of tribological properties of nanocomposite coatings.

Options	T=30°C	T=300°C	T=500°C
Sample	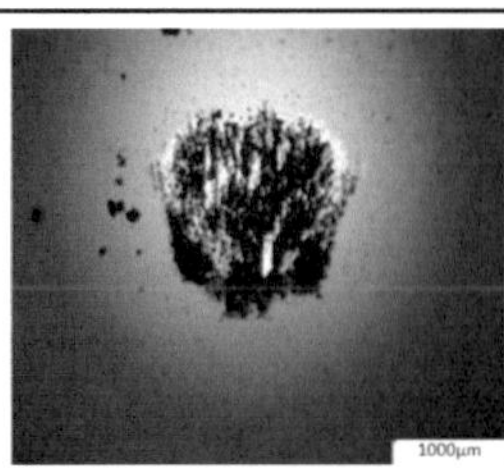	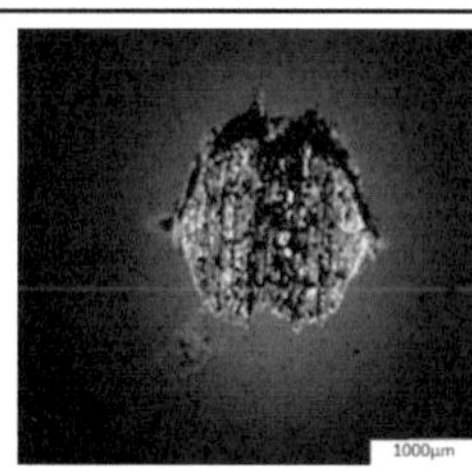	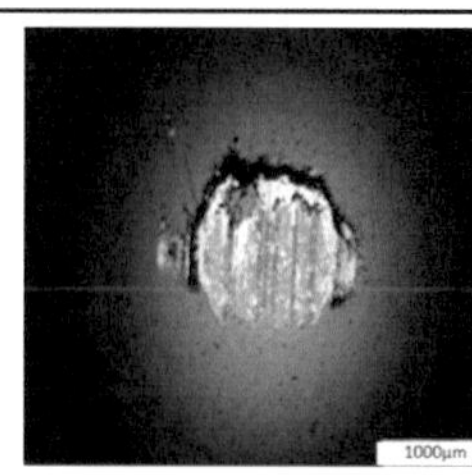
Friction track	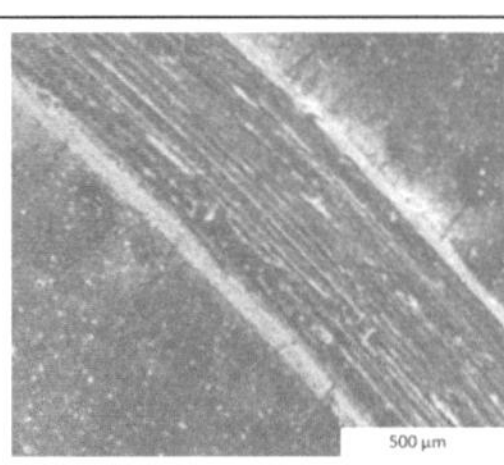	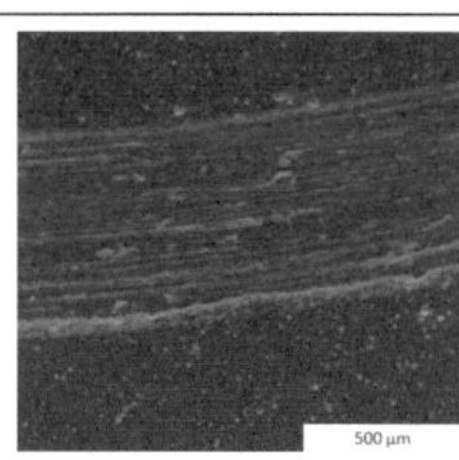	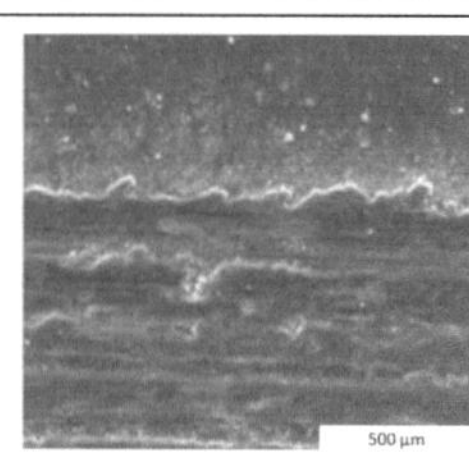

Table 2. The macrostructure of the surface of the nanocomposite coatings Ti-Zr-Si-N.

The chamber pressure is insignificant effect on the morphology, but noted that the increase in gas pressure in the chamber leads to a decrease in surface roughness. Figure 4 shows the optimal parameters of the chamber pressure and the potential applied to the substrate (U = −150 V, P = 0,8 Pa) under which the maximum peak is 16 nm.

We should also note that such high stresses characterize nitride films, which were formed under deposition with high radiation factor, which provided high adhesion to base material and development of compression stresses in the film, which was stiffly bound to the base material due to "atomic peening"- effect. [6-12]

At sliding speeds 10 cm/s is a normal abrasive oxidative wear friction. The structural-phase state coverings plays a crucial role in the processes of wear and temperature dependent. At temperatures of 30°C tests are covering adhesive interaction with the counterbody - there is a rough surface topography of the coating. At temperatures of 300°C tests for coatings based on Ti-Si-N and Zr-Ti-Si-N coating decreases the wear and wear counterface increases. With further increase in temperature to 500°C decreases the wear coating Ti-Si-N and Zr-Ti-Si-N, increases their durability. This leads to a change in the conditions of the processes occurring in the contact zone due to changes in the structure of surface layers.

Qualitative changing of phase composition was observed in films under vacuum annealing at T_{an}> 1000°C. Appearance of zirconium and titanium oxides was related to oxidation relaxation under coating surface interaction with oxygen atoms coming from residual vacuum atmosphere under annealing.

Figure 5 shows the results of RBS analysis on the samples obtained were coated with Ti-Zr-Si-N. The beam energy $^{+}$He ions is not sufficient for the analysis of the total film thickness, but the peaks of Ti and Zr are well separated and can be seen that the concentration of Ti and Zr is almost uniformly distributed over the depth of coating. [10-12]

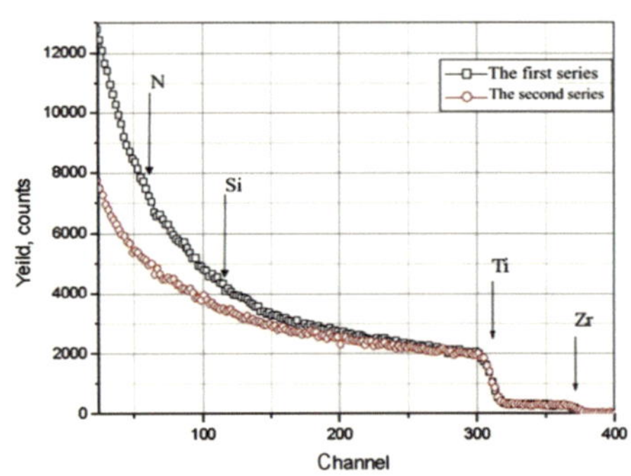

Figure 5. Energy spectra of Rutherford ion backscattering (RBS) for thin coating Zr-Ti-Si- N.

But still, Si concentration was not less than 7 at.%, while that of N might reach more than 15 at.%.

Figure 6. shows scratch properties of Zr-Ti-Si-N. The friction coefficient (μ) between two solid surfaces is defined as the ratio of the tangential force (F) required to produce sliding divided by the normal force between the surfaces (N). Normal force F_n (occasionally N) is the component, perpendicular to the surface of contact, of the contact force exerted on an object by the surface. Acoustic Emission is a naturally occurring phenomenon whereby external stimuli, such as mechanical loading, generate sources of elastic waves. Penetration Depth is a measure of how deep light or any electromagnetic radiation can penetrate into a material. It is defined as the depth at which the intensity of the radiation inside the material falls to 1/e (about 37 %) of its original value at (or more properly, just beneath) the surface.

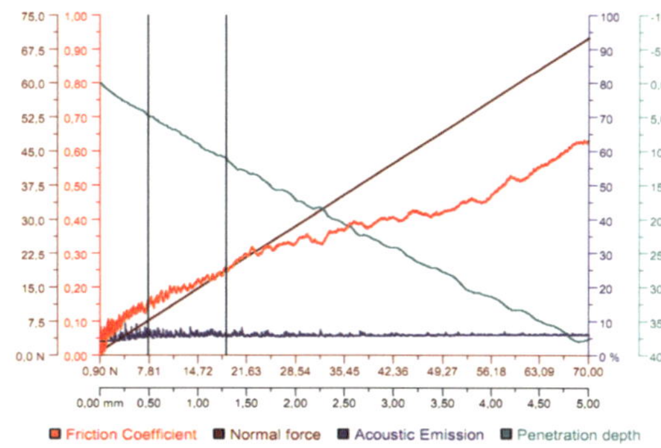

Figure 6. Scratch properties of Zr-Ti-Si-N: friction coefficient, normal force, acoustic emission, penetration depth.

Under annealing temperatures below 1000°C, coatings phase composition remained practically unchanged. One could not only changed width of diffraction lines and their shift to higher diffraction angles. The latter characterizes relaxation of compressing stresses in coatings. Changed diffraction lines were related to increased crystalline sizes (in general) and decreased micro-deformation.

Three-dimensional islands on the surface of the films with columnar structure are output on the surface of the ends of individual grains (Figure 7). It is seen that the roughness depends on the conditions of their chemical composition and the parameters of the wasp-assertion. Undulation surfaces associated with the mechanism of growth, with the formation of separate islands on the surface (Volmer-Weber mechanism)[10,11].

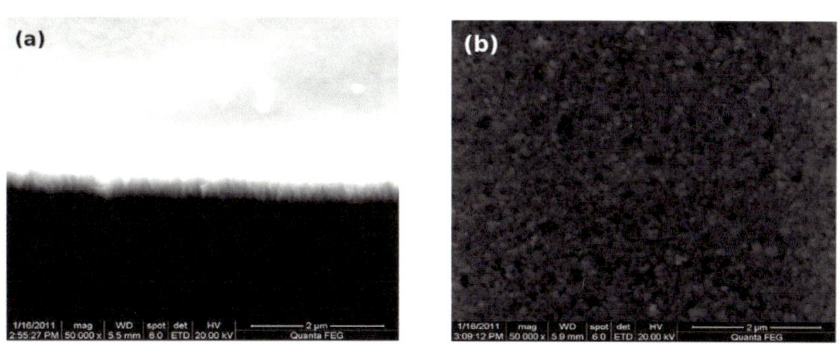

Figure 7. Coating Ti-Zr-Si-N with a columnar structure: (a) a cross-section coating, (b) surface topography of the coating.

In such a way, hardness, which was increased in the process of annealing, seems to be related to incomplete spinodal phase segregation at grain boundaries resulting from deposition of Zr-Ti-Si-N-(nanocomposite). Annealing stimulated spinodal phase segregation, forming more stable modulated film structures.

Figure 8 shows chemical composition over coating cross-section. Spectra indicate that N concentration changed from 3.16 to 4.22 wt.%, Si concentration was about 0.98 to 1.03 wt.%, Ti was 11.78 to 13.52 wt.% and that Zr = 73.90 to 77.91 wt.%. These results indicated that amount of N is essentially high, and this allowed it to participate in formation of nitrides with Zr, Ti, or (Zr, Ti)N solid solution. Si concentration was low, however, results reported by Veprek et al. indicated Si concentration as high as 6 to 7at.%, which was enough to form siliconitride phases.

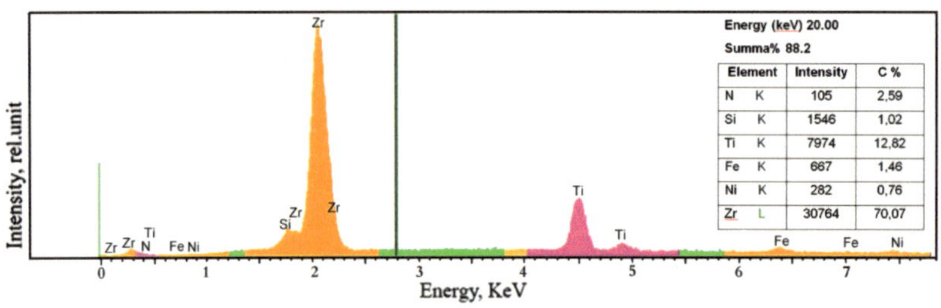

Figure 8. Data of microanalysis for point of Zr-Ti-Si-N (Ti≈12%) nanocomposite coating surface. (fifth series)

Changes occurred under macrodeformation of crystallites of basic film phase – (Zr, Ti)N solid solution. Compressing deformation of crystallite lattices increased, which seemed to be related to additional new crystalline components, which appeared in film material: oxides and siliconitrides. In the lattice itself, a period decreased corresponding to increased Ti concentration. Ordered atoms in metallic (Zr/Ti) sublattice of solid solution increased from 8.5 to 21 at.%.

In this temperature range, crystallite size increased from 15 to 25 nm, crystallite lattice microdeformation increasing non-essentially up 0.5 to 0.8 %. Table 3 summarizes substructure characteristics of (Zr, Ti)N solid solution crystallites.

Parameters of structure	After deposition	T_{an} = 300°C vacuum	T_{an} = 500°C vacuum	T_{an} = 800°C vacuum	T_{an} = 1100°C vacuum	T_{an} = 300°C air	T_{an} = 500°C air
a_0, nm	0,45520	0,45226	0,45149	0,45120	0,45064	0,45315	0,45195
□, %	-2,93	-2,40	-1,82	-1,01	-1,09	-2,15	-1,55
<□"/, %	1,4	1,0	0,85	0,5	0,8	0,95	0,88
$□_{def. pack.}$	0,057	0,085	0,107	0,155	0,150	0,090	0,128

Table 3. Changes of structure and substructure parameters occurring in ion-plasma deposited coatings of Zr-Ti-Si-N system in the course of high-temperature annealing in vacuum and in air.

In comparison with vacuum annealing, air one is characterized by a decreased of phase stability above 500°C – 600°C. Above these temperatures, one observed formation of oxides resulting in film destruction and total film destruction at 830°C.

Processes occurring in the film under annealing temperature below 600°C were similar to those occurring under vacuum annealing under the same temperature interval: they were characterized by decreased lattice period, lower values of micro- and macrodeformations accompanied by increasing concentration of deformation packing defects in metallic sublattice of solid solution.

Qualitative changing of phase composition was observed in films under vacuum annealing at T_{an}> 1000°C. Figure 9 shows characteristic diffraction curve, which was taken under 30min annealing at T_{an} = 1100°C. Under high-temperature annealing, in addition to (Zr, Ti)N nitrides (which period was close to ZrN lattice) and (Ti, Z)N (which period was close to TiN lattice), we observed diffraction peaks from zirconium oxide crystallites (ZrO_2, according to JCPDS Powder Diffraction Cards, international Center for Diffraction Data 42-1164, hexagonal lattice) and titanium oxide (TiO, JCPDS 43-1296, cubic lattice), and, probably, initial amorphous β-Si_3N_4 phase crystallites (JCPDS 33-1160, hexagonal lattice). Appearance of zirconium and titanium oxides was related to oxidation relaxation under coating surface interaction with oxygen atoms coming from residual vacuum atmosphere under annealing.

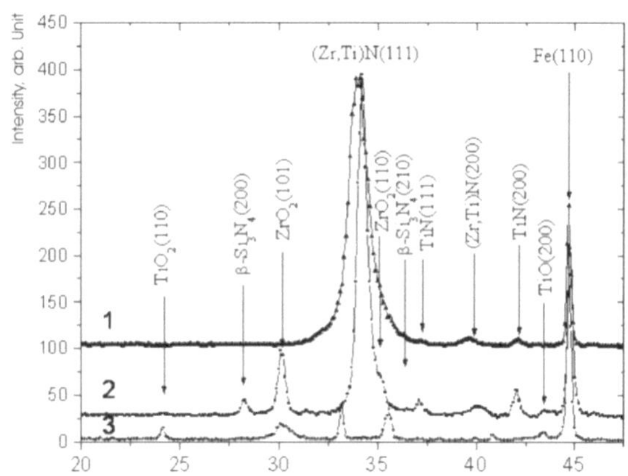

Figure 9. Region of X-ray diffraction spectra taken for the condensates of Zr-Ti-Si-N system after deposition (1); after 30 min annealing in vacuum, under T_{an} = 1180°C (2), and under T_{an} = 800°C in air (3). Three peaks, which are not designated in the curve, are for an oxide of Fe_2O_3 substrate (JCPDS 33-0664).

In solid solution, hardness increased due to increasing Ti concentration and appearance of Si_3N_4 phase. In initial state, after deposition, those samples, which phase composition included three phases (Zr,Ti)N-nc, ZrN-nc, and α-Si_3N_4), hardness was H = 40,6 ± 4 GPa; E = 392 ± 26 GPa. 500°C annealing increased H and E and decreased spread in hardness values, for example, H = 48 ± 6 GPa and E = (456 ± 78GPa), see Table 4.

Parameters	After deposition	T_{an} =300°C vacuum	T_{an} =500°C vacuum
H, Gpa	40,8±2	43,7±4	48,6±6
E, Gpa	392±26	424±56	456±78

Table 4. Changes of hardness and elastic modulus in nanocomposite coating before and after annealing.

Method X-ray scanning, demonstrated shift and broadening of diffraction peaks. Highest content of packing defects indicated shift of most closely packed planes in a fcc-sublattice (111) with respect to each other and became pronounced under vacuum annealing at T_{an} = 800 to 1100°C reaching 15.5 vol.%. As it is seen from "loading-unloading" curves and calculation results, under annealing in vacuum at 500°C, nanohardness of Zr-Ti-Si-N films was H = 46 GPa (dark circles).

When Ti and Si concentration increased and three phases nc-Zr-N, (Zr, Ti)N-nc, and α-Si_3N_4 were formed, average hardness increased to 40,8 ± 4 GPa. Figure 10 shows that in initial state, Zr-Ti-Si-N film (as received) had 40.8 GPa nanohardness. After annealing (a dark dotted curve) at 500°C in vacuum, coating nanohardness reached H = 55.3 GPa. [9,11,13]

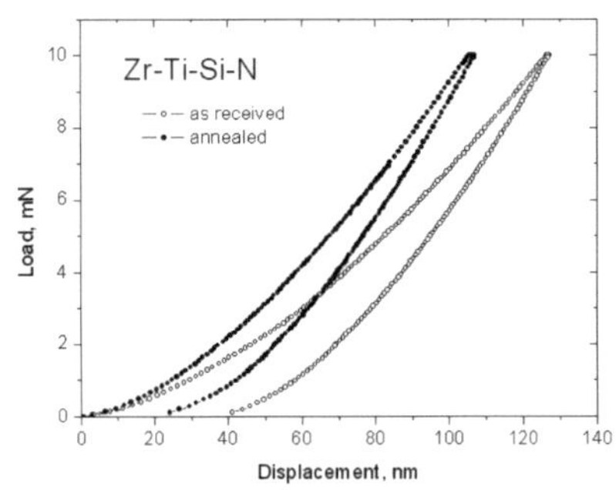

Figure 10. Load-displacement curves for as received and annealed (500°C) Zr-Ti-Si-N Effect of 500°C annealing in vacuum on nanohardness.

5. General regularities and difference of nanocomposite coatings based of Zr, Ti, Hf, V, Nb metals and their combinations

Ti-Hf-Si-N films were deposited on steel 3 substrate (20 mm diameter and 3 mm thickness) with the help of vacuum source in the HF discharge of the cathode, sintered from the Ti-Hf-Si. In order to obtain nitride, atomic N was flooded to the chamber at different pressure and substrate potential. Deposition conditions are presented in Table 5. Bulat 3T-device with generator was used for the deposition of samples. A bias potential was applied to the substrate from a HF generator, which generated impulses of convergent oscillations with ≤ 1 MHz frequency, every duration of the impulse was 60μs, their repetition frequency about 10 kHz. Due to HF diode effect the value of negative auto bias potential at substrate was about 2÷3 kV.

N°	Lattice parameter, nm	Average size of crystallite, nm	Hf content in solid solution (HfTi) coming from the size of period ***, at.%	Hardness, GPa	Nitrogen pressure in chamber, Pa	Substrate potential, V
23(100V,separated)	0.4294*	6.7**	19	42.7	0,7	-200
28(200B, non separated)	0.4430	4.0	65	37,4	0,6-0,7	-200
35 (100V, non separated)	0.4437	4.3	69	38,3	0,6÷0,7	-100
37 (200V, separated)	0.4337	5.0	33	48.6	0,6	-100
31 (200V, separated)	0.4370	3.9	45	39,7	0,3	-200

**- in textured crystallites of samples (series№23) with texture axis (220), the period is more than 0.43602 nm, which can be connected with high Hf content in them (about 40 at%).*

***- in the texture axis direction of textured crystallites the average size is larger (10.6 nm).*

****- Calculation was carried out according to Vegard rule from period values of solid solution (the influence of macrostresses on the change of diffraction lines was not taken into account).*

Table 5. Results of study of: the Ti-Hf-Si-N film depositing parameters; the lattice constant; the crystallite size; the hardness of different series of samples.

Secondary mass-spectrometers SAJW-0.5 SIMS with quadruple mass analyzer QMA-410 Balzers and SAWJ-01 GP-MS with glow discharge and quadruple mass analyzer SRS-300 (Poland, Warszawa) was used for studying of the samples chemical composition. In order to obtain complete information about samples chemical composition, 1.3 MeV ion RBS spectrometers equipped with 16 keV resolution detectors was applied. Helium ion dose was about 5 μC. Standard computer software was used for the processing of the RBS spectra, as a result the depth distribution of the concentration of compound components was plotted.

The research of the mechanical properties of the samples was carried out by the nanoindentation methods with the help of Nanoindenter G200 (MES Systems, USA) equipped with Berkovich pyramid (radius about 20 nm). An accuracy of measured indentation depth was ±0.04 nm. Measurements of the nanohardness of the samples with coating were carried out till 200 nm depth, in order to decrease influence of the substrate on the nanohardness value. The depth of indentation was substantially less than 0.1 of coating depth. XRD analysis was performed using DRON-4 and X′PertPANalitical (Holland) difractometers (step size 0.05°, speed 0.05°C, U = 40 kV, I=40 mA, emitter-copper)[11].

The cross-sections of the substrates with coatings were prepared by the ion beam. Further analysis of surface morphology, structure and chemical composition of these cross sections was carried out by the scanning ion-electron microscope Quanta 200 3D.

For determination of adhesion/cohesion strength, firmness to the scratching, and also for research of destruction mechanism, the scratch-tester REVETEST was used (CSM Instruments).

Prior to analysis of XRD data, it should be noted that for better understanding of processes occurred at near-surface region during deposition it is necessary to compare formation heats of the probable nitrides. According to standard heats of formation of such nitrides are next: ΔH_{298}(HfN) = -369.3 kJ/mole, ΔH_{298}(TiN)=-336.6 kJ/mole, $\Delta H_{298}(Si_3N_4)$ = -738.1 kJ/mole, i.e. values of the formation heats are quite large and negative. It indicates high probability of those systems formation during all stages of transport of the material from target to substrate. In addition, proximity of formation heats for TiN and HfN establish conditions for formation of the sufficiently homogenous (Ti,Hf)N solid solution.

The XRD-analysis revealed the presence of two-phase system. This system was determined as the substitutional solid solution (Ti,Hf)N because diffractions peaks of this phase are located between peaks related to mononitrides TiN (JCPDS 38-1420) and HfN (JCPDS 33-0592). The diffused peaks with less intensity at 2θ values from 40° to 60° are related to the α-Si_3N_4 phase (Figure 11).

According to Figure 11, in direct-flow mode without separation the non-textured polycrystalline coatings are formed. Rather high intensity of the peaks at XRD-patterns of (Ti,Hf)N solid solutions is attributed to relatively large concentration of hafnium, which has larger reflectance value than titanium.

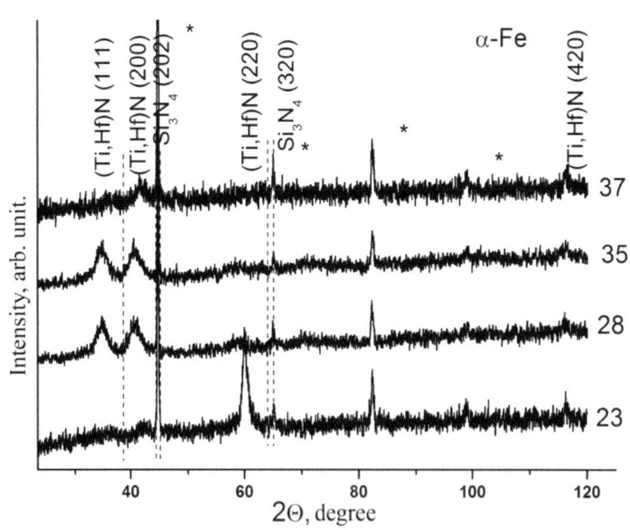

Figure 11. XRD spectra of the coatings deposited on a steel substrate at modes (1-(23) – 100V, separated, 2-(28)-200V, non-separated, 3 (35)-100V, non-separated, 4 (37)-200V, separated).

In case of beam separation the coatings have different texturation. At low substrate potential (100 V) coatings have [110] texture, and coatings consist of textured and non-textured crystallites. The volume content of textured crystallites is about 40% of total amount of the crystallites, and their lattice parameter enlarged in comparison to non-textured crystallites. We suppose that the increased lattice parameter may be caused by the inhomogeneous distribution (mainly in the lattice sites of the textured crystallites) of the hafnium atoms in coating.

At the same time, coating texture leads to increasing of the average grains size of the crystallites along the direction of particle incidence (perpendicular to the growth front). For example, in non-textured fraction of the crystallites the average grains size is about 6.7 nm, whereas in textured crystallites the value of the average grains size is substantially more, namely 10.6 nm. It should be noted that such coatings have the highest nanohardness.

The increase of the substrate potential up to 200 V caused the decrease of average grains size to 5.0 nm. The volume content of textured crystallites is also significantly decreased (less than 20%), moreover the texture axis changed from [100] to [001]. However in this case the lattice parameter is 0.4337 nm and it is larger than for the nontextured fraction in samples obtained at low substrate potential.

According to Vegard law this value of the lattice parameter corresponds to 33 at.% of Hf in metallic (Hf,Ti) solid solutions of the nitride phase (the reference data of the lattice parameters of aTiN=0.424173 nm (JCPDS 38-1420) and aHfN = 0. 452534 nm (JCPDS 33-0592) was used).

However, as a rule, the compressive stresses in coatings caused the decrease of the angles of corresponding diffraction peaks during θ-2θscan, hence calculated values of lattice parameter can be overestimated. As a result inaccuracy of the calculation of Hf concentration in solid solutions can achieve about 5-10 at. %. Therefore presented results can be considered as estimation of upper limit of the Hf concentration in solid solution.

All above mentioned results are related to samples obtained at typical pressure (0.6-0.7) Pa, whereas in a case of coating deposition at 200 V substrate potential in mode of separation (set of samples 31), the decreasing of pressure up to 0.3 Pa caused the increase of relative content of heavy Hf atoms in coatings. In addition, the average grains size of the crystallites decreased with pressure.

Indeed, the decrease in pressure should be accompanied by decrease of the probability of energy loss of atoms during collision between targets and substrate. Thus, atoms at substrates have relatively high energy which can promote secondary sputtering and radiation defect formation. So, secondary sputtering leads to decrease of relative content of heavy Hf atoms, while radiation defect formation provide the decrease of grain size with the increase of nucleus amount.

The coatings obtained under the typical pressure (0.6-0.7) Pa in case of non-separated beam (direct-flow mode) have considerably larger lattice parameter; it can be explained by the high concentration of heavy Hf atoms. [14]

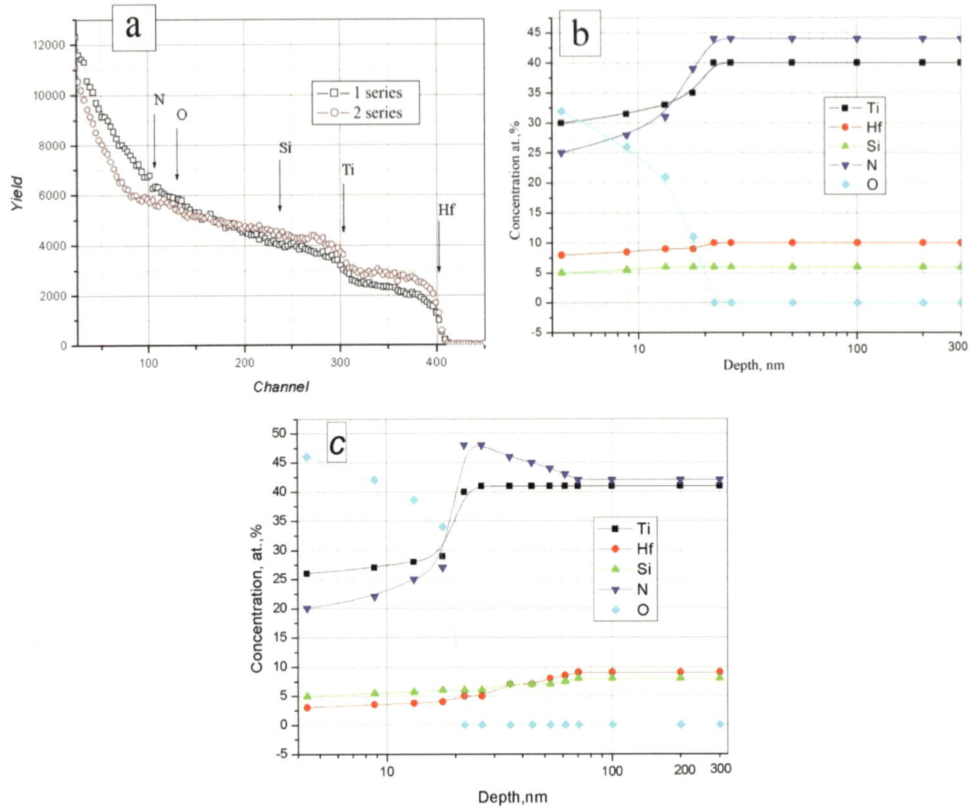

Figure 12. a) RBS spectra of He^+ with 1,3 MeV energy, obtained from steel sample with Ti-Hf-Si-N film: curve 1-potential 100V, p=0,6 Pa, curve 2 - potential 200V, p=0,7 Pa. (b) The depth profiles of elements in the Ti-Hf-Si-N coating, obtained from RBS spectrums (Figure12a). Considering that atomic density of layer is close to atomic density of titanium nitride. (c) The depth profiles in the Ti-Hf-Si-N coating obtained from spectrums (2) on Figure12a (mode 2).

Apparently, the more intensive direct-flow mode leads to the increase of the nucleus density and hence to the decrease of average grain size. In addition, more pronounced decrease of the grains size is caused by the higher substrate potential -200 V. It is obviously because increasing of radiation factor leads to the dispersion of structure. The results of the research of chemical composition of the Ti-Hf-Si-N nanostructured superhard films by the several methods are shown in Figure 12 (RBS(a), SIMS(b), GT-MS(c)). As follows from Figure 12 a, b (curves 1) chemical composition of samples from first set is $(Ti_{40}\text{-}Hf_9\text{-}Si_8)\ N_{46}$.

It is well known that RBS method is a reference for the determination of concentration of the elements with high atomic number and films thickness; also RBS is a nondestructive method. Whereas SIMS is more sensitive method (threshold of sensitivity is about 10^{-6} at.%). Therefore comparison of results obtained by the RBS, SIMS and GT-MS methods allows obtaining of more reliable data of the chemical composition and depth distribution of the concentration of compound components. This joint analysis let us to study the chemical composition along the films cross-section from the surface to the films-substrates interfaces.

http://dx.doi.org/10.5772/50567

Analysis of samples chemical composition also includes measurement of the concentration of uncontrolled oxygen from the residual chamber atmosphere.

As a result we have determined chemical composition (Ti_{40}-Hf_9-Si_8)$N_{46}$of the coatings with thickness about 1μm±0.012 μm. The second set of Ti-Hf-Si-N samples was obtained at increased bias potential (-200 V) under the pressure of 0.3 Pa.

Joint analysis of the films chemical composition by the RBS (Figure 12a curves 2), EDXS and SIMS methods allowed determining of stoichiometry of films as (Ti_{28}-Hf_{18}-Si_9) N_{45}.

The measuring of nanohardness by the triangular Berkovich pyramid (Figure 13) showed that the nanohardness of the samples from the first set is H=42.7 GPa and elastic modulus is E=390±17 GPa (Figure 13), and for the Ti-Hf-Si-N samples from the second set, the nanohardness is H=48.4±1.4 GPa and elastic modulus is E=520±12 GPa.

The XRD-analysis of the phase composition and calculation of the lattice parameter allow us to consider that the two-phase system based on substitutional solution (Ti, Hf)N and α-Si_3N_4 is formed in films.

It was determined that lattice parameter of the solid solution increased with pressure and does not depend on substrate potential. The minimal lattice parameter of the (Ti, Hf)N solid solution was observed in samples from the 23 set.

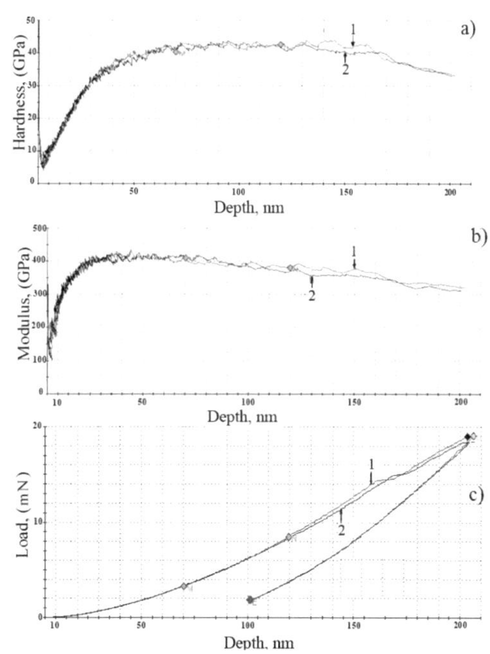

Figure 13. The dependence of hardness H(GPa) (a) on the depth of indentation, (b) the dependence of elastic modulus E(GPa) on the depth of indentation (the regions of H, E measurements are marked with points, series of probes are marked by numbers).

The calculation by the Debye-Scherer method showed that the size of nanograins of the (Ti_{28}-Hf_{18}-Si_9) N_{45} samples from the second set is 4 nm, and it is approximately 1.5 times less than for the first set of the samples. Moreover size (thickness) of amorphous (or quasiamorphous) interlayer was also less than for the first set of the samples (Table 6).

The preliminary results of the HRTEM analysis of samples with nanostructured superhard films are revealed that size of nanograined phase is about 2-5 nm, this result is in correlation with XRD data. In addition it was determined that size of α-Si_3N_4 interlayer, which is envelop the (Ti, Hf)Nnanograins, is about (0.8-1.2) nm.

The properties (hardness, elastic modulus) of Ti-Hf-Si-N samples from the first set were not changed during the storage time from 6 to 12 month.

An analysis of thermal and oxidation resistance was not performed. Therefore it is difficult to conclude that the process of spinodal segregation at the grain boundaries is fully completed. In addition the substrate temperature during the deposition was not more than 350÷400°C, and it is substantially less than full segregation temperature (550÷620°C).

Detailed study of such parameters as coefficient of friction, acoustic emission and depth penetration, were carried out with all samples.

Three-dimensional islands on the surface of the films with columnar structure are output on the surface of the ends of individual grains (Figure 14). It is seen that the roughness depends on the conditions of their chemical composition. Undulation surfaces associated with the mechanism of growth, with the formation of separate islands on the surface (Volmer-Weber mechanism).

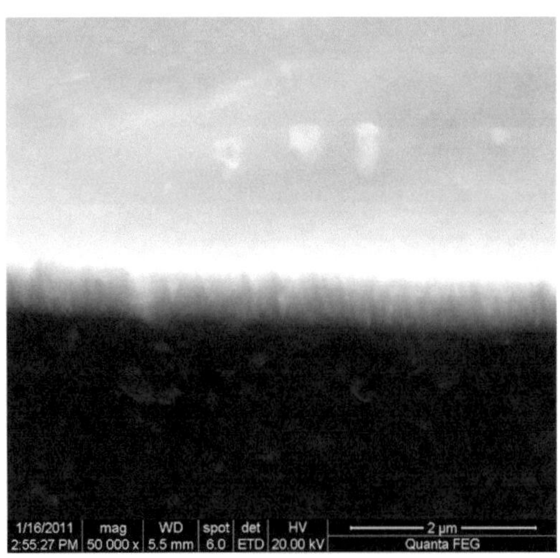

Figure 14. Cross-section of coating Ti-Hf-Si-N with a columnar structure.

http://dx.doi.org/10.5772/50567

The friction coefficientof the sample in the initial stage is equal to 0.12 (apparently due to low roughness of coatings). In the next stage (after 2.5 m of friction Figure 15a) coating starts to destruct (appearance of potholes, cracks) – it is an abrasive wear (Figure 15b). The friction coefficientincreases to 0.45 (it indicates that the coating is not of high hardness).

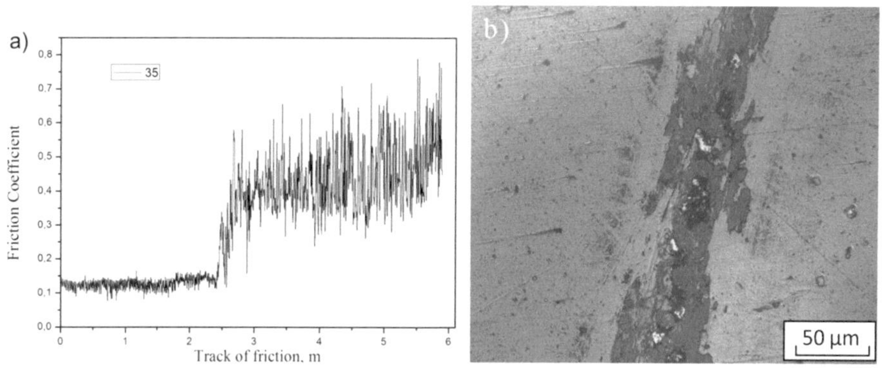

Figure 15. a) dependence of friction coefficient on friction track, b) image of the sample's №35 wear track.

Figure 16(a) represents results of tests on scratch-tester REVETEST of sample 23 with next characteristics: LC1=2,46N and LC2=10,25N.

As the criterion of adhesion strength the critical loading LC, which resulted in destruction of coating, was accepted. However, to treat the results of coatings testing many researchers use the lower (LC1) and overhead (LC2) critical loadings, which characterize ah adhesion strength. Lower critical loading (LC1) is the loading, under which initial destruction of coating occurs Figure 16 (b). And the overhead critical loading (LC2) is the loading under which the coating fully exfoliates from substrate. For samples 23 the lower critical loading for our coverage is LC1=2,46 N and overhead critical loading of LC2=10,25 N, which characterizes a good adhesion/cohesion strength. [12-14]

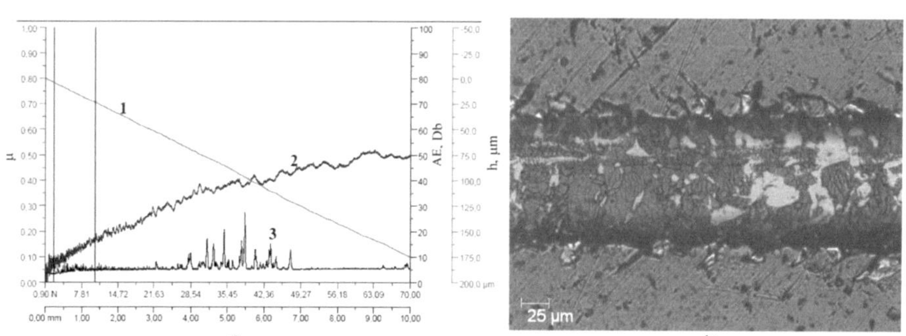

Figure 16. The results of adhesion tests of Ti-Hf-Si-N/substrate/steel coatings system on sample 23: a) 1-penetration depth, 2-Friction Coefficient (µ) and 3-dependence A_E; b) coating structure in destructions zone in load ranges 0,9 – 90 N.

6. Properties of nanostructured coatings Ti-Hf-N (Fe)

As we know, uniqueness of nanostructure nanocomposite coatings is a high volume fraction of phase boundaries and their strength, in the absence of dislocations inside the crystallites and the possibility of changing the ratio of shares of the crystalline and amorphous phases, and also the mutual solubility of metallic and nonmetallic components.

The formation of local sections of Al and C ion implantation of Al in α-Fe due to the process of segregation and the formation of helicoids were found by using microbeam ion, positron annihilation and electron microscopy in works, as increase the diffusion processes N + ions in ion-plasma modification were showed in works.

The films consisted of Ti-Hf(Fe) were deposited on steel samples with diameter of 20 and 30 mm and thick in 3 mm with vacuum-arc source in the HF (High - Frequency) discharge, where fused cathode from Ti-Hf(Fe) was used (by electron gun in an Ar atmosphere). The camera unit was filled with atomic N at various pressures and potentials on substrate for nitrides obtainment. Deposition parameters are presented in Table 6.

N°	P, nitrogen pressure in the camera, Pa	Average crystals size, nm	Hardness, GPa	Substrates potential, V
7(direct)	0,3	6.5	41.82	-200
11(separ)	0,5	4.8	47,17	-200

Table 6. The results of deposition coating Ti-Hf-N (Fe).

A scanning nuclear microprobe based on the electrostatic accelerator IAP NASU was used for analyze the properties of the coatings of Ti-Hf-N (Fe). The analysis was performed using the ions Rutherford backscattering (RBS), the characteristic X-ray emission induced by protons (PIXE and μ - PIXE) at the initial energy Ep = 1,5 MeV, the beam size (2 ÷ 4) μm, the current ≈ 10^{-5}A. PIXE analysis of the overall spectrum was performed using GUPIXWIN program, which allowed us to obtain quantitative information about the content of elements and stoichiometry. For comparison, the scanning electron-ion microscope Quanta 200 with EDS was used for elemental composition and morphology.

We used a vacuum-arc source "Bulat - 3T" with RF generator. Potential bias was applied to a substrate by HF - generator, which produced pulses of damped oscillations with a frequency ≤ 1 MHz pulse with a 60 μs, with a repetition rate of 10 kHz. The magnitude of the negative self-bias potential on the substrate by HF diode effect ranged from 2 to 3 kV. Additionally a detector with a resolution of 16 keV was used with RBS He^{+} ions with energies up to 1,3 MeV, $\theta = 170^{\circ}$. Value of helium ions μ≈5.

Mechanical properties were researched: hardness and nanohardness, elastic modulus with two devices Nanoindentor G 200 (MES System, USA) using a pyramidal Berkovich, Vickers and also indenter like "Rockwell C" with a radius of curvature of about 200 μm was used.

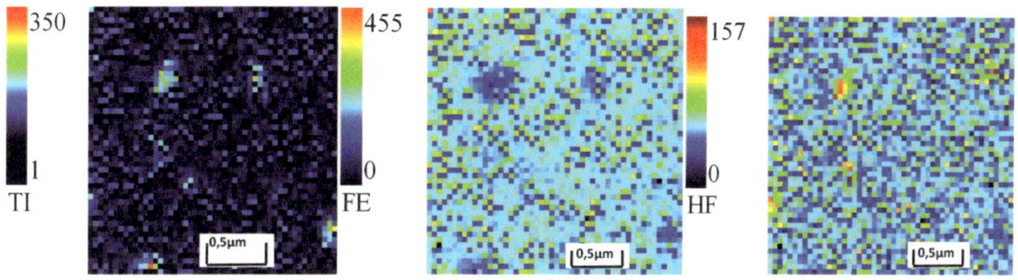

Figure 17. Maps of the distribution of elements (Ti; Hf; Fe) were obtained on samples deposited with a coating of Ti-Hf-N (Fe). In particular, see the local area the size of (2 ÷ 4) to (6 ÷ 10) μ m inclusions consisting of Hf, Ti, which sharp decreases the concentration of Fe.

As seen from these figures cover is different heterogeneity distribution of the elements Ti, Hf, Fe on the surface and the depth of coverage. Quantitative analysis and stoichiometry obtained by PIXE is shown in Figure 17. As seen from the results (integral concentration over the depth about 2 μm) a thin film of AlCformes on the surface, which is probably the result of exposure to the proton beam, and the main elements of the concentration of Fe ≈ 77%, Ti ≈ 11%, Hf≈ 11 05, Mn≈ 0,9% and Cr = 0.01%, the latter elements are apparently part of the substrate (Figure 18). [12,14,15]

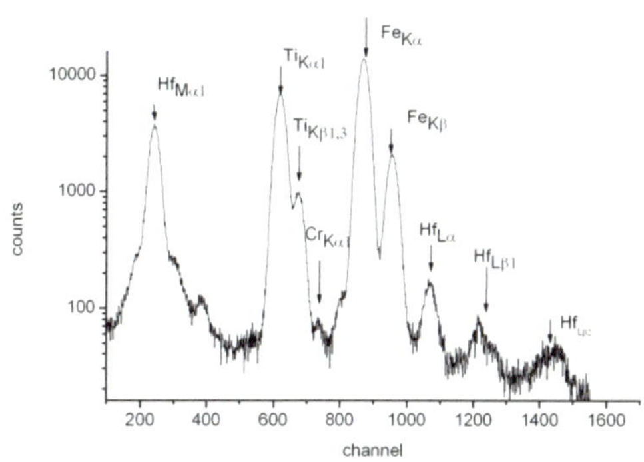

Figure 18. The mass transfers and segregations effect on formation of the super hard ≥ 48GP ananostructured coatingsTi-Hf-N (Fe).

Figure 19 (a) shows an image of the surface area coverage with the imprint of the indenter, which is equal to the value of 48,78 ± 1,2 GPa, these hardness values are very high about 50 GPa and correspond, according to modern classification as superhard coatings. The results of XRD analysis on samples, obtained with this type of coverage, show that the coating formed from at least two phases (Ti, Hf)N, (Ti, Hf)N or FeN, and the size of the nanograins certain width of diffraction peaks of Debye-Scherrer up (4,8-10,6) nm. Using foil, obtained from the coating with a TEM analysis it was found that the coating is formed by a mixture of phases, nanocrystalline (Ti, Hf)N with a grains size 3,5 ÷ 7,2 nm and quasi-amorphous, apparently, FeN. Three-dimensional islands on films surface with columnar structure come to facets surface of individual grains (Figure 19a).

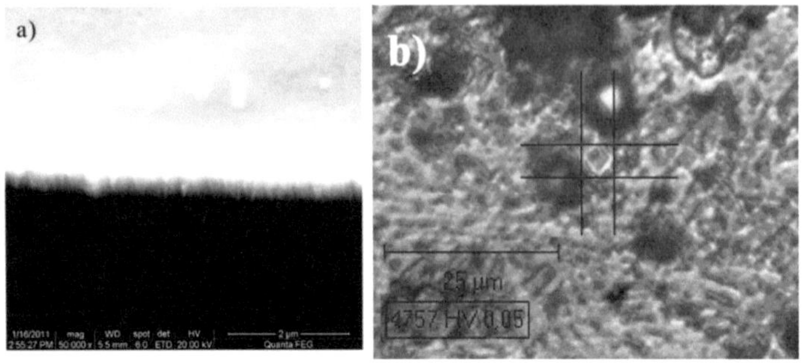

Figure 19. a,b) The mass transfers and segregations effect on formation of the superhard ≥ 48GP ananostructured coatings Ti-Hf-N (Fe).

It is seen that the surface roughness depends on the chemical composition and deposition parameters as well. Surfaces undulation associated with the mechanism of growth and the formation of separate islands on the surface (Volmer-Weber mechanism). At the same time compression microstresses were found (by measuring the XRD spectra in the geometry of up to 2 θθ, and using the method of sin2 φ) in the coating which are formed in nanograin and the corresponding value ≈ 2,6%.

Compressive stresses arise in the growth plane of the film obtained by the width of the diffractions lines peaks according to the method $\sin^2\varphi$ was about ≈ 2,78%. With the plasmas beam separation derived textured coatings with varying degrees, so for example, if the application to a substrate of high potential (-100V) - this texture with pin [110]. In the case of the formation of nanocomposites with TiN-nc and the α-Si_3N_4 (in the form of quasi-amorphous phase) of a thickness of less than 1N (about a monolayer) coatings are formed with very high hardness (superhard) 80 GPa. A necessary and sufficient condition is the end of the process of spinodal segregation at grain boundaries, but this requires a high substrate temperature during deposition (600-650°C) or a sufficiently high rate of diffusion as In our case the substrate temperature during deposition did not exceed 300°C, then apparently, the process of spinodal segregation is not completed [15].

7. Hard nanocomposite coatings with enhanced toughness

A thin coating was formed using vacuum-arc source and followed the coating surface relief formed by plasma-detonation. Its average roughness varies from 14 to 22 µm (after melting and coating deposition using vacuum-arc source). An image of X-ray energy dispersion spectrum is presented below. It indicates the following element concentrations in the thin coating: N ~ 7.0 to 7.52vol.%; Si ~ 0.7vol.%; Ti ~ 76.70 to 81vol.%. For the thick coating we found Fe ~ 0.7vol.%, and traces of Ni and Cr.

Figure 20 presents RBS data for the thick $(Cr_3C_2)_{75}$-$(NiCr)_{25}$ coating without Ti-Si-N thin one. Results for combined coating are presented below, Figure 20.

Element distribution, which was calculated according to a standard program, indicated N = 30at.%; Si ≈ 5 to 6at.%; Ti ≈ 63 to 64at.%. Spectrum of thick coating did not allow us to evaluate element concentration due to high surface roughness of the coating formed by plasma-detonation method.

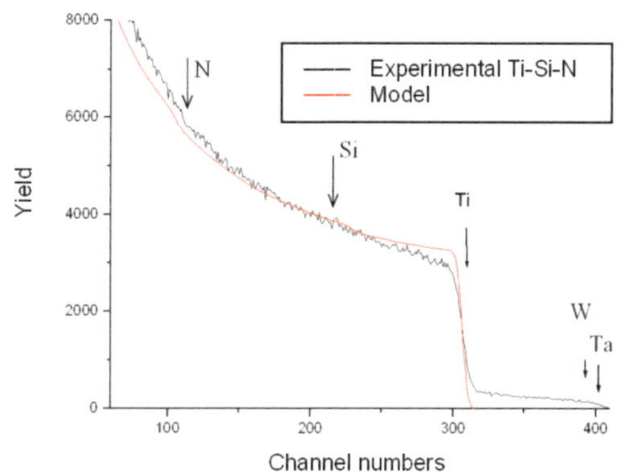

Figure 20. Energy spectra of Rutherford ion backscattering (RBS) for top thin coating Ti-Si-N/WC-Co-Cr.

Special samples were prepared for hardness measurements. Their surfaces were grinded and then polished. After grinding, thickness of $(Cr_3C_2)_{75}$-$(NiCr)_{25}$ thick coating decreased to 80 - 90μm. Thin Ti-Si-N film of about 3μm was condensed to the grinded surface. As a result, we found that hardness of different regions essentially varied within 29 ± 4 GPa to 32 ± 6GPa. Probably, it is related to non-uniformity of plasma-detonation coating surface, which hardness varied up 11.5 to 17.3 GPa. These hardness values remained after condensation of Ti-Si-N thin coating Elastic modulus also features non-ordinary behavior. [16]

Hardness of the thin coating, which was deposited to a polished steel St.45(0.45 % C) surface had maximum value of 48GPa, and its average value H_{av} was 45GPa. Variation of hardness values was lower than that found in a combined coating.

Figure 21 shows dependences of loading-unloading for various indentation depths. These dependences and calculations, which were performed according to Oliver-Pharr technique, indicated that hardness of Ti-Si-N coatings deposited to thick $(Cr_3C_2)_{75}$-$(NiCr)_{25}$ was 37.0 ± 4.0GPa under E = 483GPa.

These diffraction patterns and calculations of coating structure parameters. In the coating, basic phases are Cr_3Ni_2 for the bottom thick coating and (Ti, Si)N and TiN for the thin top coating. Diffraction patterns were taken under cobalt emission. Additionally, we found phases of pure Cr and low concentration of titanium oxide (Ti_9O_{17}) at interphase boundary between thin-thick coatings. Peaks of Ti-Si-N and TiN coincided because of low Si content. (Ti, Si)N is solid solution based on TiN (Si penetration). The phases are well distinguished at 72 to 73° angles.

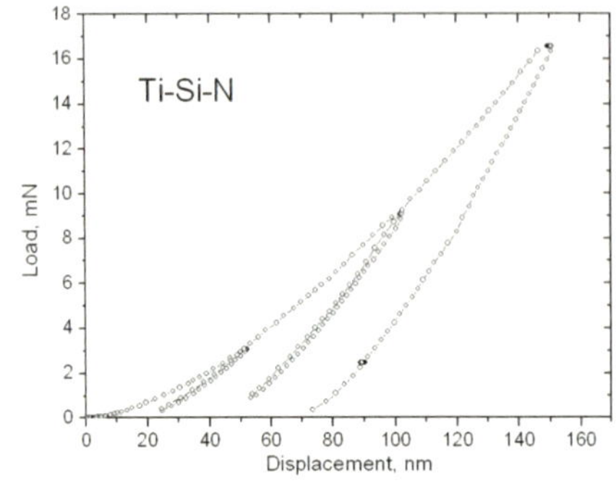

Figure 21. Loading-unloading curves for Ti-Si-N/WC-Co-Cr coating under various Berkovich indentation depths.

Figure 22 a, b shows regions of thick bottom $(Cr_3C)_{75}$-$(NiCr)_{25}$ coating and intensity distribution of X-ray emission (Figure22 c, d) for basic elements. In this coating, content of basic elements is the following: nickel and chromium - 36wt.% and 64wt.%, respectively. Also, we found carbon, oxygen, and silicon. Transversal cross-sections did not allow us to distinguish thin upper coating due to its low thickness. We found regions for pure nickel and chromium. Nickel matrix (a white region) indicated high amount of chromium inclusions with various grain sizes: small grains of < 1μm, average – of 4 to 5μm, and big – of 15 to 20μm. The white region is reach in Ni (to 90at.%). A grey region is reach in Cr (to 92at.%). In these experiments, we failed to determine composition and thickness of Ti-Si-N because of its small thickness. However, 7° angular cross-sections allowed us to find Ti-Si-N element composition and composition of the bottom thick layer $(Cr_3C_2\text{-})_{75}(NiCr)_{25}$ by 10 to 12 points. [16-18]

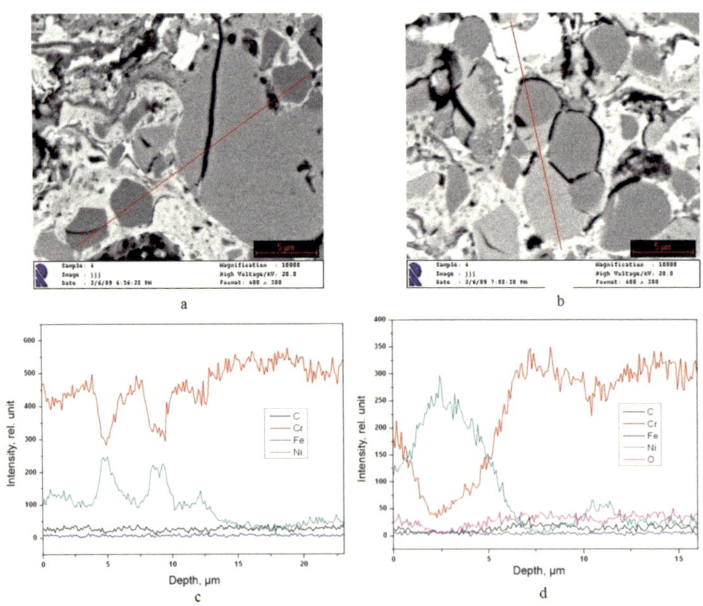

Figure 22. a, b - Regions of transversal cross-section for combined coatings (lines of element analysis are indicated) from SEM and EDS analyses. c, d - Element distribution over depth of combined coating Ti-Si-N/$(Cr_3C_2)_{75}$-$(NiCr)_{25}$ for the regions indicated in a, b.

http://dx.doi.org/10.5772/50567

8. Industrial Applications

It is well known that investigation of nanostructured objects is the most quickly progressing field of modern material science, since a superfine disperse structure is a reason for significant and, in some cases, crucial change of material properties.

Investigations of materials with superfine grain structures demonstrated that when a crystal grain size decreased below some "threshold value", material properties change crucially. This size effect manifested itself even in the case, when an average crystal grain size did not exceed 100 nm. However, it became more pronounced for materials with grain size ranged within 10 nm, and the intercrystalline (intergrain) distances were about few nanometers, contained mainly amorphous phases (nitrides, oxides, carbides, etc.). Stresses, occurring inside these interfaces, are contributed to the increase of the nanocomposite coating deformation resistance, and an absence of inside crystallite dislocations provides with improvement of the coating elasticity. [17]

In this work, we present the first results of investigation of a structure and properties of a new type nano-and microstructure Ti-N-Cr/Ni-Cr-B-Si-Fe-based protective coatings fabricated by the plasma-detonation technology and subsequent vacuum-arc deposition. The aim of this work was fabrication and investigation of the structure, physical and mechanical properties of micro-nanostructured protective coatings with thickness from 80 to 90 μm based onTi-N-Cr/Ni-Cr-B-Si-Fe. Bilayered coatings allowed us not only to protect tools from abrasive wear, but also to recover their geometrical dimensions within 60 to 250μmand more.

We selected a thick coating PG-19N-01 (Russian standard) (Cr-B-Si-Fe(W), based on Ni) since alloys based on Ni-Cr (Mo) have high corrosion resistance even in the solution of acids HCl, H_2SO_4, and HNO_3+HF, under high temperatures, and Ni is able to dissolve a great amount of doping elements (Cr, Mo, Fe, Cu). It was also known that Cr in Ni alloys and Mo in nickel-molybdenum alloys stopped dissolution of a nickel base, though Cr favored and Mo made difficult a passive character of dissolution. Moreover, hardness of (Ni-Cr-B-Si-Fe(W)) powder coating was 3 to 4 times higher than that of a substrate (1.78 ±0.12) GPa. We would like to note that Ni-Cr system is a base for many refractory nickel alloys. Therefore, chromium doping of nickel leads to essential increase of high temperature oxidation resistance. [16]

Ti-N-Cr thin coating (a solid solution) was selected taking into account the considerations that its functional properties (hardness H, elasticity modulus E, plasticity index H/E, material resistance to plastic deformation H^3/E^2, and wear resistance) were notably higher than those of Ni-Cr-B-Si-Fe(W) thick coating of (110 to 120) μmthickness.

The powder was placed inside the reaction chamber and the surface coating layer of (40 to 55)μm was melted by a plasma jet, which was doped by liquid drops coming from an eroding (doping) electrode (W). Melting, which was conducted to reduce a surface roughness from (28÷33) μm to (14÷18) μm and to obtain a more uniform element distribution in the near-surface layer was employed to achieve the necessary mechanical properties. Thus, such a combination of two layers: a thick Ni-Cr-B-Si-Fe layer of 90 μm, which was deposited using plasma-detonation technology, and a subsequently deposited Ti-Cr-N thin upper layer (with a size of the layer like units of a micron), which featured higher physical-mechanical characteristics, was selected to provide improved protective properties and restoration of worn surface regions. [18]

Figure 23a presents the SEM image of a nano-microstructured protective Ti-N-Cr coating surface region.

In the coating surface, one can see some regions with droplet fraction (they are marked with points, at which we performed microanalysis). The point 1, which was taken in the coating surface in X-ray energy dispersion (EDS) spectrum, shows N, Ti, Cr elements, and traces of Ni. Figure 23b and Table 7 show results of integral and local analyses. These results demonstrate almost the same results for N (from 0.56 to 0.98 wt.%), Ti (from 39 to 41%), and Cr (56.8 to 59.4%).We also detected Ni (0.82 to 0.98 %) in thick coating.

	Ni	Cr	Ti	N	Σ
p19_int1	0.578	40.509	58.095	0.819	100.000
p19_int2	0.487	41.867	56.797	0.850	100.000
p19_2	0.564	39.073	59.390	0.973	100.000
p19_1	0.507	40.711	57.805	0.978	100.000

Table 7. Distribution of elements (EDS) on the surface of protective coating Ti-N-Cr concentration (wt.%)

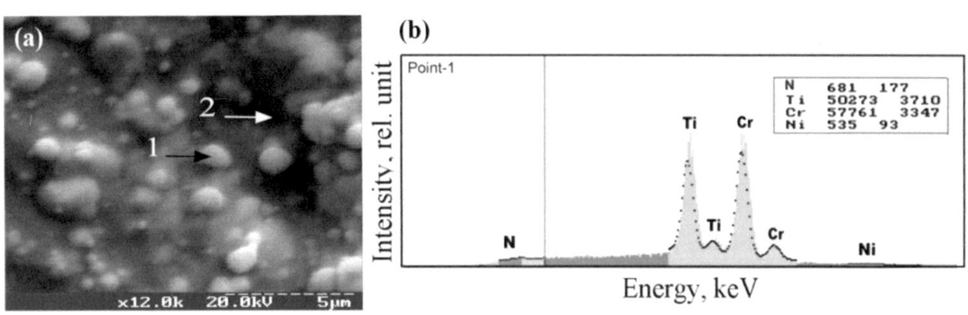

Figure 23. a) Image of the Ti-N-Cr/Ni-Cr-B-Si-Fe(W) coating surface obtained using the scanning electron microscope SEM. The photo shows the points, in which the microanalysis was taken. (b) Energy dispersion X-ray spectrum for the first of the mentioned points, see Figure23a.

Figures 24a and 24b show RBS spectra for protons (Figure 24a) and helium ions $^4He^+$ (Figure 24b). From these spectra one can see that all elements (N, C, Ti, Cr) composing Ti-N-Cr/Ni-Cr-B-Si-Fe coating were found. The fact that a "step" was present in the spectrum almost over the whole depth of analysis of this coating is worth one's attention. It indicates a uniform nitrogen distribution and formation of $(Ti\ Cr)_2N$ compounds. The compound stoichiometry was close to $Ti_{40}Cr_{40}N_{20}$ or $(Ti, Cr)_2N$.

Table 8 presented the coating composition, which was obtained using RBS and a standard program. One can also mention low W concentration (0.07 at.%) in the thin coating. But near the interface between the thick and the thin coating, this concentration increased to 0.1 at.%. We assumed that W diffuse from thick coating (from the eroding electrode). The stainless steel substrate composition demonstrated Ni_3Cr_2. Comparison of RBS, EDXS, and XRD data allowed us to state that in the thin nanostructured coating (Ti-Cr-N), oxygen was absent (less than 0.1%), and carbon was present, but its concentration was lower than XRD detection ability.

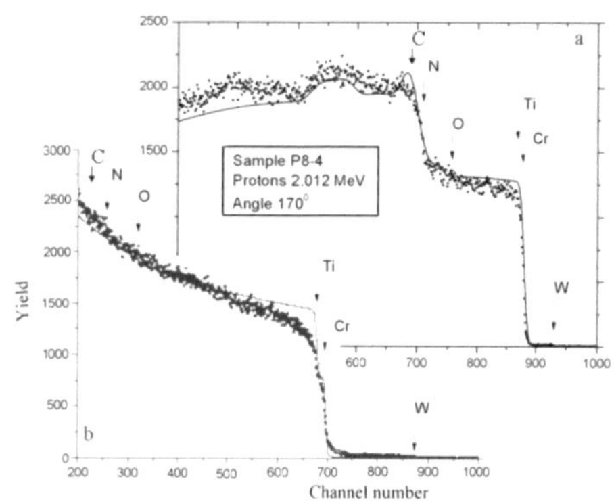

Figure 24. a, b. An energy RBS spectrum for proton scattering with the initial energy 2.012*MeV* – (a) and 2.035*MeV* – (b) taken for the sample Ti-N-Cr/Ni-Cr-B-Si-Fe(W). The arrows indicate the boundaries of kinematical factors for different elements.

Depth		Concentration (at. %)			
Nm	W	Ni	Cr	Ti	N
625	0.07	0	38.70	38.70	22,52
1251	0.07	0	38.70	38.70	22,52
2317	0.09	0	38.70	38.70	22,52
3263	0.09	0	38.70	38.70	22,52
14380	0	61.30	38.70	0	0

Table 8. Distribution of elements on the depth of protective coating Ti-N-Cr.

The general views (their cross-sections) of these coatings are presented on Figures 25 a, b.

Figure 25a shows the sample without coating. The right part of this Figure demonstrates data of micro-analysis ("A-A'" cross-section). An etched coatings layer was almost free of pores. The interface between the coating and the substrate was wavy; this indicates penetration of powder particles to the substrate.

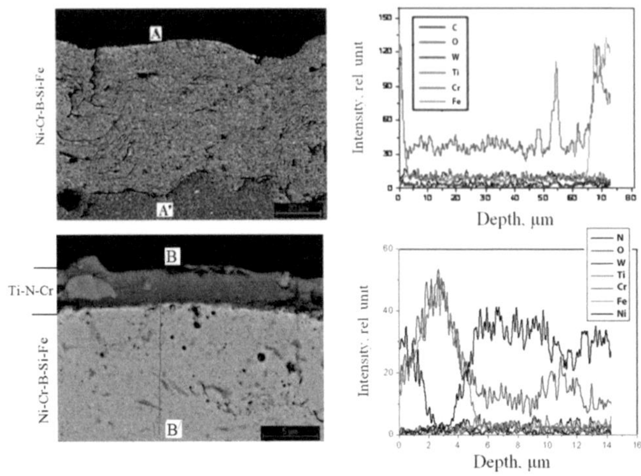

Figure 25. a) Scanning electron microscopy images for the cross-section (A-A') and distribution of the characteristic X-ray element emission over this cross-section (A-A') in a combined nanocomposite coating (a thick coating was melted by a plasma jet). (b) Scanning electron microscopy images for the cross-section and distribution of the characteristic X-ray element emission along the cross-section (B-B/) in the coating on the base of (Ti, Cr)N solid solution. The thin coating was deposited on the thick one of Ni-Cr-B-Si-Fe(W) and melted by a plasma jet.

Figure 25b shows a cross-section for Ti-N-Cr/Ni-Cr-B-Si-Fe coating. Its element distribution over cross-section depth is demonstrated in the right part of Figure 25b ("B-B/"). The thin coating was composed of Ti and Cr (N was not found, possibly due to low detector resolution). Results of XRD analysis for Ti-N-Cr/Ni-Cr-B-Si-Fe coating are presented in Table 9. Calculation of diffraction patterns (Table 9) demonstrated (Ti, Cr)N (200) and (Ti, Cr)N (220). Additionally, γ – $FeNi_3$ and $FeNi_3$ phases were found in these samples. Also we have determined that diffraction peaks were shifted, under-peak areas differed, and derived ratios of intensities. Measurements and analysis of diffraction lines, which were taken using grazing incidence diffraction, demonstrated smoothed peaks corresponding to amorphization or formation of nanocrystalline phases. [16-18]

2θ degree	Area	Intensity	Semi width	Value Angstrom	Relative Intensity	Phase	HKL
43,100	67,883	53	2,4450	2,0987	58,89	(Ti,Cr)N	200
43,640	24,025	90	0,5200	2,0740	100,00	□-(Fe,Ni)	111
50,840	6,066	27	0,4450	1,7959	30,00	□-(Fe,Ni)	200
63,020	1,693	19	0,1800	1,4750	21,11	(Ti,Cr)N	220
74,400	5,358	19	0,5500	1,2750	21,11	□-(Fe,Ni)	220
90,620	5,130	20	0,5000	1,0844	22,22	□-(Fe,Ni)	311
96,060	1,892	19	0,2000	1,0368	21,11	□-(Fe,Ni)	222

Table 9. Calculation results of diffraction patterns of Ti-N-Cr coating from the side of thin coating.

In addition to basic phases, an X-ray diffraction, which was performed at 0.5° angle, demonstrated also simple hexagonal compounds Cr_2Ti, Fe_3Ni (Fe, Ni) and various compounds of a titanium with nickel Ti_2Ni, Ni_3Ti, Ni_4Ti_3, etc. (Figure 26a). These additional phases were formed at some initial stages of the coating deposition, as a result of titanium, nickel, chromium and iron diffusion. A resulting solid solution was a small-grain dispersion mixture (grain size were calculated according to Debay-Sherer formula and reached about 2.8 to 4 nm) (Figure 26b).

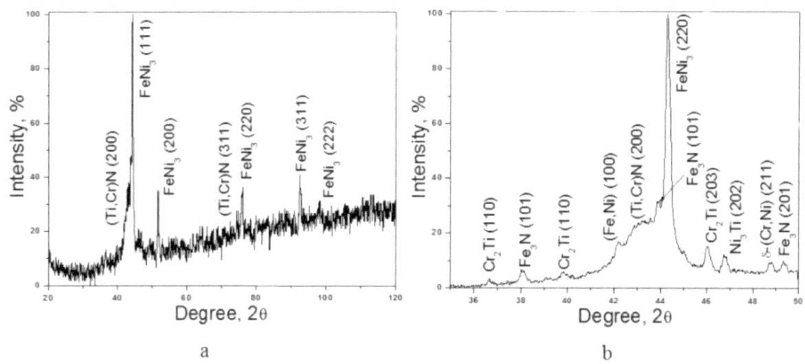

Figure 26. Fragments of the diffraction patterns taken in the incidence grazing diffraction for the whole region (a) and for the selected intensity peak (b).

TEM analysis (Figure 27a, b) demonstrated that an order of nanograin size magnitude corresponded to XRD data, namely ranged within 5 to 12 nm. (Ti, Cr)N lattice of a solid solution corresponded to NaCl, see the diffraction data in Figure 27a. The light field analysis demonstrated a uniform distribution of nanograins of various sizes (Figure 27b.), which correlated well with results reported in paper. In this report, Ti-Cr-N was deposited from two cathodes using VAD under the same coating nanohardness, which amounted from 32.8 to 42.1 GPa.

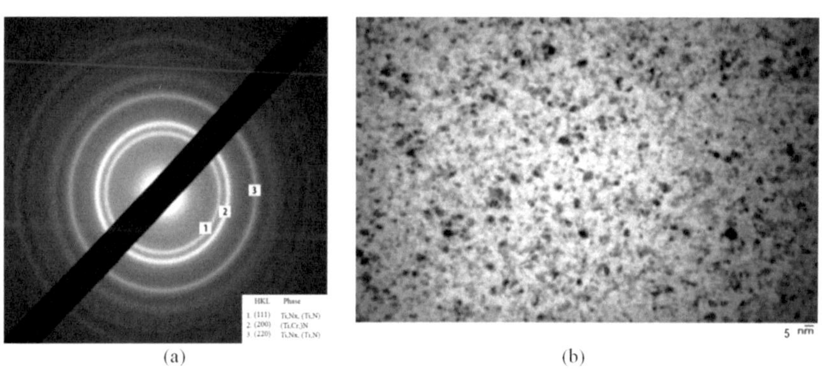

Figure 27. a) Electron diffraction patterns for nanocomposite coatings, cubic phase of solid solution (Ti,Cr)N, with NaCl type lattice. (b) Light-field image for nanocomposite films fabricated from thin coating based on (Ti,Cr)N solid solution phase.

High magnification allowed us to see a mixture of differently oriented nano-grains (Figure 28a). Interfaces of the nano-grains were not separated, and we failed to find an expressed symmetry in positions of the nano-grains and a definite orientation of atomic planes. However, the electron diffraction patterns of the studied regions demonstrated clearly visible point reflexes, which coincided with the ring electron diffraction pattern of the coating matrix (Figure 28b). Sizes of individual nano-grains reached from 2 to 3 nm, but point reflexes indicated the formation of micro-regions with identically-oriented crystalline lattices. Nevertheless, interfaces of such formations were not pronounced.

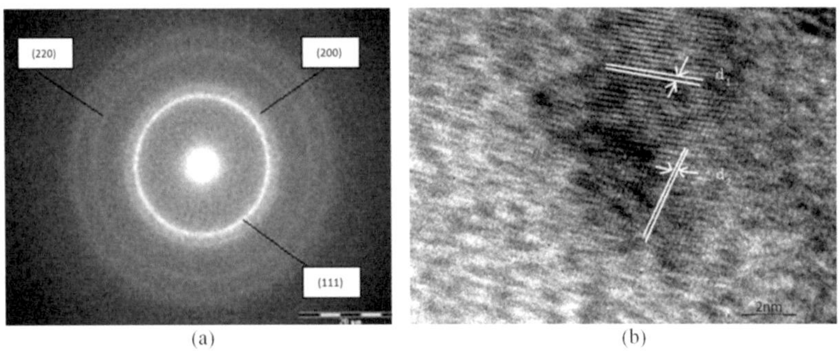

Figure 28. a) Electron diffraction from the matrix of the Ni-Cr-B-Si-Fe coating, indicating the indices of the crystallographic planes. (b) High resolution image of Ni-Cr-B-Si-Fe interface, with the presence of nanoscale domains of crystallographic orientation which coincided with the ring electron diffraction pattern.

The obtained results allowed us to conclude that an inter-metalloid Cr-Ni_3-phase having a*fcc*-lattice was formed in the coating surface. It was found that chromium compounds were formed only in a thin surface coating layer. It is evident that a main reason for the non-uniform phase formation was a non-uniform temperature profile, which was formed over the sample depth under an action of a plasma jet.

It was also found that a thin layer, which was formed in the sample surface after treatment, contained oxides, carbides and various phases of the coating elements, which badly dissolved under high temperatures. A basic coating layer featured an essentially uniform phase composition and contained a γ-phase, a solid solution on nickel base, and α-phase based on iron. The latter was found only on the coating side, which was adjacent to the substrate.

Taking into account the fact that at 700°C to 800°C temperatures, a carbide hardening phase of alloys based on Ni-Cr-B-Si-Fe coagulated faster than the intermetalloid one, we assumed that one should prefer to employ alloys with an intermetalloid type of solidification.

A partial spectrum I corresponded to stainless steel. Some asymmetry, which was observed in a quadruple duplet (various amplitudes and widths, but similar areas of resonance lines), was due to the non-uniformity in surrounding Fe atoms.

The partial spectrum II corresponded to α – Fe particles. In comparison with a standard α – Fe spectrum, these values of a Mössbauer line shift δ and a quadruple ε, which differed from zero, and a little lower value of superfine field indicated nanosized impurities ≤ 100 nm (in locally non-uniform systems). [9,11,16]

The coating surface and cross-section morphology was additionally studied using an electron scanning microscopy SEM and X-ray spectral micro-analysis (using LEO-1455 R microscope). A thin coating Ti-Cr-N based on solid solution fully repeated substrate relief.

Samples with Ti-Cr-N coatings had 6.8 and 8.4 mkg/year corrosion rate, depending on thin layer composition (stoichiometry) (Figure 29).

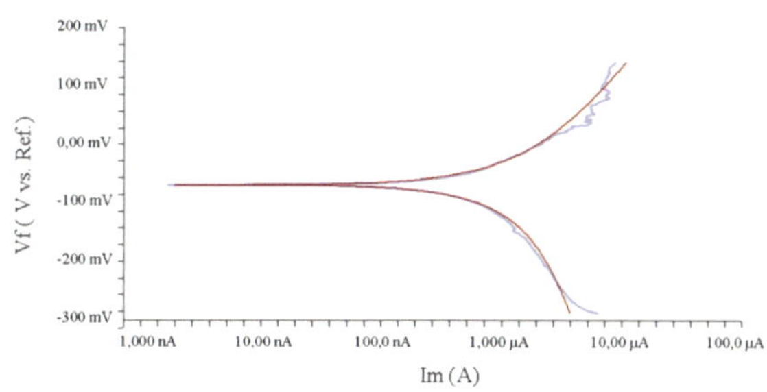

Figure 29. Experimental dependences of corrosion and Tafel curves for the sample Ti-Cr-N/Ni-Cr-B-Si-Fe(W).

The hardness *H* and elasticity modulus *E* were determined using the nanoindentation device Nanoindenter II, according to Oliver and Pharr methods and with the help of a Berkovicz indenter, see Figure 30. For a surface layer, a value of elastic recovery W_e was calculated using loading-unloading curves, according to the formula

$$W_e = \frac{h_{max} - h_r}{h_{max}} \qquad (1)$$

where h_{max} was a maximum penetration depth, and h_r was a residual depth after a load relieve.

It was obtained that the elasticity modulus of Ti-Cr-N coating had a value E_{mean}~ 440 GPa, its hardness was H_{mean} ~ 35.5G Pa, and the maximum value was 41.2 GPa (see the Table 10).

Table 10 and Figure 30a demonstrates highest hardness values obtained from nanoindentation measurements: (32,8 – 42,1) GPa for Ti-Cr-N, 6,8 GPa for Ni-Cr-B-Si-Fe, and 8,1 GPa for Ni-Cr-B-Si-Fe coatings after a plasma jet melting. We noticed a lower difference in hardness values in comparison with cases without melting. The substrate hardness was 1.78 ± 0.14 GPa. The elasticity modulus was also higher for Ti-Cr-N coatings and amounted 360 ÷ 520 GPa (Figure 30b). It was 229 ± 11 GPa for Ni-Cr-B-Si-Fe coating after the plasma jet melting.

To evaluate a material resistance to an elastic strain failure, the authors used a ratio of hardness to the elasticity modulus H/E, which was named a plasticity index. To evaluate the material resistance to a plastic deformation, they used, for example, H^3/E^2 ratio. So, to increase a resistance to an elastic strain failure and plastic deformation, a material should have a high hardness and a low elasticity modulus. As it is known from, typical ratios for ceramics and metallic ceramics did not exceed 0.2 GPa. For NiTi, due to a form memory effect, it was lower by an order of magnitude. New nanostructured materials, which were obtained in our experiments, demonstrated H^3/E^2 ratios ranging within 0.29 ± 0.03. Many materials featuring high H^3/E^2 ratios indicated a high wear resistance. And the elasticity modulus of deposited materials was close to the Young modulus of materials with high H^3/E^2 ratio, which indicated high wear resistance. Used materials resistance to plastic deformation, we used same coating nanohardness, which amounted 32.8 to 41.2 GPa one of the substrate material. These indicated high servicing characteristics under abrasive, erosion, and impact wear conditions.

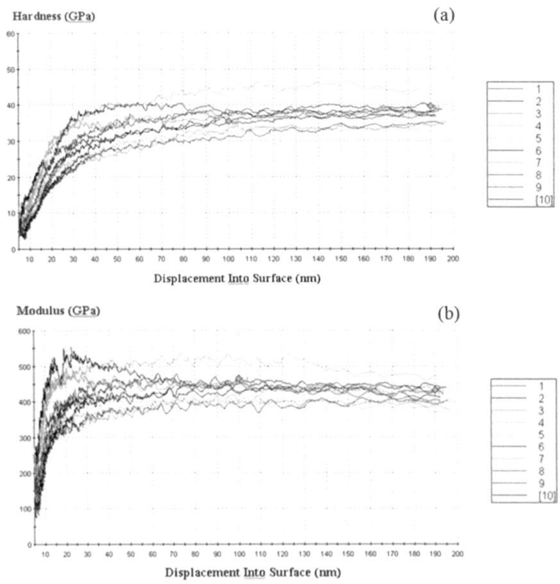

Figure 30. Curves for hardness H (a) and elasticity modulus E (b) obtained for the sample Ti-N-Cr/Ni-Cr-B-Si-Fe. The calculation results for H and E are presented in the Table 10.

Coating material	E, GPa	H, GPa
Ti-N-Cr	360 - 520	32.8 - 41.2
Ni-Cr-B-Si-Fe (W)	193±6	6.8±1.1
Ni-Cr-B-Si-Fe (W) (Melting of Plasma jets)	217±7	8.1±0.2
Substrate	–	–
(NiCr)	229±11	1.78±0.14

Table 10. Results of mechanical characteristics tests, being obtained by nanoindenter.

Therefore, we performed measurements of the wear resistance under the cylinder friction over nanocomposite combined coating surfaces without lubrication. Results of these tests are presented in Figure 31. As it is seen from the Figure 31, the maximum wear resistance of a nanostructured Ti-Cr-N coating was a factor from 27 to 30 lower than that of a steel substrate. A low wear was also observed in thick Ni-Cr-B-Si-Fe coating melted by a plasma jet. Samples coated by Ni-Cr-B-Si-Fe, which were stored in air, in a wet environment for 5 to 7 years, after repeated melting by a plasma jet demonstrated unchanged hardness, elasticity modulus, corrosion resistance, which stayed almost the same within the limits of a measurement error that undoubtedly seems to be promising for protection of steels and alloys.

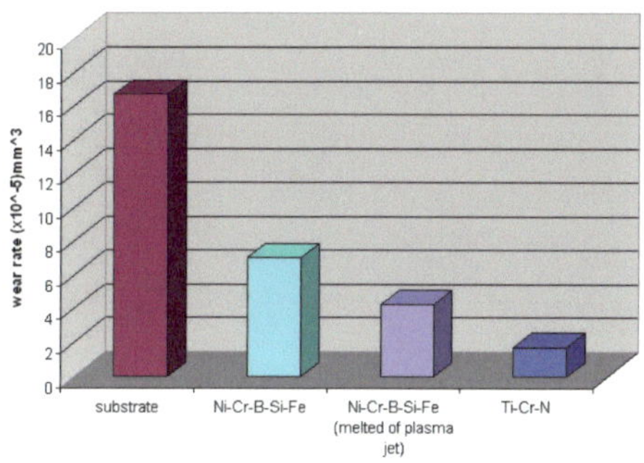

Figure 31. Histograms of dependences of wear rates for samples, which were fabricated according to scheme cylinder-plane.

9. Hard nanocomposite coatings with enhanced toughness

Hard nanocomposite coatings with enhanced toughness are coatings which are simultaneously hard and tough. Such coatings should be very elastic, exhibit a low plastic deformation, resilient properties when the plastic deformation is zero, and an enhanced resistance to cracking.

The way how to produce hard, tough and resilient coatings is indicated by the Hooke's law $\sigma = E.\varepsilon$; here, σ is the stress (load), ε is the strain (deformation). If we need to form the materialwhich exhibits a higher elastic deformation (higher value of ε) at a given value σ its Young'smodulus E must be reduced. It means that materials with the lowest value of the Young'smodulus E at a given hardness H (σ=const) need to be developed. It is a simple solution but avery difficult task.

The stress σ vs strain ε dependences for brittle, tough and resilient hard coatings areschematically displayed in Figure 32. Superhard materials are very brittle, exhibit almost n plastic deformation and very low strain ε=ε1. Hard and tough materials exhibit both elastic and plastic deformation. The material withstanding a higher strain $\varepsilon 1 \ll \varepsilon \leq \varepsilon max$ without it cracking exhibits a higher toughness. The hardness of tough materials is higher in the case when ε_{max} is achieved at higher values of σ_{max}. On the contrary, fully resilient hard coatings exhibit, compared to hard and tough materials, a lower hardness H, no plastic deformation (line 0A) and high elastic recovery We. The hardness H of hard, tough and well resilient coatings, ranging from about 15 to 25 GPa, is, however, sufficient for many applications. The main advantage of these coatings is their enhanced resistance to cracking. These are reasons why in a very near future the hard and tough, and fully resilient hard coatings will be developed. These coatings represent a new generation of advanced hard nanocomposite coatings.

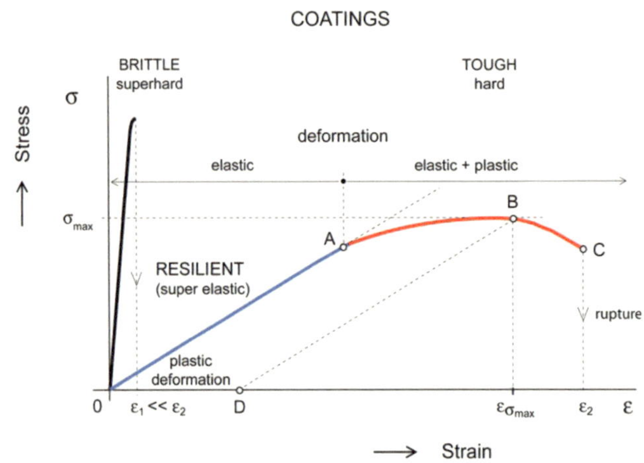

Figure 32. Schematic illustration of stress □ vs strain □ curves of superhard (brittle), hard (tough) and hard (resilient) coatings. Resilient coatings exhibit no plastic deformation (line 0A) [19].

We can conclude that a new task in the development of advanced hard nanocomposite coatings with enhanced toughness is to produce coatings with (i) a low value of the Young'smodulus E* satisfying H/E*≥ 0.1 ratio and (ii) a high value of the elastic recovery We. The coatings fulfilling these requirements can be really prepared if the element added into a base material is correctly selected as is shown in Figure 33. [20-24]

Figure 33 displays H=f(E*) dependences of five Ti-N, Ti-Al-N, Zr-N, Zr-Cu-N and Al-Cu-Nnitride coatings prepared by magnetron sputtering. Also, in this figure a straight lineH/E*=0.1, which divides the H-E* plane in two regions with H/E*>0.1 and H/E*<0.1, is displayed. From this figure it is seen that experimental points corresponding to individual nitrides are quite well distributed along mutually separated straight lines. This figure clearly shows that (1) the coating material with the same hardness H and different elemental composition can exhibit different values of the effective Young's modulus E*, (2) the value ofE* of the Me_1-Me_2-N coating depends not only on the element Me_2 added to the Me_1N binary nitride but also on the element Me_1 which forms the binary nitride, (3) not all nitrides exhibit H/E*>0.1 and (4) the coating material with the ratio H/E*>0.1 can be achieved only in the case when both elements Me_1 and Me_2 are correctly selected. The last fact represents a huge potential for new industrial applications, particularly, for the improvement of properties of the binary nitrides and the development of new advanced protective coatings, for instance, for the improvement of cutting properties and lifetime of cutting tools.

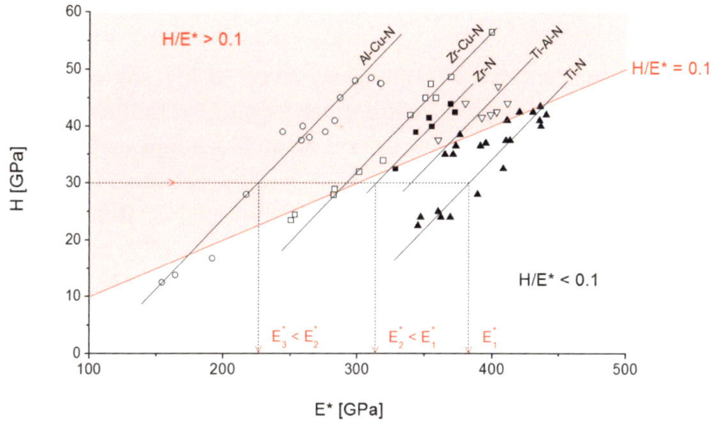

Figure 33. Control of the effective Young's modulus E* of the binary nitrides by addition of selected elements. Adapted after reference [19].

The preparation of the coatings with H/E*>0.1 is complex and difficult task because the hardness H and the effective Young's modulus E* are two mutually coupled quantities. The magnitudes of H and E* depend on deposition parameters used in the preparation of coating and are controlled not only by its elemental composition as shown above but also by its structure, phase composition and microstructure, i.e. by the energy delivered to the growing film particularly by bombarding ions and condensing atoms. At present, there are no general rules which allow predict how to prepare the coatings with H/E*>0.1. [9]

10. Trends of next development

Next research activity in the field of hard nanocomposite coatings is expected to be concentrated mainly on the solution of the following problems: (1) the development of hard coatings with enhanced toughness and increased resistance to cracking, (2) the investigation of DNG/AM composite coatings composed of small amount of nanograins dispersed in the amorphous matrix with the aim to develop new coatings with unique physical and functional properties, (3) the investigation of the electronic charge transfer between nanograins with different chemical composition and different Fermi energies in nanocomposite coatings with the aim to understand its effect on the functional properties of coating, (4) the nanocrystallization of amorphous materials at temperatures of about or less than 100°C for flexible electronics, (5) the formation of high-temperature phases at temperatures T≤500°C using superfast heating and cooling at atomic level, (6) the development of nanocomposite coatings thermally stable above 1500°C and protecting the substrate against oxidation at temperatures up to ~2000°C, (7) the formation of multilayers composed of nano-bilayers, (8) high-rate reactive deposition of hard coatings based on oxides with deposition rate a_D exceeding 10 000 nm/min, and (9) the development of new Physical Vapour Deposition (PVD) systems for the production of new advanced coatings under new physical conditions, for instance, the magnetron with molten target. [8,9, 20-25]

11. Conclusion

A current state of production and a progress achieved in investigation of properties and structures of superhardnanocomposite coatings are considered in the Chapter. The potential of various technologies employed for deposition of such coatings as Ti-Zr-Si-N, Ti-Hf-Si-N, Ti-Si-N, Ti-N, etc. is demonstrated. Investigation results obtained for micro-, nano-, and combined coatings such as Ti-Si-N / Cr_2C_3 – NiCr, Ti-N-Cr/Ni-Cr-B-Si-Fe featuring not only high nanohardness but also good corrosion resistance to NaCl, HCl, and H_2SO_4, high friction wear resistance, and very high thermal stability up to 900°C are described.

Acknowledgements

This work was supported by the SFFR of Ministry of Education and Science, Youth and Sport of Ukraine (Grant F41.1/019) "Development of physical and Technological foundation of multicomponent nano-microstructural coatings based on Ti-Hf-Si-N; Zr-Ti-Si-N with high hardness 40 GPa, the thermal stability ≥ 1000°C and high physical-mechanical properties" and(Grant № 473) "Development of basics create nanocomposite materials, coatings and layers with high physical-mechanical properties"

Author details

A. D. Pogrebnjak[1*] and V. M. Beresnev[2]

*Address all correspondence to: alexp@i.ua

1 Sumy State University, Sumy Institute for Surface Modification, Ukraine

2 Kharkov National University, Kharkov, Ukraine

References

[1] Gleiter, H. (1989). Nanocrystalline materials. *Progress in Materials Science*, 33, 223-315.

[2] Vepřek, S., & Reiprich, S. (1995). A concept for the design of novel superhard coatings. *ThinSolid Films*, 265, 64-71.

[3] Gleiter, H. (1996). Nanostructured Materials: State of the art and perspectives. *Nanostructured Materials*, 6, 3-14.

[4] Musil, J. (2000). Hard and superhardnanocomposite coatings. *Surface and Coatings Technology*, 125, 322-330.

[5] Gleiter, H. (2001). Tuning the electronic structure of solids by means of nanometer-sized microstructures. *Scripta Materialia*, 44, 1161-1168.

[6] Musil, J. (2006). Physical and mechanical properties of hard nanocomposite films prepared by reactive magnetron sputtering, Chapter 10 in Nanostructured Coatings, J.T.M. DeHosson and A. Cavaleiro (Eds.) New York, Springer Science+Business Media, LCC, 407-463.

[7] Pogrebjank, A. D., Shpak, A. P., Azarenkov, N. A., & Beresnev, V. M. (2009). Structure and properties of hard and superhardnanocomposite coatings. *Physics-Uspekhi*, 52(1), 29-54.

[8] Musil, J., Šatava, V., Zeman, P., & Čerstvý, R. (2009). Protective Zr-containing SiO2 coatings resistant to thermal cycling in air up to 1400°C. *Surface and Coatings Technology*, 203, 1502-1507.

[9] Musil, J., Hromádka, M., & Novák, P. (2011). Effect of nitrogen on tribological properties of amorphous carbon films alloyed with titanium. *Surface and Coatings Technology*, 205(2), S84-S88.

[10] Pogrebnyak, A. D., Bratushka, S. N., Malikov, L. V., Levintant, N., Erdybaeva, N. K., Plotnikov, S. V., & Gritsenko, B. P. (2009). Effect of high doses of N+, N+ + Ni+, and Mo++ W+ ions on the physicomechanical properties of TiNi. *Technical Physics*, 54(5), 667-673.

[11] Pogrebnyak, A. D., Shpak, A. P., Beresnev, V. M., Kirik, G. V., Kolesnikov, D. A., Komarov, F. F., Konarski, P., Makhmudov, N. A., Kaverin, M. V., & Grudnitskii, V. V. (2011). Stoichiometry, phase composition, and properties of superhard nanostructured Ti-Hf-Si-N coatings obtained by deposition from high-frequency vacuum-arc discharge. *Technical Physics Letters*, 37(7), 636-639.

[12] Beresnev, V. M., Sobol', O. V., Pogrebnjak, A. D., Turbin, P. V., & Litovchenko, S. V. (2010). Thermal stability of the phase composition, structure, and stressed state of ion-plasma condensates in the Zr-Ti-Si-N system. *Technical Physics*, 55(6), 871-873.

[13] Pogrebnyak, A. D., Sobol', O. V., Beresnev, V. M., Turbin, P. V., Dub, S. N., Kirik, G. V., & Dmitrenko, A. E. (2009). Features of the structural state and mechanical properties of ZrN and Zr(Ti)-Si-N coatings obtained by ion-plasma deposition technique. *Technical Physics Letters*, 35(10), 925-928.

[14] Pogrebnyak, A. D., Ponomarev, A. G., Kolesnikov, D. A., Beresnev, V. M., Komarov, F. F., Mel'nik, S. S., & Kaverin, M. V. (2012). Effect of mass transfer and segregation on formation of superhard nanostructured coatings Ti-Hf-N (Fe). *Technical Physics Letters (Rus)*, 38(13), 56-63.

[15] Pogrebnyak, A. D., Beresnev, V. M., Demianenko, A. A., Baydak, V. S., Komarov, F. F., Kaverin, M. V., Makhmudov, N. A., & Kolesnikov, D. A. (2012). Adhesive strength and superhardness, phase and element composition of nanostructured coatings formed on basis of Ti-Hf-Si-N. *Physics of the Solid State (Rus)*, 54(9), 1764-1771.

[16] Pogrebnjak, A. D., Danilionok, M. M., Uglov, V. V., Erdybaeva, N. K., Kirik, G. V., Dub, S. N., Rusakov, V. S., Shypylenko, A. P., Zukovski, P. V., & Tuleushev, Y. Zh. (2009). Nanocomposite protective coatings based on Ti-N-Cr/Ni-Cr-B-Si-Fe, their structure and properties. *Vacuum*, 83, S235-S239.

[17] Pogrebnjak, A. D., Ponomarev, A. G., Shpak, A. P., & Kunitski, Yu. A. (2012). Application of micro-nanoprobes to the analysis of small-size 3D materials, nanosystems and nanoobjects. *Uspekhi Phys-Nauk*, 182(3), 287-321.

[18] Pogrebnjak, A. D., Sobol, O. V., Beresnev, V. M., Turbin, P. V., Kirik, G. V., Makhmudov, N. A., Il'yashenko, M. V., Shypylenko, A. P., Kaverin, M. V., Tashmetov, M. Yu., & Pshyk, A. V. (2010). Nanostructured Materials and Nanotechnology IV- 34th International Conference on Advanced Ceramics and Composites, ICACC, Daytona Beach, FL, 24 January 2010 through 29 January 2010: Ceramic Engineering and Science Proceedings.

[19] Musil, J. (2012). Hard Nanocomposite Coatings: Thermal Stability and Toughness. *Surface and Coatings Technology*, (will be published).

[20] Mayrhofer, P. H., Mitterer, C., & Hultman, L. (2006). Microstructural design of hard coatings. *Progress in Materials Science*, 51, 1032-1114.

[21] Musil, J., Sklenka, J., & Čerstvý, R. (2012). Transparent Zr-Al-O oxide coatings with enhanced resistance to cracking. *Surface and Coatings Technology*, 206(8-9), 2105-2109.

[22] Musil, J., Šatava, V., & Baroch, P. (2010). High-rate reactive deposition of transparent SiO2 films containing low amount of Zr from molten magnetron target. *Thin Solid Films*, 519, 775-777.

[23] Musil, J. (2006). Physical and mechanical properties of hard nanocomposite films prepared by reactive magnetron sputtering, Chapter 10 in Nanostructured Coatings, J.T.M. DeHosson and A. Cavaleiro (Eds.) New York, Springer Science-Business Media, LCC, 407-463.

[24] Andrievski, R. A. (2005). Nanomaterials based on high-melting carbides, nitrides and borides. *Russian Chemical Reviews*, 74(12), 1061-1072.

[25] Patscheider, J. (2003). Nanocomposite hard coatings for wear protection. *MRS Bulletin*, 28(3), 180-183.

Chapter 7

Polymer Nanocomposite Hydrogels for Water Purification

Manja Kurecic and Majda Sfiligoj Smole

Additional information is available at the end of the chapter

http://dx.doi.org/10.5772/51055

1. Introduction

Contamination of water, due to the discharge of untreated or partially treated industrial wastewaters into the ecosystem, has become a common problem for many countries [1]. In various productions, such as textiles, leather, rubber, paper, plastic and other industries, the dyeing processes are among the most polluting industrial processes because they produce enormous amounts of coloured wastewaters [2-4]. In addition to their colour, some of these dyes may degrade to highly toxic products, potentially carcinogenic, mutagenic and allergenic for exposed organisms even at low concentrations (less than 1 ppm) [5]. They contaminate not only the environment but also traverse through the entire food chain, leading to biomagnifications [6-9]. The removals of such compounds particularly at low concentrations are a difficult problem.

Textile effluents are usually treated by physical and chemical processes such as sorption, oxidation, flocculation, etc. Colour removal by activated carbon, H_2O_2, sodium hyperchlorite and other chemical agents has been widely practiced in the textile industries [10]. Although activated carbon remains the most widely used adsorbent, its relatively high cost restricts its use sometimes. However, in addition to, adsorptive properties and availability are also key criteria when choosing an adsorbent for pollutant removal, thereby encouraging research into materials that are both efficient and cheap. Many non-conventional low-cost adsorbents, including natural materials, biosorbents, and waste materials from agriculture and industry have been proposed by several researchers [11-14]. Considering low cost, abundance, high sorption properties and potential ion-exchange, clay minerals are interesting materials for use as adsorbents, since they can be easily obtained and regenerated [7].

INTECH
open science | open minds

© 2012 Kurecic and Sfiligoj Smole; licensee InTech. This is a paper distributed under the terms of the Creative Commons Attribution License (http://creativecommons.org/licenses/by/3.0), which permits unrestricted use, distribution, and reproduction in any medium, provided the original work is properly cited.

2. Clay minerals

Clays are widely applied in many fields such as polymer nano-composites [15-18], catalysts [19,20], photochemical reaction fields [21], ceramics [22], paper filling and coating [23], sensors and biosensors [24], absorbents, etc. due to their high specific surface area, chemical and mechanical stabilities, and a variety of surface and structural properties [25].

The most-used clays are smectite group which refers to a family of non-metallic clays primarily composed of hydrated sodium calcium aluminium silicate, a group of monoclinic clay-like minerals with general formula of $(Ca,Na,H)(Al,Mg,Fe,Zn)_2(Si,Al)_4O_{10}(OH)_2\ nH_2O$.

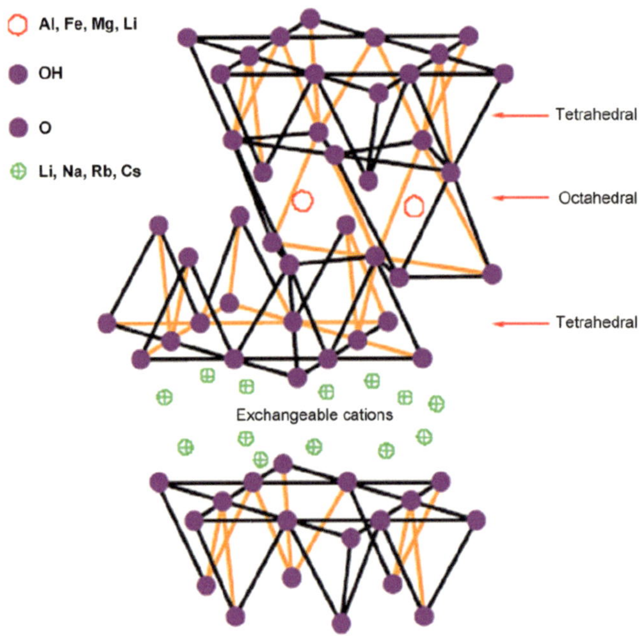

Figure 1. Structure of 2:1 layered silicate [15].

Smectite is a clay mineral having a 2:1 expanding crystal lattice (Figure 1). Its isomorphous substitution gives various types of smectite and causes a net permanent charge balanced by cations in such a manner that water may move between the sheets of the crystal lattice, giving a reversible cation exchange and very plastic properties.

Members of the smectite group include the dioctahedral minerals montmorillonite, beidellite, nontronite, bentonite, and the trioctahedral minerals hectorite (Li-rich), saponite (Mg-rich) and sauconite (Zn-rich). The basic structural unit is a layer consisting of two inward-pointing tetrahedral sheets with a central alumina octahedral sheet. The layers are continuous in the length and width directions, but the bond between layers are weak and have excellent cleavage, allowing water and other molecules to enter between the layers causing expansion in the third direction [15,16,25,26].

In the inner blocks, all corners of silica tetrahedra are connected to adjacent blocks, but in outer blocks some of the corners contain Si atoms bound to hydroxyls (Si-OH). These silanol groups at the external surface of the silicate, are usually accessible to organic species, and act as neutral adsorption sites. In addition to, some isomorphic substitutions occur in the tetrahedral sheet of the lattice of the mineral form leading to negatively charged adsorption sites which are occupied by exchangeable cations [25]. These characteristic make them powerful absorbents for organic molecules and organic cations. In order to improve the adsorption properties of clay absorbents also for organic anions, clay surface can be modified. There are many exchangeable cations on the clay surface therefore the cationic surfactants are generally used as modifiers. The characteristics of these so-called organoclays can be changed by variation of surfactant properties, such as alkyl chain length, etc. The surface properties of the clays modified by surfactants alter from organophobic to organophilic, which aids in improving clay adsorption capacities for organic compounds [27]. While crude clay minerals are effective for the adsorption of cations, organo-modified clays may adsorb negative and hydrophobic molecules [25]

2.1. Organically modified clay minerals

Surface modifications of clay minerals have received great attention because it allows the creation of new materials and new applications [28]. Organically modified clay minerals have become essential for development of polymer nanocomposites. Modified clays are also used in other applications such as adsorbents of organic pollutants in soil, water and air; rheological control agents; paints; cosmetics; refractory varnish; thixotropic fluids, etc. Several routes can be employed to modify clays and clay minerals [29]:

- adsorption,
- ion exchange with inorganic cations and organic cations,
- binding of inorganic and organic anions (mainly at the edges),
- grafting of organic compounds, reaction with acids,
- pillaring by different types of poly(hydroxo metal) cations,
- intraparticle and interparticle polymerization,
- dehydroxylation and calcination,
- delamination and reaggregation of smectites,
- lyophilisation,
- ultrasound, and
- plasma.

Ion exchange with alkylammonium ions is well-known and the preferential method to prepare organoclays. Generally, the papers describe the preparation of the organoclays in laboratory scale, with different experimental conditions, clays from several regions and suppliers, and several kinds of organic compounds [28].

The research of intercalation of organic molecules into the interlayer space of clay minerals started in the 1920s, after the introduction of X-ray diffraction in 1913 [28]. Geseking [30] found methylene blue to be very effective in replacing interlayer cations. These results indicated on the possibility of using ammonium ions of the NH_3R^+, $NH_2R_2^+$, NHR_3^+, and NR_4^+ types for better understanding of the mechanism of cation exchange in clay minerals. Different types of clay minerals were treated with the solution of hydrochlorides or hydroiodides of various amines. The clay minerals adsorbed the organic ions and increased the basal spacing more than those of the same clay minerals saturated with smaller cations such as calcium or hydrogen.

In 1944 MacEwan observed that when montmorillonite was treated with glycerol, a very sharp and intense first-order x-ray diffraction reflexion was obtained, corresponding to the basal spacing of 1.77 nm. The increase of basal spacing was due to the intercalation of glycerol into the interlayer space of the clay mineral. [31].

Studies of interactions between clay minerals and organic compounds have been presented, among others, in [32-34]. Clay–organic complexes of great industrial importance are the organoclays prepared from smectites and quaternary ammonium salts.

2.2. Organically modified clay minerals for dye adsorption

There are more than 100 000 types of dyes commercially available, with over 7 x 10^5 tons of dyestuff produced annually, which can be classified according to their structure as anionic and cationic. In aqueous solution, anionic dyes carry a net negative charge due to the presence of sulphonate (SO_3^-) groups, while cationic dyes carry a net positive charge due to the presence of protonated amine or sulphur containing groups. [35]

Reactive dyes are extensively used in the textile industry because of their wide variety of colour shades, brilliant colours, and minimal energy consumption [36]. Therefore, considerable amount of research on wastewater treatment has focused on the elimination of these dyes, essentially for three reasons: firstly, reactive dyes represent 20-30% of the total dye market; secondly, large fraction of reactive dyes (10-50%) are wasted during the dyeing process (up to 0,6 – 0,8 g dye/dm^3 can be detected in dyestuff effluent); thirdly, conventional wastewater treatment methods, which rely on adsorption and aerobic biodegradation, were found to be inefficient for complete elimination of many reactive dyes. [35]

Comprehensive research activities in the field of dye adsorption onto organically modified clays are directed to different organic modifiers in order to improve and broaden the applications of clay adsorbents [36-38]. The hexadecyltrimethylammonium (HDTMA) bentonite was synthesized by placing alkylammonium cation onto bentonite [38]. Adsorption of several textile dyes such as Everdirect Supra Yellow PG, Everdirect Supra Orange 26 CG, Everdirect Supra Rubine BL, Everdirect Supra Blue 4 BL and Everdirect Supra Red BWS on Na-bentonite and HDTMA-bentonite was investigated. While the Na-bentonite had no affinity for the dyes, the HDTMA-bentonite showed significant adsorption from aqueous solution. Wang et al. has reported that adsorption capacity for Congo Red of modified montmorillonite sharply increases from 31.1 to 299 mg of adsorbat per g of adsorbent with increasing the numbers of carbon atom of surfactant from 8 to 16 and then decreases with further increasing the number of carbon atom of surfactant from 16 to 18 [27]. He explained that alkyl chains of surfactant intercalate into the montmorillonite galleries and broad the galleries, which in turn result in an increase in the adsorption. In recent years many reports showed that the surfactant - modified clays displayed higher adsorption capacity than the original clay. Modification of bentonite clay with cetyltrimethylammonium bromide enhanced the rate at which direct dye Benzopurpin 4B is absorbed on Na-bentonite [39]. Zohra et al. explained that increase in adsorption capacity of modified clay is due to the alkyl chains in the interlamellar spaces functioning as organic solvent in portioning and electrostatic attraction with positively organoclay surface and anionic dye molecules. There have also been trials to modify montmorillonite clay with novel Gemini surfactants under microwave irradiation [40]. They have studied the adsorption behavior of methyl orange dye on MMT and three kinds of organo- MMTs modified using Gemini surfactants. All organo-MMTs displayed more excellent adsorption capacities than MMT, and as the amount or the chain length of Gemini surfactants increased, the adsorption capacity of the organo-MMTs was improved. XRD analyses were used in order to confirm the enlargement of interlayer spacing in organo-MMTs which results in higher surface area leading to the stronger adsorption capacity. In addition to, from SEM analysis it was observed that the structure of organo-MMTs was looser, which can facilitate the adsorption of the dyes on organoclays. Based on TGA results, the surface energy of organo-MMT was reduced from hydrophilic to hydrophobic, which is helpful for absorbing the organic methyl orange. With the increase of the amount or the chain length of the Gemini surfactant, the hydrophobicity of the modified MMT was higher, and it facilitated the adsorption of organic contaminants. Özcan et al. have also investigated the effect of pH on the adsorption of Reactive Blue 19 from aqueous solution onto surfactant-modified bentonite [4]. Dodecyltrimethylammonite (DTMA) bromide was used as a cationic surfactant. pH was in the range between 1-9 and it was found that the adsorption decreased with an increase of pH. Batch studies suggest that the high adsorption capacity of DTMA-bentonite in acidic solutions (pH around 1.5) is due to the strong electrostatic interaction between its adsorption site and dye anion.

In several research articles it is indicated that clay derivatives are potentially very promising sorbents for environmental and purification purposes. Although the modification of clays with surfactants increases their cost significantly, the resultant increase in adsorption capacity may still made surfactant-modified clays cost effective. The nano-clay, montmorillonite, and some modified nano-clays were used as sorbents for non-ionic, anionic and cationic dyes [41]. From the sorption differences among the different dye and clay structures, both chemical and morphological, the sorption forces that played important roles were identified. Nano-clays frequently have a sorption capacity of more than 600 mg sorbate per gram of sorbent at a liquor-to-sorbent ration of 100:1. Furthermore, a sorption of 90% at the initial dye concentration of 6g/L, or 60% based on the weight of sorbent, was observed. This indicates an extremely high dye affinity. This study showed that by modification of the nano-clay MMT, it can easily become an excellent sorbent for anionic, cationic and non-ionic dyes.

Clay minerals are in most cases used as dispersed adsorbents and as such aggravate the removal of adsorbents from clean purified water. Recently, there have been some activities to incorporate clay particles into nanocomposite hydrogels for application in wastewater technologies [42-45]. Incorporation of clay minerals in hydrogel matrix allows better manipulation with adsorbing material since clay minerals are fixed in the matrix.

3. Clay/polymer nanocomposite hydrogels

Hydrogels are 3D dense cross linked polymer network structure, containing hydrophilic and hydrophobic parts in a defined proportion. When placed in aqueous medium, they intensively swell. By swelling they increase their initial volume for several times without either dissolving or considerably changing their shape, because hydrophilic chains contact one to the other by cross-linking [42,46,47]. The response of hydrogel is dependent on the presence of hydrophilic functional groups such as –OH, -COOH. These groups make the hydrogel hydrophilic and due to the capillary action and the difference in the osmotic pressure, water diffuses into the hydrogel.

Polymerization methods, the presence of functional groups and the nature of cross-linking agents are important parameters that control the swelling ability of hydrogel [43].

Owing to their advantageous properties, such as swellability in water, hydrophilicity, biocompatibility and lack of toxicity, hydrogels have been utilized in a wide range of hygienic, agricultural, medical and pharmaceutical applications and in such applications, water absorbency and water retention properties are essential [46,48]. The most well-established hydrogel applications are superasorbing hydrogels in diapers and hydrogels for contact-lenses, just to mention few [49-53].

Recently hydrogels have gain particular interest in wastewater treatment due to their high adsorption capacities, especially regeneration abilities and reuse for continuous processes [54]. But pure hydrogels often have some limitations such as low mechanical stability and gel strength. In the initial phase of nanocomposite hydrogels development, various clay minerals were widely added to polymer hydrogel matrix in order to improve weak mechanical stability of hydrogels. Sodium montmorillonite (NaMMT) or attapulgite were used as reinforcing filler in the preparation of hydrogels to improve mechanical properties or swelling ability (55-58). Liang et al. [59] used organically modified montmorillonite to prepare hydrogels that exhibited higher swelling degree and enhanced thermal response compared to conventional poly(N-isopropylacrylamide) (PNIPAM) hydrogels. Hydrogels were also prepared with ionic monomers and montmorillonite [60]. However, the transparency, swelling degree and mechanical property did not improve simultaneously, in particular at relatively high NaMMT loading, because of the poor dispersion of clay mineral particles and structural inhomogeneity of the hydrogel network caused by the crosslinker N,N′-methylene-bis-acrylamide (BIS) [61]. To overcome this problem, a special type of an inorganic-organic thermo-responsive PNIPAM nanocomposite hydrogel was developed by Haraguchi, containing laponite XLG without any chemical crosslinker. The exfoliated laponite particles acted as multifunctional crosslinker, and the polymer chains were anchored to the particles and entangled to form a network [49,50,53]. According to this mechanism several researches prepared various nanocomposite hydrogels by using laponite as multifunctional crosslinker [56,61]. The resulting hydrogels exhibited not only excellent optical and ultrahigh mechanical properties but also large swelling ratios and rapid shrinking capability [49,50,53,61,62].

Despite of undesired low mechanical properties, which can be improved by introduction of clay minerals in the hydrogel matrix, hydrogels have many predominant properties including low interfacial tension and a variety of functional groups which can trap ionic dyes from wastewater and provide high adsorption capacities [63]. Introduction of clay materials into hydrogel combines improvement of elasticity and permeability of the hydrogels with high ability of the clays to adsorb different substances. Application of low-priced and biodegradable adsorbents is a good tool to minimize the environmental impact caused by dye manufacturing and textile effluents. Consequently, research concerning development of hydrogels with clay particles for adsorbing dyes and metal ions is exponentially increasing [43,42,54,64].

Shirsath et al. synthesized polymer nanocomposite hydrogels using metal hybrid polymer along with clay. Ultrasonic irradiation was used to initiate the emulsion polymerization to form hydrogel through the generation of free radicals. The high shear gradients generated by the acoustic cavitations process help to control the molecular weight of hydrogels formed in aqueous solutions. Ultrasound was found to be an effective method for polymerization of monomers and for the production of hydrogel in the absence of a chemical initiator [43].

3.1. Clay/polymer nanocomposite synthesis

Hydrogels are usually crosslinked during polymerization via condensation polymerization or free radical polymerization (thermal polymerization, radiation polymerization, photo-polymerization or plasma polymerization) [65-67]. Photo-polymerization, in addition to its environmental-friendly aspects, offers a number of advantages, such as ambient temperature operations, location and time-control of the polymerization process and minimal heat production, in comparison with other techniques [68]. Photo-polymerization can be induced by ultraviolet (100-400 nm), visible (400-700 nm) or infrared (780-20000 nm) radiation. Light quanta are absorbed by molecules via electronic excitation [66]. During photo-polymerization process, photo-initiators are generally used having high absorption capacities at specific wavelengths of light thus enabling them to produce radically initiated species [69].

The preparation of a clay mineral-polymer composite with light for initiation of polymerization requires suitable monomers and a suitable photoiniciating system. In the research PNIPAM/clay nanocomposite hydrogels were synthesized using aqueous dispersion of organo-clay (O-MMT) particles, modified by distearyldimethyl ammonium chloride (Nanofil 8, Süd Chemie, Germany) in varied concentrations (0, 0.25, 0.5, 0.75, 1 wt% regarding the monomer content). Aqueous dispersions of different concentration of O-MMT were kept under constant stirring for 2 hours at room temperature after the addition of 1% NIPAM and 1 wt% (regarding the monomer content) BIS. After this period, 1 wt% (regarding the monomer content) Irgacure 2959 photoiniciator was added and the dispersion was kept under the same conditions for additional 1 hour. Prepared dispersions were pored in glass Petri dishes, bubbled with nitrogen for 5 minutes and covered. Petri dishes were placed on the sample holder in the middle of a UV chamber (Luzchem). Polymerization was carried out in UV chamber using 6 UVA lamps (centred at 350 nm) placed on top of the chamber with the distance to the sample 15 cm. Time of polymerization was 2 hours. After polymerization the hydrogels were washed with deionised water for 4 days (daily exchange of water). After washing, the hydrogels were dried at 40°C until a constant mass was reached. Preparation scheme is presented in Figure 2.

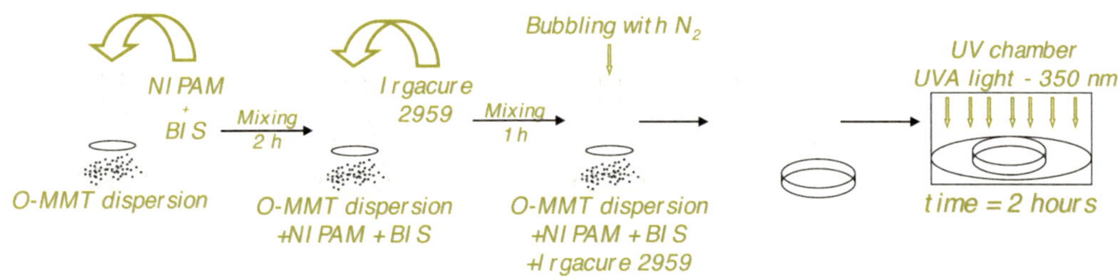

Figure 2. Nanocomposite hydrogel preparation process.

3.2. Clay/polymer nanocomposite hydrogel structure

Wide angle (WAXS) or small angle X-ray scattering (SAXS) are generally used methods for characterization of nanocomposite structure. These techniques enable determination of the spaces between structural layers of the silicate utilizing Bragg's law: $\sin\theta = n\lambda/2d$, where λ corresponds to the wave length of the X-ray radiation used in the diffraction experiment, d the spacing between diffraction lattice planes and θ is the measured diffraction angle [70]. By monitoring the position, shape and intensity of the basal reflection from the distributed silicate layers, the nanocomposite structure may be identified [71].

Depending on the nature of the components used (clay mineral, organic cation and polymer matrix) and the method of preparation, three main types of composites may be obtained when a clay mineral is combined with polymer. Then the polymer is unable to be intercalated, a phase-separated composite is obtained, whose property stay in the same range as those of traditional micro composites. Beyond this classical family composite, two further types of nanocomposites can be obtained. An intercalated structure in which a single (and sometimes more than one) extended polymer chain is intercalated between the silicate layers results in well-ordered multilayers morphology built up of alternating polymeric and inorganic layers. When the silicate layers are completely and uniformly dispersed in a continuous polymer matrix, an exfoliated or delaminated structure is obtained [70] Intercalated structures can be identified using SAXS or WAXS analyses [46,72].

To analyse the effect of monomer, crosslinker and photoinitiator content on composite structure formation we have prepared dispersion of MMT particles, monomer, crosslinker and photoiniciator to study the influence of reagents on MMT particles intercalation by measuring distances between silicate galleries of clay particles using small angle X-ray scattering (SAXS). Figure 3 shows two x-ray diffraction curves of O-MMT particles which were dispersed in monomer, crosslinker and photoinitiator water dispersion.

On the diffraction curve of O-MMT particles dispersed in water, the characteristic maximum for O-MMT particles (q= 2,8 nm^{-1}) is observed. According to the Braggs` law it corresponds to the distance between silicate layers d_{001} = 2.03 nm. By the addition of monomer, crosslinker and photoinitiator into the O-MMT dispersion, the characteristic discrete maximum on the diffraction curve is shifted to a lover angle value (q=1,65nm^{-1}) which corresponds to the distance between silicate layers d_{001} = 3,81 nm^{-1}. According to the pronounced change in silicate layers distances, we are concluding that monomer molecules have intercalated between silicate layers.

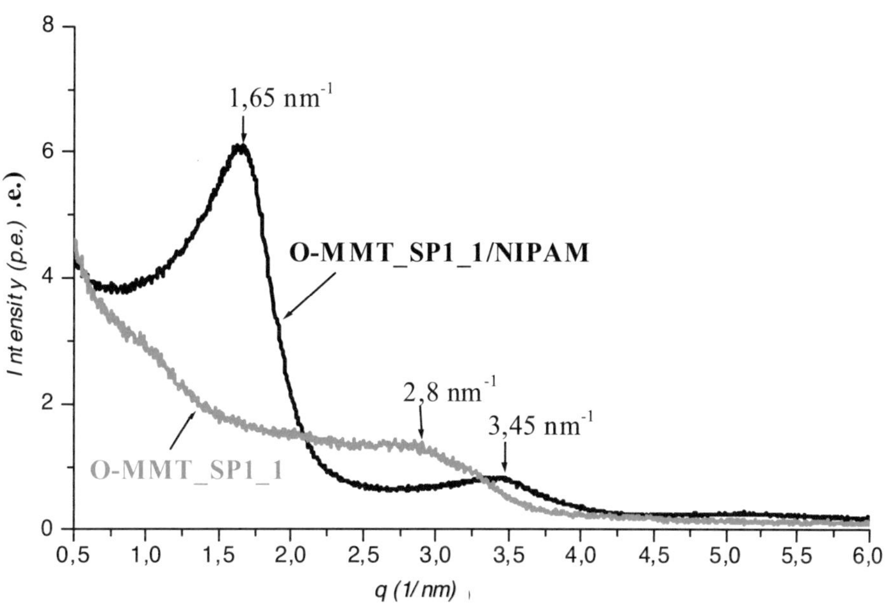

Figure 3. SAXS-pattern of O-MMT particles and O-MMT particles dispersed in monomer, crosslinker and photoinitiator solution.

Thereby silicate layers are pushed apart which increases the distance between them, however the repetitive silicate multi layer structure is still preserved, allowing the interlayer spacing to be determined. By addition of monomer, crosslinker and photoinitiator into O-MMT aqueous dispersion, we obtained O-MMT dispersion with intercalated structure of O-MMT particles as shown in Figure 4.

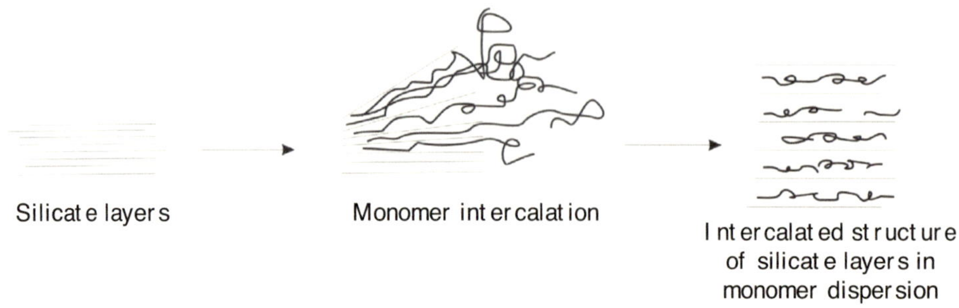

Figure 4. Intercalation of monomer, crosslinker and photoinitiator molecules between clay minerals silicate layers.

Since nanocomposite material is formed when the complete exfoliation of silicate platelets is possible, in-situ polymerized hydrogels were also analyzed using SAXS. Figure 5 presents x-ray spectra of composite hydrogels with different concentrations of O-MMT particles (0,25; 0,5; 0,75; 1 wt%).

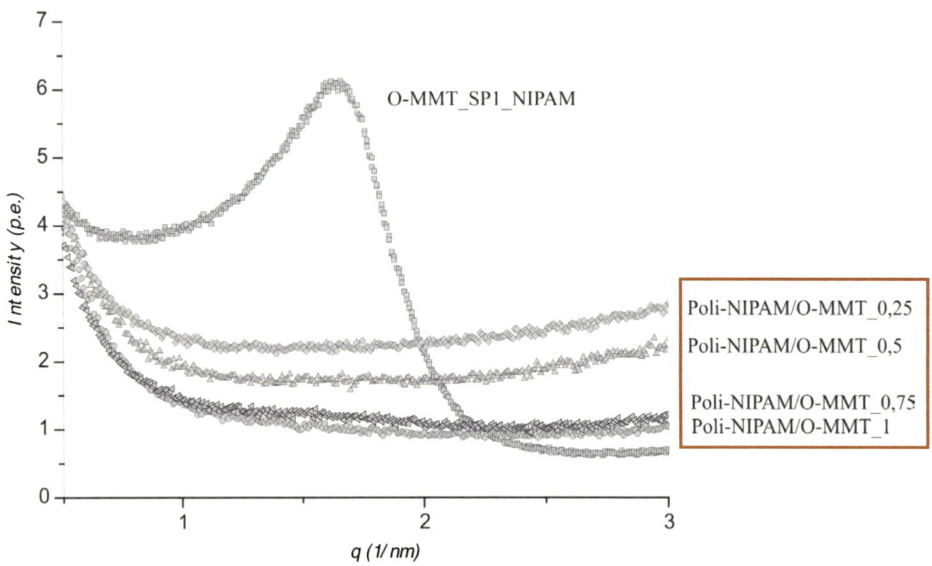

Figure 5. SAXS pattern of O-MMT particles dispersed in monomer, crosslinker and photoinitiator solution (O-MMT_SP1_NIPAM) and nanocomposite hydrogels (Poli-NIPAM/O-MMT) with different clay content.

The discrete maximum at q=1.65 nm^{-1} characteristic for O-MMT particles dispersed in monomer solution, disappears on small angle scattering curves of nanocomposite hydrogels. This phenomenon indicates that monomer molecules between platelets galleries polymerize and crosslink due to UV irradiation. Polymer formation causes movement of clay platelets apart and thereby the exfoliated structure of polymerized O-MMT/NIPAM nanocomposite hydrogel is formed.

http://dx.doi.org/10.5772/51055

In contrast to the intercalated structure, the extensive layer separation associated with exfoliated structures disrupts the coherent layer stacking and results in a featureless diffraction patters. Thus, for exfoliated structures no more diffraction peaks are observed in X-ray diffractograms either because of a much too large spacing between the layers, (i.e. exceeding 8 nm in the case of ordered exfoliated structure) or because the nanocomposites did not present ordering [70,71,73].

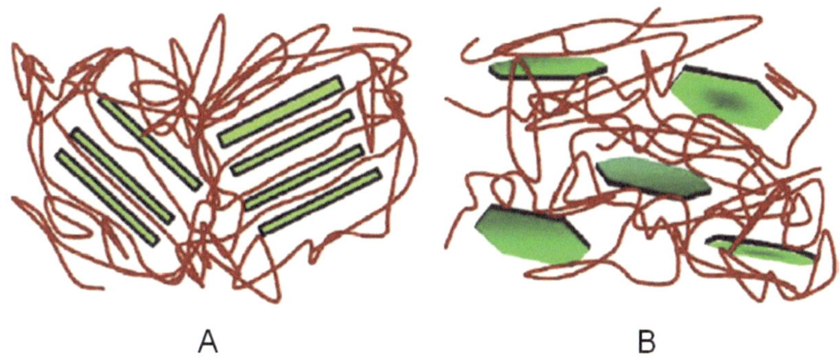

Figure 6. Scheme of intercalated and exfoliated nanocomposite structure [71].

3.3. Clay/polymer nanocomposite hydrogel swelling and gel fraction

The weight ratio of the dried hydrogels in rinsed and unrinsed conditions can be assumed as a measure of crosslinking degree or gel fraction. Therefore the gel fraction of sample can be calculated as follows [74]:

$$\text{Gel fraction}\ (\%) = \frac{W_f - W_c}{W_i - W_c}\quad 100$$

Where W_f and W_i are the weight of the dried hydrogel after and before rinsing, respectively and W_c is the weight of organoclay incorporated into the sample.

To perform gel fraction measurement, pre-weighed hydrogel sample was dried under vacuum at room temperature until no change in mass was observed.

A typical dependency of the gel fraction on the clay concentration in hydrogels is given in Figure 7.

Gel fraction of samples is increased by increasing the amount of clay. The relationship is almost linear. The gel fraction data reveal that presence of clay within the three dimensional networks of hydrogels causes an increase in crosslinking, thus creates more entangled structure. By adding O-MMT to the hydrogel, strong interactions are developed between functional groups of organoclay and polymer chains.

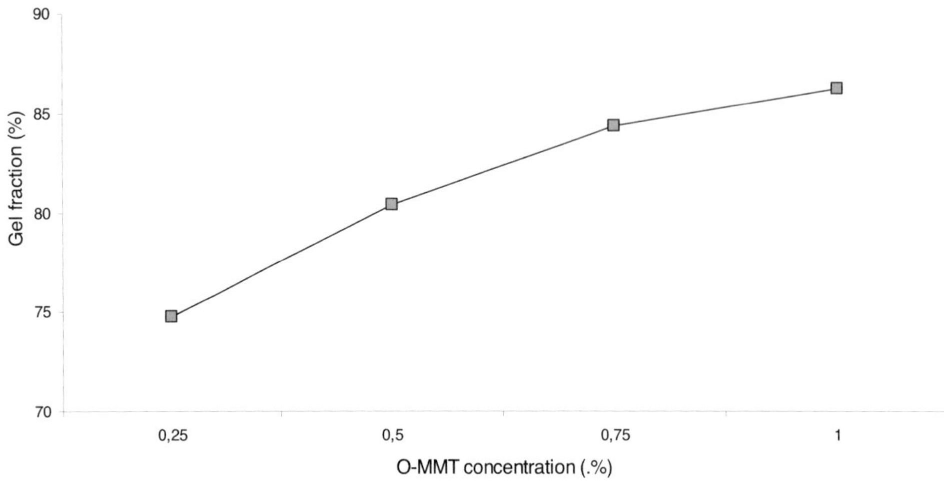

Figure 7. Gel fraction of nanocomposite hydrogels regarding the O-MMT clay concentration.

When pre-weighed samples of nanocomposite hydrogels are kept in water, the compact (dry) network structure of the polymer matrix relaxed and swells due to the diffusion of water molecules into the matrix until equilibrium is reached. At this stage, pressure inside the hydrogel matrix increases due to the presence of large amount of water molecules. Cross-linked structure prevents the dissolution of hydrogels [43].

For determination of an equilibrium swelling degree (EDS), we used pre-weighted hydrogel samples and immersed them into deionised water. Samples were removed from water every hour, wiped with filter paper in order to remove surface water, weighted and placed back into the water for further swelling. The equilibrium was reached when no mass difference was determined. EDS was calculated using the equation [46]:

$$EDS(\%)=\frac{W_s-W_d}{W_d}\quad 100 \tag{1}$$

where *Ws* and *Wd* are the masses of the gel in swollen and dried states, respectively.

When hydrogel is exposed to water, water molecules diffuse into hydrogel structure and consequently hydrogel swells. Hydrogel ability to swell or uptake water is one of its key characteristics.

Figure 8 demonstrates the equilibrium degree of swelling (EDS) of NIPAM hydrogel and NIPAM/clay nanocomposite hydrogels as a function of the amount of clay. Decreasing trend of equilibrium swelling degree by increasing the quantity of organoclay is observed.

By comparing the equilibrium swelling degree and gel fraction values (Figure 7) a relationship between these properties is observed, i.e. more gel fraction leads to less swelling. Denser hydrogel structure which is formed by increasing clay particles concentration affects the water uptake and decreases the swelling degree. Water uptake represents the migration of water molecules into preformed gapes between polymer chains [75]. Denser hydrogel structure diminish the accessibility of water molecules to hydrophilic parts of polymer molecules, therefore less water can penetrate into the hydrogel structure.

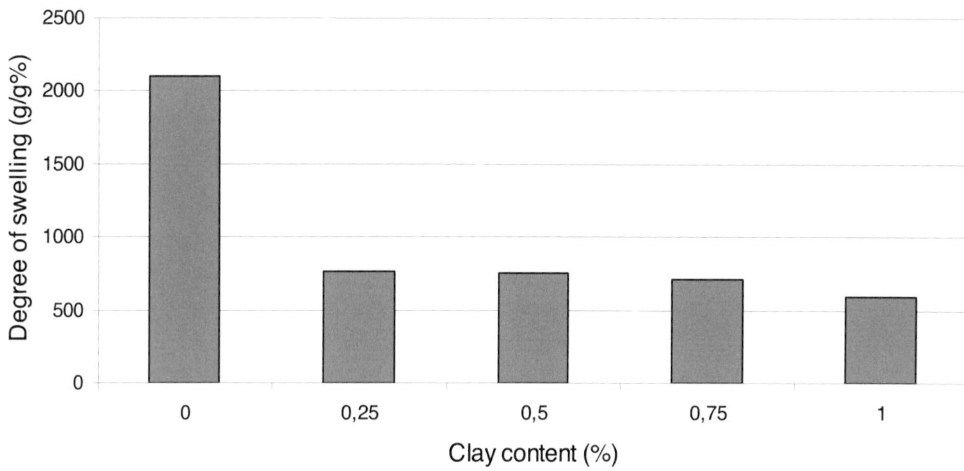

Figure 8. Degree of swelling for hydrogel and nanocomposite hydrogels with different clay content.

3.4. Clay/polymer nanocomposite hydrogel morphology

Figure 9 shows SEM micrographs of a cross-section and a pore surface of nanocomposite hydrogels with different clay content. Samples were lyophilized after the equilibrium swelling has been reached at room temperature in order to preserve natural nanocomposite hydrogel structure in swollen state. Nanocomposite hydrogel cross-section (Figure 9 A/I, B/I, C/I, D/I) shows very porous structure with several pores and wide pore size distribution. The pore structure has a sponge-like shape with spherical opens and interconnected cells. This porous microstructure is essential for a large active surface of hydrogel and assures the capillary effect of water uptake. Comparing the hydrogels pore structure regarding the clay content we observed a drastic change in pore size for nanocomposite hydrogel with 1 wt% O-MMT particles in the hydrogel matrix. At the concentration of 0.25 % O-MMT particles the pore size is approximately 200 μm, while the pore size for nanocomposite hydrogel with 1% O-MMT particles is 100 μm. This pore size reduction confirms that silicate platelets represent additional crosslinking points in nanocomposite hydrogel structure.

Figures 9 A/II, B/II; C/II and D/II show pore surfaces of nano-hydrogels. On the surface incorporated clay particles are observed.

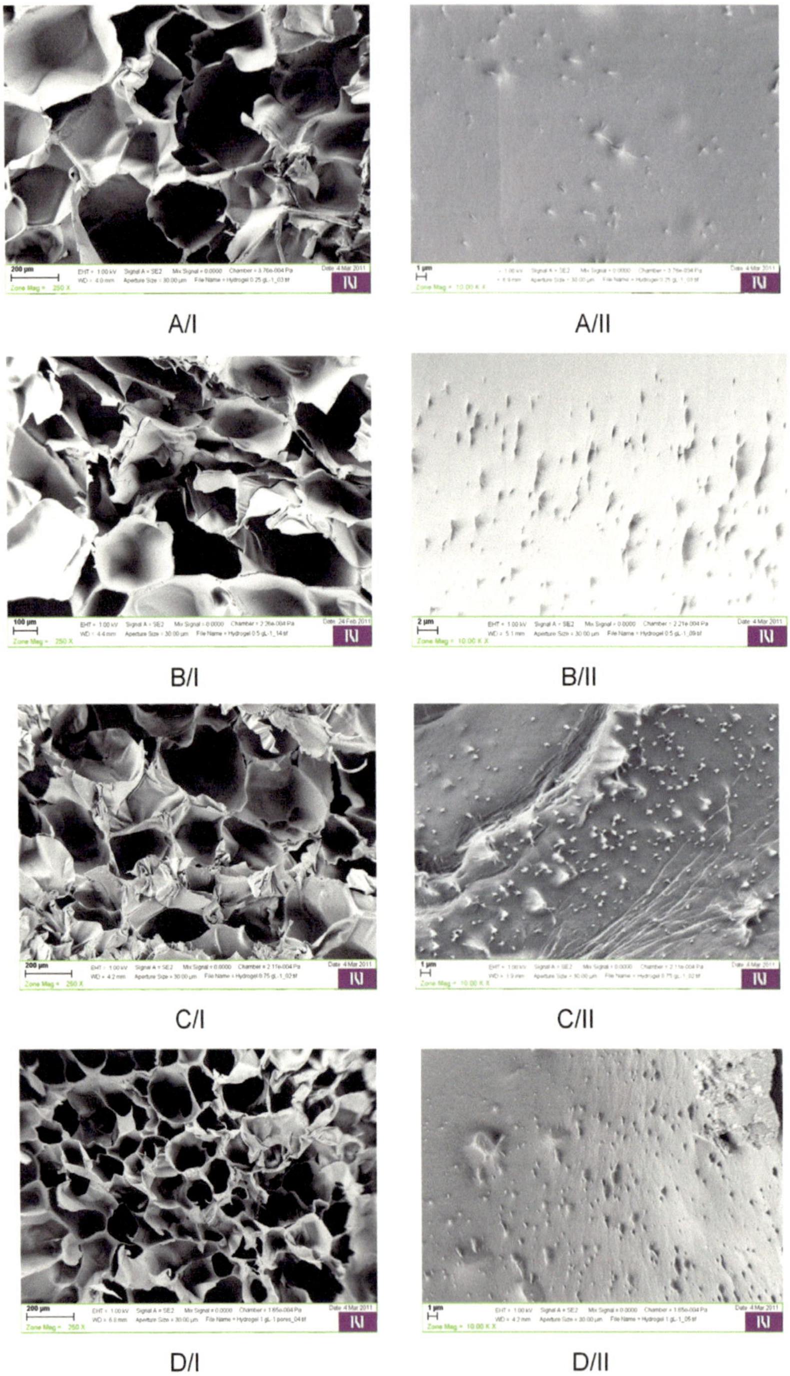

Figure 9. SEM of freeze-dried nanocomposite hydrogels with different clay content (A: hydrogel with 0.25% O-MMT; B: hydrogel with 0.5% O-MMT; C: hydrogel with 0.75% O-MMT; D: hydrogel with 1% O-MMT).

4. Adsorption properties of clay/polymer nanocomposite hydrogels

Adsorption studies are the key for evaluating the effectiveness of an adsorbent. Montmorillonite particles used for preparation of nanocomposite hydrogels are organically modified and therefore contain positively charged nitrogen atoms that attract opposite negatively charged anions with electrostatic attraction. Binding efficiency was studied with determination of adsorption degree. To study the effects of different experimental parameters, such as, pH, dye concentration, clay structure on the adsorption of anionic dye Acid Orange 33 onto clay/polymer nanocomposite hydrogel, UV spectroscopy was used. A Carry 50 spectrophotometer (Varian) was used for analyses. The dye concentration was determined at a wavelength corresponding to the maximum absorbance. The adsorption degree was calculated using following equation [43]:

$$\text{Adsorption degree } (\%) = \frac{C_0 - C_e}{C_0} \quad 100$$

Where C_0 and C_e are the initial and equilibrium concentrations of Acid Orange 33 dye (mg/L), respectively.

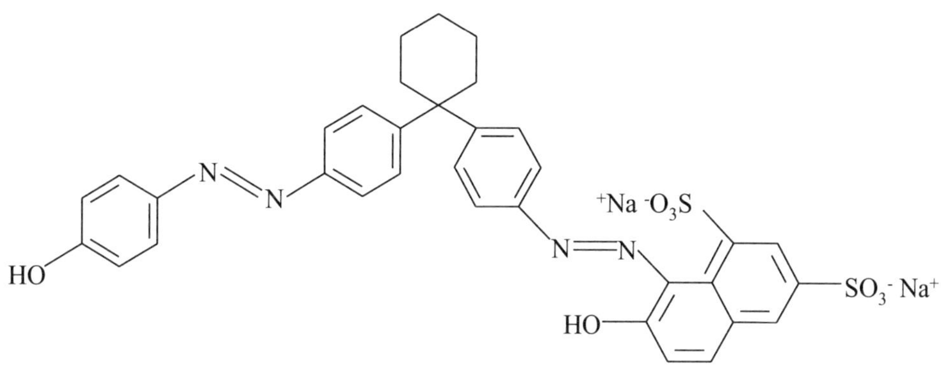

Figure 10. Chemical structure of Acid Orange 33.

Acid Orange 33 is an anionic dye used for dyeing wool, silk and PA, since it contains negatively charged SO_3^- groups in the structure (Figure 10).

4.1. Adsorption degree: pH dependence

pH of the solution is one of the main parameters that control the adsorption process. The effect of pH solution depends on the ions present in the reaction mixture and electrostatic interactions at the adsorption surface [43]. To determine the effect of different pH on Acid Orange 33 dye removal, the adsorption was carried out at different pH values of dye solution (pH= 3-9). pH was adjusted using acid/base buffer solutions. Figure 11 presents the effect of pH on the dye adsorption at an initial dye concentration of 100 mg/L.

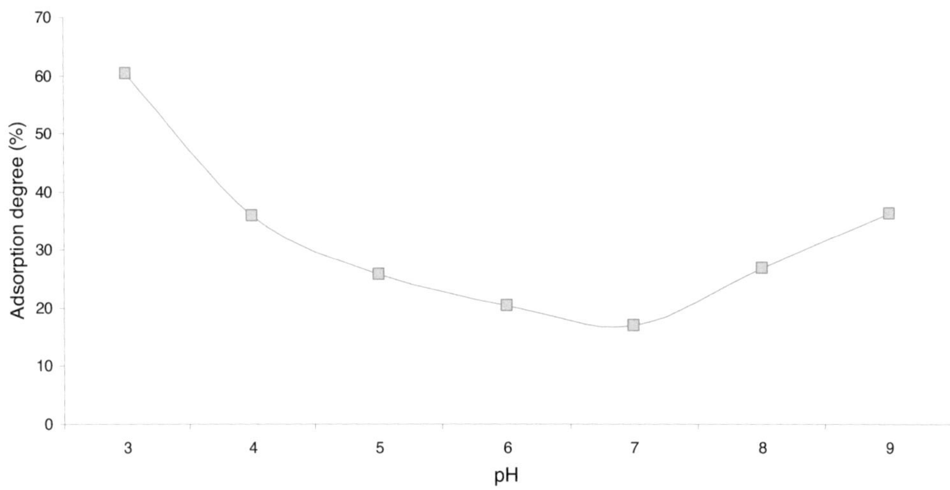

Figure 11. Adsorption degree vs. pH.

In Figure 11 the highest dye adsorption degree is observed at pH = 3 (around 60%). We assume that this high dye adsorption onto clay/polymer nanocomposite hydrogel at low pH values is due to the neutralization of the negative charge of $-SO_3^-$ anion, which influences the protonation and thereby increases the electrostatic attraction between the negatively charged $-SO_3^-$ anion and the positively charged adsorption site. The reason for high adsorption capacity at low pH is due to the strong electrostatic interaction between the cationic surfactant head groups of clay minerals incorporated in the hydrogel matrix and dye anions [4].

By increasing the pH to higher, neutral values (pH=4-7) we observe a decrease in adsorption degree. This is due to the decease of positive charge on the clay surface and the number of negatively charged sites increases. The negatively charged surface sites on clay do not favour the adsorption of anionic dye due to the electrostatic repulsion [76].

In alkaline pH region (pH=8-9) we observe another slight increase in adsorption degree, which is lower regarding the adsorption at pH = 3. Barkaralingam et al. reported that in alkaline medium a competition between OH^- ions and dye anions will be expected [76], however a significant colour adsorption is still observed as the pH of dye solution increases from 7 to 9. He suggested that a second mechanism is operating at these conditions. The mechanism of colour removal at higher pH values can be explained by formation of covalent bonds between the external surface –OH groups of Si and Al atoms of adsorbent and negatively charged dye molecules [10].

The maximum adsorption degree of Acid Orange 33 is at pH =3, which was therefore selected for all further adsorption experiments.

http://dx.doi.org/10.5772/51055

4.2. Adsorption degree - adsorption time dependence

The effect of adsorption time on the adsorption capacities of Acid Orange 33 is shown in Figure 12. The adsorption capacity increased rapidly within the first 60 minutes, after that it increased slowly until the adsorption equilibrium was reached. Under experimental conditions (1 wt% O-MMT, 0,1g/L Acid Orange 33 and pH3), the equilibrium time for the adsorption of Acid Orange 33 onto clay/polymer nanocomposite is 360 minutes. The rapid adsorption observed during the first 60 minutes is probably due to the abundant availability of active sites on the clay surface, and with the gradual occupancy of these sites, the adsorption becomes less efficient [77].

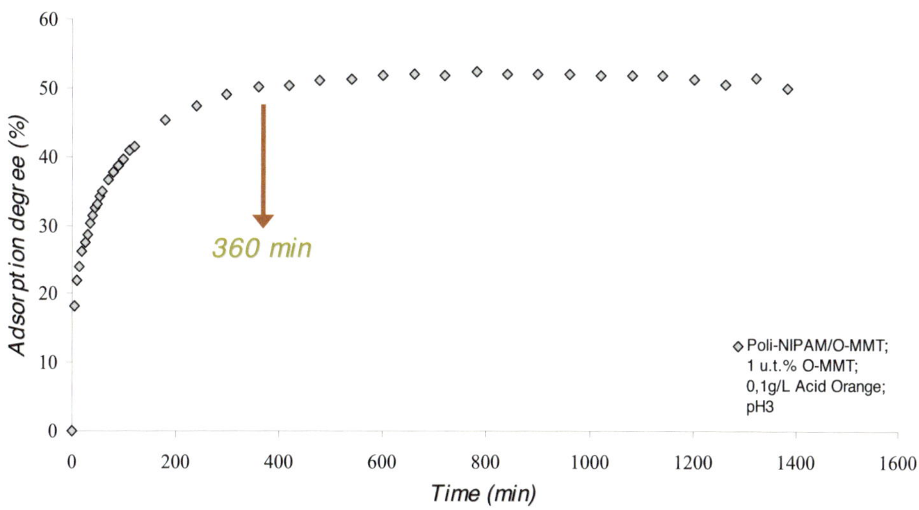

Figure 12. Degree of swelling vs. time.

4.3. Adsorption degree - initial dye concentration dependence

Generally, the removal of dye is dependent on the initial concentration of the dye in the solution. Results shown in Figure 13 indicate that the equilibrium dye uptake by nanocomposite hydrogel increases with increasing initial dye concentration. This is because at higher initial dye concentration, the availability of the number of dye molecules is higher, which can easily penetrate through hydrogel matrix. However, the removal efficiency is increasing only slightly after the initial dye concentration 0.05 g/L. This can be due to the saturation of hydrogel sites [43] or due to the fact that the formation of dye molecules agglomerates makes it almost impossible for them to diffuse deeper into the nanocomposite hydrogel [78].

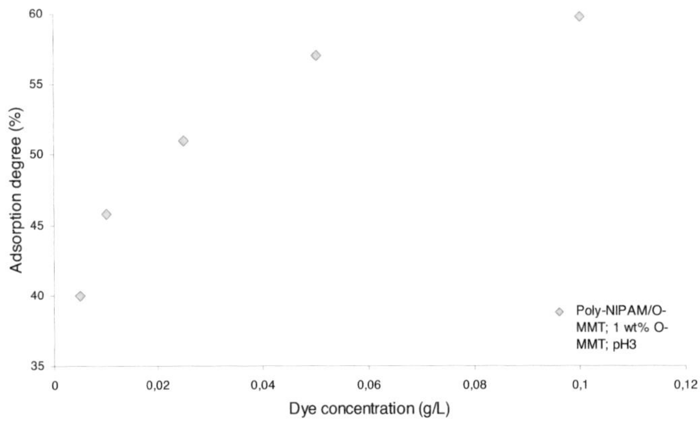

Figure 13. Adsorption degree vs. initial dye concentration.

4.4. Adsorption degree - O-MMT clay particles concentration dependence

The effect of the amount of active sites was studied by using hydrogels with different concentration (0.25 – 5%) of O-MMT particles incorporated in hydrogel matrix. The adsorption degree was measured in a solution containing 100 mg/L Acid Orange 33 dye at pH3. Figure 14 shows that the removal of the dye from the solution increases with an increase in the quantity of O-MMT particles incorporated into the hydrogel matrix. This indicates that the presence of higher quantity of O-MMT particles provides a larger number of active sites, which are positively charged and are capable to absorb more Acid Orange 33 dye molecules due to electrostatic forces. From the results we can conclude that the adsorption degree is significantly increasing from 30 to 59,9% when the concentration of nanoparticles increases (from 0.25 to 1% of O-MMT), but with additional increase of O-MMT particles in hydrogel matrix there is no change in the adsorption degree (the adsorption degree is 60,3% for the hydrogel containing 5% of O-MMT). We assume that at higher clay concentrations agglomerates are formed inside the hydrogel matrix, and therefore the expected additional active sites are not formed [78].

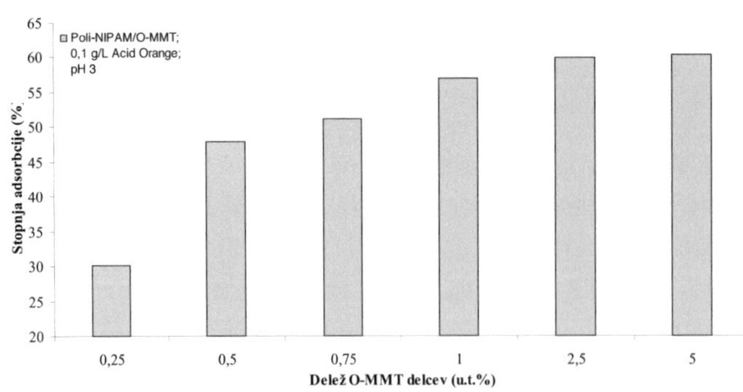

Figure 14. Adsorption degree vs. clay concentration.

5. Conclusions

Hydrogel nanocomposites have been prepared and their potential to be used as adsorbent materials for the removal of dyes which is a serious problem, especially in the textile industry was studied.

It was shown that incorporation of O-MMT particles into the hydrogel matrix induced adsorption capability for acid dye Acid Orange 33. The results obtained from absorption study show that:

The adsorption capacity is maximal at the pH value 3 of the dye solution and is decreasing with increasing pH.

The equilibrium time for the adsorption of Acid orange 33 onto clay/polymer nanocomposite is 360 minutes.

Equilibrium dye uptake by nanocomposite hydrogel increases with increasing initial dye concentration.

As the content of O-MMT particles in hydrogel matrix increases, adsorption capacity is increasing. At O-MMT content higher than 1wt% the adsorption capacity remains unchanged.

Author details

Manja Kurecic* and Majda Sfiligoj Smole

*Address all correspondence to: manja.kurecic@uni-mb.si

University of Maribor, Faculty of Mechanical Engineering, Department for Textile Materials and Design, Slovenia

References

[1] Akar, S. T., & Uysal, R. (2010). Untreated clay with high adsorption capacity for effective removal of C.I. Acid Red 88 from aqeous solutions: Batch and dynamic flow mode studies. *Chemical Engineering Journal*, 162, 591-598.

[2] Ojstrsek, A., & Fakin, D. (2011). Colour and TOC reduction using biofilter packed with natural zeolite for the treatment of textile wastewater. *Desalination and Water Treatment*, 33, 147-155.

[3] Bennani, Karim. A., Mounir, B., Hackar, M., Bakasse, M., & Yaacoubi, A. (2009). Removal of Basic Red 46 dye from aqueous solution by adsorption onto Moroccan clay. *Journal of Hazardous Materials*, 168, 304-309.

[4] Özcan, A., Ömeroğlu, Ç., Erdoğan, Y., & Özcan, A. S. (2007). Modification of bentonite with a cationic surfactant: An adsorption study of textile dye vReactive Blue 19. *Journal of Hazardous Materals*, 140, 173-179.

[5] Rafatullah, M., Sulaiman, O., Hashim, R., & Ahmad, A. (2010). Adsorption of methylene blue on low-cost adsorbents: A review. *Journal of Hazardous Materials*, 177, 70-80.

[6] Gürses, A., Doğar, Ç., Yalçin, M., Açikyildiz, , Bayrak, R., & Karaca, S. (2006). The adsorption kinetics of the cationic dye, methylene blue, onto clay. *Journal of Hazardous Materials*, B131, 217-228.

[7] Gil, A., Assis, F. C. C., Albeniz, S., & Korili, S. A. (2011). Removal of dyes from wastewaters by adsorption on pillared clays. *Chemical Engineering Journal*, 168, 1032-1040.

[8] Lin, S., Juang, R., & Wang, Y. (2004). Adsorption of acid dye from water onto pristine and acid-activated clays in fixed beds. *Journal of Hazardous Materials*, B113, 195-200.

[9] Li, S. (2010). Removal of crystal violet from aqueous solution by sorption into semi-interpenetratined networks hydrogels constituted of poly(acrylic acid-acrylamide-methacrylate) and amylose. *Boiresource Technology*, 101, 2197-2202.

[10] Baskaralingam, P., Pulikesi, M., Ramamurthi, V., & Sivanesan, S. (2006). Equilibrium studies for the adsorption of Acid dye onto modified hectorite. *Journal of Hazardous Materials*, B136, 989-992.

[11] Namasivayam, C., & Arasi, D. J. S. E. (1997). Removal of Congo Red from wastewater by adsorption onto waste red mud. *Chemosphere*, 34(2), 401-417.

[12] Namasivayam, C., & Yamuna, R. T. (1992). Removal of Congo Red from aqueous solutions by biogas waste slurry. *Journal of Chemical Technology and Biotechnology*, 53(22), 153-157.

[13] Namasivayam, C., & Kanchana, N. (1992). Waste banana pith as adsorbent for colour removal from wastewaters. *Chemosphere*, 25(11), 1691-1705.

[14] Namasivayam, C., Muniasamy, N., Gayathri, K., Rani, M., & Ranganathan, K. (1996). Removal of dyes from aqueous solutions by cellulosic waste orange peel. *Bioresource Technology*, 57(1), 37-43.

[15] Pavlidou, S., & Papaspyrides, C. D. (2008). A review on polymer-layered silicate nanocomposites. *Progress in polymer science*, 33, 1119-1198.

[16] Singa Ray, S., & Okamoto, M. (2003). Polymer/layered silicate nanocomposites: a review from preparation to processing. *Progress in Polymer Science*, 28, 1539-1641.

[17] Fischer, H. (2003). Polymer nanocomposites: from fundamental research to specific applications. *Material Science and Engineering*, C23, 763-772.

[18] Zeng, Q. H., Yu, A. B., Lu, G. Q., & Paul, D. R. (2005). Clay-based polymer nanocomposites: research and commercial development. *Journal of Nanoscience and Nanotechnology*, 5, 1574-1592.

[19] Gil, A., Gandia, L. M., & Vicente, M. A. (2000). Recent advances in the synthesis and catalyticapplications of pillared clays. *Catalysis Reviews, Science and Engineering*, 42, 145-212.

[20] De Stefanis, A., & Toklinson, A. A. G. (2006). Towards designing pillared clays for catalysts. *Catalysis Today*, 114, 126-141.

[21] Shichi, T., & Takaqi, K. (2000). Clay minerals as photochemical reaction fields. *Journal of Photochemistry and Photobiology C: Photochemistry Reviews*, a, 113-130.

[22] Burst, J. F. (1991). Application of clay minerals in ceramics. *Applied Clay Science*, 5, 421-443.

[23] Bundy, W. M., & Ishley, J. N. (1991). Kaolin in paper filling and coating. *Applied Clay Science*, 5, 397-420.

[24] Mousty, C. (2004). Sensors and biosensors based on clay-modified electrodes- new trends. *Applied Clay Science*, 27, 159-177.

[25] Liu, P., & Zhang, L. (2007). Adsorption of dyes from aqueous solutions of suspensions with clay nano-adsorbents. *Separation and Purification Technology*, 58, 32-39.

[26] Meunier, A. (2005). Clays. Heidelberg: Springer-Verlag Berlin.

[27] Wang, L., & Wang, A. (2008). Adsorption properties of Congo Red from aqueous solution onto surfactant-modified montmorillonite. *Journal of Hazardous materials*, 160, 173-180.

[28] Betega de Paiva, L., Morale, A. R., & Díaz, F. R. V. (2008). Organocalys: Properties, preparation and applications. *Applied Clay Science*, 42, 8-24.

[29] Lagaly, G., & Ziesmer, S. (2003). Colloid chemistry of clay minerals: the coagulation of montmorillonite dispersions. *Advances in Colloid and Interface Science*, 100-102, 100-102.

[30] Gieseking, J. E. (1939). The mechanism of cation exchange in the montmorillonite-beidellite-nontronite type of clay minerals. *Soil Science*, 47, 1-14.

[31] MacEwan, D. M. C. (1944). Identification of the montmorillonite group of minerals by X-rays. *Nature*, 154, 577-578.

[32] Theng, B. K. G. (1974). The chamistry of Clay-Organic Reactions. London: Adam Hilger.

[33] Lagaly, G. (1984). Clay-organic interactions. *Transactions of the Royal Society of London*, 211, 315-332.

[34] Yariv, S., & Cross, H. (2002). Ograno-Clay Complexes and Interaction. Marcel Dekker.

[35] Errais, E., Duplay, J., Darragi, F., M'Rabet, I., Aubert, A., Huber, F., & Morvan, G. (2011). Efficient anionic dye adsorption on natural intreated clay: Kinetic study and thermodynamic parameters. *Desalination*, 275, 74-81.

[36] Ojstrsek, A., Doliska, A., & Fakin, D. (2008). Analysis of reactive dyestuffs and their hydrolysis by capillary electrophoresis. *Analytical Science*, 24, 1581-1587.

[37] Wang, L., & Wang, A. (2008). Adsorption properties of Congo Red from aqueous solution onto surfactant-modified montmorillonite. *Journal of Hazardous Materials*, 160, 173-180.

[38] Koswojo, R., Utomo, R. P., Ju, Y., Ayucitra, A., Soetaredjo, F. E., Sunarso, J., & Ismadji, S. (2010). Acid Green 25 removal from wastewater by organo-bentonite from Pacitan. *Applied clay science*, 48, 81-86.

[39] Ceyhan, Ö., & Baybas, D. (2001). Adsorption of some textile dyes by hexadecyltrimethylammonium bentonite. *Turkish Journal of Chemistry*, 25, 193-200.

[40] Zohra, B., Aicha, K., Fatima, S., Nourredine, B., & Zoubir, D. (2008). Adsorption of Direct Red 2 on bentonite modified by cetyltrimethylammonium bromide. *Chemical Engineering Journal*, 136, 295-305.

[41] Liu, B., Wang, X., Yang, B., & Sun, R. (2011). Rapid Modification of montmorillonite with novel cationc Gemini surfactants and its adsorption for methyl orange. *Materials Chemistry and Physics*, 130, 1220-1226.

[42] Yang, Y. Q., Han, S. Y., Fan, Q. Q., & Uqbolue, S. C. (2005). Nanoclay and modified nanoclay as sorbents for anionic, cationic and nonionic dyes. *Textile research journal*, 75, 622-6.

[43] Taleb, M. F. A., Hegazy, D. E., & Ismail, S. A. (2012). Radiation synthesis characterization and dye adsorption of alginate-organophilic montmorillonite nanocomposite. *Carbohydrate Polymers*, 87, 2263-2269.

[44] Shirsath, S. R., Hage, A. P., Zhou, M., Sonawane, S. H., & Ashokkumar, M. (2011). Ultrasound assisted preparation of nanoclay Bentonite-FeCo nanocomposite hybtrid hydrogel: A potential responsive sorbent for removal of organic pollutant from water. *Desalination*, 281, 429-437.

[45] Dalaran, M., Emik, S., Güçlü, G., İyim, T. B., & Özgümüş, S. (2011). Study on a novel polyampholyte nanocomposite seperabsorbent hydrogels: Synthesis, characterization and investigation of removal of indigo carmine from aqueous solution. *Desalination*, 297, 170-182.

[46] Li, S., Zhang, H., Feng, J., Xu, R., & Liu, X. (2011). Facile preparation of poly(acrylic acid-acrylamide) hydrogels by frontal polymerization and their use in removal of cationic dyes from aqueous solution. *Desalination*, 280, 95-102.

[47] Janovak, L., Varga, J., Kemeny, L., & Dekany, I. (2009). Swelling properties of copolymer hydrogels in the presence of montmorillonite and alkylammonium montmorillonite. *Applied clay science*, 42, 260-270.

[48] Osada, Y. (2001). (editor). Gels Handbook. San Diego: Academic Press.

[49] Ramírez, E., Burillo, S. G., Barrera-Díaz, , Roa, G., & Bilyeu, B. (2011). Use of pH-sensitive polymer hydrogels in lead removal from aqueous solution. *Journal of Hazardous Materials*, 192, 432-439.

[50] Haraguchi, K. (2007). Nanocomposite hydrogels. *Current Opinion in Solid State and Materials Science*, 11, 47-54.

[51] Haraguchi, K. (2007). Nanocomposite Gels: New Advanced Functional Soft Materials. *Macromolecular Symphosium*, 256, 120-130.

[52] Nguyen, K. T., & West, J. L. (2001). Photopolymerizable hydrogels for tissue engineering applications. *Biomaterials*, 23, 4307-4314.

[53] Bulut, Y., Akcay, G., Elma, D., & Serhatli, I. E. (2009). Synthesis of clay-based superabsorbent composite and its sorption capability. *Journal of Hazardous Matererials*, 171, 717-723.

[54] Haraguchi, K., & Li, H. J. (2006). Mechanical properties and structure of polymer-clay nanocomposite gels with high clay content. *Macromolecules*, 39, 1898-1905.

[55] Kaplan, M., & Kasgoz, H. (2011). Hydrogels nanocomposite sorbents for removal of basic dyes. *Polymer Bulletin*, 67, 1153-1168.

[56] Xia, X. H., Yih, J., D`Souza, N. A., & Hu, Z. B. (2003). Swelling and mechanical behavior of poly(N-isopropylacrylamide)/Na-montmorilonite layered silicates composite gels. *Polymer*, 44, 3389-3393.

[57] Zhou, S. H., Yang, J. G., & Wu, C. P. (2003). Synthesis and swelling properties of poly (N,N`-diethylacrylamide)-clay nanocomposites. *Acta Polymer Sinica*, 3, 326-329.

[58] Bignoti, F., Satore, L., Penco, M., Ramorino, G., & Peroni, I. (2004). Effect of montmorillonite on the properties of thermosensitive poly(N-isopropylacrylamide) composite hydrogels. *Journal of Applied Polymer Science*, 93, 1964-1971.

[59] Xiang, Y. Q., Peng, Z. Q., & Chen, D. J. (2006). A new polymer/clay nano-composite hydrogel with imporved response rate and tensile mechanical properties. *European Polymer Journal*, 42, 2125-2132.

[60] Liang, L., Liu, J., & Gong, X. (2000). Thermosensitive poly(N-isopropylacrylamide)-clay nanocomposites with enhanced temperature response. *Langmuir*, 16, 9895-9899.

[61] Xu, K., Wang, J. H., Xiang, S., Chen, Q., Zhang, W. D., & Wang, P. X. (2007). Study on the synthesis and performance of hydrogels with ionic monomers and montmorillonite. *Applied Clay Science*, 38, 139-145.

[62] Zhang, Q., Li, X., Zhao, Y., & Chen, L. (2009). Preparation and performance of nanocomposite hydrogels based on different clay. *Applied Clay Science*, 46, 346-350.

[63] Nie, J. J., Du, B. Y., & Oppermann, W. (2005). Swelling, elasticity and spatial inhomogeneity of poly(N-isopropylacrylamide)/clay nanocomposite hydrogels. *Macromolecules*, 38, 5729-5736.

[64] Yi, J., Ma, Y., & Zhang, L. (2008). Synthesis and decoloring properties of sodium humate/poly (N-isopropylacrylamide) hydrogels. *Bioresource Technology*, 99(13), 5362-5367.

[65] Kasgöz, H., & Durmus, A. (2008). Dye removal by a novel hydrogel-clay nanocomposite with enhanced swelling properties. *Polymers for Advanced Technologies*, 19, 838-845.

[66] Nguyen, K. T., & West, L. J. (2002). Photopolymerizable hydrogels for tissue engineering applications. *Biomaterials*, 23, 4307-4314.

[67] He, D., Susanto, H., & Ulbricht, M. (2009). Photo-irradiation for preparation, modification and stimulation of polymeric membranes. *Progress in Polymer Science*, 34, 62-98.

[68] Tamirisa, P. A., Koskinen, J., & Hess, D. W. (2006). Plasma polymerized hydrogel thin films. *Thin Solid Films*, 515, 2618-2624.

[69] Hongyan, H., Ling, L., & James, Lee. L. (2006). Photopolymerization and structure formation of methacrylic acid based hydrogels in water/ethanol mixture. *Polymer*, 47, 1612-1619.

[70] Kurecic, M., Sfiligoj-Smole, M., & Stana-Kleinschek, K. (2012). UV polymerization of poly(N-isopropylacriylamide) hydrogel. *Materials and technology*, 46, 69-73.

[71] Alexandre, M., & Dubois, P. (2000). Polymer-layered silicate nanocomposites: preparation, properties and uses of a new class of materials. *Materials science and engineering:R:Reports*, 28, 1-63.

[72] Pavlidov, S., & Papaspyrides, C. D. (2008). A review on polymer-layered silicate nanocomposites. *Progress in Polymer science*, 33, 1119-1198.

[73] Haraguchi, K., & Li, H. J. (2006). Mechanical properties and structure of polymer-clay nanocom- poiste gels with high clay content. Macromolecules, 39, 1898-1905.

[74] Vaia, R. A., & Giannelis, E. P. (1997). Polymer melt intercalation in organicallymodified layered silicates: model predictions and experiment. *Macromolecules*, 30, 8000-9.

[75] Kokabi, M., Sirousazar, M., & Hassan, Z. M. (2007). PVA-clay nanocomposite hydrogel for wound dressing. *European Polymer Journal*, 43, 773-781.

[76] Can, V., Abdurrahmanoglu, S., & Okay, O. (2007). Unusual swelling behavior of polymer-clay nanocomposite hydrogels. *Polymer*, 48, 5016-5023.

[77] Baskaralingam, P., Pulikesi, M., Elango, D., Ramamurthi, V., & Sivanesan, S. (2006). Adsorption of acid dye onto organobentonite. *Journal of Hazardous Materials*, B128, 138-144.

[78] Errais, E., Duplay, J., Darragi, F., M'Rabet, I., Aubert, A., Huber, F., & Morvan, G. (2011). Efficient anionic dye adsorption of natural intreated clay:Kinetic study and thermodynamic parameters. *Desalination*, 275, 74-81.

[79] Kurecic, M. (2011). Synthesis of ion exchange nanocomposite hydrogels inside the PP membrane pores. PhD thesis.

Chapter 8

Ecologically Friendly Polymer-Metal and Polymer-Metal Oxide Nanocomposites for Complex Water Treatment

Amanda Alonso, Julio Bastos-Arrieta,
Gemma.L. Davies, Yurii.K. Gun'ko, Núria Vigués,
Xavier Muñoz-Berbel, Jorge Macanás, Jordi Mas,
Maria Muñoz and Dmitri N. Muraviev

Additional information is available at the end of the chapter

1. Introduction

The physical characteristics of nanomaterials, those with a size smaller than 100 nm, are known to be substantially dependent on their size scale. The increase of interest in nanotechnology studies has been due to the incorporation of nanoparticles (NPs) into commercial available products.

Thus, the development of methods for the synthesis of NPs with a narrow size distribution, the techniques of separation and preparation of customized of engineered nanoparticles is one of the most important points of research to focus in.

By taking into account some parameters during NPs preparation, such as: time, temperature, stirring velocity and concentrations of reactants and stabilizing reagents, one can obtain the ideal distribution and morphology of these novel materials [1].

In this regard,polymeric supports play a very important role for several reasons including, the ease of their preparation in the most appropriate physical forms (e.g., granulated, fibrous, membranes, etc.), the possibility to produce the macroporous matrices with highly developed surface area and some others. However, the immobilization of NPs on the appropriate polymeric support represents a separate task [2] and thus, the incorporation of poly-

INTECH
open science | open minds

© 2012 Alonso et al.; licensee InTech. This is a paper distributed under the terms of the Creative Commons Attribution License (http://creativecommons.org/licenses/by/3.0), which permits unrestricted use, distribution, and reproduction in any medium, provided the original work is properly cited.

mers as support for NPs synthesis is another way to control the growth of NPs as well as preventing the NPs aggregation and their release.

Embedded NPs in polymer matrix have gained interest in the past years, because of the unique applications of the final nanocomposite materials, for example, optical, magnetic, sensors and biosensors [3–11].

Several parameters of the polymeric matrices may be considered for their use on the synthesis of nanocomposites as it is discussed in the following sections.

1.1. Ion Exchange Polymers

One simple consideration of the ion exchange process is the equivalent exchange of ions between two or more ionized species located in different phases, at least one of which is an ion exchanger. The process takes place without the formation of chemical bonds by certain equilibrium between charges of ions and, in the case of polymers, functional groups.

Depending on the functional group charge, ion exchangers are called cation exchangers if they bear negative charged functional groups and carry exchangeable cations. Anion exchangers carry anions due to the positive charge of their fixed groups. Chemical interactions could be different for different polymeric and inorganic ion exchanges. However, most of the thermodynamic and kinetic approaches, as well as the practical methods and technologies are essentially the same [12]. Table 1 shows the most common physical and chemical parameters specific for polymeric ion exchangers.

Physical properties	Chemical Properties
Physical structure and morphology	Cross Linking degree
Surface Area	Ionic form(functional groups)
Particle Size	Ion Exchange Capacity(IEC)
Partial Volume in swollen state	Type of matrix

Table 1. General Properties of Ion Exchange Materials [13].

Probably, the first extensive investments in the development of ion exchangers and ion exchange processes were done bearing in mind the potential application for isotope separation in nuclear industry, although nowadays one of the most common applications of these materials is in water purification processes among other applications of interest as is shown in Table 2.

Functional groups define chemical properties, by bearing on surface a negative or positive charge. Due to this fact, different dissociation properties of groups leads to a difference among strong and weak exchangers; which are recognized similar to that of strong and weak electrolytes. Table 3 shows some of the most common types of cation and anion exchangers.

Application	Brief Description
Water Treatment	Preparation of pure and ultrapure deionized water Potable Water preparation Water Softening
Food Industry	Removing off tasters and odors. Deacidification of fruit juice Recovery of glutamic acid
Nuclear Industry	Separation of uranium isotopes Final storage of radioactive waste. Waste decontamination
Pulp and Paper Industry	Removal of inorganic salts from liquors Detoxification of by-products transferred to bio-cultivation

Table 2. Different applications of ion exchange materials [12,13].

Cation Exchange groups (Negatively charged)	Anion Exchange Groups (Positively Charged)
Sulphonic SO_3^-	Quaternary Amonium $-(N\text{-}R_3)^+$
Carboxilic: COO^-	Phosphonium: $-(P\text{-}R_4)^+$
Arsenate: AsO_3^-	Sulphonium: $-(S\text{-}R_3)^+$
Phosphate $-PO_3^{2-}$	

Table 3. Functional Groups in Ion Exchange Polymers [13].

Ion exchange capacity (IEC) is the main feature of ion exchange materials. An ion exchanger can be considered as a "reservoir" containing exchangeable counterions. The counterion content, in a given amount of material, is defined essentially by the amount of fixed charges which must be compensated by the counterions and thus, is essentially constant [6,7].

Some facts must be considered to define IEC: availability of functional groups for exchange reaction, the macrostructure architecture, swelling degree and the size of the ions to be exchanged.

Depending on the final application of an ion exchange polymer (e.g., granulated resins), a pre-treatment stage is advised in order to obtain an enhanced and optimal exchange capacity. For cationexchanger resins an acid stage pretreatment (usually HCl 0.1 M) is the more advisable, as for the anionexchangerresin would be a basic stage pre-treatment (usually NaOH 0.1 M).

Regarding the applications of the ion exchange resins, they have been used in pharmaceutics and food industry which is determined by another advantage of these materials: while being chemically active, they are highly stable in both physical and chemical senses and, as a result, do not contaminatethe final product.

Concretely, the use of the functional ion exchange polymers as supports for the synthesis of metal nanoparticles (MNPs) and metal oxide nanoparticles (MONPs) [14] has in this sense, one most important advantage dealing with the possibility to synthesize the NPs of interest directly at the "point of use", i.e. on the supporting polymer by an "in-situ" reaction. For instance, in the case of the metal catalyst nanoparticles, this results in the formation of catalytically-active polymer-metal nanocomposites [6,7,15].

Overall, this chapter focuses on the feasible application of ion exchange polymeric matrices (i.e., both cation and anion granulated exchangers resins) used as support for MNPs and MONPs synthesized by using the developed Intermatrix Synthesis (IMS) methodology [5]-[8,10, 14,16]. This methodology has been shown by the enhancement of the accessibility of the nanomaterial for de desirable final functionality due to the final NPs distribution. The overall combination of certain features of these matrices: cross linking degree, the different solubility in organic solvents, their stability and insolubility in water, etc. offer a wide range of possible applications and functionalities, depending on the embedded NPs nature as well of the type of polymer for a specific final use. For instance, examples of water treatment and catalysis have been presented on several publications of the authors [14,6].Concretely, this chapter is based on the use of these materials for the bacteria elimination on water treatment applications.

2. Intermatrix synthesis of metal nanoparticles

The IMS method represents one of the most efficient and simple techniques for the "in-situ" preparation of metal-polymer nanocomposites. The general principles of IMS apply to all types of polymer matrices and NPs.In general:

- Polymer molecules serve as nanoreactors and provide a confined medium for the synthesis (thus controlling particle size and distribution).
- Polymer molecules stabilize and isolate the generated NPs, thus preventing their aggregation.

In the case of ion exchange matrices, the functional groups that can immobilize metal ions and metal ion complexes are the key points for IMS because they are homogeneously distributed in the ion exchange matrix and behave as combinations of single isolated nanoreactors generating homogeneous nanocomposites.

The major part of our work in this field has been done with the polymers bearing negative charged functional groups (i.e., cation exchanger polymers containing both carboxylated and sulfonated [5,8,10] functional groups), which first have to be loaded with the desired metal ions (MNP precursors) followed by their chemical reduction to zero-valent state (MNPs) by using an appropriate reducing agent. Several recent publications by the authors describe the IMS of MNPs with the most favorable distribution near the surface of nanocomposite for instance for catalytic applications or for killing bacteria by a contact mechanism.This distribution is due to the coupling of the classical IMS methodology with the Donnan Exclusion effect (DEE) [6,7,9,16].

The polymeric matrix bears a charge due to the presence of well-dissociated functional groups. This means that a reducing agent negatively charged (e.g., brohydride BH_4^-) presents the same charge than the support, therefore they cannot penetrate inside the polymer because of the action of electrostatic repulsion. This is known as Donnan Exclusion Effect. It refers to the impossibility to penetrate deeply in a matrix when there is a coincidence between the charge of the outside ions (e.g. from the reducing agent) and the ones of the functional groups on the polymer surface. Thus, an equilibrium between ion concentration (either functional groups or from metal or reducing agent solution) and electrostatic repulsion takes place (Figure 1).

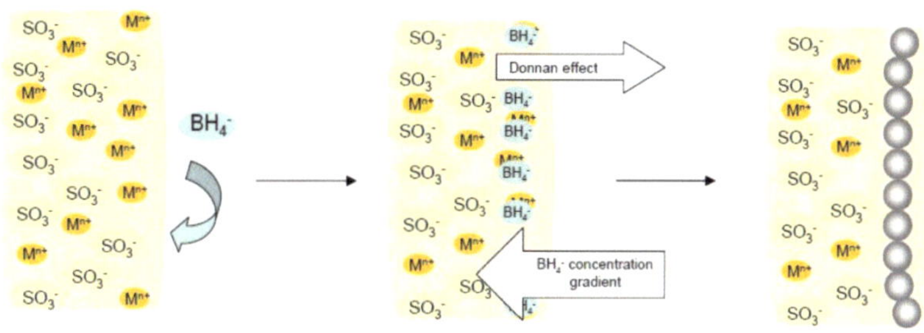

Figure 1. Scheme of IMS on a sulfonated exchange polymer with Donnan Exclusion Effect to obtain MNPs mainly on the polymer surface.

2.1. Traditional and novel versions of IMS technique

The difference between the traditional and the novel version of the IMS technique developed in this study become clear after comparison of the respective reaction schemes, which can be written for the case of formation of Ag-NPs in (a) strong acidic polymers (e.g. containing sulfonated $-SO_3^-$ functional groups)and; (b) strong basic polymers (e.g. containing quaternary ammonium $-NR_4^+$ functional groups)as follows [6,7]:

a) IMS in cation exchanger polymers (traditional version):

1) Metal-loading stage

$$R-SO_3^-Na^+ + Ag^+NO_3^- \rightarrow R-SO_3^-Ag^+ + Na^+NO_3^- \qquad (1)$$

2) Metal-reduction stage

$$2R-SO_3^-Ag^+ + 2\,NaBH_4 + 6H_2O \rightarrow 2R-SO_3^-Na^+ + 2Ag^0 + 7H_2 + 2B(OH)_3 \qquad (2)$$

b) IMS in anion exchange polymers (novel version):

1) Reduce-loading stage

$$R-(R'_4-N)^+Cl^- + NaBH_4 \rightarrow (R_4-N)^+BH_4^- + NaCl \quad (3)$$

2) Metal-loading-reduction stage

$$2\left[R-(R'_4-N)^+BH_4^-\right] + 2Ag^+NO_3^- + 6H_2O \rightarrow 2(R_4-N)^+NO_3^- + 2Ag^0 + 7H_2 + 2B(OH)_3 \quad (4)$$

As it is seen from the above reactions, the main difference between (a) and (b) versions of IMS consists in the first stage of the process. In the first case, the functional groups of the polymer are loaded with the desired metal ions, while in the second case, the loading is carried out with the desired reducing ions. The second stage, in the first case, consists on the reduction of metal ions with ionic reducing agent, located in the external solution. As far as the charge sign of reducer anions coincide with that of the polymer matrix, they cannot deeply penetrate inside the polymer due to the action of the DEE and as the result, the reduction process appears to be "localized" near the surface of the polymer.

As the result, the second stage of this version permits to couple the metal-loading and the metal-reduction processes in one step. The metal loading is carried out by using external solution containing metal ions bearing the charge of the same sigh as that of the functional groups of the polymer, what does not allow them to deeply diffuse inside the polymer matrix (due to DEE process). Again, the reduction of metal ions and therefore, the formation of MNPs have to proceed near the surface of the polymer. For obvious reasons the second version of IMS technique (version b) can be classified as a sort of the symmetrical reflection of version (a) as it is shown in Figure 2.

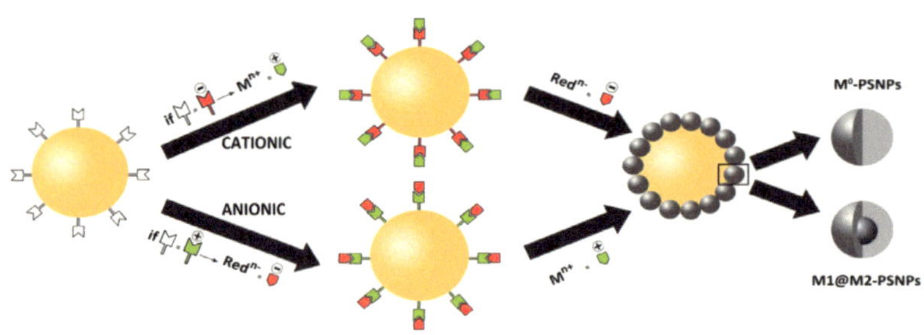

Figure 2. Scheme of IMS steps for the synthesis of NPs by using either a cation or anion granulated exchanger polymer as a matrix, throwing to symmetrical version of IMS influenced by DEE.

In both versions of IMS methodologies shown, DEE plays a very important role as it appears to be responsible for the desired nonhomogeneous distribution of MNPs inside the polymeric nanocomposite. The action of this effect is observed in both cases within the second stage of IMS process (see equations 2 and 4). The following two "driving forces" acting in the opposite directions are responsible for the DEE: 1) the electric field determined by the charge of the polymer matrix and 2) the concentration of the ionic component in the external solution [17,18].

The first force rejects the ions of the same charge as that of the functional groups of the polymer while the second one drives these ions to move into the polymeric matrix. The first force can be hardly varied as it has a constant value determined by the ion exchange capacity of the polymer and the degree of dissociation of its functional groups. The second force can be easily varied by changing the concentration of respective component in the external solution, what has to result in the changes in the composition of the final nanocomposite (MNPs content).

The possibility to use the second approach follows from the above reaction schemes (see reactions 2 and 4). Indeed, after finishing the metal reduction (IMS version a) or the metal-loading-reduction stages (IMS version b) the functional groups of the polymer appear to be converted back into the initial ionic form (Na-form in the first and Cl-form in the second case). This means that in both cases IMS reactions of MNP can be repeated without any additional pretreatment of the ion exchanger. This has to result in the accumulation of a higher amount of the metal (or MNPs) inside the polymer when the same metal precursor is used or, what is more relevant in this work, the possibility of the formation of core-shell NPs (e.g. Ag@Co-NPs, what means a Co-core coated by a Ag-shell). This approach leads to a wide range of possibilities for the applications of core-shell MNPs or MONPs [6,7,10,16,19–22]. Thus, the following Figure 3 shows the range of types of MNPs and MONPs (either monometallic or core-shell bimetallic structures) synthesized by the mentioned methodology presenting different properties for the applications of interest.

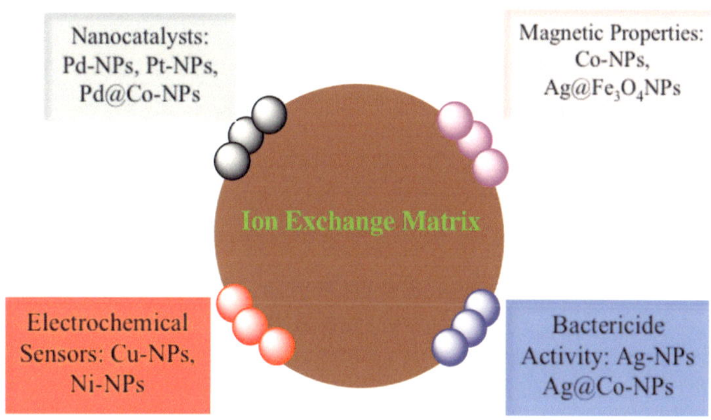

Figure 3. Scheme of IMS feasibility to produce different MNPs or MONPs with an accessible distribution of NPs on polymer surface for different final applications.

Among all the MNPs and MONPs possibilities, in this chapter, we focused on the synthesis and characterization (Section 3) of monocomponent NPs based on magnetite (Fe_3O_4), cobalt (Co) and silver (Ag) as well as bicomponent core-shell NPs based on Ag@Fe_3O_4 and Ag@Co. In these cases, the core is always composed of a magnetic element for its interest regarding the final application and the prevention of MNPs leaching, as it will be discussed in Section 4.

3. Characterization of nanocomposite materials

In order to understand the enhancement of the properties of the polymeric nanocomposites obtained by IMS technique, a proper material characterization is mandatory and it is presented in this section. The application of these novel materials is presented in later on, and deals with their bactericidal activity for water treatment application.

3.1. Synthesis and sharacterization of sulfonated granulated nanocomposites containing Fe_3O_4- and Ag@Fe_3O_4-NPs

Fe_3O_4 and Ag@Fe_3O_4 nanocomposites were developed by an extension of IMS technique by the coupling of the general precipitation technique of Fe_3O_4-NPs(23)with the IMS methodology as shown in the following reactions:

$$8(R-SO_3^-Na^+) + Fe^{2+} + 2Fe^{3+} \rightarrow (R-SO_3^-)_2(Fe^{2+}) + 2[(R-SO_3^-)_3(Fe^{3+})] + 8Na^+ \quad (5)$$

$$(R-SO_3^-)_2(Fe^{2+}) + 2[(R-SO^{3-})_3(Fe^{3+})] + 8NaOH \rightarrow 8(R-SO_3^-Na^+) + Fe_3O_4 + 4H_2O \quad (6)$$

The synthesis of Ag@Fe_3O_4-NPs was performed by the subsequent reduction reaction of Ag^+ onto Fe_3O_4-NPs surface and within the matrix as follows:

$$R-SO_3^-Na^+ + Fe_3O_4 + Ag^+ \rightarrow R-SO_3^-Ag^+ + Na^+ + Fe_3O_4 \quad (7)$$

$$R-SO_3^-Ag^+ + Fe_3O_4 + NaBH_4 + 3H_2O \rightarrow R-SO_3^-Na^+ + 7/2H_2 + B(OH)_3 + Ag@Fe_3O_4 \quad (8)$$

The main characterization techniques for nanomaterials have been used in these systems.

3.1.1. X-Ray Diffraction, XRD

XRD technique was used to determine the crystalline structure of the particles. Figure 4 shows the XRD graphs of Fe_3O_4-NPs as a reference (synthesized by liquid phase method [24] and the sample corresponding to Fe_3O_4-NPs stabilized in a sulfonated polymeric matrix represented as C100E code (from Purolite S.A).

The position and relative intensity of all diffraction peaks from the Fe_3O_4-nanocomposite sample are in good agreement with those for the Fe_3O_4 powder. The relative intensity is lower for the nanocomposite sample due to the "diluting" polymer effect.

3.3.2. Microscopy characterization

The microscopy techniques (e.g., Scanning Electron Microscopy, SEM, and Transmission Electron Microscopy, TEM) allow the characterization of both surface and inside area of the nanocomposites. For instance, the NPs metal concentration profiles, along the cross-sectioned polymeric beads Figure 5), was examined by using SEM technique coupled with Energy Dispersive Spectroscopy (EDS). EDS analysis demonstrated that Ag and Fe elements were mostly found on the edge of the bead.In general, this non-homogeneous distribution of the NPs may be attributed to the Donnan Exclusion Effect as shown before for NPs in most of the polymers.

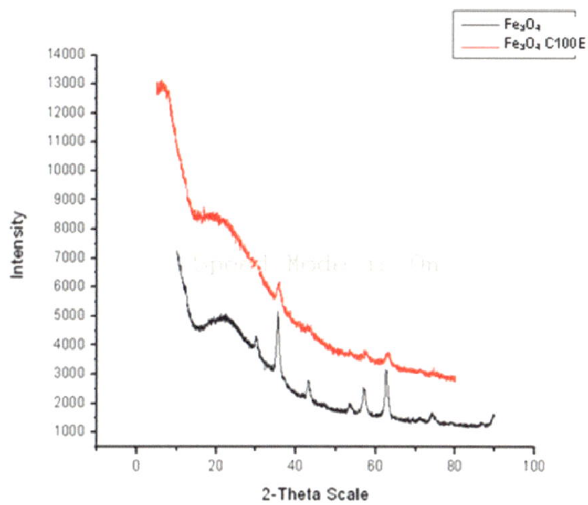

Figure 4. X-ray diffraction patterns of Fe_3O_4-NPs (black) and, Fe_3O_4-NPs stabilized on sulfonated polymer (red).

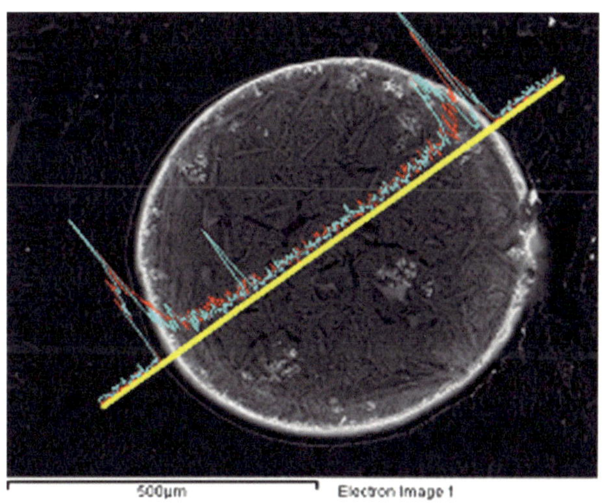

Figure 5. SEM image of a cross-sectioned Ag@Fe_3O_4-sulfonated nanocomposites resin and EDS metal content distribution profile (Ag in blue line and, Fe in red line).

3.1.3.Thermogravimetric Analysis, TGA

TGA technique is used to determine polymer degradation temperatures in polymer or composite materials [24]. Figure 6 shows the TGA curves for Fe_3O_4- and Ag@Fe_3O_4-NPs stabilized in sulfonated resin as well as the corresponding raw polymer.

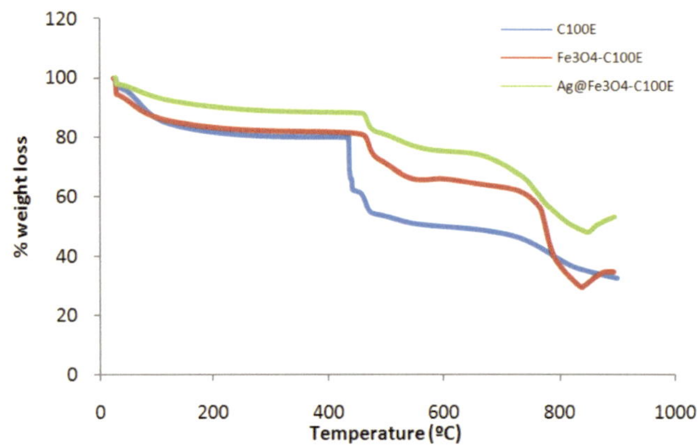

Figure 6. TGA curves of %weight loss vs temperature of sulfonated samplesFe_3O_4- and Ag@Fe_3O_4.

As seen in Figure 6, TGA curves for all samples are characterized by four weight-loss regions, which can be described as follows:

1. The weight loss between 30 and 400ºC can be mainly attributed to adsorbed water molecules, both "free" and strongly "bound" to surface groups from the polymer and the nanoparticles, where applicable.

2. A significant weight loss at 450ºC for all samples. This loss is particularly important for the raw polymer (NPs-free) in comparison with the NPs-modified polymers and can be associated with the loss of the functional groups including free sulfonic functionalities.

3. A third gradual weight-loss is observed between 500ºC and 700ºC and may be attributed to the degradation of the polymer side chains. Again, this loss is less important for the nanocomposite samples.

4. Finally, the weight changes at temperatures higher than 700ºC may be caused by further thermodegradation of the polymer, but is noteworthy that for Ag@Fe_3O_4-C100E and Fe_3O_4-C100E nanocomposites, there is weight gain, probably due to the oxidation of the magnetic material from Fe_3O_4 to Fe_2O_3.

As can be seen, lines are almost parallel and only Fe_3O_4-C100E sample shows a quite different behaviour close to 800ºC [24].

3.1.4. Magnetic characterization

As mentioned, the bicomponent core-shell NPs are based on a magnetic core which lead to obtaind magnetic properties to the nanocomposite. The characterization of the magnetic properties of the nanocomposites was preceded by using a vibrating sample magnetometer (VSM), as shown in Figure 7 by the representation of the magnetization curves of the samples when a magnetic field is applied. Also, the magnetic behavior of Fe_3O_4-NPs as powder structure (without polymeric support) was analysed.

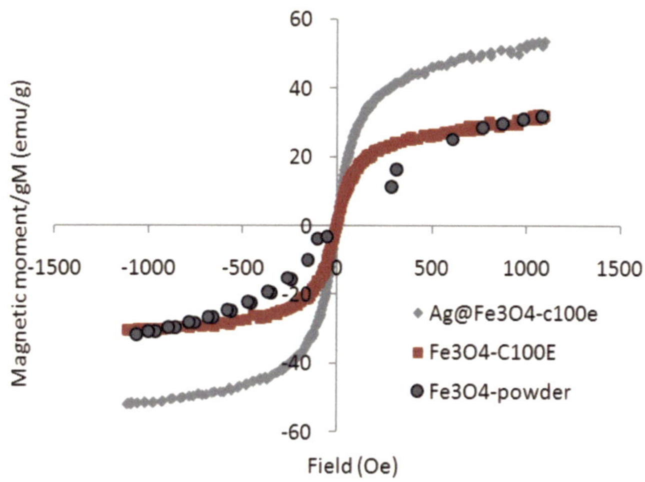

Figure 7. Magnetization curves for Fe_3O_4- and Ag@Fe_3O_4-NPs in sulfonated polymer and for powered Fe_3O_4-NPs.

As shown, superparamagnetic behaviour was observed in all the Fe_3O_4-based nanocomposites and powder. When comparing the magnetization values of the Fe_3O_4-nanocomposite with powdered Fe_3O_4-NPs, similar magnetization values were obtained at aprox. 30 emu/g. Besides, Ag@Fe_3O_4-nanocomposites showed higher magnetic saturation than for those Fe_3O_4 ones. [22–25] This result, as an example, shows the advantage of the nanocomposites containing Ag@Fe_3O_4-NPs since they show the combination of the properties from both components: bactericidal activity from Ag as well as magnetic properties from Fe_3O_4core.

3.2. Amine-based granulated nanocomposites containing Fe_3O_4- and Ag@Fe_3O_4-NPs(14)

On the other hand, and with the goal of expanding the IMS technique applications, Fe_3O_4- and Ag@Fe_3O_4-NPs were also synthesized in anion exchanger polymers. The granulated resin (in this case, A520E from Purolite), containing quaternary ammonium functional groups ($-NR_3^{\ +}$), was used as polymeric matrix.

As already introduced, the synthesis of MNPs in anionic exchanger polymers is the "mirror image" methodof the traditional IMS (see Figure 2 and Eqs 3-4). Thus, for the case of the combination of the precipitation technique with the IMS for the synthesis of Fe_3O_4-NPs in quaternary ammonium based polymers, the use of an initial positively charged element is needed to modify the charge of the raw polymer and lead to the procedure of the synthesis.

The following equations show this synthetic procedure. Initially, the raw material was pre-treated with 1.0 M trisodium citrate, $((CH_2)_2COH)(COONa)_3$, to compensate the positive charge of the polymer (where cit = citrate).

$$3(R'-NR_3^+)(Cl^-) + (Na^+)_3(cit^{3-}) \rightarrow (R'-NR_3^+)_3(cit^{3-}) + 3NaCl \tag{9}$$

Afterwards, the polymer was used for the Fe_3O_4-NPssynthesis :

$$4(R'-NR_3^+)_3(cit^{3-}) + Fe^{2+} + 2Fe^{3+} \rightarrow (R'-NR_3^+)_4(cit^{3-})_4(Fe^{2+}, 2Fe^{3+}) + 8R'-NR_3^+ \tag{10}$$

$$\left(R'-NR_3^+\right)_4\left(cit^{3-}\right)_4\left(Fe^{2+}, 2Fe^{3+}\right) + 8NaOH \rightarrow 4\left(R'-NR_3^+\right)\left(cit^{3-}\right)\left(Na^+\right)_2 + Fe_3O_4 + 4H_2O \tag{11}$$

$Ag@Fe_3O_4$-NPs were obtained after the loading with $NaBH_4$ followed by the use of $AgNO_3$.

Schematically, Figure 8 describes the synthetic procedure.

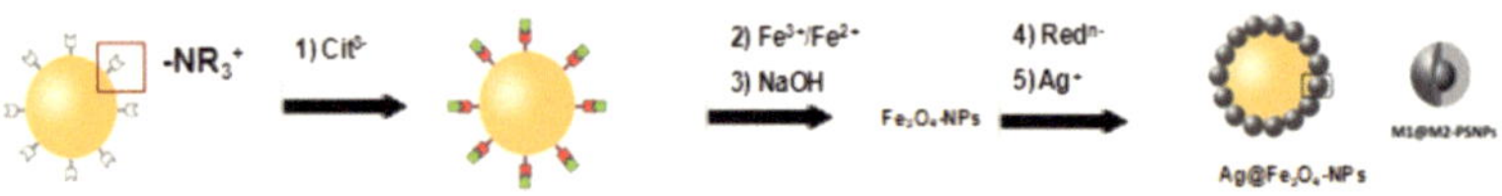

Figure 8. Synthetic methodologies for the synthesis of monocomponent Ag-NPs or Fe_3O_4-NPs and bicomponent $Ag@Fe_3O_4$-NPs stabilized in A520E support

Next, the microscopy characterization and the evaluation of the magnetic properties of these materials are shown.

3.2.1. Microscopy characterization

As described before, SEM technique was used to evaluate the NPs distribution. Thus, Figure 9a shows the metal profile of $Ag@Fe_3O_4$-NPs stabilized on an anionexchanger polymer. By EDS ScanLine is observed that Ag and Fe co-localize in $Ag@Fe_3O_4$-nanocomposite matrix. In addition, the particles structure in the nanocomposite was analysed by TEM. Figure 9bshows a magnified TEM image of the edge of the cross sectioned area from Figure 9a with a dispersed distribution of the NPs. By these results, the wide range of systems that can be studied based on IMS technique and the success on their formation is clearly shown.

3.2.3. Magnetic characterization

The magnetic properties of $Ag@Fe_3O_4$-nanocomposites were determined with a Superconducting Quantum Interference Device (SQUID) and compared with those obtained by polymeric structures only containing Fe_3O_4-NPs.

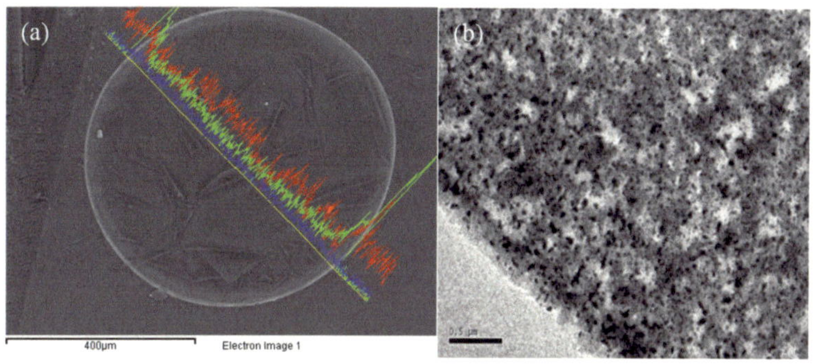

Figure 9. a)SEM images of cross sectioned Ag@Fe_3O_4-nanocomposite(on A520E matrix).EDS LineScan shows Ag (blue), Fe (red) and O (green). b) TEM images of crosssectioned area of the nanocomposite.

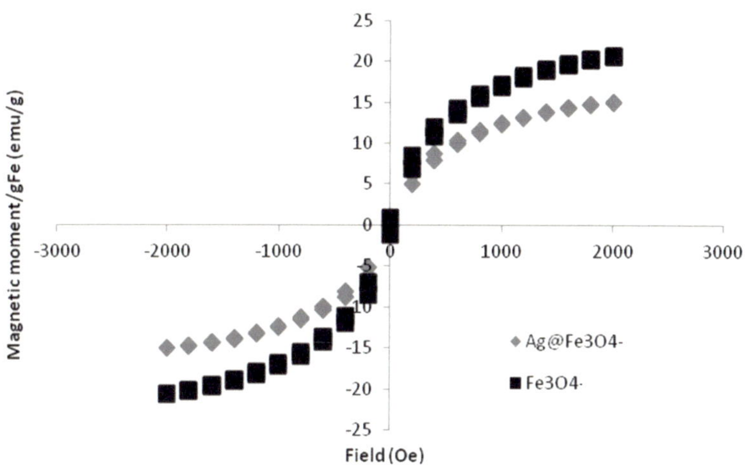

Figure 10. SQUID magnetization curves of Fe_3O_4- and Ag@Fe_3O_4- nanocomposites.

Similar magnetic hysteresis curves and saturation values were obtained when comparing both nanocomposites, suggesting that the presence of Ag did not affect the magnetic properties of the material. This was especially relevant when considering the final application of the nanocomposites, for example, for water purification.

It is generally known that Ag-NPs are much more toxic than the bulk Ag metal, limiting their application to real live environments. Thus, the possibility of collecting Ag@Fe_3O_4-NPs accidentally released from the polymeric matrix with a simple magnetic trap would be extremely desirable for water purification. Further studies about the use of Ag-based nanocomposites (and containing a magnetic core) for water treatment applications is detailed in the next section.

4. Ecological concerns regarding uncontrollable release of NPs to the environment

4.1. Environmental and safety concerns and uncontrollable release of NPs

The use of engineered nanoparticles in the environment as a consequence of the development of nanotechnology is a serious case of concern of environmental biologists worldwide. However, a few studies have already demonstrated the toxic effects of NPs on various organisms, including mammals. Nanotechnology is still in discovery phase in which novel materials are first synthesized in small scale in order to identify new properties and further applications [26–31].

Perception and knowledge are important parts of public understanding of nanotechnology. They can be influential for achievable benefit obtained and the possible risks and hazards.

Therefore, detail understanding of their sources, release interaction with environment, and possible risk assessment would provide a basis for safer use of engineered nNPs with minimal or no hazardous impact on environment. Thus, ecotoxicology of NPs will be closely related to their intrinsic properties as shown in Table 4 and Figure 11.

Physicochemical properties	Toxicological findings	Biological Effects
Size	Affects reactivity and permeability of cells and organs.	Increase biodistribution of NPs in environmental system.
Surface /volume ratio	Higher reactivity	Inflammatory effects
Chemical Composition	Increase in UVA absorption, higher activation of reactive oxygen species in cell media.	Cancerigen, cell proliferation reduced.
Aggregation state	More pronounced cytotoxic effects	Cytotoxicity
Surface Charge	Charged NPs present higher deposition degree in tissues.	Bioacummulation in brain, lungs and others.

Table 4. Biological effects due to physicochemical properties of nanomaterials 27.

Increase surface activity, mobility, and diffusion and adsorption ability are some other effects [26].

A further comprehension of the structure- function relationships in nanomaterials matter could lead to new protocols for nanomaterials manufacture wherein high precision, low waste methods are included [1,32,33].

Some criteria could be taken into account referring to NPs release and effects study:

1. NPs effects should be scale dependent and not the same in larger scale or agglomerates. This means that effects may be quiet different to adopt specific and more appropriate regulations.

2. These differences are based on size, surface chemistry and other specific interactions depending on the scale. Thus, the same material may have different regulations through the different sizes presented.

3. Effects must be conclusive to those products which commercialization is imminent. So, the NPs presented in the final product may be the ones, which the studies should focus on.

4. For the novelty of some materials, data to extrapolate environmental effects are difficult to obtain; so enhanced simulation system are needed.

The wide application of engineered NPs and their entry into the environment, the study of their impact on the ecosystem and a growing concern in society regarding the possible adverse effects of manufactured nanoparticles has been raised in recent years [26–31,34–39].

Therefore, it is required to study their release, uptake, and mode of toxicity in the organisms. Furthermore, to understand the long-term effect of NPs on the ecosystem, substantial information is required regarding their persistence and bioaccumulation.

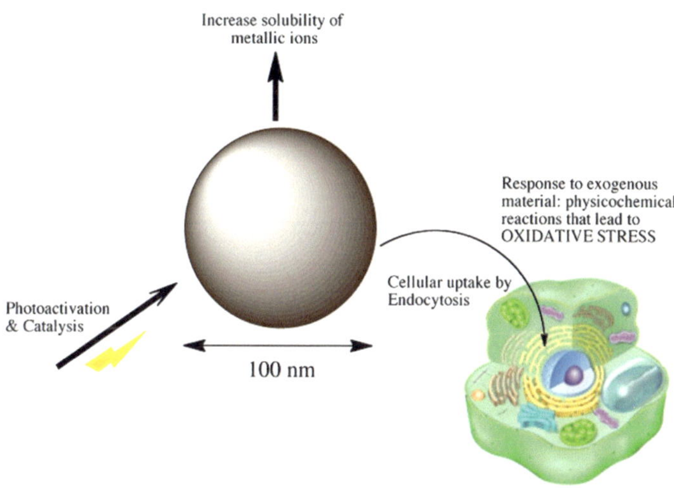

Figure 11. Effects on cellular activity due to the release of metal contain of NPs to the environment.

4.2. Safe polymer-metal nanocomposites

Table 5 presents some of green chemistry principles could be applied to the synthesis of nanomaterials, including nanocomposites.

General Principle	Toxicological findings
Safer nanomaterials	Find the influence of morphology, functionality and other features of nanomaterials that lead to the properties of interest , avoiding and understanding whatever parameter which leads to the incorporation of toxic nature to the material.
Reduced environmental impact	Analyse degradation and routes of incorporation to the environment, looking for a design of harmless products. One possibility is avoid the use of known hazardous precursors for the nanomaterials.
Waste reduction	Optimize solvent use by applying alternative purification techniques and media reactions.
Overall process safety	Make use of benign precursors and solvents in the designing and enhancement of the synthesis and even suggest greener alternative procedures and reagents for existing methodologies.
Materials efficiency	Think about new strategies that incorporate raw materials in products by bottom-up strategy. Also the application of catalytic procedures to enhance selectivity and yield of the overall process.
Energy efficiency	Design room temperature synthetic routes, with real time monitoring to optimize energy consumption.

Table 5. Advisable enhancements for nanomaterial synthesis methodology [1].

The environmental safety of materials, which consist of or contain nanosize components, becomes one of the most important emerging topics of the Nanotechnology within the last few years. The main concerns dealing with the rapid development and commercialization of various nanomaterials are associated with [32,40,41]:

1. the approved higher toxicity of many nanomaterials (NMs) in comparison with their larger counterparts,
2. the absence of the adequate analytical techniques for detection of NMs in the environment
3. the absence of the legislation normative for permitted levels of various NMs in water and air.

In this regard the increase of the safety of NMs is of particular importance. One way to prevent risk is the development of the environmentally-safe polymer-metal nanocomposite materials that consist in a functional polymer with immobilized MNPs distributed mainly by the surface of the polymer with a higher stability to prevent release of the MNPs.

The material represents what makes them maximally accessible for the bacteria to be eliminated. Core-shell MNPs contain a superparamagnetic core coated with the functional metal shell, which provides the maximal bactericide activity. The MNPs are strongly captured inside the polymer matrix that prevents their escape into the medium under treatment. The superparamagnetic nature of MNPs provides an additional level of the material safety as MNPs leached from the polymer matrix can be easily captured by the magnetic traps to completely prevent any post-contamination of the treated medium.

4.2.1. Characterization of MNPs: key factor to ensure the safety of new technologies.

Nevertheless, the lack of specific characterization techniques of environmental effects of MNPs, existing and described methodologies should be modified to obtain valid results.

Some parameters must be taken into account in order to understand the relation between NPs behaviour and their physical and chemical structure [1,32,37,41,42].

Without detailed material physicochemical characterization, toxicity studies become difficult to interpret, and inter-comparison of studies becomes near impossible. Factors such as agglomeration state, surface chemistry, material source, preparation method, and storage take on a significance that has often been overlooked, potentially leading to inappropriate conclusions being drawn. Table 6 presents some approaches to the nanotoxicity evaluation [39,42].

This becomes particularly significant where hazard is dependent on structural and surface properties, as changes in these properties may lead to significant differences between the released (or basic) material, and the material people are exposed to. With no specific characterization techniques for nanotoxicity, the actual techniques are being modified and enhanced to determine and evaluate NPs effects.

5. Ecological safe MNPs or MONPs nanocomposites for bactericidal applications for water treatment.

Due to their relevant optical, electrical and thermal properties; Ag-NPs are being incorporated into several commercially available products such as biological and chemical sensors; as well as into bactericidal processes. The antibacterial features of Ag-NPs are one of the top topics of investigation into noble metals research.

Products as wound dressings and biomedical devices with Ag-NPs continuously release Ag in low levels that leads to protection against bacteria.

Considering the unusual properties of nanometric scale materials in contrast with those from macro counterparts, Ag-NPs are widely used for the more efficient antimicrobial activity compared with Ag^+ ions. The incorporation of magnetic cores to the preexisting nanocomposite materials increases the applicability of these in a macro scale for the easiest separation and the enhanced performance [43–47].

5.1. Bactericidal activity test for sulfonated nanocompositescontaining Ag@Co-NPs

In general, the bactericidal activity was determined as the relationship between the number of viable bacteria before and after the treatment in percentage terms (% cell viability) at several extractions/treatment times in all the tests Eq. 12)where t_f corresponds to the extraction time and t_o to the initial time).

$$\%CellViability = \frac{\left(CFU/_{mL}\right)_{t_f}}{\left(CFU/_{mL}\right)_{t_0}} x100 \tag{12}$$

The relationship between the Ag metal content in the sulfonated polymeric matrix and its antibacterial activity was then evaluated by following both batch and flow protocols.

The capacity of the nanocomposites to inhibit bacterial proliferation was evaluated by using the Minimum Inhibitory Concentration (MIC) test as a batch protocol, by using E. Coli. MIC is definedas the concentration of an antimicrobial agent that completely inhibits the microorganisms' proliferation in the sample. [23] Parallelly, the MIC_{50} corresponds to the antimicrobial concentration which inhibits just the 50%. In this case, the MIC of each material was determined by introducing an increasing amount of nanocomposite (in individual wells, from Microtiter plates with 96 wells,containing 10^5 CFU/mL of E. coli suspension in LB medium. After overnight incubation, bacterial proliferation was evaluated by measuring the optical density of each well at 550 nm (this wavelength is indicative of bacterial proliferation). The bactericidal activity of the Ag, Co and Ag@Co nanocomposites (in sulfonated polymeric granulated matrices) was determined as showin in Figure 12. As a result, the MIC_{50} values are expressed as number of nanocomposite beads in 200 μL of culture medium (beads/200 μL).

Assay / Technique	Aim	NPs applicability.
Synchrotron radiation based techniques	Distribution of NPs in different systems, analyze Oxidative Stress precursors, chemical speciation	Nanoscale zerovalent iron, TiO_2, ZnO, CeO_2. NPs
Colony forming efficiency Test	Cytotoxicity	Cobalt NPs
Transmission electron Microscopy (TEM)	Intracellular location, morphology.	Fullerene derivatives, ultrafine particles, metal NPs as AgNPs
Light Microscopy	Morphological observations	Single – wall carbon nanotubes, metal nanoparticles.
Neutral red Assay	Cell viability	Carbon nanotubes, Ag-NPs, Ti-NPs, TiO_2-NPs

Table 6. Overview of different techniques and assays for nanotoxicity evaluation.

The raw sulfonated material did not present inhibitory activity in the concentration range under test. However, it became antibacterial when modified with NPs providing a quite higher value of MIC_{50} (between 13-16 beads/200 μL) compared with that of Ag@Co- nanocomposites with the same Ag content (MIC_{50} around 4 beads/200 μL). The reason for the enhancement inhibition of bacteria proliferation recorded by Ag@Co-NPs in sulfonated matrices is still controversial. However, thanks to the better knowledge of Ag@Co-granulated nanocomposites obtained by further characterization with different techniques, it is possible to link some physic-chemical parameters with the final bactericidal activity of the materials.This best result is in agreement with the reported value for organo-silver compounds incorporated in microspheres (~ 0.125 mM) [24].

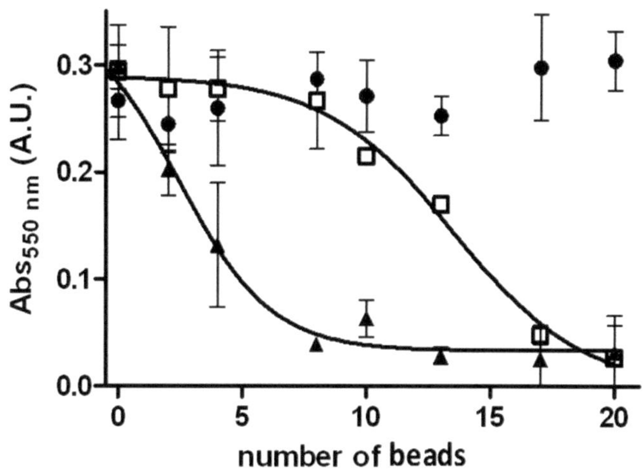

Figure 12. Variation of the absorbance at 550 nm with the number of polymer beads for (●) the raw material, (□) Ag- and, (▲) Ag@Co-sulfonated (C100E) nanocomposites (3 replicates).

In the flow method, nanocomposite-based filters containing Ag@Co-NPs or without NPs were set in a filtering column support of the experimental set-up and connected to a peristaltic pump that allowed the control of the flow rate. This set-up can operate by a single pass, when bacterial suspensions passed through the filter containing nanoparticles only once. The number of viable cells was determined at regular times. 10^3 CFU/mL of E. coli suspensions were forced to pass through the filter at a flow rate of ranging 1.0mL/min and the bactericidal activity of the material was evaluated.

Culture medium samples after passing through the column were extracted once a week under sterile conditions and the number of viable cells was determined. Figure 14 compares the % cell viability versus treatment time for sulfonated granulated material modified with Ag- or Ag@Co-NPs. Also, raw material response is shown.

The cell viability in the suspension after being treated by sulfonated nanocomposites for 60 min of continuous operation was found to decrease near to 0 %. Also, little differences between Ag and Ag@Co stabilized in different polymeric matrices are observed. It should be emphasized that, in this case, control samples showed also a decrease of % cell viability after 60 min of treatment. Therefore, these nanocomposites showed good performance and stability even under continuous operation.

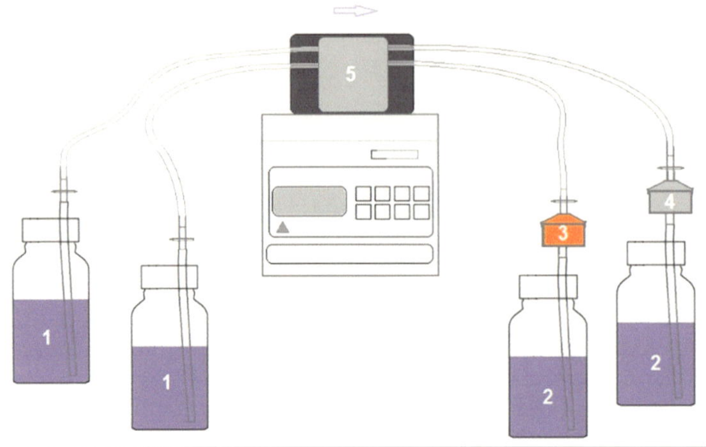

Figure 13. Scheme of the flow experimental one-step where:1. Initial Bacterial suspension, 2. Treated solution, 3. Nanocomposite filter (Ag@Co-NPs), 4. Control filter (without NPs), 5. Pump.

As it aforementioned, the Ag@Co-nanocomposites bactericidal activity was evaluated in granulated polymers. The nanocomposite showed high bactericidal activity with a cell viability close to 0 % for bacterial suspensions with an initial concentration below 10^5 CFU/mL) and only the more concentrated suspensions (over 10^5 CFU/mL) required recirculation to guarantee a complete bacterial removal.

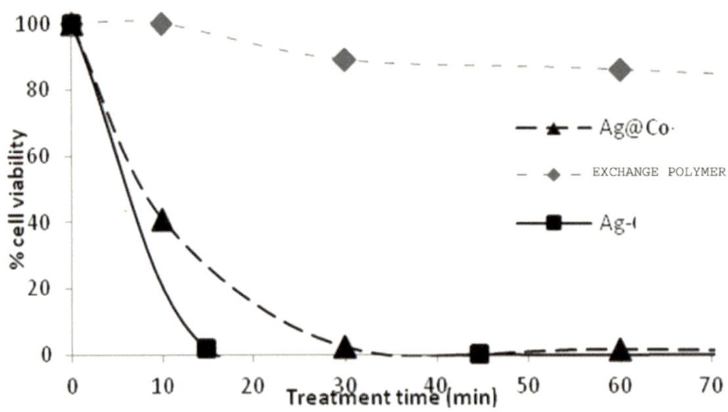

Figure 14. Representation of the variation of the % of cell viability with the treatment time for the Ag-, Ag@Co- and raw sulfonated granulated nanocomposite.

Also, the materials lifetime was tested obtaining high activities for different kinds of bacteria and applied in long term experiments. It was observed in all cases that bimetallic Ag@Co-NPs in any type of support showed higher bactericidal activity in comparison with monometallic Ag- or Co-NPs. However, the presence of Co showed high toxicity [14].

5.3. Bacterial applications test for solfonated granulated resins containing Ag@Fe_3O_4 nanocomposites.

Hence, the described Ag@Fe_3O_4-nanocomposites were tested and compared for antibacterial applications. In general, their antibacterial activity was evaluated by quantifying cell viability (% cell viability) at several extractions/treatment times after incubation with the E.coli bacteria by following the batch protocol as shown in Figure 15. It is determined the kinetics in terms of % of cell viability per mg of Ag for the samples to compare the activity for Ag- or Ag@Fe_3O_4-NPs on C100E polymers.

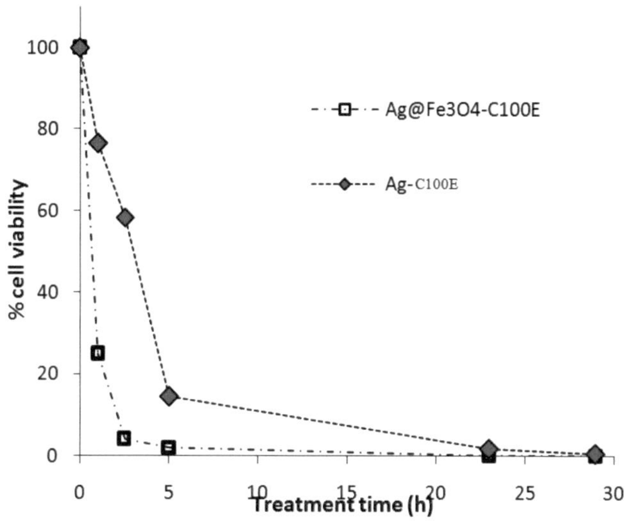

Figure 15. Cell viability versus treatment time for Ag@Fe_3O_4- nanocomposites in sulfonated polymers.

All the Ag@Fe_3O_4 samples showed initially, a fast decrease in cell viability what corresponds to a decrease of more than the 90 % after 2.5 h of treatment.

5.4. Bactericidal applications test for ammine based nanocomposite containing Ag@Fe_3O_4-NPs

The capacity of the nanocomposites to inhibit bacterial proliferation was evaluated by using the MIC test [24] by using E. Coli as described before.

The MIC of both Ag- and Ag@Fe_3O_4-A520E was determined and compared with that obtained by the raw material without NPs or containing Fe_3O_4-NPs as shown in Figure 16.

Ag- and Ag@Fe_3O_4-A520E nanocomposites showed high bactericidal activity with a deep decrease of the absorbance magnitude at 550 nm (Abs_{550} when increasing the number of nanocomposite beads in the suspension. Conversely, raw material and Fe_3O_4-nanocomposite did not present significant bactericidal activity at this concentration range, with a constant Abs_{550} value around 0.4 a.u. in all cases.

This result indicated that Ag-NPs were responsible of the bactericidal activity recorded and it was not affected by the presence of magnetite.

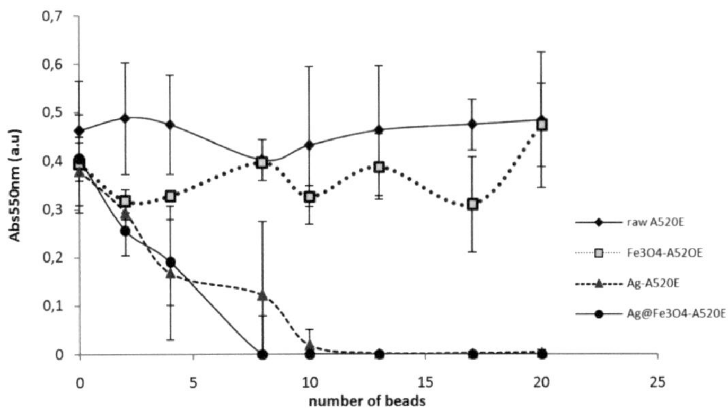

Figure 16. Variation of the absorbance at 550 nm with the number of polymer beads for the Ag- and Fe_3O_4-based A520E nanocomposites. Raw material is also analysed (3 replicates).

6. Conclusions

The following conclusions could be derived from the results and discussion shown in this chapter.

- It was shown that the Intermatrix Synthesis (IMS) method is applicable to all the ion-exchange materials tested and useful to any kind of NPs composition since, also, the coupling of the IMS with the co-precipitation technique was succeed to obtain magnetite-based nanocomposites.
- IMS methodology coupling with Donnan Exclusion Effect was observed for the NPs structures synthesized on the ion exchange polymers. All NPs were highly stabilized on the surface of the polymer and showed magnetic properties what allows their recovery by applying a magnetic trap.
- Also, the development of the IMS route to the synthesis of NPs on anion exchanger polymers was obtained showing comparable results than the materials formed by using cationexchangers.
- The rapid growth of interest in engineered NPs has presented many challenges for ecotoxicology, not least being the effort required to analyse and understand the NPs themselves.

- Given the importance of potable water to people, it is a clear need for the development of innovative new technologies and materials whereby challenges associated with the provision of safe potable water can be addressed.
- Ag@Co and Ag@Fe_3O_4-NPs stabilized in both anion and cationexchanger polymeric resins were applied for the water purification against E.coli bacteria suspensions. Both NPs structure showed high bactericidal activity being, however, less citotoxic the materials containing Fe_3O_4-NPs.
- Thus, an efficient material based on Ag@Fe_3O_4-NPs stabilized in anion and cation exchanger polymeric resins was obtained for water treatment applications by showing high bactericide activity as well as low citotoxity for animal cells.

Acknowledgements

We are sincerely grateful to all our associates cited throughout the text for making this publication possible. Part of this work was supported by Research Grant MAT2006-03745, 2006–2009 from the Ministry of Science and Technology of Spain, which is also acknowledged for the financial support of Dmitri N. Muraviev.

We also thank ACC1Ó for VALTEC 09-02-0057 Grant within "Programa Operatiu de Catalunya" (FEDER). AGAUR is also acknowledged for the support of A.Alonso with the predoctoral FI and BE grants. J. Bastos also thanks the Autonomous University of Barcelona for the personal grant.

Author details

Amanda Alonso[1], Julio Bastos-Arrieta[1], Gemma.L. Davies[2], Yurii.K. Gun'ko[2], Núria Vigués[3], Xavier Muñoz-Berbel[4], Jorge Macanás[5], Jordi Mas[3], Maria Muñoz[1] and Dmitri N. Muraviev[1]

*Address all correspondence to: Dimitri.Muraviev@uab.cat

1 Analytical Chemistry Division, Department of Chemistry, Autonomous University of Barcelona, 08193 Bellaterra, Barcelona, Spain

2 Trinity College Dublin, Dublin 2, , Ireland

3 Department of Genetics and Microbiology, Autonomous University of Barcelona, 08913, Bellaterra, Barcelona, Spain

4 Centre Nacional de Microelectrònica (IMB-CNM, CSIC), 08913, Bellaterra, Barcelona, Spain

5 Department of Chemical Engineering, Universitat Politècnica de Catalunya (UPC), 08222, Terrassa, Spain

References

[1] Dahl, J.A., Maddux, B.L.S., & Hutchison, J.E. (2007). Toward greener nanosynthesis. *Chemical reviews [Internet]*, 107(6), 2228-69, http://www.ncbi.nlm.nih.gov/pubmed/17564480.

[2] Campelo, J. M., Luna, D., Luque, R., Marinas, J. M., & Romero, A. A. (2009). Sustainable preparation of supported metal nanoparticles and their applications in catalysis. *ChemSusChem [Internet]*, 2(1), 18-45, http://www.ncbi.nlm.nih.gov/pubmed/19142903.

[3] Erkey, C. (2009). *The Journal of Supercritical Fluids Preparation of metallic supported nanoparticles and films using supercritical fluid deposition*, 47, 517-22.

[4] Han, B. J., Liu, Y., & Guo, R. (2009). *Reactive Template Method to Synthesize Gold Nanoparticles with Controllable Size and Morphology Supported on Shells of Polymer Hollow Microspheres and Their Application for Aerobic Alcohol Oxidation in Water*, 1112-7.

[5] Han, B. J., Liu, Y., & Guo, R. (2009). *Reactive Template Method to Synthesize Gold Nanoparticles with Controllable Size and Morphology Supported on Shells of Polymer Hollow Microspheres and Their Application for Aerobic Alcohol Oxidation in Water*, 1112-7.

[6] Macanás, J., Bastos-arrieta, J., Shafir, A., Alonso, A., Mu, M., & Muraviev, D.N. (2012). *Article in press*, 10-3.

[7] Muraviev, D. N., Alonso, A., Shafir, A., & Vallribera, A. (2012). *Article in press.*

[8] Muraviev, D., Macanas, J., Farre, M., Munoz, M., & Alegret, S. (2006). Novel routes for inter-matrix synthesis and characterization of polymer stabilized metal nanoparticles for molecular recognition devices. *Sensors and Actuators B: Chemical [Internet], [cited 2011 May 9]*, 118(1-2), 408-17, http://linkinghub.elsevier.com/retrieve/pii/S092540050600325X.

[9] Parrondo, J., & Mun, M. (2007). *Cation-exchange membrane as nanoreactor: Intermatrix synthesis of platinum- copper core- shell nanoparticles.*, 67, 1612-21.

[10] Ruiz, P., Muñoz, M., Macanás, J., & Muraviev, D.N. (2011). Reactive & Functional Polymers Intermatrix synthesis of polymer-stabilized PGM @ Cu core- shell nanoparticles with enhanced electrocatalytic properties. *Reactive and Functional Polymers [Internet]*, 71(8), 916-24, http://dx.doi.org/10.1016/j.reactfunctpolym.2011.05.009.

[11] Taylor, P., Muraviev, D. N., Pividory, M. I., Luis, J., & Soto, M. (2012). *Solvent Extraction and Ion Exchange Extractant Assisted Synthesis of Polymer Stabilized Platinum and Palladium Metal Nanoparticles for Sensor Applications Extractant Assisted Synthesis of Polymer.*

[12] Barbaro, P., & Liguori, F. (2009). Ion exchange resins: catalyst recovery and recycle. *Chemical reviews [Internet]*, 109(2), 515-29, http://www.ncbi.nlm.nih.gov/pubmed/19105606.

[13] Zegorodni, A. (2006). Ion exchange material. *Elsevier [Internet], [cited 2012 May 27]*, http://www.sciencedirect.com/science/book/9780080445526.

[14] Alonso, A., Muñoz-Berbel, X., Vigués, N., Rodríguez-Rodríguez, R., Macanás, J., Mas, J., et al. (2012). Intermatrix synthesis of monometallic and magnetic metal/metal oxide nanoparticles with bactericidal activity on anionic exchange polymers. *RSC Advances [Internet], [cited 2012 Apr 27]*, 2(3), 4596-9, http://xlink.rsc.org/?DOI=c2ra20216.

[15] Kung, H.H., & Kung, M.C. (2004). Nanotechnology: applications and potentials for heterogeneous catalysis. *Catalysis Today [Internet], Nov [cited 2011 Aug 3]*, 97(4), 219-24, http://linkinghub.elsevier.com/retrieve/pii/S0920586104004262.

[16] Alonso, A., Macanás, J., Shafir, A., Muñoz, M., Vallribera, A., Prodius, D., et al. (2010). Donnan-exclusion-driven distribution of catalytic ferromagnetic nanoparticles synthesized in polymeric fibers. *Dalton transactions (Cambridge, England: 2003) [Internet], Mar [cited 2011 Feb 24]*, 39(10), 2579-86, http://www.ncbi.nlm.nih.gov/pubmed/20179851.

[17] Ayyad, O., Muñoz-Rojas, D., Oró-Solé, J., & Gómez-Romero, P. (2009). From silver nanoparticles to nanostructures through matrix chemistry. *Journal of Nanoparticle Research*, 12(1), 337-45.

[18] Lira-Cantú, M., & Gómez-Romero, P. (1997). Cation vs. anion insertion in hybrid materials based on conducting organic polymers for energy storage applications. *Ionics*, 3(3-4), 194-200.

[19] Ruiz, P., Muñoz, M., Macanás, J., Turta, C., Prodius, D., & Muraviev, D. N. (2010). Intermatrix synthesis of polymer stabilized inorganic nanocatalyst with maximum accessibility for reactants. *Dalton transactions (Cambridge, England: 2003) [Internet], Feb [cited 2011 Feb 24]*, 39(7), 1751-7, http://www.ncbi.nlm.nih.gov/pubmed/20449418.

[20] Ruiz, P., Muñoz, M., Macanás, J., & Muraviev, D. N. (2010). Intermatrix Synthesis of Polymer−Copper Nanocomposites with Tunable Parameters by Using Copper Comproportionation Reaction. *Chemistry of Materials [Internet], Dec [cited 2011 Feb 24]*, 22(24), 6616-23, http://pubs.acs.org/doi/abs/10.1021/cm102122c.

[21] Muraviev, D. N., Ruiz, P., Muñoz, M., & Macanás, J. (2008). Novel strategies for preparation and characterization of functional polymer-metal nanocomposites for electrochemical applications. *Pure and Applied Chemistry [Internet], cited 2011 Feb 24]*, 80(11), 2425-37, http://iupac.org/publications/pac/80/11/2425/.

[22] Muraviev, D.N., Macanás, J., Ruiz, P, & Muñoz, M. Synthesis, stability and electrocatalytic activity of polymer-stabilized monometallic Pt and bimetallic Pt/Cu core-shell nanoparticles. *Physica Status Solidi (a) [Internet]. 2008 Jun [cited 2012 May 23]*, 205(6), 1460-4, http://doi.wiley.com/10.1002/pssa.200778132.

[23] Ferreira, T. A. S., Waerenborgh, J. C., Mendonça, M. H. R. M., Nunes, M. R., & Costa, F. M. (2003). *Structural and morphological characterization of FeCo 2 O 4 and CoFe 2 O 4 spinels prepared by a coprecipitation method*, 5, 383-92.

[24] Alonso, A. (2012). Development of polymeric nanocomposites with enhanced distribution of catalytically active or bactericide nanoparticles. *PhD Thesis. Universitat Autònoma de Barcelona.*

[25] Poulter, N., Muñoz-Berbel, X., Johnoson, A.L., Dowling, A.J., Waterfield, N., & Jenkis, T.A. (2009). *Chemical Communications*, 7312-7314.

[26] Toksha, B. G., Shirsath, S. E., Patange, S. M., & Jadhav, K. M. (2008). *Structural investigations and magnetic properties of cobalt ferrite nanoparticles prepared by sol- gel auto combustion method*, 147, 479-83.

[27] Search, H., Journals, C., Contact, A., & Iopscience, M. (2003). Address IP. *Applications of magnetic nanoparticles in biomedicine*, 167.

[28] Amara, D., Felner, I., Nowik, I., & Margel, S. (2009). *Colloids and Surfaces A: Physicochemical and Engineering Aspects Synthesis and characterization of Fe and Fe 3 O 4 nanoparticles by thermal decomposition of triiron dodecacarbonyl*, 339, 106-10.

[29] Æ, PCÆFVDK, & Hofmann, MBÆT. (2008). Nanoparticles: structure, properties, preparation and behaviour. *Environmental media*, 326-43.

[30] Arora, S., Rajwade, J. M., & Paknikar, K. M. (2012). Nanotoxicology and in vitro studies: The need of the hour. *Toxicology and Applied Pharmacology [Internet]*, 258(2), 151-65, http://dx.doi.org/10.1016/j.taap.2011.11.010.

[31] Blaser, S. A., Scheringer, M., Macleod, M., & Hungerbühler, K. (2007). Estimation of cumulative aquatic exposure and risk due to silver: Contribution of nano-functionalized plastics and textiles. *Europe*.

[32] Cushen, M., Kerry, J., Morris, M., Cruz-romero, M., & Cummins, E. (2011). Nanotechnologies in the food industry e Recent developments, risks and regulation. *Trends in Food Science & Technology [Internet]*, http://dx.doi.org/10.1016/j.tifs.2011.10.006.

[33] Chen-fang, M., & Li-lan, H. (2007). The consumer' s attitude toward genetically modified foods in Taiwan. *Food Quality and Preference*, 18(40), 662-74.

[34] Innovation, N., & Island, R. (2010). Ion Release Kinetics and Particle Persistence in Aqueous Nano-Silver Colloids. *Environmental Science & Technology*, 2169-75.

[35] Ry, G. V. L. O. W., Rez, P. A. L. V. A., Ysios, D. I. A. N., Ysiou, D., & Biswas, P. R. A. Assessing the Risks of Manufactured. *Environmental Science & Technology*.

[36] Maynard, A.D. (2006). *Nanotechnology: Nanotechnology is seen as a transformative technology, which has the*, 1(2), 22-33.

[37] Wagner, B., Marconi, F., Kaegi, R., Odzak, N., & Box, P. O. (2008). *Toxicity of Silver Nanoparticles to Chlamydomonas reinhardtii*, 8959-64.

[38] Tiede, K., Hassellöv, M., Breitbarth, E., Chaudhry, Q., Boxall, A.B.A., Hutton, S., et al. (2009). *Considerations for environmental fate and ecotoxicity testing to support environmental risk assessments for engineered nanoparticles*, 1216, 503-9.

[39] Simonet, B. M., & Valcárcel, M. (2009). *Monitoring nanoparticles in the environment*, 17-21.

[40] Siegrist, M., Cousin-eve, M., Kastenholz, H., & Wiek, A. (2007). *Public acceptance of nanotechnology foods and food packaging: The influence of affect and trust. Appetite*, 49, 459-66.

[41] Press AIN. (2007). Nanotechnologies: What we do not know. *Technology in Society*, 29, 43-61.

[42] Nowack, B. (2008). *Exposure Modeling of Engineered Nanoparticles in the Environment*, 41(0), 4447-53.

[43] Savolainen, K., Pylkkänen, L., Norppa, H., Falck, G., Lindberg, H., Tuomi, T., et al. (2010). Nanotechnologies, engineered nanomaterials and occupational health and safety- A review. *Safety Science [Internet]*, 48(8), 957-63, http://dx.doi.org/10.1016/j.ssci.2010.03.006.

[44] Midander, K., Cronholm, P., Karlsson, H.L., Elihn, K., Leygraf, C., & Wallinder, I.O. (2009). Surface Characteristics, Copper Release, and Toxicity of Nano- and Micrometer-Sized Copper and Copper (II) Oxide Particles: A Cross-Disciplinary Study. *Small* [3], 389-99.

[45] Siegrist, M., Stampfli, N., Kastenholz, H., & Keller, C. (2008). Perceived risks and perceived benefits of different nanotechnology foods and nanotechnology food packaging. *Appetite*, 51, 283-90.

[46] Hu-song, B. J., Zhong-shu, L., Song-guo, W., & Wan-jun, L. (2008). Synthesis of Hierarchically Structured Metal Oxides and their Application. *Heavy Metal Ion Removal*, 2977-82.

[47] Lv, Y., Liu, H., Wang, Z., Liu, S., Hao, L., Sang, Y., et al. (2009). *Silver nanoparticle-decorated porous ceramic composite for water treatment*, 331, 50-6.

[48] Rajh, T., Chen, L. X., Lukas, K., Liu, T., Thurnauer, M. C., & Tiede, D. M. (2002). *Surface Restructuring of Nanoparticles: An Efficient Route for Ligand-Metal Oxide Crosstalk*, 10543-52.

[49] Theron, J., Walker, J. A., & Cloete, T. E. (2008). *Nanotechnology and Water Treatment: Applications and Emerging Opportunities*, 43-69.

[50] Zhon-shu, BL., Hu-song, J., Liang-pu, H., Cao-min, A., Song-guo, W., & Wan-jun, L. (2006). Self-Assembled 3D Flowerlike Iron Oxide Nanostructures and Their Application. *Water Treatment*, 2426-31.

Chapter 9

Impact Response of Nanofluid-Reinforced Antiballistic Kevlar Fabrics

Roberto Pastore, Giorgio Giannini,
Ramon Bueno Morles, Mario Marchetti and
Davide Micheli

Additional information is available at the end of the chapter

http://dx.doi.org/10.5772/50411

1. Introduction

In the last decades the research on composite materials have been acquiring importance due to the possibility of increasing the material mechanical performances while contemporary decreasing both mass and volume of the structures. Mass lowering is a "must" especially in military and space applications, since aircraft aerodynamic profile needs to be optimized and because of the high costs of launch and launcher and payload mass constraints [1]. The need to face up to the well know problem of the so called "space debris" has led many aerospace researchers to look for advanced lightweight materials for ballistic applications. Among all innovative materials, a promising branch of such research focuses on the polymeric composite materials with inclusions of nanostructures [2]. The present work fits in a more general research project, the aim of which is to realize, study and characterize nanocomposite materials. These latter are currently manufactured in the SASLab of Astronautic Engineering Department of University of Rome "Sapienza" *(www.saslab.eu)* by mixing the nanoparticles within polymeric matrixes in such a way to obtain a material as homogeneous as possible, in order to have a final composite with improved physical characteristic [3]. The goal of the present study is to perform a ballistic characterization of the nanocomposites by means of an in-house built electromagnetic accelerator. The realization of such experimental apparatus, and mostly the optimization with a view to space debris testing planes, is quite complex since the fundamental machine parameters have high non-linearity theoretical behavior [4]. Hereafter experimental preliminary results of a prototypal device are presented and discussed. An intriguing issue of nanoscience research for aerospace applications is to produce a new thin, flexible, lightweight and inexpensive material that have an equivalent

INTECH
open science | open minds

© 2012 Pastore et al.; licensee InTech. This is a paper distributed under the terms of the Creative Commons Attribution License (http://creativecommons.org/licenses/by/3.0), which permits unrestricted use, distribution, and reproduction in any medium, provided the original work is properly cited.

or even better ballistic properties than the existing Kevlar fabrics. A shear thickening fluid (STF) is a material with remarkable properties [5]. STFs are very deformable materials in the ordinary conditions and flow like a liquid as long as no force is applied. However they turn into a very rigid solid-like material at high shear rates. Shear thickening is a non-newtonian fluid behavior defined as the increase of viscosity with the increase in the applied shear rate. This phenomenon can occur in micro/nano colloidal dispersions. More concentrated colloidal suspensions have been shown to exhibit reversible shear thickening resulting in large, sometimes discontinuous, increases in viscosity above a critical shear rate. Two main causes of reversible shear thickening have been proposed: the order-disorder transition and the "hydrocluster" mechanism. This transition from a flowing liquid to a solid-like material is due to the formation and percolation of shear induced transient aggregates, or hydroclusters, that dramatically increase the viscosity of the fluid. Support for such mechanism has been demonstrated experimentally through rheological, rheo-optics and flow-SANS experiments as well as computer simulation. It has been reported in the literature that shear thickening has been observed for a wide variety of suspensions such as clay-water, calcium carbonate-water, polystyrene spheres in silicon oil, iron particles in carbon tetrachloride, titanium dioxide-resin, silica-polypropylene glycol, and silica-ethylene glycol. The phenomenon of shear thickening of suspensions in general has no useful applications in industrial production. Recently Wagner's group and U.S. Army research lab developed a body armor using shear thickening fluid and Kevlar fabric [6]. These research results demonstrate that ballistic penetration resistance of Kevlar fabric is enhanced by impregnation of the fabric with a colloidal STF. Impregnated STF/fabric composites are shown to provide superior ballistic protection as compared with simple stacks of neat fabric and STF. Comparisons with fabrics impregnated with non-shear thickening fluids show that the shear thickening effect is critical to achieving enhanced performance. Many researchers have used various techniques to prepare the STFs. Acoustic cavitations technique is one of the efficient ways to disperse nanoparticles into the liquid polymers. In this case, the application of alternating acoustic pressure above the cavitations threshold creates numerous cavities in the liquid. Some of these cavities oscillate at a frequency of the applied field (usually 20 kHz) while the gas content inside these cavities remains constant. However, some other cavities grow intensely under tensile stresses while yet another portion of these cavities, which are not completely filled with gas, starts to collapse under the compression stresses of the sound wave. In the latter case, the collapsing cavity generates tiny particles of 'debris' and the energy of the collapsed one are transformed into pressure pulses. It is noteworthy that the formation of the debris further facilitates the development of cavitation. It is assumed that acoustic cavitations in liquids develop according to a chain reaction. Therefore, individual cavities on real nuclei are developing so rapidly that within a few microseconds an active cavitations region is created close to the source of the ultrasound probe. The development of cavitations processes in the ultrasonically processed melt creates favorable conditions for the intensification of various physical-chemical processes. Acoustic cavitations accelerate heat and mass transfer processes such as diffusion, wetting, dissolution, dispersion, and emulsification. SASLab objective in this research field is to synthesize a STF in a single step reaction through high power ultrasound technique, fabricate STF/fabric composite and characterize

http://dx.doi.org/10.5772/50411

it for ballistic resistance applications. The STF is a combination of silicon dioxide (silica) nanoparticles suspended in a liquid polymer. This mixture of flowable and hard components at a particular composition results in a material with remarkable properties. The STF is prepared by ultrasound irradiation of silica nanoparticles dispersed in liquid polyethylene glycol polymer. The as-prepared STFs are then tested for their rheological properties. Kevlar fabrics are soaked in STF/ethanol solution to make STF/fabric composite. Ballistic tests are performed on the neat fabrics and STF/fabric composite targets. The results show that STF impregnated fabrics have better penetration resistance as compared to neat fabrics, without affecting the fabric flexibility. That indicates that the STF addition to the fabric may enhance the fabric performance and thus can be used for ballistic applications.

2. Materials manufacturing and characterization

In this section the procedures adopted for the shear thickening nanofluids realization and the Kevlar-reinforced fabrics manufacturing are basically described, providing rheological and morphological (SEM *TESCAN-Vega LSH,* Large Stage High Vacuum scanning electron microscope) characterization of the materials under testing too. Silica nanoparticles (n-SiO_2, *Sigma Aldrich* Fumed Silica powder 0.007μm) and Polyethylene glycol (PEG, *Sigma Aldrich* Poly(ethylene glycol) average mol wt 200) have been chosen as nanofiller and carrier fluid respectively, to follow the tracks of the most remarkable results in STF applications for absorbing impact energy [6-8]. Ethanol (*Sigma Aldrich* Ethanol puriss. p.a., ACS reagent) was used as solvent for nanopowder disentanglement and dispersion within the polymeric matrix. SEM images of the as-received silica nanoparticles are shown in Figure 1 below.

Figure 1. Low (left) and medium (right) magnification SEM photos of the silica nanoparticles adopted; the BSE image enhances the high degree of the as-received material homogeneity.

Several are the parameters of the procedure for the solution preparation and, moreover, their combination greatly affects the efficacy of the later fabrics impregnation (i.e. the effective nanofluid amount intercalated between the fibers) as well as the ballistic behavior of the final manufactured test article. Mixing tools are fundamental to achieve a correct preparation of the nanostructured solution, since a suitable nanoparticles dispersion inside the liquid matrix and the subsequent mixture homogenization are absolutely not trivial tasks, in particular in case (as the present) of substantial filler weight percentages. Nanopowders very high "surface area" (surface/weight ratio, in the case of the adopted nanoparticles the value 390 m^2/g is reported in the data sheet) give rise to so huge volumetric gaps between the host fluid and the filler (dry nanopowder volume may be one order of magnitude greater than the matrix volume). Beside the dilution in organic solvent (which must be used in controlled excess to avoid the whole mixture degradation) a first mechanical mixing (performed by *Velp Scientifica* Magnetic Stirrer BS Type 0÷2000rpm) to gradually introduce the nanoparticles in solution is needed. Then, 20kHz ultrasonication technique (by *Sonics & Materials* VC 750 Ultrasonic Liquid Processor) is adopted to exfoliate the micrometric agglomerates in which the nanoparticles are typically entangled, in order to increase the mixture homogeneity as well as to reduce the presence of internal air voids. During this step an increase by several tens degrees of the solution temperature may occur, due to the relatively high energy quantities exchanged: as each case requires (i.e. depending on parameters as solvent amount, evaporation rate, compound thermal stability, etc.) the sonication can be carried out in thermostatic environment or not. Of course, the timing procedure strictly depends on the material quantities utilized, which in turn are linked to the characteristics (surface dimensions and absorption rate) of the specific fabric typology treated. Schematically, the method for the preparation of by about 120g of nanofluid loaded at 20wt%, an amount estimated to perform the full treatment of eleven 16×16cm layers of reference batavia Kevlar fabric (see below) consists of the following steps: 60g of PEG mechanical 500rpm mixing in 200÷300ml of solvent for about 10 minutes, gradual addition of 12g of silica nanoparticles, high energy ultrasonication (50% of mixer maximum power) for about 30 minutes, low energy ultrasonication (25% of mixer maximum power) for about 4 hours in low temperature (0÷5 C) environment, and low energy ultrasonication for about 2 hours in warming temperature (up to 50 C). The result of such procedure is an homogeneous solution of volume reduced to 80÷120ml, mainly due to the evaporation of a certain amount of solvent as well as to the nanoparticles/polymeric macromolecules coupling inside the solution (testified by an evident chromatic transition from opaque to quasi-transparent solution). Whereas the next step for the fabric-reinforced manufacture should be the fabric impregnation in the solution (followed by the total evaporation of the solvent in excess), in order to obtain directly a fluid with non-newtonian (bulk) properties the complete solvent evaporation is required (6÷8 hours at 70÷80 C are typically enough). In Figure 2 and Figure 3 the morphological and rheological characterizations of several nanosilica wt% filled PEG solutions are respectively reported: in the SEM images the n-SiO_2/PEG chemical interaction is highlighted, while the viscosity/shear measurements (performed by parallel-plate rheometer) give evidence of the STF fashion as from 10wt% of nanosilica inclusion (showing the typical knee [5,9,10] at shear of about 10Hz) and a quasi-solid behavior for 15wt% and over loading.

Three textile materials have been treated with the different wt% loaded solutions realized. In Table 1 below their main characteristics are listed. XP Kevlar is highlighted to pick out its reference as starting best material in terms of density and claimed ballistic properties: this advanced *DuPont* material is produced by not trivial polymeric/fiber intercalation treatment, resulting in high compact thin lightweight paper-like flexible structure. B Kevlar is a conventional typology of aramidic fiber woven, while hybrid KN material results from an experimental try to couple spongy waste Kevlar to commercial Nylon fabric.

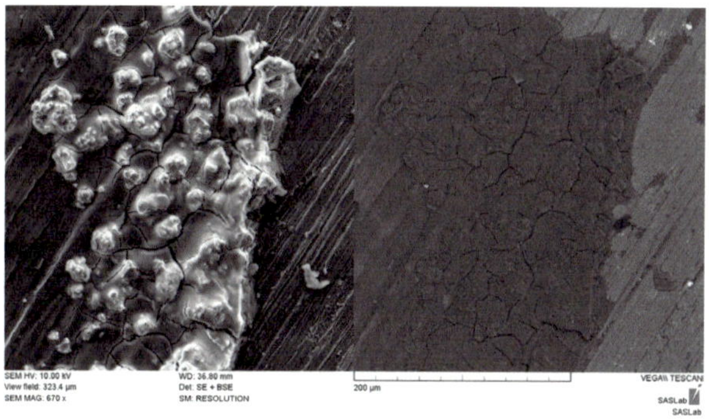

Figure 2. SEM photos of a drop-sample from a solution realized with n-SiO_2 at 20wt% inside PEG matrix after the ethanol evaporation. The SE image (left) shows the coupled morphology of the two chemical species, the correspondent BSE one (right) enhances the excellent mixture uniformity degree and the very low amounts of inner voids.

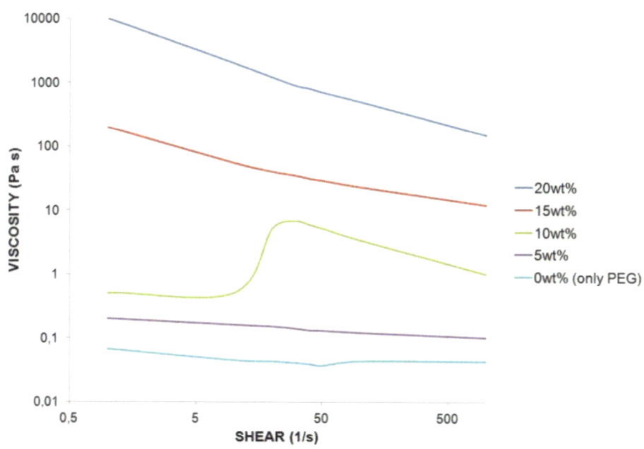

Figure 3. Viscosity/shear behavior of mixtures loaded with different weight percentages of nanosilica (measurements performed by parallel plate rheometer, shear 1 1/s to 1000 1/s, T 25 C): an evident phase transition (non newtonian behavior) is detectable around shear values of 10 Hz for the solution loaded at 10wt%.

The fabric impregnation with n-SiO_2/PEG mixtures diluted in solvent has to take place in relatively prompt way, in order to avoid unevenness treatment of the several layers due to potential physical/chemical changes of the post-sonicated solution (further solvent evaporation, filler sedimentation, cluster formation, etc.). For each kind of fabric and of mixture concentration the suitable fluid amount needed to achieve the maximum absorption is preliminary estimated. That is necessary because the highly diluted solutions saturate the fabrics with an effective n-SiO_2/PEG absorption lesser than how potentially possible, as clear from halfway imbibitions and weight control operations. Such evaluation is performed by wetting drop by drop a layer of fixed dimensions until the first saturation, waiting for solvent evaporation in oven at 70÷80 C (considered run out when the weight reduction is less than 1% for measurements taken one hour apart), impregnating again and so on, until the dry layer weight has stabilized. For each kind of material the absorption properties (i.e. the weight increase) must be strictly linked to the results of the ballistic test, thus are reported in details in the experimental section. From a qualitative point of view, the following general considerations can be pointed out by visual inspection as well as SEM morphology investigations (Figures 4,5): very high concentration (>20wt%) mixtures reinforced fabric show a so poor manufacturing degree (Figures 4a-b), with the presence of a clotty gloss weakly attached to the layer surfaces; XP fabric is basically refractory to the treatment due to its above mentioned chemical composition, as clear from so low absorption rates and structure degradation phenomena (cfr. Figures 4c-d); B fabric shows the best behavior in terms of fibers-nanofluid interaction, resulting in highly uniform woven bulk structure (Figures 4e-f); KN fabric is treated only on the Kevlar side, which presents a tridimensional woven mat (Figure 5a) that assists the absorption mechanism (Figures 5c-d), while the hydrophobic Nylon backside surface (Figure 5b) doesn't show any kind of interaction with the fluid.

material	symbol	areal density (kg/m^2)
DuPont™ Kevlar XP	**XP**	0.51
Saatilar batavia 4/4	**B**	0.62
hybrid Kevlar-Nylon	**KN**	0.65

Table 1. Main characteristics of the three Kevlar-based fabric tested.

The procedure of fabric impregnation consists in the simply dipping within a bowl filled with the suitable solution amount, then the layers are squeezed (Figure 6a) and put inside the oven for the solvent evaporation (typically 6÷8h at 70÷80 C, see Figure 6b). Finally, the treated layers are enveloped with polyethylene sheets (Figure 6c-d) in order to minimize the loss of material not perfectly stuck on the surfaces and to avoid unwanted interaction at the interfaces between neat/treated surfaces (lubrication or degradation of fluid incompatible fabrics). Fabric-reinforced flexibility has been discovered essentially unchanged comparing with neat material, even in the case of treatments with high concentration nanofluid mixtures.

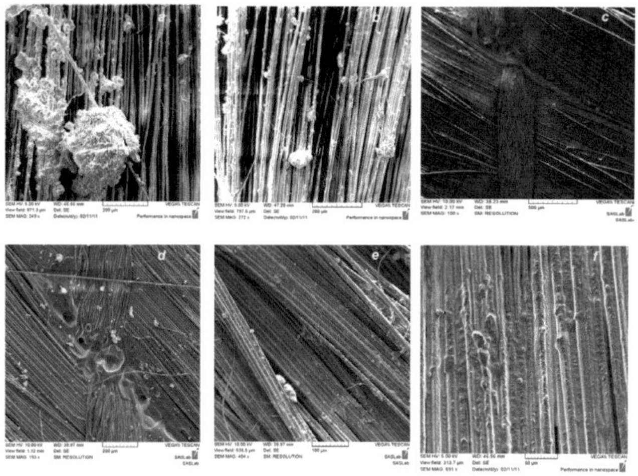

Figure 4. SEM images of neat/STF-reinforced fabrics: a)-b) B fabric treated with 50wt% STF solution; c) XP fabric neat morphology; d) XP fabric treated with 10wt% STF solution; e)-f) B fabric treated with 10wt% STF solution.

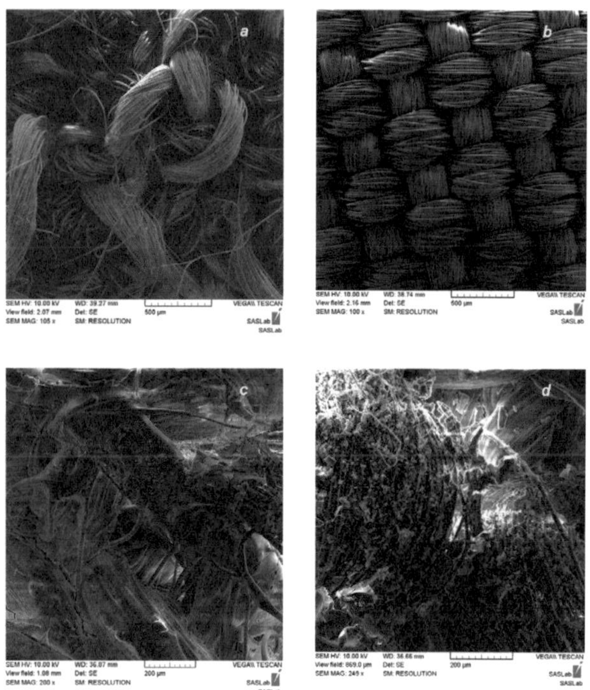

Figure 5. SEM images of neat/STF-reinforced fabrics: a) KN fabric neat morphology, Kevlar side; b) KN fabric neat morphology, Nylon side; c) KN fabric treated on Kevlar surface with 10wt% STF solution; d) KN fabric treated on Kevlar surface with 20wt% STF solution.

Figure 6. Different step of Kevlar-reinforced based antiballistic panels manufacturing: a) fluid application onto layers surface, b) solvent evaporation in oven, c) enveloping procedure, d) the panels ready for the ballistic test.

3. Experimental set-up

The ballistic characterization of the above described manufactured materials has been performed by means of an in-house built device called Coil Gun (CG), that is a typology of the more general electromagnetic accelerators equipment class. The idea to use intense electromagnetic pulses to exploit the intriguing matter/field interaction for propulsion applications is not so new, the first scientific researches in this area being carried out since many decades ago [4,11,12]. Several important results have been achieved in terms of ballistic performances, at the present time [13,14], anyway, the technological challenge is to reduce the system devices cost, weight and dimensions in order to compete with the conventional ballistic facilities. The CG basic background is the well known phenomenon of attraction suffered by a ferromagnetic body toward the middle of an hollow coil when a fixed current flows through this latter. As schematically depicted in Figure 7, the current flow produces an axial magnetic field inside the coil with maximum value (proportional to current intensity and coil turns number) around the coil central zone; the magnetic field decreases of about one half nearby the two coil's ends and goes rapidly to zero outside. A ferromagnetic object located not so far from one end of the coil suffers a strong magnetization (usually several order of magnitude greater than the magnetic

http://dx.doi.org/10.5772/50411

induction, due to the high magnetic permeability of ferromagnetic materials), thus resulting in an axial force depending in intensity and sign from the first derivative of the magnetic field [15]. In the case of a continuous steady current, the object's equilibrium position is clearly the center of the coil (i.e. where the body's center of mass fits to that of the coil), that is reached by friction after some (very fast) oscillations back and forth around the equilibrium center. If, on the contrary, an high current pulse is provided in such a way (i.e. with a characteristic time-constant) that the intensity falls to zero when the object is just coming to the coil's middle zone, then the backward recalling force is cut off and the object may move fast forward (and outside the coil) without kinetic energy loss. One simple way to obtain a pulse of current is to produce a capacitor discharge: a CG system thus works by exploiting a capacitor discharge across an inductance, that is via an RLC circuit discharge (Figure 8). In other words, the aim of the CG is to shoot a ferromagnetic bullet by converting the electrostatic energy stored in a capacitance into projectile's kinetic energy, thanks to the switch to magnetic energy inside an inductance coiled round the projectile's barrel.

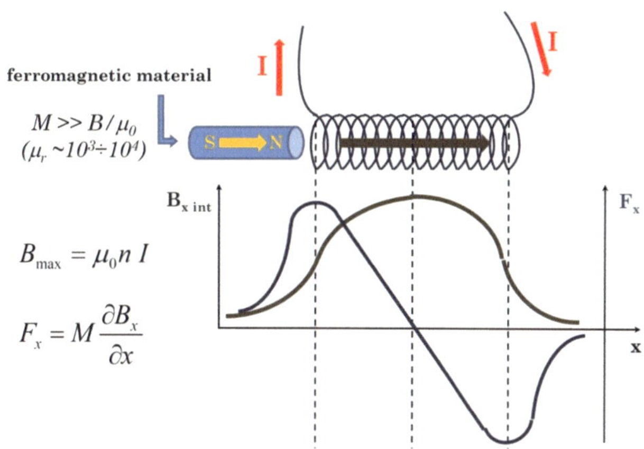

Figure 7. Coil Gun basic background: schematic representations and expressions of the magnetic field inside the coil due to the current flow, the induced magnetization of the ferromagnetic body, and the force acting on this latter.

The in-house built CG is shown in Figure 9: the main parts of the CG are the coil inductor, the projectile's barrel (that acts as support for the coil wrapping), the capacitors bank, the switch system, and the rectifier diodes. The coil inductor increases the acceleration of the projectile during its passage across itself. Its dimensions and turns number are crucial parameters; in fact, since the greater is the inductor turns number the higher is the inductance, then the electric discharge impulse rises, and above all decay time could result too much higher compared to the velocity of the projectile within the inductor. In such a case the efficiency of the CG could be compromised. The greatest efficiency is obtained when the impulse is shorter than the time took by the projectile to cross the half coil inductors length. If this condition is not satisfied then the inductors will apply an attractive force on the projectile. This force will act in the opposite direction with respect to the projectile motion, thus decreasing the projectile acceleration. The diodes connected to the coil in the opposite polarity with respect to the capacitors are necessary to dump the negative voltage semi-wave oscillation caused by the capacitors discharge and inductors charge process. The dimensioning of the inductor and the capacitors must be computed in order to obtain the maximum efficiency. This means that the coil inductor should have the lowest time charge constant while the capacitors the fast discharge time constant. This is the fundamental condition required in order to avoid the forward-back projectile magnetic strength effect. In fact, once the projectile has overcome the half coil length, the back magnetic action strength starts to act on the projectile decreasing the initial forward acceleration imparted to the projectile. Since the capacitance discharge acts across the coil inductors, the best compromise can be found taking into account contemporary both the capacitance discharge constant time and the inductor charge one. Such a compromise can be obtained by reducing the coil inductor turns' number, as well as the capacitors' capacitance. Preliminary numerical simulations [16,17] have indicated that by a suitable arrangement of high capacitance ($4\times10^3\mu F$) capacitors as discharge trigger for a typical bullet/barrel system (mass projectile ~10g, gun length ~40cm), it's possible to reach values of 1÷2km/s for the bullet's speed, thanks to an effective coil propulsion force of by about 10^3kN. By now the highest measured speed was near below 90m/s with capacitors of $12\times10^3\mu F$; next implementation will surely give the opportunity to come nearer the computed values. In such a case the device will be really appropriate for ballistic aerospace testing, by providing faithful results about the interaction between materials and space debris (~8km/s).

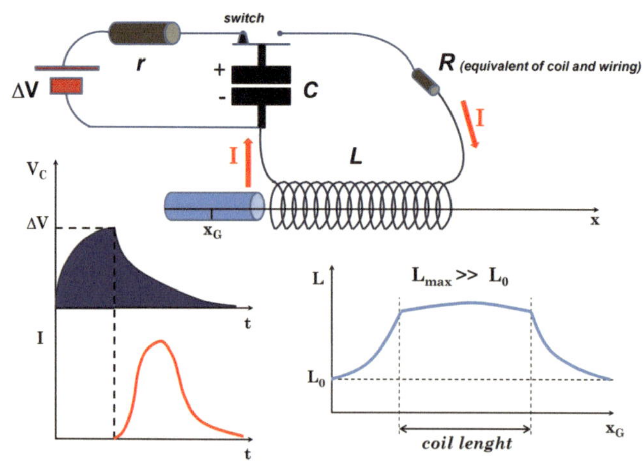

Figure 8. Coil Gun working: schematic representation of the RLC circuit (the charging phase concerns the r-C circuit, r being the capacitor loading resistance) with temporal behavior of capacitor voltage and intensity of current inside the coil. The qualitative variation of the circuit inductance highly dependent on the projectile's position during the discharge is also highlighted.

http://dx.doi.org/10.5772/50411

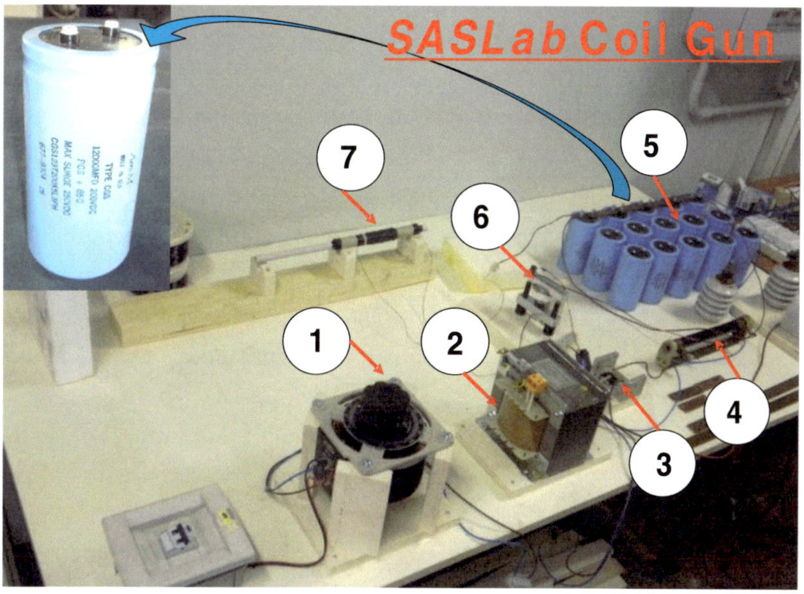

Figure 9. Picture of the Coil Gun system in-house realized (*SASLab-DIAEE* of *Sapienza* University): 1. Resistive Variac (V_{input} 220 V, V_{output} 0-250 V CA); 2. Transformer (V_{input} 250 V, V_{output} 1200 V CA, 1A); 3. Rectifier Diodes to convert AC in DC supply; 4. Resistor (1kΩ, 50W); 5. Capacitors (12000 μF, 200 V DC); 6. High power SCR ($3000V_{max}$, 300A); 7. CG inductor (copper coil winded on aluminum barrel).

It has to be pointed out, on the other hand, that any numerical approach toward the system optimization deals with a so complex analytical problem. In fact, the system of second order differential equations for the two time laws I(t) and x(t)

$$\begin{aligned} & L\,\frac{d^2 I(t)}{dt^2} + R\frac{dI(t)}{dt} + \frac{1}{C}I(t) = 0 \\ & m\ddot{x}_G = F_x = M\frac{\partial B_x}{\partial x}[B_x = B_x[x; I(t)]] \\ & I(0) = 0 \quad V_c(o) = \Delta V \\ & x_G(o) = x_0 \quad \ddot{x}_G(0) = 0 \end{aligned} \tag{1}$$

at first sight of relatively simple resolution (starting from the trivial expression of I(t) for RLC discharge), actually hides a tremendous non linear coupling due to the really appreciable variation of L during the discharge. The inductance of an hollow coil, as well known, can raise by several order of magnitude if a magnetic material is located inside the coil's core: since the body is magnetized (with M not constant too, since it depends on the magnetic field, i.e. on the current intensity) the circuit inductance is thus highly unsettled during the projectile's motion, with obvious consequences on the system time evolution. The modeling has thus to take into account for the physical changes of the fundamental parameters

$$\begin{aligned} M &= M[I(t)] \\ L &= L\,[x(t); M(I(t))] \end{aligned} \tag{2}$$

so that the system (1) should be solved with an heuristic recursive approach, starting from the experimental measurements of velocity hereafter reported (Table 2). These latters are obtained by varying the charging applied voltage (at one's pleasure) and the CG macroscopic parameters (within the technical limits), that are: the iron bullets (S - length 8cm, diameter 6.3mm, mass 17.2g; L - length 16cm, diameter 6.3mm, mass 36.6g; see Figure 10a), the coils (A - wire diameter 2.1mm, length 15cm, 58 coils, 8 turns, L_0=1.54mH; B - wire diameter 3.2mm, length 14cm, 40 coils, 7 turns, L_0=0.55mH; see Figure10b-c), and the bank capacitors (C=12000μF) configuration (C1 - 5 C series, C_{eq} = 2400μF; C2 - 2 C1 parallel, C_{eq} = 4800μF; C3 - 3 C1 parallel, C_{eq} = 7200μF). The gun barrel is kept fixed (aluminum tube: length 31.6cm, outer diameter 10mm, inner diameter 7.6mm), the speed measurements are recorded by means of a ballistic chronograph (*ProChrono* chronograph, minimum speed 17m/s, precision 0.5m/s; see Figure 10d), the results are averaged over five shots for each arrangement.

Figure 10. CG set-up pictures: a) short (S) and long (L) iron bullets; b) different copper coils tested; C) inductance static measurements by *Agilent* LCR Meters; d) ballistic *ProChrono* chronograph for bullet's speed measurements.

http://dx.doi.org/10.5772/50411

BULLET SPEED (m/s)								
d.d.p. (Volts)	*COIL A*						*COIL B*	
	C1		C2		C3		C3	
	S	L	S	L	S	L	S	L
350	-	-	<17	-	22.2±0.7	<17	33.1±1.1	27.1±1.1
400	-	-	19.5±0.7	-	23.5±0.6	17.1±1.1	37.1±1.1	32.1±1.1
460	-	-	23.3±0.6	-	24.5±0.6	17.6±0.6	40.0±1.5	35.2±1.0
500	-	-	25.2±0.8	-	25.1±0.4	19.1±0.8	43.5±1.3	38.4±1.0
550	<17	-	27.7±0.6	-	25.0±0.6	22,7±1.0	45.9±1.2	42.8±1.3
600	<17	-	28.5±0.6	<17	24.2±0.5	25.0±0.6	49.2±1.2	48.5±1.2
660	<17	-	30.0±0.6	<17	23.6±0.6	28.1±0.7	53.1±1.2	55.6±1.1
700	18.1±1.8	-	31.3±0.7	<17	23.0±1.0	29.5±0.9	55.3±1.5	61.6±1.2
750	19.5±1.3	-	31.5±0.8	19.3±1.3	22.3±1.1	31.5±0.8	57.2±1.5	68.3±1.1
800	21.7±1.2	<17	32.1±0.6	27.1±1.2	20.5±1.2	32.6±0.6	58.8±1.4	74.9±1.0
850	24.6±0.9	<17	32.4±0.5	29.4±1.1	19.1±1.7	33.3±0.5	59.5±1.3	80.3±0.7
900	26.5±0.8	<17	32.2±0.8	32.7±1.1	<17	34.2±0.5	59.2±1.3	84.1±0.5
950	29.9±1.0	25.4±1.3	31.5±1.0	35.3±0.9	<17	34.5±0.3	58.1±1.0	86.0±0.4
1005	31.5±0.8	31.5±1.0	31.2±1.0	37.1±0.7	<17	34.0±0.5	55.4±1.1	85.7±0.6
1050	33.1±0.7	35.2±0.5	30.5±1.1	37.5±0.5	-	33.3±0.6	50.3±1.5	82.5±0.8
1095	34.2±1.0	38.9±0.7	29.3±1.1	37.3±0.6	-	31.5±0.7	42.2±1.4	78.5±0.8

Table 2. Speed measurements at different input voltages for the several CG arrangements tested.

The several trends of the bullet's speed depending on the input voltage for the different arrangements highlight the intriguing, complex and highly non linear behavior of the system with respect to its physical main parameters. A first preliminary analysis of the experimental results listed in Table 2 and outlined in Figure 11 stresses in fact the not obvious dependence between the several variables involved, mainly between the time charge/discharge circuit response and the mass of the bullet. The bell-shaped curves demonstrate that the energetic system balance, defined by

$$\eta = \frac{K}{E_C} \qquad \left(E_C = \frac{CV^2}{2} \quad , \quad K = \frac{mv^2}{2} \right) \tag{3}$$

with obvious physical meaning of the symbols, doesn't follow a trivial trend (speed increasing for higher voltages). It's clear, for example, that for coil A, with the bullets adopted and around the maximum voltage that can be applied to not more than five capacitor in series (each one can be charged up to 250V), the system efficiency raise up by decreasing the total capacitance, thus supplying less energy to the system (E_c): from (3) one can find a best efficiency $\eta \sim 2\%$ for LAC1 configuration at the maximum voltage.

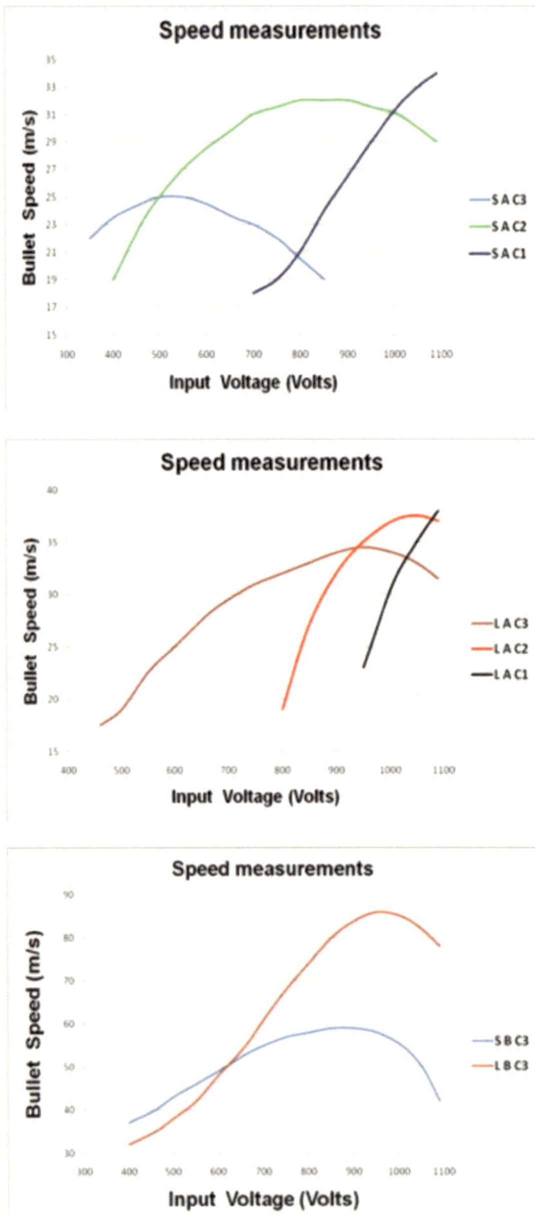

Figure 11. Graphic representations of the speed measurements reported in Table 2.

http://dx.doi.org/10.5772/50411

For coil B in C3 configuration, on the other hand, the heavier L bullets results faster than the lighter S over almost the whole voltage range, with maximum efficiency η ~ 4.5% around 850V. Such results, of course, are connected to the time employed by the bullets to reach the coil center, over which they are recalled back, as explained above. To get higher speed and efficiency all the parameters have to be accurately matched: the results obtained for coil B and bullet L suggest that a suitable arrangement of capacitors bank may let one able to raise the bullet's speed up to 100m/s only with the present Coil Gun stage. Anyway, to perform an as precise as possible material ballistic characterization, a good test reproducibility rather than higher bullets kinetic energies has been addressed by now. With such aim, two fixed configurations with the lowest statistical dispersions in terms of bullets velocities were chosen: LAC3 and LBC3 both at 950V, in what follow related to the low energy (~22J) and high energy (~135J) test respectively. By watching at the experimental error values in Table 2, in fact, it's clear that it's worth operating around maximum points of the curves of Figure 11 in order to avoid ballistic characterization mismatches as much as possible. Furthermore, the use of the longer bullet ensures more stable conditions about the relevant error source due to the bullet initial position: in fact, the coil inductance increases appreciably even with a not magnetized (metallic) body inside its core (changes up to 6÷8 times were found by inductance static measurements performed by inserting both bullets, partially and totally, inside the coil; cfr. Figure 10c), thus resulting in speed changes for same coil/bullet/capacitance arrangements at same input voltages (Figure 12) that cannot be neglected.

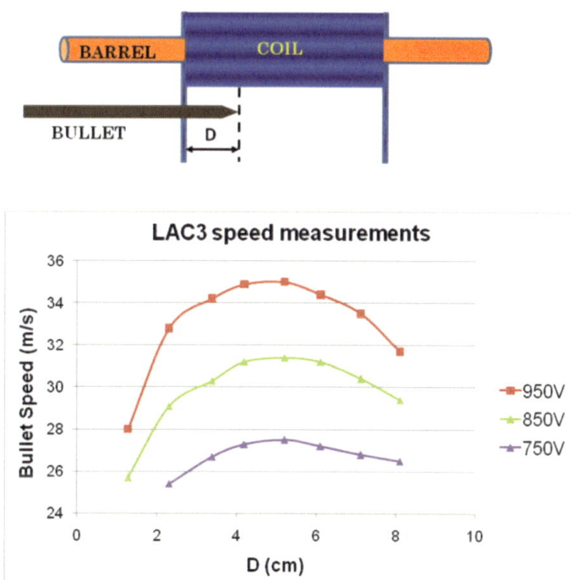

Figure 12. Bullet velocity versus initial position: the measurements performed in the same arrangement (LAC3) for three different input voltages showed that the CG efficiency may considerably change by few millimeters shift of bullet initial positioning.

4. Results and discussion

In this section the most significant results of the ballistic characterization of the nanofluid-reinforced Kevlar-based fabric by means of the CG device are reported and discussed. The choice of the test panel configurations (layers type and number, target surface, alternation and coupling between neat and treated layers, assembling modality, etc.) has been first suggested by the experimental set up best solutions, beyond the purpose to obtain performances similar or even better than the best reference samples in terms of weight/resistance ratio. Preliminary characterization test of the experimental apparatus have indicated the most suitable ensemble of physical parameters (bullet typology, shot energies, gun-target distance, etc.) in order to achieve the best compromise between test efficacy and reproducibility. In particular, the combined constraints of gun and projectile's direction stability during the shot to obtain a 90 central impact on the targets have suggested to keep the sample surface dimensions inside 20×20cm (such value was further lowered cause the big quantity of fluid required to carry out a significant number of experimental test). In Figure 13 the experimental set up adopted is schematically depicted: the samples are fixed by elastic clamps to the wood support, where a plasticine witness is centrally positioned (Figure 14) to estimate the different panel performances in terms of absorbed energy.

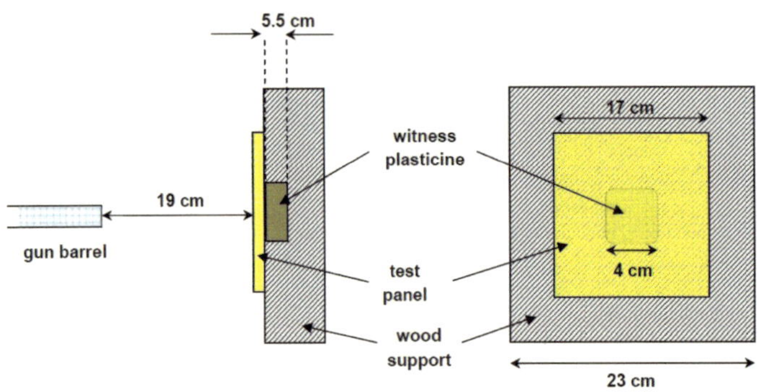

Figure 13. Panels ballistic characterization test by CG: schematic lateral (left) and frontal (right) views of the adopted set-up.

About the layers number, the investigation has been oriented by the reference fabric (neat XP) performances: the number of layers was set to eleven, such to obtain a precise quantitative evaluation of its ballistic properties with the operating experimental conditions. As a consequence, the prototype panel configurations have been established by taking into account most of all the panel total weight requirements, in order to carry out a reliable comparison. For the same reason the results obtained with high (>20wt%) STF concentration have not been taken into consideration, since the critical manufacturing issues due to overfilled solutions produce not homogeneous structures and thus a poor test reproducibility (beyond a not practically useful material). Direct measurements and evaluations for several panel configurations are summarized in Table 3 below, the results being averaged over five shots for each ballistic test. For each panel typology are indicated symbol, layers sequence (following the nomenclature of Table 1 for the fabric type, and subscribing the STF concentration for the treated materials), and total weight (P); for the two low/high energy test above defined the measured penetration depth in the plasticine witness (L), and the computed relative absorbed energy in percentage (ΔE) and efficiency (Q) are then reported. The penetration depth is measured by depth gauge (precision 0.05mm, see Figure 15), the relative absorbed energy and efficiency are simply defined as follow

$$\Delta E = \frac{L_0 - L}{L_0}, \quad Q = \frac{\Delta E}{P} \tag{4}$$

where L_0 indicates the penetration depth into the plasticine witness without sample (low energy test: $L_0 = 9.2$ mm; high energy test: $L_0 = 50.8$ mm). The absorbed energy is an index of the intrinsic ballistic effectiveness of the tested material, while the efficiency factor represents, by weighting on the material density, a balanced evaluation of the global material properties in view of applications in which both lightness and resistance are contemporaneously required.

Figure 14. Picture of the plasticine witness located in the square hole of the sample's support in front of the CG barrel.

By analyzing the numerical results of Table 4 the effectiveness of the fabric reinforcement treatment by the STF solutions realized is evident at first sight: the performances of every blank typology (b_0, B and KN) are improved by the corresponding STF-reinforced configurations in terms of absorbed energy. Moreover, the higher is the nanosilica concentrations within the STF solutions, the greater is the percentage of absorbed energy, as shown in Figure 16, thus enhancing the direct role of reinforcing element played by the n-SiO_2 based nanofluids when coupled to B and KN type fabrics. That is supported by the observation of the zone perforated by the bullets: a more collective resistance work done by the fibers of the reinforced fabric comparing to the neat one is detectable in the panels backside images after the shot (Figure 17), as well as in the SEM photos of the impact point (Figure 18).

panel typology			low energy test (~22J)			high energy test (~135J)		
symbol	*configuration*	*P (g)*	*L (mm)*	*ΔE*	*Q*	*L (mm)*	*ΔE*	*Q*
XP	*11 XP*	175 ± 2	1.2 ± 0.2	87%	**0.50**	10.8 ± 0.5	79%	**0.45**
b_0	*1 XP / 9 B / 1 XP*	284 ± 5	5.8 ± 0.3	37%	**0.13**	36.0 ± 1.1	29%	**0.10**
b_0*	*1 XP / 9 B_{15} / 1 XP*	319 ± 8	5.1 ± 0.3	45%	**0.14**	29.3 ± 1.0	42%	**0.13**
B	*5 XP / 3 B / 1 XP*	180 ± 4	5.5 ± 0.4	40%	**0.22**	34.1 ± 0.4	33%	**0.18**
B_{10}	*5 XP / 3 B_{10} / 1 XP*	192 ± 6	5.3 ± 0.3	42%	**0.22**	31.5 ± 0.5	38%	**0.20**
B_{15}	*5 XP / 3 B_{15} / 1 XP*	196 ± 4	4.8 ± 0.3	48%	**0.24**	27.6 ± 0.6	46%	**0.23**
B_{20}	*5 XP / 3 B_{20} / 1 XP*	199 ± 5	4.1 ± 0.2	55%	**0.28**	22.8 ± 1.1	55%	**0.28**
KN	*5 XP / 4 KN / 1 XP*	184 ± 3	5.2 ± 0.3	43%	**0.24**	33.1 ± 0.7	35%	**0.19**
KN_{10}	*5 XP / 4 KN_{10} / 1 XP*	199 ± 6	4.3 ± 0.2	53%	**0.27**	26.0 ± 1.0	49%	**0.25**
KN_{15}	*5 XP / 4 KN_{15} / 1 XP*	203 ± 4	3.2 ± 0.3	65%	**0.32**	18.2 ± 0.5	64%	**0.32**
KN_{20}	*5 XP / 4 KN_{20} / 1 XP*	205 ± 4	1.9 ± 0.2	79%	**0.39**	10.5 ± 0.6	79%	**0.39**

Table 3. Table 3. Ballistic characterization results (the horizontal blocks enclose the results for single groups of blank/treated materials).

Such improvements are confirmed by the Q-factor trend, even if in a less effective way: the more concentrated solution raise the fabric saturation level in terms of weight increasing, thus lowering the panel global efficacy. In particular the Q values founded for the b_0 type panels give evidence of the drawbacks due to an overall utilization of the STF treatment, because the structure heaviness may offset the gain in impact resistance capability. For this reason the mix configurations with the first 5 layers made of neat XP were designed: as best reference material, the XP Kevlar has confirmed the better weight/resistance trade-off in these experimental conditions. On the other hand, as mentioned in the first section, the poor coupling between STF and XP doesn't give any contribution to the fabric's resistance, rather degrading the fibers structure. As far as the impact energy is concerned, an interesting property of the STF-based samples can be noticed: the quasi-unchanged ΔE and Q values of the reinforced panels (mainly of those with higher nanofiller wt%) in the two different energy range test, with respect to the correspondent lowering discovered in the neat samples. That is highlighted by the crossing around 20wt% of the two curve pairs (B type and KN type) in the graphic of Figure 16, while in the starting points (0wt%, i.e. neat panels) the ΔE values are well spaced, even for the best configuration (horizontal reference lines).

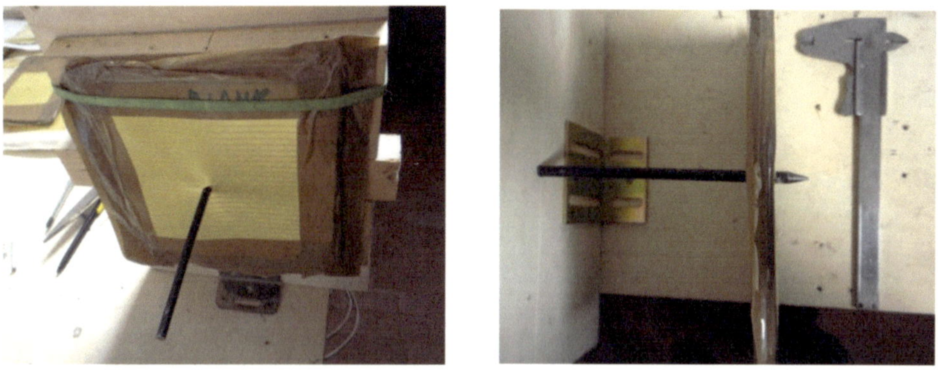

Figure 15. Pictures of a panel sample just after the ballistic test.

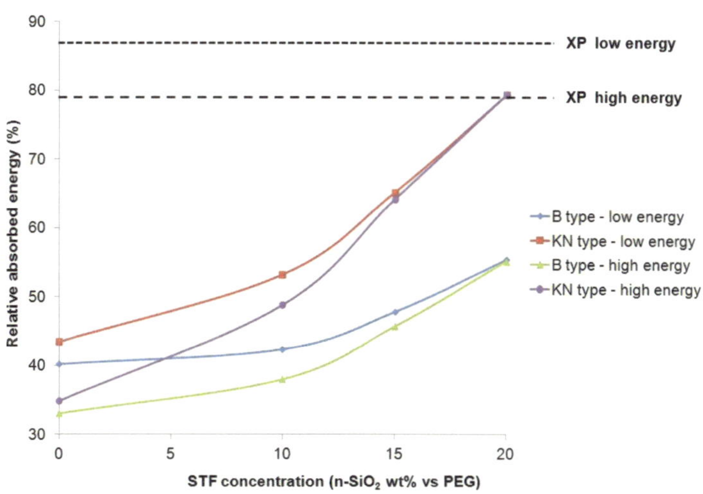

Figure 16. Absorbed energy dependence on nanoparticles wt% inside the several typologies of STF-reinforced fabric panels.

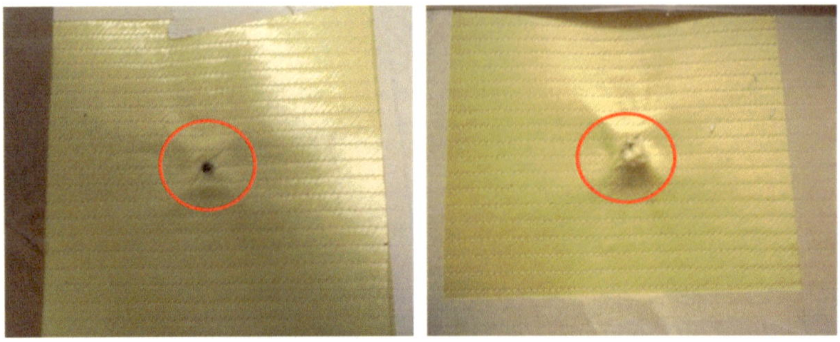

Figure 17. Backside (exit wound) pictures of a B (left) and a B_{20} (right) configuration panel samples after the low energy test.

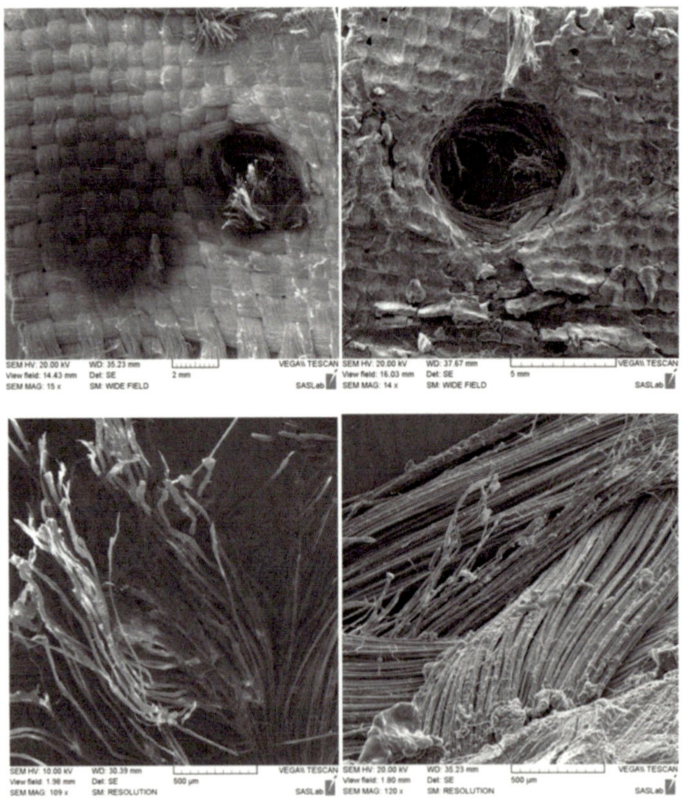

Figure 18. Low (up) and medium (down) magnification SEM photos of the impact zone in B (left) and B_{20} (right) configuration panel samples after the low energy ballistic test.

Such results suggest an intriguing fashion of the impact resistance mechanism by the STF-treated materials. In a very simplified scenario, a conventional structure puts up resistance by a constant friction, so that the L quantities in (4) only depend on the intrinsic material properties: by this way the absorbed energy may be written as

$$\Delta E = \frac{E_{INC} - E_F}{E_{INC}} \tag{5}$$

where E_{INC} and E_F are the incident bullet's energy and the friction physical work respectively: if this latter is approximately constant (i.e. not dependent on the impact energy), the relative absorbed energy is clearly reduced by increasing the bullet's kinetic energy. In the energy range investigated this latter description seems to be reasonable for what concerns the untreated samples. On the contrary, the experimental results obtained for the treated fabric-based panels indicate a more complex mechanism of interaction (cfr. Figure 19), able to raise the friction effect (i.e. the fabric response upon impact) at higher incident energies. This feature may be so promising, mainly for the globally good performances of the treated KN-based type panels. In this case, in fact, the nanofluid/fabric suitable coupling (due to the particular fabric morphology) enhances the STF behavior: that makes this kind of structure (at the highest concentration wt%) competitive with the reference one, even matching it in the high energy range in terms of impact absorption. The next steps of the research has to be then addressed to the further upgrade of the treated KN-based structures, in particular for what concerns the manufacturing reliability of high percentage STF-filled materials.

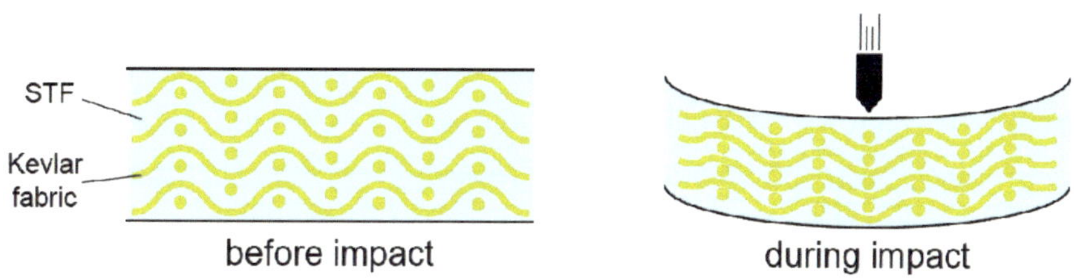

Figure 19. Schematic representation of elastic-like mechanical behavior of STF-reinforced Kevlar fabric upon a projectile impact.

5. Conclusions

In the present study the possibility to employ nanoparticles-based shear thickening fluid for improving the antiballistic properties of Kevlar fabrics has been investigated. Nanosilica particles have been used to realize the reinforcing solutions, and an electromagnetic accelerator device called Coil Gun has been designed, realized and characterized to perform ballistic impact tests on several type of Kevlar-based panels. The fabric samples realization modality consists essentially of two phases: the nanofluid preparation and the fabric impregnation treatment. The first one has to be accurately defined, in order to be confident of the basic effectiveness of the proposed reinforcing material: with such an aim, the combination of the main parameters (material amount percentages, mixing techniques, solvent influence, etc.) affecting the nanofluid preparation have been analyzed, and the several solutions characterized in terms of their rheological properties (viscosity/shear). The second one is strictly dependent on the physical/chemical coupling at the nanofluids/fabric interface: the macroscopic indications provided by the manufacturing procedure, as well as the morphological characterization analyses of the fabric surfaces, have suggested to employ two particular typologies of Kevlar fabric reinforced with nanofluid solutions with concentrations up to 20wt%. The in-house built Coil Gun has been carefully characterized in terms of its main parameters (bullets velocity, energy efficiency, system stability, etc.): two particular configuration (low/high energy) have been established for the fabric ballistic characterization, so that two different impact energy ranges have been investigated and, at the same time, the maximum test reproducibility has been achieved. The results obtained have outlined a better resistance upon impact provided by the highest concentrations of nanofluid-reinforced materials against the corresponding unreinforced ones, thus suggesting further implementation of such nano-reinforced fabrics for antiballistic applications. In particular, a not conventional impact response mechanism seems to be dependent on the nanofluids employment, enhancing their effectiveness for energy increasing: such result can make the treated fabrics able to reach, and eventually overcome, the performances of the best commercial Kevlar-based material (here taken as reference). With such an objective, several technical improvements have to be supplied to the present state of art. Firstly, the manufacturing technique has to be optimized in order to realize fabrics reinforced by higher concentrations of nanofluid solutions: the experimental results have shown, in fact, a clear influence of the nanosilica percentage of inclusion on the fabric absorbing energy capability. This goal, of course, has to be addressed without lack of material homogeneity and flexibility, in order to realize prototype materials of practical application. Secondly, a Coil Gun implementation in terms of efficiency is needed, in order to explore different (higher) energy ranges with the same degree of test reproducibility. Such step will be needful for achieving a deeper knowledge of the underlying impact response mechanism showed by the nanofluid-based material, thus giving the opportunity for their further optimization in terms of antiballistic performances.

Author details

Roberto Pastore, Giorgio Giannini, Ramon Bueno Morles, Mario Marchetti and Davide Micheli

Astronautic, Electric and Energetic Engineering Department, University of Rome "Sapienza", Rome - Italy

References

[1] Mangalgiri, P. D. (1999). Composite materials for aerospace applications. *Bulletin of Materials Science*, 22(3), 657-664.

[2] Njuguna, J., & Pielichowski, K. (2003). Polymer Nanocomposites for Aerospace Applications. *Advanced Engineering Materials*, 5(11), 769-778.

[3] Micheli, D., Apollo, C., Pastore, R., & Marchetti, M. X. (2003). Band microwave characterization of carbon-based nanocomposite material, absorption capability comparison and RAS design simulation. *Composites Science and Technology*, 70(2), 400-409.

[4] Haghmaram, R., & Shoulaie, A. (2004). Study of Traveling wave Tubular Linear Induction Motors. *International Conference on power System Technology, POWERCON, November 21-24, 2004, Singapore.*

[5] Hassan, T. A., Rangari, V. K., & Jeelani, S. (2010). Synthesis, processing and characterization of shear thickening fluid (STF) impregnated fabric composites. *Materials Science and Engineering A*, 527, 2892-2899.

[6] Egres, R.G, Lee, Y.S., Kirkwood, J.E., Kirkwood, K.M., Wetzel, E.D., & Wagner, N.J. (2004). Liquid armor": protective fabrics utilizing shear thickening fluids. *IFAI 4th International Conference on Safety and Protective Fabrics, Pittsburgh, PA.*, 26-27.

[7] Wetzel, E. D., & Wagner, N. J. (2004). Stab Resistance of Shear thickening Fluid (STF)-Kevlar Composite for Body Armor Applications. *24th Army Science Conference, December 2, Orlando, FL.*

[8] Decker, M. J., Halbach, C. J., Nam, C. H., Wagner, N. J., & Wetzel, E. D. (2007). Stab resistance of shear thickening fluid (STF)-treated fabrics. *Composite Science and Technology*, 67, 565-578.

[9] Maranzano, B.J., & Wagner, N.J. (2001). The effect of interparticle interactions and particle size on reversible shear thickening: hard-sphere colloidal dispersions. *Journal of Rheology*, 45-1205.

[10] Lee, Y. S., & Wagner, N. J. (1983). Dynamic properties of shear thickening colloidal suspension. *Rheologica Acta*, 42, 199-208.

[11] Lell, P., Igenbergs, E., & Kuczera, H. (2003). An electromagnetic accelerator. *Journal of Physiscs E*, 16, 325-330.

[12] McNab, IR. (2003). Launch to space with an electromagnetic railgun. *IEEE Transaction on magnetic*, 39, 295-304.

[13] Schmidt, E., & Bundy, M. (2005). Ballistic Launch to Space. Vancouver BC, Canada. *International symposium on Ballistics*, 14-18.

[14] Shope, S., Alexander, J., Gutierrez, W., Kaye, R., Kniskern, M., Long, F., Smith, D., Turman, B., Marder, B., Hodapp, A., & Waverik, R. (2003). Results of a study for a

long range coilgun naval bombardment system. *Sandia National Laboratories, Albuquerque, NM 87185.*

[15] Purcell, E.M. (1965). *Electricity and Magnetism*, Mc Graw-Hill book Company, U.S.A.

[16] Micheli, D., Apollo, C., Pastore, R., & Marchetti, M. (2010). Ballistic characterization of nanocomposite materials by means of "Coil Gun" electromagnetic accelerator. *XIX International Conference on Electrical Machines- ICEM, Rome, Italy.*

[17] Micheli, D., Pastore, R., Apollo, C., & Marchetti, M. (2011). *Coil Gun electromagnetic accelerator for aerospace material anti-ballistic application.*, 1826-4697.

Chapter 10

Graphene/Semiconductor Nanocomposites: Preparation and Application for Photocatalytic Hydrogen Evolution

Xiaoyan Zhang and Xiaoli Cui

Additional information is available at the end of the chapter

http://dx.doi.org/10.5772/51056

1. Introduction

1.1. What is graphene?

Graphene is a flat monolayer of sp^2-bonded carbon atoms tightly packed into a two-dimensional (2D) honeycomb lattice. It is a basic building block for graphitic materials of all other dimensionalities (see Fig.1 from ref. [1]), which can be wrapped into 0D fullerene, rolled into 1D nanotubes or stacked into 3D graphite. It has high thermal conductivity (~5,000 W $m^{-1}K^{-1}$) [2], excellent mobility of charge carriers (200,000 cm^2 V^{-1} s^{-1}) [3], a large specific surface area (calculated value, 2,630 m^2 g^{-1}) [4] and good mechanical stability [5]. Additionally, the surface of graphene is easily functionalized in comparison to carbon nanotubes. Thus, graphene has attracted immense attention [1,6-8] and it shows great applications in various areas such as nanoelectronics, sensors, catalysts and energy conversion since its discovery in 2004 [9-14].

To date, various methods have been developed for the preparation of graphene via chemical or physical routes. Novoselov in 2004 firstly reported the micromechanical exfoliation method to prepare single-layer graphene sheets by repeated peeling [1]. Though the obtained graphene has high quality, micromechanical exfoliation has yielded small samples of graphene that are useful for fundamental study. Then methods such as epitaxial growth and chemical vapor deposition have been developed [15-20]. In epitaxial growth, graphene is produced by decomposition of the surface of silicon carbide (SiC) substrates *via* sublimation of silicon atoms and graphitization of remaining C atoms by annealing at high temperature (1000-1600°C). Epitaxial graphene on SiC(0001) has been demonstrated to exhibit high mobilities, especially multilayered films. Recently, single layered SiC converted graphene over a

INTECH
open science | open minds

© 2012 Zhang and Cui; licensee InTech. This is a paper distributed under the terms of the Creative Commons Attribution License (http://creativecommons.org/licenses/by/3.0), which permits unrestricted use, distribution, and reproduction in any medium, provided the original work is properly cited.

large area has been reported and shown to exhibit outstanding electrical properties [21]. Kim et al. [17] reported the direct synthesis of large-scale graphene films using chemical vapor deposition on thin nickel layers under flowing reaction gas mixtures (CH_4:H_2:Ar = 50:65:200 standard cubic centimeters per minute), and successful transferring of them to arbitrary substrates without intense mechanical and chemical treatments. However, the graphene obtained from micromechanical exfoliation and chemical vapor deposition has insufficient functional groups, which makes its dispersion and contact with photocatalysts difficult [22]. Among the various preparation methods, the reduction of exfoliated graphene oxide (GO) was proven to be an effective and reliable method to produce graphene owing to its low cost, massive scalability, and especially that the surface properties of the obtained graphene can be adjusted via chemical modification [23]. Thus, the development of functionalized graphene-based nanocomposites has aroused tremendous attraction in many potential applications including energy storage [24], catalysis [25], biosensors [26], molecular imaging [27] and drug delivery [28].

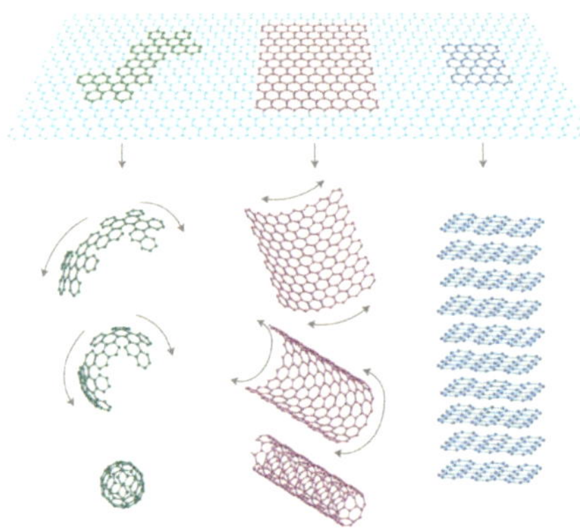

Figure 1. Mother of all graphitic forms. (from ref. [1])

1.2. What is photocatalytic hydrogen evolution?

Photocatalytic water splitting is a chemical reaction for producing hydrogen by using two major renewable energy resources, namely, water and solar energy. As the feedstocks for the reaction, water is clean, inexpensive and available in a virtually inexhaustible reserve, whereas solar energy is also infinitely available, non-polluting and appropriate for the endothermic water splitting reaction. Thus, the utilization of solar energy for the generation of hydrogen from water has been considered as an ultimate solution to solve the crisis of energy shortage and environmental degradation [29]. The following is the dissociation of the water molecule to yield hydrogen and oxygen:

$$H_2O \rightarrow 1/2O_2(g) + H_2(g); \Delta G = +237 \text{ kJ/mol} \quad (1)$$

http://dx.doi.org/10.5772/51056

This simple process has gathered a big interest from an energetic point of view because it holds the promise of obtaining a clean fuel, H_2, from a cheap resource of water [30,31]. As shown in Reaction (1), its endothermic character would require a temperature of 2500 K to obtain ca. 5% dissociation at atmospheric pressure, which makes it impractical for water splitting [32]. The free energy change for the conversion of one molecule of H_2O to H_2 and $1/2O_2$ under standard conditions corresponds to $\Delta E° = 1.23$ eV per electron transfer according to the Nernst equation. Photochemical decomposition of water is a feasible alternative because photons with a wavelength shorter than 1100 nm have the energy (1.3 eV) to split a water molecule. But, the fact is that only irradiation with wavelengths lower than 190 nm works, for that a purely photochemical reaction has to overcome a considerable energy barrier [33]. The use of a photocatalyst makes the process feasible with photons within solar spectrum since the discovery of the photoelectrochemical performance for water splitting on TiO_2 electrode by Fujishima and Honda [34].

To use a semiconductor and drive this reaction with light, the semiconductor must absorb radiant light with photon energies of larger than 1.23 eV (≤ wavelengths of 1000 nm) to convert the energy into H_2 and O_2 from water. This process must generate two electron-hole pairs per molecule of H_2 (2 × 1.23 eV = 2.46 eV). In the ideal case, a single semiconductor material having a band gap energy (E_g) large enough to split water and having a conduction band-edge energy (E_{cb}) and valence band-edge energy (E_{vb}) that straddles the electrochemical potentials $E°_{(H+/H2)}$ and $E°_{(O2/H2O)}$, can drive the hydrogen evolution reaction and oxygen evolution reaction using electrons/holes generated under illumination (see Fig. 2) [29,35].

$$H_2O + 2(h^+) \longrightarrow \tfrac{1}{2} O_2 + 2H^+ \qquad \text{(HER)}$$

$$2H^+ + 2e^- \longrightarrow H_2 \qquad \text{(OER)}$$

$$H_2O \longrightarrow \tfrac{1}{2} O_2 + H_2 \qquad \Delta G = +237.2 \text{ kJ/mol}$$

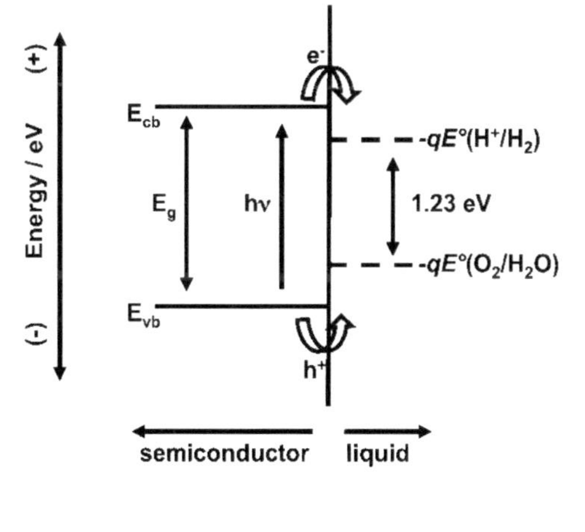

Figure 2. The mechanism of photocatalytic hydrogen evolution from water (see ref. [35])

To date, the above water splitting can be photocatalyzed by many inorganic semiconductors such as titanium dioxide (TiO_2), which was discovered in 1971 by Fujishima and Honda [34, 36]. Among the various types of widely-investigated semiconductor material, titanium dioxide (TiO_2) has been considered the most active photocatalyst due to its low cost, chemical stability and comparatively high photocatalytic efficiency [37, 38].

Frequently, sacrificial agents such as methanol [39-41], ethanol [42-44] or sulfide/sulfite [45-47] are often added into the photocatalytic system with the aim to trap photogenerated holes thus improving the photocatalytic activity for hydrogen evolution. The reaction occurred in this case is usually not the water photocatalytic decomposition reaction [48]. For example, overall methanol decomposition reaction will occur in a methanol/water system, which has a lower splitting energy than water [49]. The reaction proposed by Kawai [50] and Chen [51] was as follows:

$$CH_3OH(l) \leftrightarrow HCHO(g) + H_2(g) \qquad \Delta G_1{}^\circ = 64.1\ kJ/mol \tag{2}$$

$$HCHO(g) + H_2O(l) \leftrightarrow HCO_2H(l) + H_2(g) \qquad \Delta G_2{}^\circ = 47.8\ kJ/mol \tag{3}$$

$$HCO_2H(l) \leftrightarrow CO_2(g) + H_2(g) \qquad \Delta G_3{}^\circ = -95.8\ kJ/mol \tag{4}$$

With the overall reaction being

$$CH_3OH(l) + H_2O(l) \leftrightarrow CO_2(g) + 3H_2(g) \qquad \Delta G^\circ = 16.1\ kJ/mol \tag{5}$$

Consequently, it is easier for methanol decomposition in comparison to water decomposition in the same conditions.

2. Synthesis and Characterization of Graphene/Semiconductor Nanocomposite Photocatalysts

Considering its superior electron mobility and high specific surface area, graphene can be expected to improve the photocatalytic performance of semiconductor photocatalysts such as TiO_2, where graphene can act as an efficient electron acceptor to enhance the photoinduced charge transfer and to inhibit the recombination of the photogenerated electron-holes [52,53]. Thus, graphene-based semiconductor photocatalysts have also attracted a lot of attention in photocatalytic areas [7,8]. A variety of semiconductor photocatalysts have been used for the synthesis of graphene (or reduced graphene oxide) based composites. They mainly include metal oxides (e.g. TiO_2 [42-46], ZnO [61-66], Cu_2O [67], Fe_2O_3 [68], NiO [69], WO_3 [70],), metal sulfides (e.g. ZnS [71], CdS [72-77], MoS_2 [78]), metallates (e.g. Bi_2WO_6 [79],

$Sr_2Ta_2O_7$ [80], $BiVO_4$ [81], $InNbO_4$ [82] and g-Bi_2MoO_6 [83]), other nanomaterials (e.g. CdSe [84], Ag/AgCl [85,86], C_3N_4 [87,88]). The widely used synthetic strategies to prepare graphene-based photocatalysts can be divided into four types, which are sol-gel, solution mixing, in situ growth, hydrothermal and/or solvothermal methods. In fact, two or more methods are usually combined to fabricate the graphene-based semiconductor nanocomposites.

2.1. Sol -gel process

Sol-gel method is a wet-chemical technique widely used in the synthesis of graphene-based semiconductor nanocomposites. It is based on the phase transformation of a sol obtained from metallic alkoxides or organometallic precursors. For example, tetrabutyl titanate dispersed in graphene-containing absolute ethanol solution would gradually form a sol with continuous magnetic stirring, which after drying and post heat treatment changed into TiO_2/graphene nanocomposites [52,55]. The synthesis process can be schematically illuminated in Fig. 3(A) (from ref. [55]). The resulted TiO_2 nanoparticles closely dispersed on the surface of two dimensional graphene nanosheets (see Fig. 3(B) from ref. [55]). Wojtoniszak et al. [89] used a similar strategy to prepare the TiO_2/graphene nanocomposite via the hydrolysis of titanium (IV) butoxide in GO-containing ethanol solution. The reduction of GO to graphene was realized in the post heat treatment process. Farhangi et al. [90] prepared Fe-doped TiO_2 nanowire arrays on the surface of functionalized graphene sheets using a sol-gel method in the green solvent of supercritical carbon dioxide. In the preparation process, the graphene nanosheets acted as a template for nanowire growth through surface -COOH functionalities.

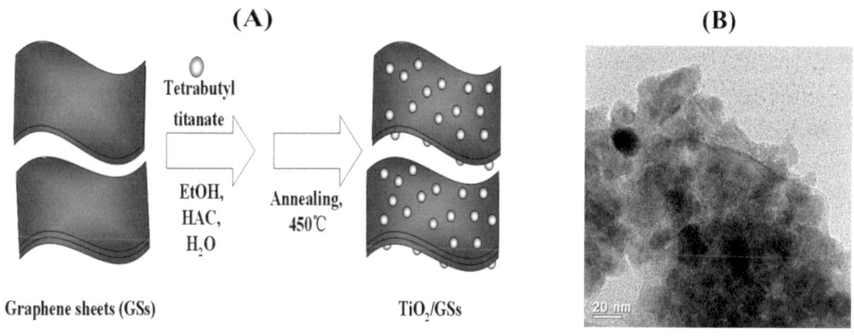

Figure 3. Schematic synthesis procedure (A) and typical TEM image of the TiO_2/graphene nanocomposites (B). (from ref. [55])

2.2. Solution mixing method

Solution mixing is a simple method to fabricate graphene/semiconductor nanocomposite photocatalysts. The oxygenated functional groups on GO facilitate the uniform distribution of photocatalysts under vigorous stirring or ultrasonic agitation [91]. Graphene-based nanocomposites can be obtained after the reduction of GO in the nanocomposite.

For example, Bell et al. [92] fabricated TiO_2/graphene nanocomposites by ultrasonically mixing TiO_2 nanoparticles and GO colloids together, followed by ultraviolet (UV)-assisted photocatalytic reduction of GO to graphene. Similarly, GO dispersion and N-doped $Sr_2Ta_2O_7$ have been mixed together, followed by reduction of GO to yield $Sr_2Ta_2O_{7-x}N_x$/graphene nanocomposites under xenon lamp irradiation [80]. Graphene-CdSe quantum dots nanocomposites have also been synthesized by Geng et al. [84]. In this work, pyridine-modified CdSe nanoparticles were mixed with GO sheets, where pyridine ligands were considered to provide π-π interactions for the assembly of CdSe nanoparticles on GO sheets. They thought that pyridine ligands could provide π-π interactions for the assembly of CdSe nanoparticles capped with pyridine on GO sheets. Paek et al. [93] prepared the SnO_2 sol by hydrolysis of $SnCl_4$ with NaOH, and then the prepared graphene dispersion was mixed with the sol in ethylene glycol to form the SnO_2/graphene nanocomposite. Most recently, Liao et al. [88] fabricated GO/g-C_3N_4 nanocomposites via sonochemical approach, which was realized by adding g-C_3N_4 powder into GO aqueous solution followed by ultrasonication for 12 h and then drying at 353 K.

2.3. Hydrothermal/solvothermal approach

The hydrothermal/solvothermal process is another effective method for the preparation of semiconductor/graphene nanocomposites, and it has unique advantage for the fabrication of graphene-based photocatalysts. In this process, semiconductor nanoparticles or their precursors are loaded on the GO sheets, where GO are reduced to graphene simultaneously with or without reducing agents or in the following step.

For example, Zhang et al. [54] synthesized graphene-TiO_2 nanocomposite photocatalyst by hydrothermal treatment of GO sheets and commercial TiO_2 powders (Degussa P25) in an ethanol-water solvent to simultaneously achieve the reduction of GO and the deposition of P25 on the carbon substrate. In order to increase the interface contact and uniform distribution of TiO_2 nanoparticles on graphene sheets, a one-pot hydrothermal method was applied using GO and $TiCl_4$ in an aqueous system as the starting materials [94]. Wang et al. [95] used a one-step solvothermal method to produce graphene-TiO_2 nanocomposites with well-dispersed TiO_2 nanoparticles by controlling the hydrolysis rate of titanium isopropoxide. Li and coworkers [74] synthesized graphene-CdS nanocomposites by a solvothermal method in which graphene oxide (GO) served as the support and cadmium acetate ($Cd(Ac)_2$) as the CdS precursor. Reducing agents can also be added into the reaction system. Recently, Shen et al. [96] added glucose as the reducing agent in the one-pot hydrothermal method for preparation of graphene-TiO_2 nanocomposites. Ternary nanocomposites system can also be obtained by a two-step hydrothermal process. Xiang et al. [42] prepared TiO_2/MoS_2/graphene hybrid by a two-step hydrothermal method.

Furthermore, some solvothermal experiments can result in the semiconductor nanoparticles with special morphology on graphene sheets. Shen et al. [97] reported an ionic liquid-assisted one-step solvothermal method to yield TiO_2 nanoparticle-graphene composites with a dendritic structure as a whole. Li et al. [78] synthesized MoS_2/graphene hybrid by a one-step solvothermal reaction of $(NH_4)_2MoS_4$ and hydrazine in a N, N dimethylformamide (DMF) solution of GO. During this process, the $(NH_4)_2MoS_4$ precursor was reduced to MoS_2 on GO sheets and the GO simultaneously to RGO by reducing agent of hydrazine. The existence of graphene can change the morphology of the resulted MoS_2 in the graphene/MoS_2 nanocomposite in comparison to pure MoS_2 (see Fig. 4 from ref. [78]). Ding et al. [98] reported graphene-supported ultrathin anatase TiO_2 nanosheets with exposed (001) high-energy facets by a simple solvothermal method. In this process, anatase TiO_2 nanosheets directly grew from titanium (IV) isopropoxide onto the GO support during the solvothermal growth of TiO_2 nanocrystals in isopropyl alcohol solvent, and then GO was reduced to graphene via a post thermal treatment under N_2/H_2 to finally obtain the graphene-TiO_2 nanocomposite.

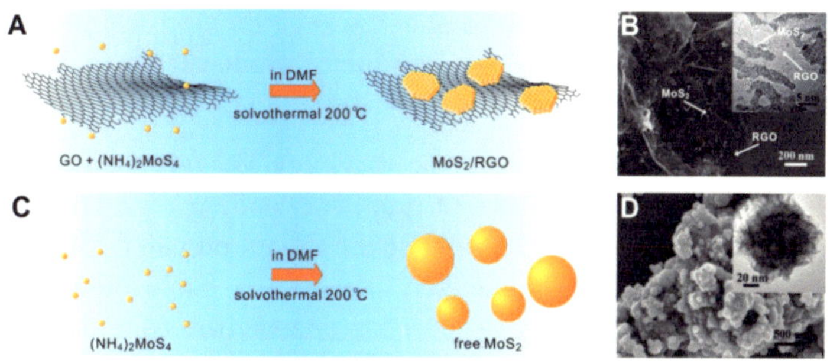

Figure 4. Synthesis of MoS_2 in solution with and without graphene sheets. (A) Schematic solvothermal synthesis with GO sheets. (B) SEM and (inset) TEM images of the MoS_2/graphene hybrid. (C) Schematic solvothermal synthesis without any GO sheets, resulting in large, free MoS_2 particles. (D) SEM and (inset) TEM images of the free particles. (from ref. [78])

2.4. *In situ* growth strategy

In situ growth strategy can afford efficient electron transfer between graphene and semiconductor nanoparticles through their intimate contact, which can also be realized by hydrothermal and/or solvothermal method. The most common precursors for graphene and metal compound are functional GO and metal salts, respectively. The presence of epoxy and hydroxyl functional groups on graphene can act as the heterogeneous nucleation sites and anchor semiconductor nanoparticles avoiding the agglomeration of the small particles [99]. Zhu et al. [100] reported a one-pot water-phase approach for synthesis of graphene/TiO_2 composite nanosheets using $TiCl_3$ as both the titania precursor and the reducing agent. Lambert et al. [101] also reported the *in situ* synthesis of nanocomposites of petal-like TiO_2-GO by the hydrolysis of TiF_4 in the presence of aqueous dispersions of GO, followed by post chemical or thermal treatment to produce TiO_2-graphene hybrids. With the concentration of graphene oxide high enough and stirring off, long-range ordered assemblies of TiO_2-GO sheets were obtained because of self-assembly. Guo et al. [102] synthesized TiO_2/graphene nanocomposite sonochemically from $TiCl_4$ and GO in ethanol-water system, followed by a hydrazine treatment to reduce GO into graphene. The average size of the TiO_2 nanoparticles was controlled at around 4-5 nm on the sheets, which is attributed to the pyrolysis and condensation of the dissolved $TiCl_4$ into TiO_2 by ultrasonic waves.

3. Applications of Graphene-based Semiconductor Nanocomposites for Photocatalytic Hydrogen Evolution

Hydrogen is regarded as an ultimate clean fuel in the future because of its environmental friendliness, renewability, high-energy capability, and a renewable and green energy carrier [103-105]. Using solar energy to produce hydrogen from water splitting over semiconductor is believed to be a good choice to solve energy shortage and environmental crisis [106,107]. Various semiconductor photocatalysts have been reported to have the performance of photocatalytic hydrogen evolution from water. However, the practical application of this strategy is limited due to the fast recombination of photoinduced electron-holes and low utilization efficiency of visible light. Because of the superior electrical property of graphene, there is a great interest in combining semiconductor photocatalysts with graphene to improve their photocatalytic H_2 production activity [8,54].

Zhang et al. firstly reported the photocatalytic activity of TiO_2/graphene nanocomposites for hydrogen evolution [55]. The influences of graphene loading contents and calcination atmosphere on the photocatalytic performance of the sol-gel prepared TiO_2-graphene composites have been investigated, respectively. The results show that the photocatalytic performance of the sol-gel prepared TiO_2/5.0wt%graphene nanocomposites was much higher than that of P25 for hydrogen evolution from Na_2S-Na_2SO_3 aqueous solution under UV-Vis light irradiation. Yu and his coworkers studied the photocatalytic performance of graphene/TiO_2 nanosheets composites for hydrogen evolution from methanol/water solution (see Fig. 5 from ref. [108]). They investigated the effect of TiO_2 precursor on the photocatalytic performance of the synthesized nanocomposites under UV light irradiation. Enhanced photocatalytic H_2 production was observed for the prepared graphene/TiO_2 nanosheets composite in comparison to that of graphene/P25 nanoparticles composites as shown in Figure 6 (see ref. [108]).

Figure 5. TEM images of the graphene/TiO_2 nanosheets nanocomposite. (from ref. [108])

Fan et al. [58] systematically studied the influence of different reduction approaches on the efficiency of hydrogen evolution for P25/graphene nanocomposites prepared by UV-assisted photocatalytic reduction, hydrazine reduction, and a hydrothermal reduction method. The photocatalytic results show that the P25/graphene composite prepared by the hydrothermal method possessed the best performance for hydrogen evolution from methanol aqueous solution under UV-Vis light irradiation, followed by P25/graphene-photo reduction and P25/graphene-hydrazine reduction, respectively. The maximum value exceeds that of pure P25 by more than 10 times. Figure 7 shows the morphology and XRD patterns of the one-pot hydrothermal synthesized TiO_2/graphene composites [94]. It can be observed that TiO_2 nanoparticles dispersed uniformly on graphene sheets as shown in Figure 7(A). The TiO_2/graphene nanocomposites are composed mainly anatase TiO_2 confirmed from the XRD results as shown in Figure 7(B).

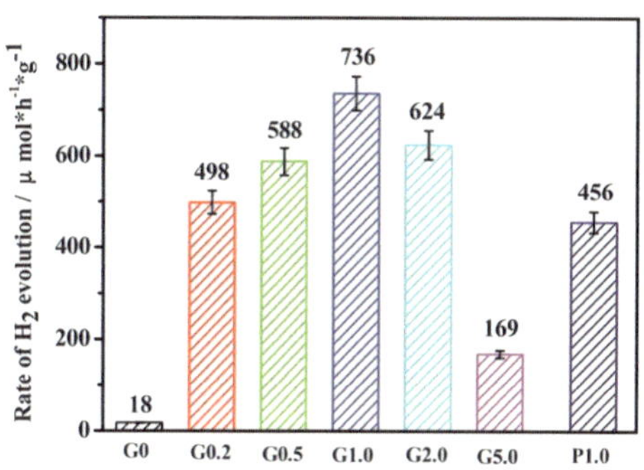

Figure 6. Comparison of the photocatalytic activity of the G0, G0.2, G0.5, G1.0, G2.0, G5.0 and P1.0 samples for the photocatalytic H_2 production from methanol aqueous solution under UV light irradiation. (Gx, x is the weight percentage of graphene in the graphene/TiO_2 nanosheets nanocomposites; P1.0 is the graphene/P25 nanocomposite with 1.0wt% graphene.) (from ref. [108])

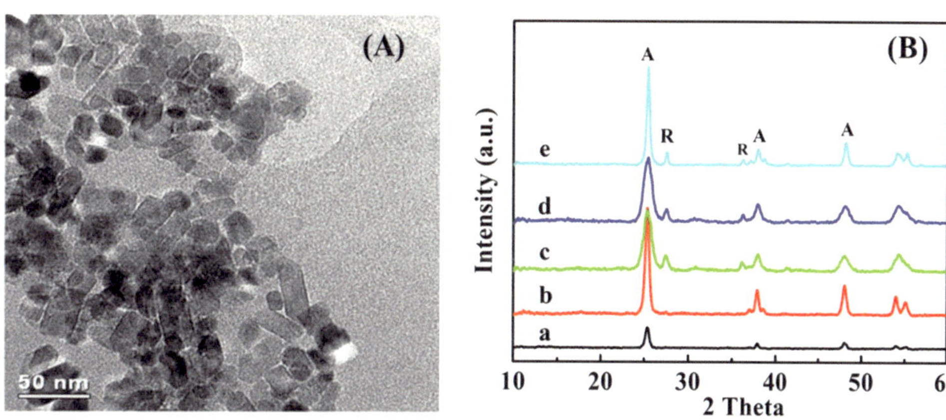

Figure 7. Typical TEM image (A) and XRD patterns (B) of the one-pot hydrothermal synthesized TiO_2/graphene nanocomposites. (from ref. [94])

The CdS/graphene nanocomposites have also attracted many attentions for photocatalytic hydrogen evolution. Li et al. [74] investigated the visible-light-driven photocatalytic activity of CdS-cluster-decorated graphene nanosheets prepared by a solvothermal method for hydrogen production (see Fig. 8). These nanosized composites exhibited higher H_2-production rate than that of pure CdS nanoparticles. The hydrogen evolution rate of the nanocomposite with graphene content as 1.0 wt % and Pt 0.5 wt % was about 4.87 times higher than that of pure CdS nanoparticles under visible-light irradiation.

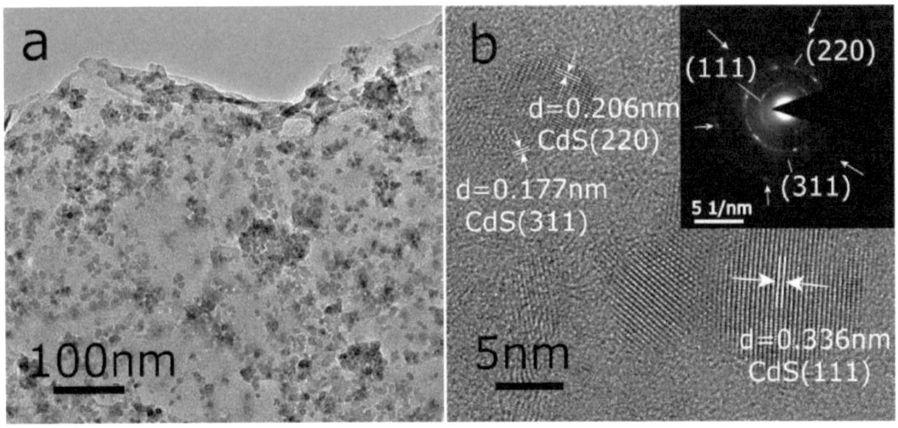

Figure 8. a) TEM and (b) HRTEM images of sample GC1.0, with the inset of (b) showing the selected area electron diffraction pattern of graphene sheet decorated with CdS clusters. (GC1.0 was synthesized with the weight ratios of GO to $Cd(Ac)_2$ $2H_2O$ as 1.0%). (see ref. [74])

4. Mechanism of the Enhanced Photocatalytic Performance for H_2 Evolution

It is well-known that graphene has large surface area, excellent conductivity and high carriers mobility. The large surface of graphene sheet possesses more active adsorption sites and photocatalytic reaction centers, which can greatly enlarge the reaction space and enhance photocatalytic activity for hydrogen evolution [74,110].

Excellent conductivity and high carriers mobility of graphene sheets facilitate that graphene attached to semiconductor surfaces can efficiently accept and transport electrons from the excited semiconductor, suppressing charge recombination and improving interfacial charge transfer processes. To confirm this hypothesis, the impedance spectroscopy (EIS) of the graphene/TiO_2 nanocomposite films was given as shown in Figure 9 (see ref. [108]). In the EIS measurements, by applying an AC signal to the system, the current flow through the circuit can be modeled to deduce the electrical behavior of different structures within the system. Figure 9 shows the conductance and capacitance as a function of frequency for FTO electrodes coated with TiO_2 and reduced graphene oxide (RGO)-TiO_2 with different RGO content (0.5, 1.0, and 1.5 mg) using a custom three-electrode electrochemical cell with a gold wire counter electrode and Ag/AgCl reference electrode in 0.01M H_2SO_4 electrolyte in a frequency range from 1 mHz to 100 kHz. Information about the films themselves is obtained from the region between 1 mHz and 1 kHz. At frequencies below 100 Hz, the conductivity is the films themselves, and at ultralow frequencies (1 mHz), the conductivity is dominated by the interface between the film and the FTO. So it can be seen that the RGO in the nanocomposites films not only enhances conductivity within the film but also the conduction between the film and the FTO substrate. The same results are obtained from the inset Nyquist plots, where the radius of each arc is correlated with the charge transfer ability of the corresponding film; the larger the radius the lower the film's ability to transfer charge. The luminescence decay spectra in Figure 10 (see ref. [109]) indicate the electron transfer from photoexcited CdS nanoparticles into modified graphene (mG), thereby leading to decrease of emission lifetime from CdS to CdS-mG, further confirming that graphene can improve the charge separation and suppress the recombination of excited carriers.

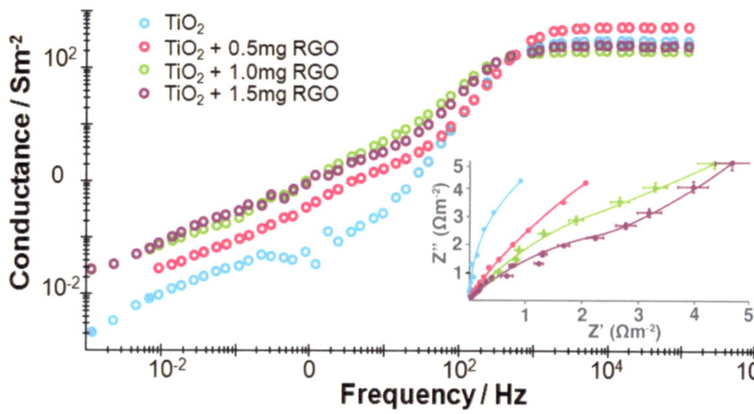

Figure 9. EIS conductance plot of TiO_2 and RGO- TiO_2 films. (Inset) Nyquist plots of the same films. (see from ref. [109])

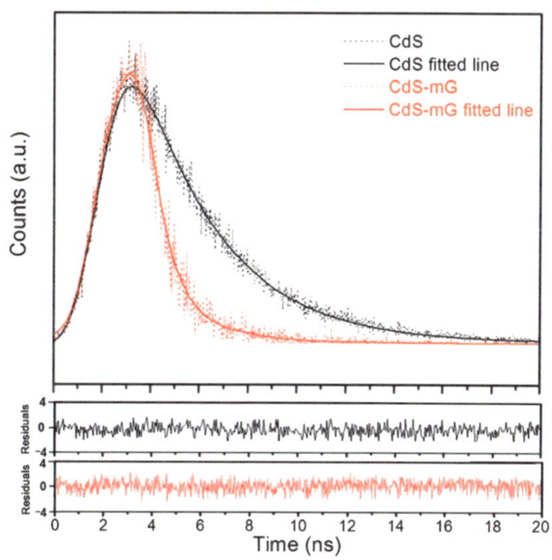

Figure 10. Time-resolved fluorescence decays of the CdS and CdS-mG solution at the 20 ns scanning range. Excited wavelength is at 355 nm, and emission wavelength is 385 nm. Bold curves are fitted results. (mG is modified graphene) (see ref. [110])

Figure 11 shows (a) the schematic illustration for the charge transfer and separation in the graphene/TiO_2 nanosheets system under UV light irradiation and (b) the proposed mechanism for photocatalytic H_2-production under UV light irradiation. Normally, the photogenerated charge carriers quickly recombine with only a small fraction of the electrons and holes participating in the photocatalytic reaction, resulting in low conversion efficiency [110,111]. When graphene was introduced into TiO_2 nanocomposite, the photogenerated electrons on the conduction band (CB) of TiO_2 tend to transfer to graphene sheets, suppressing the recombination of photogenerated electron-holes.

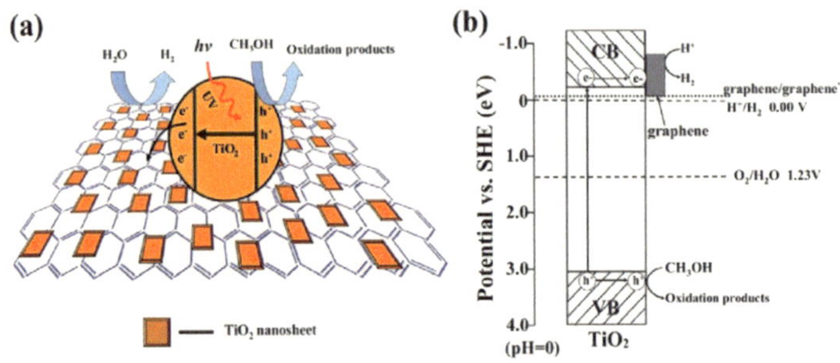

Figure 11. a) Schematic illustration for the charge transfer and separation in the graphene-modified TiO_2 nanosheets system under UV light irradiation; (b) proposed mechanism for photocatalytic H_2-production under UV light irradiation. (from ref. [108])

Moreover, a red shift of the absorption edge of semiconductor photocatalyst upon modified by graphene (or reduced graphene oxide) was observed (see Fig. 12 from ref. [58]) by many researchers from the diffuse reflectance UV-Vis spectroscopy, which was proposed to be ascribed to the interaction between semiconductor and graphene (or reduced graphene oxide) in the nanocomposites [55,58,73,108,112]. Therefore, it can be inferred that the introduction of graphene in semiconductor photocatalysts is effective for the visible-light response of the corresponding nanocomposite, which leads to more efficient utilization of the solar energy.

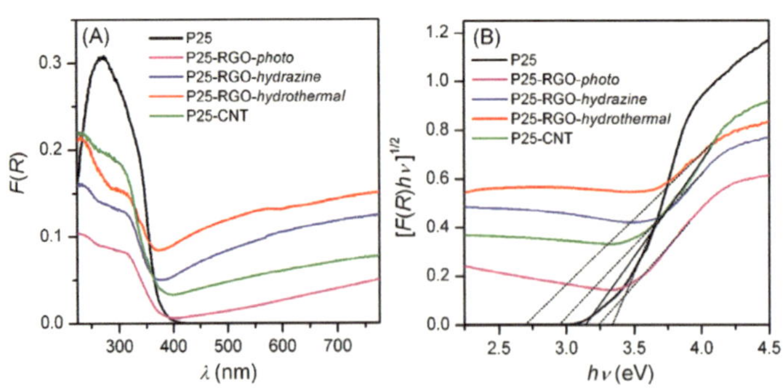

Figure 12. A) Diffuse reflectance UV-Vis spectra of P25, P25-RGO nanocomposites (P25/RGO = 1/0.2) prepared by different methods, and P25-CNT composite (P25/CNT = 1/0.3). (B) Corresponding plot of transformed Kubelka-Munk function versus the energy of the light. (see from ref. [58])

The above results suggest an intimate interaction between semiconductor photocatalysts and graphene sheets is beneficial for the visible light absorption and separation of photogenerated electron and hole pairs, leading to enhanced photocatalytic performance for hydrogen evolution.

5. Summary and Perspectives

In summary, graphene can be coupled with various semiconductors to form graphene-semiconductor nanocomposites due to its unique large surface area, high conductivity and carriers mobility, easy functionalization and low cost. The unique properties of graphene have opened up new pathways to fabricate high-performance photocatalysts. In this chapter, we have summarized the various fabrication methods such as solution mixing, sol gel, in situ growth, and hydrothermal/solvothermal methods that have been developed for fabricating the graphene-based semiconductor photocatalysts. These composites have shown potential applications in energy conversion and environmental treatment areas.

Although great progress has been achieved, challenges still exist in this area and further developments are required. The first challenge is that the quality-control issues of graphene still need to be addressed. Graphene oxide is believed to be a better starting material than pure graphene to form nanocomposite with semiconductor photocatalysts. However, reduction of graphene oxide into graphene usually can bring defects and impurity simultaneously. Thus, new synthesis strategies have to be developed to fabricate high-performance graphene-semiconductor composites. The second one is the semiconductor photocatalysts. The introduction of graphene into the nanocomposites mainly acts to promote the separation of charge carriers and transport of photogenerated electrons. The performance of photocatalysts is highly dependent on the semiconductor photocatalysts and their surface structures such as the morphologies and surface states. Therefore, the development of novel photocatalysts is required. Furthermore, the underlying mechanism of the photocatalytic enhancement by the graphene-based semiconductor nanocomposites is partly unclear. For example, whether graphene can change the band gap of the semiconductor photocatalysts, and whether graphene can truly sensitize semiconductor photocatalysts. Nevertheless, there are still many challenges and opportunities for graphene-based semiconductor nanocomposites and they are still expected to be developed as potential photocatalysts to address various environmental and energy-related issues.

Acknowledgements

This work is supported by the National Science Foundation of China (No. 21273047) and National Basic Research Program of China (Nos. 2012CB934300, 2011CB933300), the Shanghai Science and Technology Commission (No. 1052nm01800) and the Key Disciplines Innovative Personnel Training Plan of Fudan University.

Author details

Xiaoyan Zhang and Xiaoli Cui*

*Address all correspondence to: xiaolicui@fudan.edu.cn

Department of Materials Sciences, Fudan University, Shanghai, 200433, China

References

[1] Novoselov, K. S., Geim, A. K., Morozov, S. V., Jiang, D., Zhang, Y., Dubonos, S. V., Grigorieva, I. V., & Firsov, A. A. (2004). *Science*, 306, 666.

[2] Balandin, A. A., Ghosh, S., Bao, W. Z., Calizo, I., Teweldebrhan, D., Miao, F., & Lau, C. N. (2008). *Nano Lett.*, 8, 902-907.

[3] Bolotin, K. I., Sikes, K. J., Jiang, Z., Klima, M., Fudenberg, G., Hone, J., Kim, P., & Stormer, H. L. (2008). *Solid State Commun.*, 146, 351-355.

[4] Stoller, M. D., Park, S., Zhu, Y., An, J., & Ruoff, R. S. (2008). *Nano Lett.*, 8, 3498-3502.

[5] Lee, C., Wei, X., Kysar, J. W., & Hone, J. (2008). *Science*, 321, 385-388.

[6] Pyun, J. (2011). *Angew. Chem. Int. Ed.*, 50, 46.

[7] Xiang, Q. J., Yu, J. G., & Jaroniec, M. (2012). *Chem. Soc. Rev.*, 41(2), 782-796.

[8] An, X. Q., & Yu, J. C. (2011). *RSC Advances*, 1, 1426-1434.

[9] Wei, D., & Liu, Y. (2010). *Adv. Mater.*, 22, 3225.

[10] Allen, M. J., Tung, V. C., & Kaner, R. B. (2010). *Chem. Rev.*, 110, 132.

[11] Chen, J. S., Wang, Z., Dong, X., Chen, P., & Lou, X. W. (2011). *Nanoscale*, 3, 2158.

[12] Yi, J., Lee, J. M., & Park, W. I. (2011). *Sens. Actuators*, B155, 264.

[13] Fan, Y., Lu, H. T., Liu, J. H., Yang, C. P., Jing, Q. S., Zhang, Y. X., Yang, X. K., & Huang, K. J. (2011). *Colloids Surf.*, B83, 78.

[14] Johnson, J., Behnam, A., Pearton, S. J., & Ural, A. (2010). *Adv. Mater.*, 22, 4877.

[15] Lin, Y. M., Dimitrakopoulos, C., Jenkins, K. A., Farmer, D. B., Chiu, H. Y., & Grill, A. (2010). *Science*, 327, 662.

[16] de Heer, W. A., Berger, C., Wu, X., First, P. N., Conrad, E. H., Li, X., Li, T., Sprinkle, M., Hass, J., Sadowski, M. L., Potemski, M., & Martinez, G. (2007). *Solid State Commun.*, 143, 92.

[17] Kim, K. S., Zhao, Y., Jang, H., Lee, S. Y., Kim, J. M., Kim, K. S., Ahn, J. H., Kim, P., Choi, J. Y., & Hong, B. H. (2009). *Nature*, 457, 706-710.

[18] Sutter, P. W., Flege, J. I., & Sutter, E. A. (2008). *Nature Mater.*, 7, 406-411.

[19] Reina, A., Jia, X. T., Ho, J., Nezich, D., Son, H., Bulovic, V., Dresselhaus, M. S., & Kong, J. (2009). *Nano Lett.*, 9(1), 30-35.

[20] Dato, A., Radmilovic, V., Lee, Z., Phillips, J., & Frenklach, M. (2008). *Nano Lett.*, 8, 2012-2016.

[21] Wu, X., Hu, Y., Ruan, M., Madiomanana, N. K., Hankinson, J., Sprinkle, M., Berger, C., & de Heer, W. A. (2009). *Appl. Phys. Lett.*, 95, 223108.

[22] Cui, X., Zhang, C., Hao, R., & Hou, Y. (2011). *Nanoscale*, 3, 2118.

[23] Park, S., & Ruoff, R. S. (2009). *Nat. Nanotechnol.*, 4, 217.

[24] Wang, K., Ruan, J., Song, H., Zhang, J., Wo, Y., Guo, S., & Cui, D. (2011). *Nanoscale Res. Lett.*, 6, 8.

[25] Wang, Q., Guo, X., Cai, L., Cao, Y., Gan, L., Liu, S., Wang, Z., Zhang, H., & Li, L. (2011). *Chem. Sci.*, 2, 1860-1864.

[26] Park, S., Mohanty, N., Suk, J. W., Nagaraja, A., An, J., Piner, R. D., Cai, W., Dreyer, D. R., Berry, V., & Ruoff, R. S. (2010). *Adv. Mater.*, 22, 1736.

[27] Cai, W. B., & Chen, X. Y. (2007). *Small*, 3, 1840.

[28] Akhavan, O., Ghaderi, E., & Esfandiar, A. (2011). *J. Phys. Chem.*, B115, 6279.

[29] Osterloh, F. E. (2008). *Chem. Mater.*, 20, 35-54.

[30] Serpone, N., Lawless, D., & Terzian, R. (1992). *Sol. Energy*, 49, 221.

[31] Hernández-Alonso, M. D., Fresno, F., Suárez, S., & Coronado, J. M. (2009). *Energy Environ. Sci.*, 2, 1231-1257.

[32] Kodama, T., & Gokon, N. (2007). *Chem. Rev.*, 107, 4048.

[33] Gonzalez, M. G., Oliveros, E., Worner, M., & Braun, A. M. (2004). *Photochem. Photobiol.*, C5(3), 225.

[34] Fujishima, A., & Honda, K. (1972). *Nature*, 238(5358), 37.

[35] Walter, M. G., Warren, E. L., McKone, J. R., Boettcher, S. W., Mi, Q. X., Santori, E. A., & Lewis, N. S. (2010). *Chem. Rev.*, 110(11), 6446-6473.

[36] Fujishima, A., & Honda, K. B. (1971). *Chem. Soc. Jpn.*, 44(4), 1148.

[37] Ni, M., Leung, M. K. H., Dennis, Y. C., Leung, K., & Sumathy, . (2007). *Renewable and Sustainable Energy Reviews*, 11, 401-425.

[38] Hoffmann, M. R., Martin, S. T., Choi, W., & Bahnemann, D. W. (1995). *Chem. Rev.*, 95, 69-96.

[39] Wang, L., & Wang, W. Z. (2012). *Int. J. Hydrogen Energy*, 37, 3041-3047.

[40] Yang, X. Y., Salzmann, C., Shi, H. H., Wang, H. Z., Green, M. L. H., & Xiao, T. C. (2008). *J. Phys. Chem.*, A112, 10784-10789.

[41] Sreethawong, T., Laehsalee, S., & Chavadej, S. (2009). *Catalysis Communications*, 10, 538-543.

[42] Xiang, Q. J., Yu, J. G., & Jaroniec, M. (2012). *J. Am. Chem. Soc.*, 134, 6575.

[43] Wang, Y. Q., Zhang, Z. J., Zhu, Y., Li, Z. C., Vajtai, R., Ci, L. J., & Ajayan, P. M. (2008). *ACS Nano*, 2(7), 1492-1496.

[44] Eder, D., Motta, M., & Windle, A. H. (2009). *Nanotechnology*, 20, 055602.

[45] Yang, H. H., Guo, L. J., Yan, W., & Liu, H. T. (2006). *J. Power Sources*, 159, 1305-1309.

[46] Yao, Z. P., Jia, F. Z., Tian, S. J., Li, C. X., Jiang, Z. H., & Bai, X. F. (2010). *Appl. Mater. Interfaces*, 2(9), 2617-2622.

[47] Jang, J. S., Hong, S. J., Kim, J. Y., & Lee, J. S. (2009). *Chemical Physics Letters*, 475, 78-81.

[48] Chen, T., Feng, Z. C., Wu, G. P., Shi, J. Y., Ma, G. J., Ying, P. L., & Li, C. (2007). *J. Phys. Chem.*, C111, 8005-8014.

[49] Lin, W. C., Yang, W. D., Huang, I. L., Wu, T. S., & Chung, Z. J. (2009). *Energy & Fuels*, 23, 2192-2196.

[50] Kawai, T., & Sakata, T. (1980). *J. Chem. Soc. Chem. Commun.*, 15, 694-695.

[51] Chen, J., Ollis, D. F., Rulkens, W. H., & Bruning, H. (1999). *Water. Res.*, 33, 669-676.

[52] Zhang, X. Y., Li, H. P., & Cui, X. L. (2009). *Chinese Journal of Inorganic Chemistry*, 25(11), 1903-1907.

[53] Lightcap, I. V., Kosel, T. H., & Kamat, P. V. (2010). *Nano Lett.*, 10, 577.

[54] Zhang, H., Lv, X. J., Li, Y. M., Wang, Y., & Li, J. H. (2010). *ACS Nano*, 4, 380.

[55] Zhang, X. Y., Li, H. P., Cui, X. L., & Lin, Y. H. (2010). *J. Mater. Chem.*, 20, 2801.

[56] Du, J., Lai, X. Y., Yang, N. L., Zhai, J., Kisailus, D., Su, F. B., Wang, D., & Jiang, L. (2011). *ACS Nano*, 5, 590.

[57] Zhou, K. F., Zhu, Y. H., Yang, X. L., Jiang, X., & Li, C. Z. (2011). *New J. Chem.*, 35, 353.

[58] Fan, W. Q., Lai, Q. H., Zhang, Q. H., & Wang, Y. (2011). *J. Phys. Chem.*, C115, 10694-10791.

[59] Liang, Y. T., Vijayan, B. K., Gray, K. A., & Hersam, M. C. (2011). *Nano Lett.*, 11, 2865.

[60] Zhang, L. M., Diao, S. O., Nie, Y. F., Yan, K., Liu, N., Dai, B. Y., Xie, Q., Reina, A., Kong, J., & Liu, Z. F. (2011). *J. Am. Chem. Soc.*, 133, 2706.

[61] Li, B. J., & Cao, H. Q. (2011). *J. Mater. Chem.*, 21, 3346.

[62] Williams, G., & Kamat, P. V. (2009). *Langmuir*, 25, 13869.

[63] Lee, J. M., Pyun, Y. B., Yi, J., Choung, J. W., & Park, W. I. (2009). *J. Phys. Chem.*, C113, 19134.

[64] Akhavan, O. (2010). *ACS Nano*, 4, 4174.

[65] Xu, T. G., Zhang, L. W., Cheng, H. Y., & Zhu, Y. F. (2011). *Appl. Catal.*, B101, 382.

[66] Akhavan, O. (2011). *Carbon*, 49, 11.

[67] Xu, C., Wang, X., & Zhu, J. W. (2008). *J. Phys. Chem.*, C112, 19841.

[68] Morishige, K., & Hamada, T. (2005). *Langmuir*, 21, 6277.

[69] Wang, D. H., Kou, R., Choi, D., Yang, Z. G., Nie, Z. M., Li, J., Saraf, L. V., Hu, D. H., Zhang, J. G., Graff, G. L., Liu, J., Pope, M. A., & Aksay, I. A. (2010). *ACS Nano*, 4, 1587.

[70] Guo, J. J., Li, Y., Zhu, S. M., Chen, Z. X., Liu, Q. L., Zhang, D., Moonc, W-J., & Song, D-M. (2012). *RSC Advances*, 2, 1356-1363.

[71] Hu, H., Wang, X., Liu, F., Wang, J., & Xu, C. (2011). *Synth. Met.*, 161, 404.

[72] Cao, A. N., Liu, Z., Chu, S. S., Wu, M. H., Ye, Z. M., Cai, Z. W., Chang, Y. L., Wang, S. F., Gong, Q. H., & Liu, Y. F. (2010). *Adv. Mater.*, 22, 103.

[73] Jia, L., Wang, D. H., Huang, Y. X., Xu, A. W., & Yu, H. Q. (2011). *J. Phys. Chem. C*, 115, 11466-11473.

[74] Li, Q., Guo, B., Yu, J., Ran, J., Zhang, B., Yan, H., & Gong, J. (2011). *J. Am. Chem. Soc.*, 133, 10878.

[75] Nethravathi, C., Nisha, T., Ravishankar, N., Shivakumara, C., & Rajamathi, M. (2009). *Carbon*, 47, 2054.

[76] Wu, J. L., Bai, S., Shen, X. P., & Jiang, L. (2010). *Appl. Surf. Sci.*, 257, 747.

[77] Ye, A. H., Fan, W. Q., Zhang, Q. H., Deng, W. P., & Wang, Y. (2012). *Catal. Sci. Technol.*, 2, 969-978.

[78] Li, Y. G., Wang, H. L., Xie, L. M., Liang, Y. Y., Hong, G. S., & Dai, H. J. (2011). *J. Am. Chem. Soc.*, 133, 7296-7299.

[79] Gao, E. P., Wang, W. Z., Shang, M., & Xu, J. H. (2011). *Phys. Chem. Chem. Phys.*, 13, 2887.

[80] Mukherji, A., Seger, B., Lu, G. Q., & Wang, L. Z. (2011). *ACS Nano*, 5, 3483.

[81] Ng, Y. H., Iwase, A., Kudo, A., & Amal, R. (2010). *J. Phys. Chem. Lett.*, 1, 2607.

[82] Zhang, X. F., Quan, X., Chen, S., & Yu, H. T. (2011). *Appl. Catal.*, B105, 237.

[83] Zhou, F., Shi, R., & Zhu, Y. F. (2011). *J. Mol. Catal. A: Chem.*, 340, 77.

[84] Geng, X. M., Niu, L., Xing, Z. Y., Song, R. S., Liu, G. T., Sun, M. T., Cheng, G. S., Zhong, H. J., Liu, Z. H., Zhang, Z. J., Sun, L. F., Xu, H. X., Lu, L., & Liu, L. W. (2010). *Adv. Mater.*, 22, 638.

[85] Zhang, H., Fan, X. F., Quan, X., Chen, S., & Yu, H. T. (2011). *Environ. Sci. Technol.*, 45, 5731.

[86] Zhu, M. S., Chen, P. L., & Liu, M. H. (2011). *ACS Nano*, 5, 4529.

[87] Xiang, Q. J., Yu, J. G., & Jaroniec, M. (2011). *J. Phys. Chem.*, C115, 7355-7363.

[88] Liao, G. Z., Chen, S., Quan, X., Yu, H. T., & Zhao, H. M. (2012). *J. Mater. Chem.*, 22, 2721.

[89] Wojtoniszak, M., Zielinska, B., Chen, X. C., Kalenczuk, R. J., & Borowiak-Palen, E. (2012). *J. Mater. Sci.*, 47(7), 3185-3190.

[90] Farhangi, N., Chowdhury, R. R., Medina-Gonzalez, Y., Ray, M. B., & Charpentier, P. A. (2011). *Appl. Catal. B: Environmental*, 110, 25-32.

[91] Zhang, Q., He, Y. Q., Chen, X. G., Hu, D. H., Li, L. J., Yin, T., & Ji, L. L. (2011). *Chin. Sci. Bull.*, 56, 331.

[92] Bell, N. J., Yun, H. N., Du, A. J., Coster, H., Smith, S. C., & Amal, R. (2011). *J. Phys. Chem.*, C115, 6004.

[93] Paek, S. M., Yoo, E., & Honma, I. (2009). *Nano Lett.*, 9, 72.

[94] Zhang, X. Y., Sun, Y. J., Li, H. P., Cui, X. L., & Jiang, Z. Y. (2012). *Inter. J. Hydrogen Energy*, 37, 811.

[95] Wang, P., Zhai, Y. M., Wang, D. J., & Dong, S. J. (2011). *Nanoscale*, 3, 1640.

[96] Shen, J. F., Yan, B., Shi, M., Ma, H. W., Li, N., & Ye, M. X. (2011). *J. Mater. Chem.*, 21, 3415.

[97] Shen, J. F., Shi, M., Yan, B., Ma, H. W., Li, N., & Ye, M. X. (2011). *Nano Res.*, 4(8), 795-806.

[98] Ding, S. J., Chen, J. S., Luan, D. Y., Boey, F. Y. C., Madhavi, S., & Lou, X. W. (2011). *Chem. Commun.*, 47, 5780.

[99] Li, N., Liu, G., Zhen, C., Li, F., Zhang, L. L., & Cheng, H. M. (2011). *Adv. Funct. Mater.*, 21, 1717.

[100] Zhu, C. Z., Guo, S. J., Wang, P., Xing, L., Fang, Y. X., Zhai, Y. M., & Dong, S. J. (2010). *Chem. Commun.*, 46, 7148.

[101] Lambert, T. N., Chavez, C. A., Hernandez-Sanchez, B., Lu, P., Bell, N. S., Ambrosini, A., Friedman, T., Boyle, T. J., Wheeler, D. R., & Huber, D. L. (2009). *J. Phys. Chem.*, C113, 19812.

[102] Guo, J., Zhu, S., Chen, Z., Li, Y., Yu, Z., Liu, Q., Li, J., Feng, C., & Zhang, D. (2011). *Ultrason. Sonochem.*, 18, 1082.

[103] Yu, J. G., Zhang, J., & Jaroniec, M. (2010). *Green Chem.*, 12, 1611.

[104] Yu, J. G., & Ran, J. R. (2011). *Energy Environ. Sci.*, 4, 1364.

[105] Qi, L. F., Yu, J. G., & Jaroniec, M. (2011). *Phys. Chem. Chem. Phys.*, 13, 8915.

[106] Li, Q. Y., & Lu, G. X. (2007). *J. Mol. Catal.*, (China), 21, 590-598.

[107] Yu, C. L., & Yu, J. C. (2009). *Catal. Lett.*, 129, 462-470.

[108] Bell, N. J., Ng, Y. H., Du, A., Coster, H., Smith, S. C., & Amal, R. (2011). *J. Phys. Chem.*, C115, 6004-6009.

[109] Lv, X. J., Fu, W. F., Chang, H. X., Zhang, H., Cheng, J. S., Zhang, G. J., Song, Y., Hub, C. Y., & Li, J. H. (2012). *J. Mater. Chem.*, 22, 1539-1546.

[110] Yu, J. C., Yu, J. G., Ho, W. K., Jiang, Z. T., & Zhang, L. Z. (2002). *Chem. Mater.*, 14, 3808.

[111] Yu, J. G., Wang, W. G., Cheng, B., & Su, B. L. (2009). *J. Phys. Chem. C*, 113, 6743.

[112] Zhang, H., Lv, X., Li, Y., Wang, Y., & Li, J. (2010). *ACS Nano*, 4, 380-386.

Chapter 11

New Frontiers in Mechanosynthesis: Hydroxyapatite – and Fluorapatite – Based Nanocomposite Powders

Bahman Nasiri–Tabrizi, Abbas Fahami,
Reza Ebrahimi–Kahrizsangi and Farzad Ebrahimi

Additional information is available at the end of the chapter

http://dx.doi.org/10.5772/50160

1. Introduction

Mechanosynthesis process is a solid state method that takes advantage of the perturbation of surface-bonded species by pressure or mechanical forces to enhance the thermodynamic and kinetic reactions between solids. Pressure can be applied via conventional milling equipment, ranging from low-energy ball mills to high-energy stirred mills. In a mill, the reactants are crushed between the balls and wall (horizontal or planetary ball mill, attritor, vibratory ball mill), or between rings or ring and wall (multi–ring media mill) (Bose et al., 2009). These processes cause the creation of defects in solids; accelerate the migration of defects in the bulk, increase the number of contacts between particles, and renew the contacts. In these circumstances, chemical interaction occurs between solids (Avvakumov et al., 2002). This procedure is one of the most important fields of solid state chemistry, namely, the mechanochemistry of inorganic substances, which is intensively developed; so that, a large number of reviews and papers published on this subject in the last decades (Silva et al., 2003; Suryanarayana, 2001; De Castro & Mitchell, 2002). The prominent features of this technique are that melting is not essential and that the products have nanostructural characteristics (Silva et al., 2003; Suryanarayana, 2001; De Castro & Mitchell, 2002). In the field of bioceramics, high efficiency of the mechanochemical process opens a new way to produce commercial amount of nanocrystalline calcium phosphate-based materials. A review of scientific research shows that the mechanosynthesis process is a potential method to synthesis of nanostructured bioceramics (Rhee, 2002; Silva et al., 2004; Suchanek et al., 2004; Tian et al., 2008; Nasiri–Tabrizi et al., 2009; Gergely et al., 2010; Wu et al. 2011; Ramesh et al., 2012).

INTECH
open science | open minds

© 2012 Nasiri–Tabrizi et al.; licensee InTech. This is a paper distributed under the terms of the Creative Commons Attribution License (http://creativecommons.org/licenses/by/3.0), which permits unrestricted use, distribution, and reproduction in any medium, provided the original work is properly cited.

On the other side, bioceramics play a vital role in several biomedical applications and have been expanding enormously the recent years (Adamopoulos & Papadopoulos, 2007). Among different forms of bioceramics, particular attentions have been placed to calcium phosphates-based powders, granules, dense or porous bodies, and coatings for metallic or polymeric implants due to their excellent biocompatibility and osteointegration properties (Marchi et al., 2009). It is well known that hydroxyapatite (HAp: $Ca_{10}(PO_4)_6(OH)_2$) is a major mineral component of bones and teeth (Zhou & Lee, 2011). Therefore, synthetic HAp has been extensively utilized as a bioceramic for maxillofacial applications owing to its excellent osteoconductive properties (Adamopoulos & Papadopoulos, 2007). Besides this field, in a variety of other biomedical applications calcium phosphates have been used as matrices for controlled drug release, bone cements, tooth paste additive, and dental implants (Rameshbabu et al., 2006). Nevertheless, HAp intrinsic poor mechanical properties (strength, toughness and hardness), high dissolution rate in biological system, poor corrosion resistance in an acid environment and poor chemical stability at high temperatures have restricted wider applications in load-bearing implants (Fini et al., 2003; Chen et al., 2005).

According to the literature (Jallot et al., 2005), the biological and physicochemical properties of HAp can be improved by the substitution with ions usually present in natural apatites of bone. In fact, trace ions substituted in apatites can effect on the lattice parameters, the crystallinity, the dissolution kinetics and other physical properties (Mayer & Featherstone, 2000). When OH^- groups in HAp are partially substituted by F^-, fluoride-substituted HAp (FHAp: $Ca_{10}(PO_4)_6(OH)_{2-x}F_x$) is obtained. If the substitution is completed, fluorapatite (FAp: $Ca_{10}(PO_4)_6F_2$), is formed. When fluoride consumed in optimal amounts in water and food, used topically in toothpaste, and mouth rinses, it increases tooth mineralization and bone density, reduces the risk and prevalence of dental caries, and helps to promote enamel remineralization throughout life for individuals of all ages (Palmer & Anderson, 2001). It is found that the incorporation of fluorine into HAp induced better biological response (Rameshbabu et al., 2006). On the other hand, the incorporation of bioinert ceramics and addition of appropriate amount of ductile metallic reinforcements into calcium phosphate-based materials has demonstrated significant improvement in structural features as well as mechanical properties. Therefore, improvements on structural, morphological, and mechanical properties of HAp ceramics have been tried by a number of researches (Cacciotti et al., 2009; Schneider et al., 2010; Farzadi et al., 2011; Pushpakanth et al., 2008; Rao & Kannan, 2002; Viswanath & Ravishankar, 2006; Gu et al., 2002; Ren et al., 2010). These studies have shown that such characteristics of HAp might be exceptionally strengthened by various methods such as making nanocomposites, use of different sintering techniques, and adding dopants. In the field of nanocomposites, an ideal reinforcing material for the HAp-based composites, which satisfies all of the requirements, has not yet been found. Thus, synthesis and characterization of novel nanostructured calcium phosphate-based ceramics provided the key target for current research. In most researches (Enayati–Jazi et al., 2012; Rajkumar et al., 2011; Choi et al., 2010), calcium phosphate-based nanocomposites were prepared using multiple wet techniques which ordinarily comprise of several step processes. Over the past decades, the mechanochemical synthesis has been extended for the production of a wide range of nanostructured materials (Suryanarayana, 2001), particularly for the synthesis of nanocrystalline calcium phosphate-based ceramics (Rhee, 2002; Suchanek et al., 2004; Tian et al., 2008; Nasiri–Tabrizi et al., 2009; Gergely et al., 2010; Wu et al. 2011; Ramesh et al., 2012). The advantages of this procedure remains on the fact that melting is not necessary and the powders are nanocrystalline (Silva et al., 2007).

In this chapter, a new approach to synthesis of HAp- and FAp-based nanocomposites via mechanochemical process is reported. The effect of high-energy ball milling parameters and subsequent thermal treatment on the structural and morphological features of the nanocomposites were discussed in order to propose suitable conditions for the large scale synthesis of HAp- and FAp-based nanocomposites. Powder X-ray diffraction (XRD), Fourier transform infrared (FT-IR) spectroscopy, and energy dispersive X-ray spectroscopy (EDX) techniques are used to provide evidence for the identity of the samples. Transmission electron microscopy (TEM), Field-Emission Scanning Electron Microscope (FE-SEM), and scanning electron microscopy (SEM) are also utilized to study of the morphological features of the nanocomposites. Literature reported that the size and number of balls had no significant effect on the synthesizing time and grain size of FAp ceramics, while decreasing the rotation speed or ball to powder weight ratio increased synthesizing time and the grain size of FAp (Mohammadi Zahrani & Fathi, 2009). On the other hand, our recent experimental results confirm that the chemical composition of initial materials and thermal annealing process are main parameters that affect the structural features (crystallinity degree, lattice strain, crystallite size) of the products via mechanochemical method (Nasiri-Tabrizi et al., 2009; Honarmandi et al., 2010; Ebrahimi-Kahrizsangi et al., 2010; Fahami et al., 2011; Ebrahimi-Kahrizsangi et al., 2011; Fahami et al., 2012). Consequently, the present chapter is focused on the mechanochemical synthesize of HAp- and FAp-based nanocomposites. In the first part of this chapter, an overview of recent development of ceramic-based nanocomposites in biomedical applications and mechanochemical process are provided. The other sections describe the application of these procedures in the current study. The effects of milling media and atmosphere to prepare novel nanostructured HAp-based ceramics are studied. Moreover, mechanochemical synthesis and characterization of nanostructured FAp-based bioceramics are investigated.

2. Recent developments of ceramic-based nanocomposites for biomedical applications

Over the past decades, innovations in the field of bioceramics such as alumina, zirconia, hydroxyapatite, fluorapatite, tricalcium phosphates and bioactive glasses have made significant contribution to the promotion of modern health care industry and have improved the quality of human life. Bioceramics are mainly applied as bone substitutes in biomedical applications owing to their biocompatibility, chemical stability, and high wear resistance. However, the potential of bioceramics in medical applications depends on its structural, morphological, mechanical, and biological properties in the biological environment. The first successful medical application of calcium phosphate bioceramics in humans is reported in 1920 (Kalita et al., 2007). After that the first dental application of these ceramics in animals was described in 1975 (Kalita et al., 2007). In a very short period of time, bioceramics have found various applications in replacements of hips, knees, teeth, tendons and ligaments and repair for periodontal disease, maxillofacial reconstruction, augmentation and stabilization of the jawbone and in spinal fusion (Kalita et al., 2007).

Today, many specialty ceramics and glasses have been developed for use in dentistry and medicine, e.g., dentures, glass-filled ionomer cements, eyeglasses, diagnostic instruments, chemical ware, thermometers, tissue culture flasks, fiber optics for endoscopy, and carriers for enzymes and antibodies (Hench, 1998). Among them, calcium phosphate-based bioceramics have been utilized in the field of biomedical engineering due to the range of properties that they offer, from tricalcium phosphates (α/β-TCP) being resorbable to HAp being bioactive (Ducheyne & Qiu, 1999). Hence, different phases of calcium phosphate-based bioceramics are used depending upon whether a resorbable or bioactive material is desired. The phase stability of calcium phosphate-based bioceramics depends significantly upon temperature and the presence of water, either during processing or in the use environment. It is found that at body temperature; only two calcium phosphates are stable in contact with aqueous media, such as body fluids. These stable phases are $CaHPO_4.2H_2O$ (dicalcium phosphate, brushite) and HAp at pH<4.2 and pH>4.2, respectively (Hench, 1998). At higher temperatures, other phases, such as α/β-TCP and tetracalcium phosphate ($Ca_4P_2O_9$) are present. The final microstructure of TCP will contain β or α–TCP depending on their cooling rate. Rapid cooling from sintering temperature gives rise to α–TCP phase only, whereas slow furnace cooling leads to β–TCP phase only. Any moderate cooling rate, in between these two results mixed phase of both β and α–TCP (Nath et al., 2009).

One of the primary restrictions on clinical use of bioceramics is the uncertain lifetime under the complex stress states, slow crack growth, and cyclic fatigue that result in many clinical applications. Two creative approaches to these mechanical limitations are use of bioactive ceramics as coatings, and the biologically active phase in composites. Because of the anisotropic deformation and fracture characteristics of cortical bone, which is itself a composite of compliant collagen fibrils and brittle HAp crystals, the Young's modulus varies ~ 7–25 GPa, the critical stress intensity ranges ~ 2–12 $MPa.m^{1/2}$, and the critical strain intensity increases from as low as ~ 600 $J.m^{-2}$ to as much as 5000 $J.m^{-2}$, depending on orientation, age, and test condition. On the contrary, most bioceramics are much stiffer than bone and many exhibit poor fracture toughness (Hench, 1998). Therefore, the only materials that exhibit a range of properties equivalent to bone are composites. For this reason, many attempts have been made to improve the mechanical properties as well as structural features through the incorporation of ceramic second phases (Viswanath & Ravishankar, 2006; Evis, 2007; Nath et al., 2009; Ben Ayed & Bouaziz, 2008). These studies have shown that the mechanical properties of HAp and fluoridated HAp might be exceptionally strengthened by composite making technique.

It is found that (Kong et al., 1999) the following conditions should be satisfied to be effective as a reinforcing agent for a ceramic matrix composite material. First, the strength and the elastic modulus of the second phase must be higher than those of the matrix. Second, the interfacial strength between the matrix and the second phase should be neither too weak nor too strong. Indeed, for an appropriate interfacial strength, no excessive reaction should occur between the matrix and the second phase. Third, the coefficient of thermal expansion (CTE) of the second phase should not differ too much from that of the matrix in order to prevent micro-cracks formation in densification process. Fourth, in the case of biomaterials, the biocompatibility of the reinforcing agent is another crucial factor that should be considered. Nevertheless, an ideal reinforcing material for the calcium phosphate-based composites, which satisfies all of requirements, has not yet been found. So, some attempts have been made to develop HAp- and fluorhydroxyapatite-based composites such as HAp–Al_2O_3 (Viswanath & Ravishankar, 2006), HAp–ZrO_2 (Evis, 2007), HAp–TiO_2 (Nath et al., 2009), FHAp–Al_2O_3 (Adolfsson et al., 1999), FHAp–ZrO_2 (Ben Ayed & Bouaziz, 2008), poly(lactide-co-glycolide)/β-TCP (Jin. et al., 2010), polyglycolic acid (PGA)/β–TCP (Cao & Kuboyama, 2010), and HAp–CNT (Lee et al., 2011) composites. These experimental studies exhibited that interfacial reactions occurred during the high temperature processing of composites due to the large interfacial area available for the reactions. Interfacial reactions result in the formation of new phases, influence densification, mechanical properties and even degrade the biological properties of the composite in some cases which often limit their performance (Viswanath & Ravishankar, 2006). Hence, control over nanocomposite characteristics is a challenging task.

3. Mechanosynthesis of ceramic-based nanocomposites

To date, several approaches, including wet chemical methods (Mobasherpour et al., 2007; Kivrak & Tas, 1998), hydrothermal processes (Liu et al., 2006), solid–state reaction (Silva et al., 2003), and sol–gel method (Balamurugan et al., 2002), have been developed for synthesis of nanobioceramics. Among them, mechanochemical process has been extended for the production of a wide range of nanostructured materials (Suryanarayana, 2001; De Castro & Mitchell, 2002). According to literature (Bose et al., 2009), mechanochemical synthesis was originally designed for the production of oxide dispersion-strengthened (ODS) alloys. Over the past 20 years, however, the number of available mechanochemical synthesis has grown, such that Nowadays it is used for the fabrication of a wide range of advanced materials, both metallic and nonmetallic in composition. In mechanosynthesis, the chemical precursors typically consist of mixtures of oxides, chlorides and/or metals that react either during milling or during subsequent thermal treatment to form a composite powder consisting of the dispersion of ultrafine particles within a soluble salt matrix. The ultrafine particle is then recovered by selective removal of the matrix phase through washing with an appropriate solvent.

Mechanochemical approach is a very effective process for synthesizing nanocomposites with various classes of compounds: metals, oxides, salts, organic compounds in various combinations. For example, Khaghani-Dehaghani et al. (Khaghani-Dehaghani et al., 2011) synthesized Al_2O_3–TiB_2 nanocomposite by mechanochemical reaction between titanium dioxide, acid boric and pure aluminum according to the following reactions:

$$2H_3BO_3 \rightarrow B_2O_3 + 3H_2O \tag{1}$$

$$3TiO_2 + 3B_2O_3 + 10Al \rightarrow 3TiB_2 + 5Al_2O_3 \tag{2}$$

Titanium diboride has an attractive combination of high Vickers hardness, electrical conductivity, excellent chemical resistance to molten nonferrous metals and relatively low specific gravity (Gu et al., 2008). However, titanium diboride has poor fracture toughness and impact strength. Thus, the composites of TiB_2 such as Al_2O_3–TiB_2 improve those mechanical properties. These nanocomposites are useful in variety of applications such as cutting tools, wear-resistant substrates, and lightweight armor (Mishra et al., 2006). Results reveal that the Al_2O_3–TiB_2 nanocomposite was successfully synthesized after 1.5 h of milling. Also, the determined amounts of structural features demonstrate that after 20 h of milling the steady state was obtained. Increasing milling time up to 40 h had no significant effect other than refining the crystallite size. The SEM and TEM observations show that increase of milling time was associated with decrease of powder particles, so that a fine structure was produced after 40 h of milling. Figure 1 shows the morphological features of the Al_2O_3–TiB_2 nanocomposite powders after 40 h of milling by SEM and TEM. It is clear that the particles exhibited high affinity to agglomerate. The agglomerates include fine particles of TiB_2 and Al_2O_3.

Thermodynamic studies, based on thermodynamic databases, show that the change in Gibbs free energy of the reduction of boron oxide and titanium oxide with aluminum (Eqs. (3) and (4)) is favorable at room temperature.

$$\begin{gathered} 4Al + 3TiO_2 \rightarrow 2Al_2O_3 + 3Ti \\ \Delta G^{\circ}_{298K} = -495.488\ kJ,\ \Delta H^{\circ}_{298K} = -516.306\ kJ \end{gathered} \tag{3}$$

$$\begin{gathered} 2Al + B_2O_3 \rightarrow Al_2O_3 + 2B \\ \Delta G^{\circ}_{298K} = -389.053\ kJ,\ \Delta H^{\circ}_{298K} = -403.338\ kJ \end{gathered} \tag{4}$$

It is well known if a reaction is highly exothermic, the impact of the milling balls can initiate a mechanically induced self-sustaining reaction (MSR) (Xia et al., 2008). MSR was usually observed in highly exothermic reactions. The ignition of MSR takes place after a certain activation time, during which the powder mixtures reach a critical state due to the physical and chemical changes caused by ball milling (Takacs, 2002; Takacs et al., 2006). That certain activation time depends mainly on the exothermicity of the process, the milling conditions and the mechanical properties of the raw materials. Takacs (Takacs, 2002) showed that a reaction can propagate in the form of a self sustaining process, if $\Delta H/C$, the magnitude of the heat of reaction divided by the room temperature heat capacity of the products, is higher than about 2000 K. The calculations on the system Al–B_2O_3–TiO_2 show that the value of $\Delta H/C$ is about 5110 K. Therefore, the proposed reactions occurred through an expanded MSR reaction in milled samples which led to the formation of Al_2O_3–TiB_2 nanocomposite after short milling times.

Figure 1. a) SEM micrograph and (b) TEM image of Al_2O_3–TiB_2 nanocomposite after 40 h of milling (Khaghani-Dehaghani et al., 2011).

3.1. Mechanochemical synthesize of hydroxyapatite nanostructures

HAp and its isomorphous modifications are valuable and prospective materials in biomedical applications. Therefore, a large number of studies was performed on this subject in the last decade (Rhee, 2002; Silva et al., 2004; Suchanek et al., 2004; Tian et al., 2008; Nasiri-Tabrizi et al., 2009; Gergely et al., 2010; Wu et al. 2011; Ramesh et al., 2012). Generally, the fabrication methods of HAp nanostructures can be classified into two groups: wet and dry (Rhee, 2002). The advantage of the wet process is that the by-product is almost water and as a result the probability of contamination during the process is very low. On the other hand, the dry process has benefit of high reproducibility and low processing cost in spite of the risk of contamination during milling. Furthermore, the dry mechanochemical synthesis of HAp presents the advantage that melting is not necessary and the powder obtained is nanocrystalline. The calcium and phosphorous compounds used as the starting materials in the dry process are dicalcium phosphate anhydrous ($CaHPO_4$), dicalcium phosphate dihydrate ($CaHPO_4.2H_2O$), monocalcium phosphate monohydrate ($Ca(H_2PO_4)_2.H_2O$), calcium pyrophosphate ($Ca_2P_2O_7$), calcium carbonate ($CaCO_3$), calcium oxide (CaO), and calcium hydroxide ($Ca(OH)_2$), etc.

Otsuka et al. (Otsuka et al., 1994) investigated the effect of environmental conditions on the crystalline transformation of metastable calcium phosphates during grinding. Based on the results, the mixture of $CaHPO_4$ and $Ca(OH)_2$ transformed into low-crystallinity HAp after grinding in air. Nevertheless, under N_2 atmosphere, a mixture of initial materials did not transform into HAp. After that, Toriyama et al. (Toriyama et al., 1996) proposed a method to prepare powders and composite ceramic bodies with a matrix comprising HAp. The powders Was produced by the utilization of a simple and economic mechanochemical method. The composite ceramic bodies were easily obtained by simple firing of the powders at a suitable temperature (1250 C). After sintering, the obtained products exhibited a flexural strength of more than 100 MPa in standard samples. This value is significantly higher than that usually attainable with commercially available powders (60 MPa). In another research

(Yeong et al., 2001), nanocrystalline HAp phase has been produced by high-energy mechanical activation in a dry powder mixture of CaO and $CaHPO_4$. The initial stage of mechanical activation resulted in a significant refinement in crystallite and particle sizes, together with a degree of amorphization in the starting powder mixture. A single-phase HAp of high crystallinity was attained by >20 h of mechanical activation. The resulting HAp powder exhibits an average particle size of ~ 25 nm. It was sintered to a density of 98.20% theoretical density at 1200 C for 2 h. The hardness increases almost linearly with rising sintering temperature from 900 to 1200 C, where it reaches a maximum of 5.12 GPa. This is followed by a slight decrease, to 4.92 GPa, when the sintering temperature is raised to 1300 C. Afterward, Rhee (Rhee, 2002) synthesized HAp powder by mechanochemical reaction between $Ca_2P_2O_7$ and $CaCO_3$. The two powders were mixed in acetone and water, respectively, and the single phase of HAp was observed to occur only in the powder milled in water, without the additional supply of water vapor during heat-treatment at 1100 C for 1 h. The results indicated that the mechanochemical reaction could supply enough amount of hydroxyl group to the starting powders to form a single phase of HAp. Therefore, the powder of high crystalline HAp can be obtained by the simple milling in water and subsequent heat-treatment. With the development of nanostructured materials using mechanochemical processes, nanocrystalline powders of HAp was produced in 2003 by Silva et al. (Silva et al., 2003). To produce nanocrystalline powders of HAp, five different experimental procedures in a pure dry process were utilized. For four different procedures, HAp was obtained after a couple of hours of milling (in average 60 h of milling, depending in the reaction procedure). In the preparation of nanocrystalline HAp, commercial oxides $Ca_3(PO_4)_2.xH_2O$, $Ca(OH)_2$, $CaHPO_4$, P_2O_5, $CaCO_3$ and $(NH_4)H_2PO_4$ were used in the HAp preparation. This milling process, presents the advantage that melting is not necessary and the powder obtained is nanocrystalline with crystallite size in the range of 22 nm to 39 nm. Subsequently, Silva et al. (Silva et al., 2004) synthesized nanocrystalline powders of HAp using three different experimental procedures (HAPA: $Ca(H_2PO_4)_2$ + $Ca(OH)_2$; HAPB: $Ca(H_2PO_4)_2$ + $CaCO_3$; and HAPC: $CaHPO_4$ + $CaCO_3$). Nanocrystalline HAp was obtained after 5, 10 and 15 h of milling in the reactions HAPA and HAPB, but it is necessary 15 h of milling in the reaction HAPC to obtain HAP. Moreover, in order to improve the mechanical properties of HAp calcium phosphate ceramics, with titanium (CaP-Ti) and zirconium (CaP-Zr), were prepared by dry ball milling using two different experimental procedures: CaP-Ti1: $Ca(H_2PO_4)_2$ + TiO_2; CaP-Ti2: $CaHPO_4$ + TiO_2; and CaP-Zr1: $Ca(H_2PO_4)_2$ + ZrO_2, CaP-Zr2: $CaHPO_4$ + ZrO_2. The calcium titanium phosphate phase, $CaTi_4P_6O_{24}$, was produced in the reaction CaP-Ti1. In the reactions CaP-Ti2, CaP-Zr1 and CaP-Zr2, it was not observed the formation of any calcium phosphate phase even after 15 h of dry mechanical alloying.

Nanocrystalline HAp powders were synthesized by the mechanochemical–hydrothermal method using emulsion systems consisting of aqueous phase, petroleum ether (PE) as the oil phase and biodegradable Tomadol 23–6.5 as the nonionic surfactant (Chen et al., 2004). $(NH_4)_2HPO_4$ and $Ca(NO_3)_2$ or $Ca(OH)_2$ were used as the phosphorus and calcium sources, respectively. The calcium source and emulsion composition had significant effects on the stoichiometry, crystallinity, thermal stability, particle size, and morphology of final products.

Disperse HAp crystals with a 160 nm length were formed in an emulsion system containing 10 wt% PE, 60 wt% water, and 30 wt% surfactant. The HAp particles had needle morphology with a specific surface area of 190 m^2/g. According to obtained results, HAp nanopowders with specific surface areas in the range of 72–231 m^2/g were produced. In the same year, Mochales et al. (Mochales et al., 2004) investigated the possibility of mechanochemistry to synthesize calcium deficient HAp (CDHA) with an expected molar calcium to phosphate (Ca/P) ratio ± 0.01. To optimize the experimental conditions of CDHA preparation from dicalcium phosphate dihydrate (DCPD) and calcium oxide by dry mechanosynthesis reaction, the kinetic study was carried out with two different planetary ball mills (Retsch or Fritsch Instuments). Results obtained with the two mills led to the same conclusions although the values of the rate constants of DCPD disappearance and times for complete reaction were very different. Certainly, the origin of these differences was from the mills used, thus the influence of instrumental parameters such as the mass and the surface area of the balls or the rotation velocity were examined on the mechanochemical reaction kinetics of DCPD with CaO. Results exhibited that the DCPD reaction rate constant and the inverse of the time for complete disappearance of CaO both vary linearly with (i) the square of the rotation velocity, (ii) the square of eccentricity of the vial on the rotating disc and (iii) the product of the mass by the surface area of the balls. The consideration of these four parameters allows the transposition of experimental conditions from one mill to another or the comparison between results obtained with different planetary ball mills. Gonzalez et al. (Gonzalez et al., 2006) studied the mechanochemical transformation of two mixtures: $Ca(OH)_2$–$(NH_4)_2HPO_4$ and $Ca(OH)_2$–P_2O_5, milled in a mortar dry grinder for different periods of time. Mechanical grinding and thermal treatment was a successful method to obtained biphasic mixtures of HAp/β-TCP. Amorphization, for both reactant mixtures, was observed after prolonged milling, 17.5 h for $Ca(OH)_2$–$(NH_4)_2HPO_4$ mixture and 5 h for the $Ca(OH)_2$–P_2O_5 mixture. The composition of the milled powders varied in the range of $1.50 < Ca/P < 1.67$ for different milling periods. Calcination of milled powders of both mixtures at 800 C led to the formation of HAp and β-TCP, with an average particle size of 200 nm. Further, the Ca/P ratio affects the proportion of HAp and β-TCP phases obtained after thermal treatment. Also, Kano et al. (Kano et al., 2006) developed a novel mechanochemical process to prepare HAp fine particles. For this aim, a non-thermal process for dechlorinating of Polyvinyl chloride (PVC) was utilized. This process was composed of two steps: The first step was to grind the PVC waste with an active grinding additive such as CaO, leading to transformation of organic chlorine into water soluble chloride mechanochemically. The second step is to remove the formed chloride from the milled product by washing with water. When the filtrate was mixed with solution which contains phosphate ion $PO_4{}^{2-}$, HAp fine particles formed which has sorption ability for heavy metals such as Pb^{2+}. El Briak-BenAbdeslam et al. (El Briak-BenAbdeslam et al., 2008) investigated the influence of water addition on the kinetics of the mechanochemical reaction of dicalcium phosphate dihydrate with calcium oxide. The DCPD disappearance rate constant k and the final reaction time t_f were determined in each case and correlated with the water content present in the slurry. Results showed that the addition water (i) slowed down the reaction rate and (ii) increased the powder contamination by mill material (hard porcelain) due to ball and vial erosion; and that (iii) wet milling did not generate the expected products, in contrast to dry grinding, because porcelain induced HAp decomposition with the formation of β-TCP and silicon-stabilized tricalcium phosphate. Consequently, dry mechanosynthesis appears preferable to wet milling in the preparation of calcium phosphates of biological interest.

3.1.1. *Single-crystal hydroxyapatite nanoparticles*

A new approach to mechanochemical synthesis of HAp nanostructures was developed in 2009 by Nasiri-Tabrizi et al. (Nasiri–Tabrizi et al., 2009). Single-crystal HAp nanorods and nanogranules synthesized successfully by a mechanochemical process using two distinct experimental procedures.

$$6CaHPO_4 + 4Ca(OH)_2 \rightarrow Ca_{10}(PO_4)_6(OH)_2 + 6H_2O \quad (5)$$

$$4CaCO_3 + 6CaHPO_4 \rightarrow Ca_{10}(PO_4)_6(OH)_2 + 4CO_2 + 2H_2O \quad (6)$$

The feasibility of using polymeric milling media to prepare HAp nanoparticles is described. By controlling the temperature and milling time during mechanical activation (45-min milling steps with 15-min pauses), powders with three different crystallite size, lattice strain and crystallinity degrees are produced. Figure 2 presents the XRD patterns of reactions 5 and 6, respectively. The XRD patterns show that the product of reaction 5 is HAp. The extra peaks ($CaHPO_4$, ■) occurred in $2\theta = 26.59$ and 30.19 , consecutively. In reaction 6, the extra peaks are not observed after 40, 60 and 80 h of milling and the only detected phase is HAp, as shown in Figure 2(b). Therefore, during milling process, $CaHPO_4$ is a compound that should be avoided if the purpose is to achieve pure HAp without any extra phase presentation. In order to determine crystallite size and lattice strain in activated samples, the full width at half maximum (FWHM) of each peak is usually considered. Furthermore, the fraction of crystalline phase (Xc) in the HAp powders is evaluated by Landi equation (Landi et al., 2000).

According to obtained data, the crystallite size decreases and the lattice strain increases with increase of milling time. However, the rate of both variations, i.e. increasing lattice strain and decreasing crystallite size, decreases by increasing the milling time. Furthermore, the obtained data show that by choosing the total milling time to 80 h for reaction 5, the crystallinity degree increases first and reaches to a maximum at 60 h of milling, and then by further increasing the milling time to 80 h, the crystallinity degree decreases. Moreover, the increase of HAp crystallinity compared to the increase of milling time was not linear. The fraction of crystalline phase in the HAp powders from reaction 6 indicates that by increasing the milling time from 40 to 80 h, the crystallinity degree decreases mostly after 60 h and reaches to a minimum at 80 h of milling time. Based on these results, we conclude that the chemical composition of initial materials and the milling time are important parameters that affect the structural properties of product via mechanochemical process.

The morphological features of the synthesized HAp products were further examined by TEM technique. Figures 3 and 4 show the TEM micrographs of nanorods and nanogranules, respectively. Figure 3a shows that the sample possesses a mostly rod-like structure after 60 h milling time in polymeric milling vial for reaction 5. In Figure 3b, it can be seen that the morphology of nanocrystalline HAp after 80 h milling time, similar to 60 h, is also the rod shape; although, few particles appear to be close to a spherical shape. Using HAp nanorods as raw materials is an effective way to obtain dense bioceramics with high mechanical properties. Hence, this product may be used as strength enhancing additives for the preparation of the HAp ceramics or biocompatible nanocomposites.

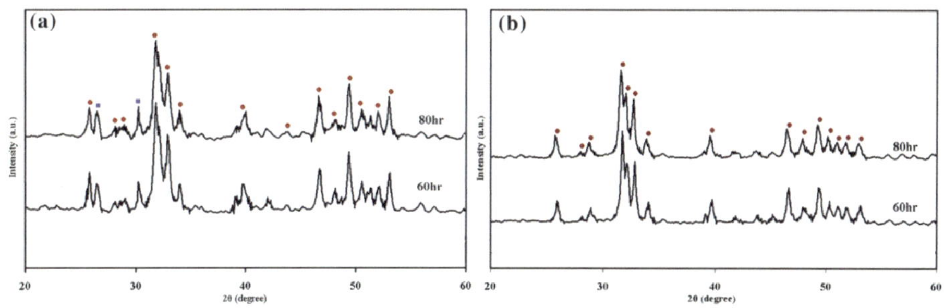

Figure 2. XRD patterns of samples milled for 60 and 80 h, (a) reaction 5 and (b) reaction 6. (Nasiri-Tabrizi et al., 2009).

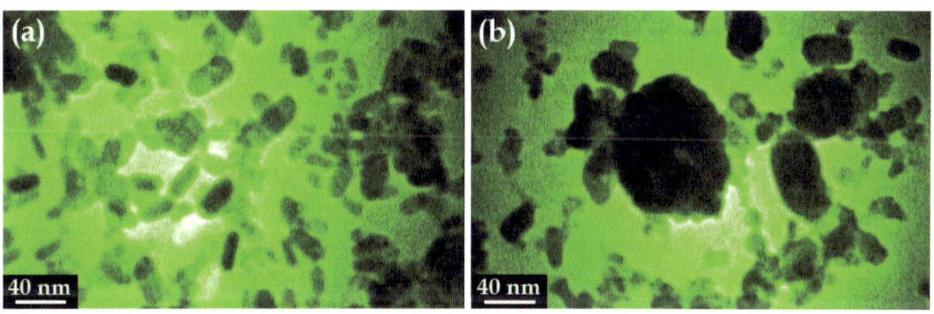

Figure 3. Typical TEM micrograph of nanorods HAp after 60 h (a) and 80 h (b) milling time for reaction 5 (Nasiri-Tabrizi et al., 2009).

In reaction 5, more agglomeration also occurs by increasing milling time from 60 h to 80 h. In fact, the obtained product nearly had a uniform geometry distribution just after 60 h milling time. Although, it may appear some ellipse or round like shapes from this image, it is due to the axis orientation of nanorods with respect to the image plane. In other words, if the rod axis is perpendicular or oblique on the image plane, the rod may be seen as a full circle or ellipse, respectively. Despite of previous research that a perfect spherical shape rarely observed in the mechanically alloyed powders, nanosphere particles were successfully obtained. In Figure 4, it can be seen that the morphology of nanocrystalline HAp for reaction 6, either after 60 or 80 h milling time, is absolutely spherical granules with a reasonable smooth geometry.

Therefore, we reach to an important conclusion that using polyamide-6 milling vial leads to the spherical granules HAp. Since spherical geometry compared to irregular shape is important for achieving osseointegration (Komlev et al., 2001; Nayar et al., 2006; Hsu et al., 2007), the latest product is well preferred for medical applications. Similar to previous reaction, the obtained product after 60 h has a better uniform geometry distribution than one after 80 h milling time. It should be noted that the HAp particles out of reaction 5 are in average length of 17 ± 8 nm and 13 ± 7 nm after 60 and 80 h milling time, respectively. Similarly, the HAp particles out of reaction 6 are in average diameter of 16 ± 9 nm and 15 ± 8 nm after 60 and 80 h milling time. Based on obtained data, the maximum particle distribution is below the crystallite size which is estimated from the line broadening of the given X-ray diffraction peak.

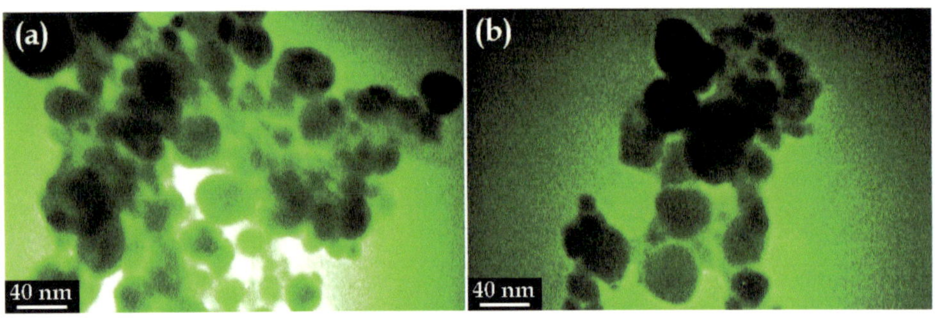

Figure 4. Typical TEM micrograph of nanospheres HAp after 60 h (a) and 80 h (b) milling time for reaction 6 (Nasiri-Tabrizi et al., 2009).

Thus, after 80 h milling time, we ascertain that this method gives rise to the single-crystal HAp with their average size below 20 nm and 23 nm for reactions 6 and 7, respectively. In fact, a novel method for the synthesis of nanosize single-crystal HAp is developed in both spherical and rod-like particles.

3.1.2. Milling media effects on structural features of hydroxyapatite

Honarmandi et al. (Honarmandi et al., 2010) investigated the effects of milling media on synthesis, morphology and structural characteristics of single-crystal HAp nanoparticles. Typical TEM images of nanosize HAp particles produced through reactions 5 and 6 after being milled in both metallic and polymeric vials have been shown in Figure 5.

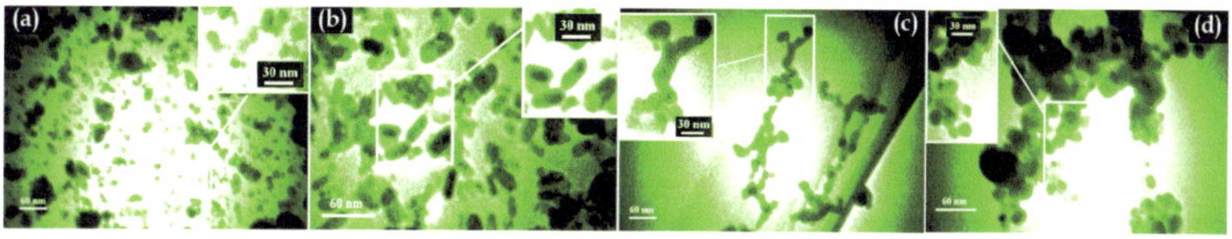

Figure 5. Morphologies of HAp synthesized through reactions 5 after being milled for 60 h in (a) metallic vials and (b) polymeric vials; through reactions 6 after being milled for 60 h in (c) metallic vials and (d) polymeric vials.

The results reveal that the single-crystal HAp nanoparticles have been successfully produced in metallic and polymeric vials through two different experimental procedures. Transmission electron microscopy images illustrate the wide morphology spectrums of the single-crystal HAp nanoparticles which are ellipse-, rod- and spherical-like morphologies each of which can be applied for specific purpose. After 60 h milling, this method results in the single-crystal HAp with their average sizes below 21 and 24 nm in the tempered chrome steel and polyamide-6 vials, respectively. According to TEM images the obtained single-crystal HAp in polymeric vials have more production efficiency and better uniform geometry distribution than products in metallic vials. In metallic vial, intense agglomeration happens during mechanochemical process as shown in Figure 6. Therefore, an important conclusion reaches that the polyamide-6 vial is more suitable than the tempered chrome steel vial for the synthesis of single-crystal HAp nanoparticles with appropriate morphology.

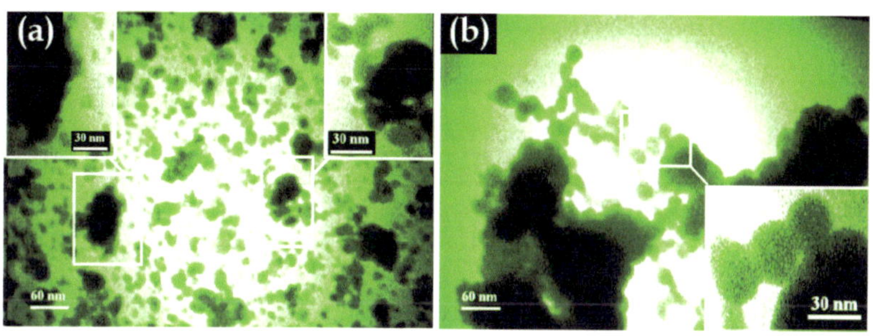

Figure 6. TEM images of agglomerated products which is obtained after 60 h milling in metallic vials through a (a) reaction 5 and (b) reaction 6 (Honarmandi et al., 2009).

3.1.3. Milling atmosphere effect on structural features of hydroxyapatite

In recent years, various mechanochemical processes were utilized to synthesis HAp nanostructures. For instance, Gergely et al. (Gergely et al., 2010) synthesized HAp by using recycled eggshell. The observed phases of the synthesized materials were dependent on the mechanochemical activation method (ball milling and attrition milling). Attrition milling proved to be more efficient than ball milling, as resulted nanosize, homogenous HAp even after milling. SEM micrographs showed that the ball milling process resulted in micrometer sized coagulated coarse grains with smooth surface, whereas attrition milled samples were characterized by the nanometer size grains. Wu et al. (Wu et al. 2011) synthesized HAp from oyster shell powders by ball milling and heat treatment. The wide availability and the low cost of oyster shells, along with their biological– natural origin are highly attractive properties in the preparation of HAp powders for biomedical application. Chemical and microstructural analysis has shown that oyster shells are predominantly composed of calcium carbonate with rare impurities. Solid state reactions between oyster shell powders (calcite polymorph of $CaCO_3$) and calcium pyrophosphate ($Ca_2P_2O_7$) or dicalcium phosphate dihydrate ($CaHPO_4.2H_2O$, DCPD) were performed through ball milling and subsequently heat treatment. The ball milling and heat treatment of $Ca_2P_2O_7$ and oyster shell powders in air atmosphere produced mainly HAp with a small quantity of β-TCP as a by product. However, oyster shell powder mixed with DCPD and milled for 5 h followed by heat-treatment at 1000 C for 1 h resulted in pure HAp, retaining none of the original materials.

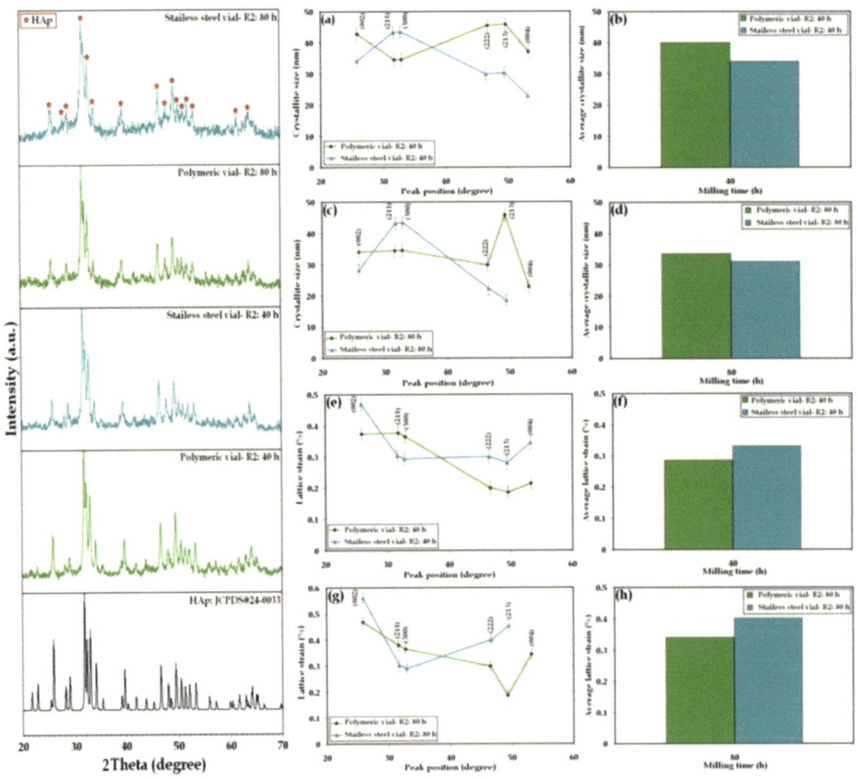

Figure 7. XRD patterns, crystallite size, lattice strain and their average of samples milled for 40 and 80 h in polymeric and metallic vials under argon atmosphere.

Mechanosynthesis of calcium phosphates can be performed under air or inert gas atmosphere. In most papers and patents, grinding under air atmosphere was selected. So far, only a few papers were devoted to mechanosynthesis of calcium phosphates under inert gas atmosphere (Nakano et al, 2001). To understand the effect of inert gas atmosphere, the mechanochemical synthesis under argon atmosphere was investigated by our research group. The starting reactant materials are $CaCO_3$ and $CaHPO_4$. The initial powders with the desired stoichiometric proportionality were mixed under a purified argon atmosphere (purity> 99.998 vol %). Figure 7 shows the XRD patterns of the powder mixture after 40 and 80 h of milling in the polymeric and metallic vials under argon atmosphere. The XRD patterns of obtained powders exhibit that the production of mechanical activation is single phase HAp. Also, Figure 7 illustrates the determined amounts of crystallite size; lattice strain and their average for experimental outcomes after 40 and 80 h of milling in polymeric and metallic vials under argon atmosphere.

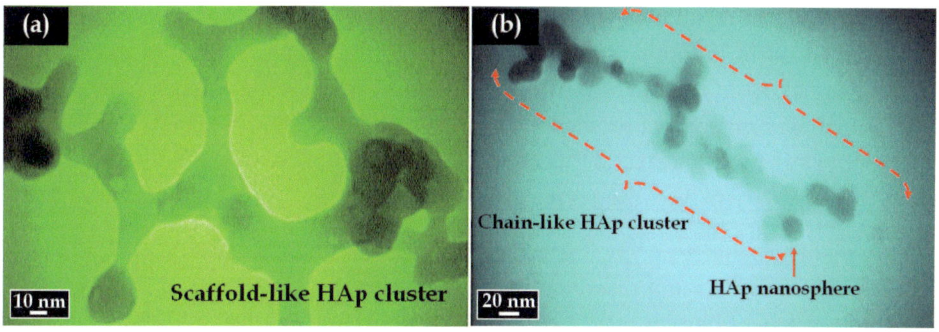

Figure 8. Typical TEM micrograph of nanocrystalline HAp after 80 h of milling under argon atmosphere in (a) polymeric and (b) metallic vials.

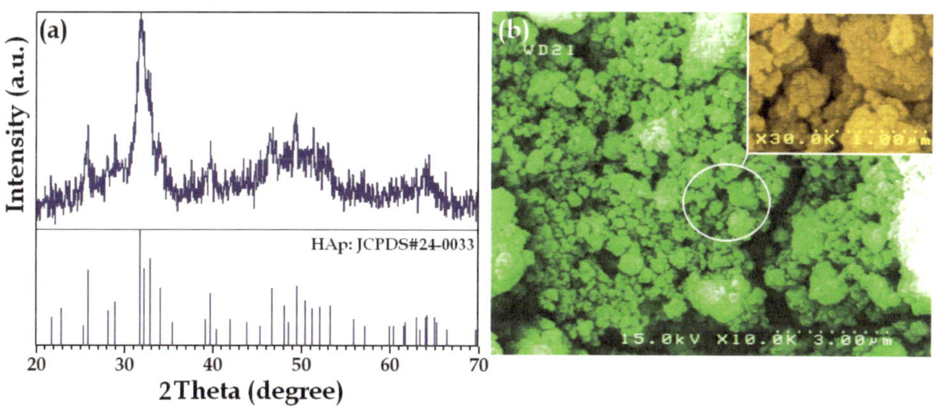

Figure 9. a) XRD profile and (b) FE-SEM images of nanocrystalline HAp with low degree of crystallinity after 2 h of milling in polymeric vial under air atmosphere.

Using the (0 0 2) plane (Figure 7a), the crystallite size of HAp is around 43 and 34 nm after 40 h of milling in polymeric and metallic vials, respectively. For comparison, the mean values determined from the use of six planes simultaneously, i.e. (0 0 2), (2 1 1), (3 0 0), (2 2 2), (2 1 3), and (0 0 4) planes. The calculated data indicates that the average crystallite size of HAp is around 40 and 34 nm, respectively. Moreover, using the (0 0 2) plane the crystallite size of HAp is around 34 and 28 nm after 80 h of milling in polymeric and metallic vials, respectively. However, the average crystallite size of HAp is around 34 and 31 nm after 80 h of milling in polymeric and metallic vials, respectively. The evaluation of the lattice strain of HAp reveals that the average of lattice strain partially increased from 0.286 % to 0.340 % after 80 h of milling in polymeric vial. A similar trend was observed in the average lattice strain of HAp after 80 h of milling. According to Figure 7, the average crystallite size decreases and the average lattice strain increases with increase of milling time from 40 up to 80 h. The TEM micrographs of synthesized powder after 80 h of milling in polymeric and metallic vials under argon atmosphere are shown in Figure 8. The TEM micrographs show that HAp particles can attach at crystallographically specific surfaces and form scaffold- and chain-like cluster composed of many primary nanospheres. In is found that (Pan et al., 2008) the living organisms build the outer surface of enamel by an oriented assembly of the rod-like crystal and such a biological construction can confer on enamel protections against erosion. It should be noted that, comparison of the physical, mechanical and biocompatibility between classical HAp ceramics and the novel nanostructures will be carried out in our laboratory.

Whilst the main advantages of the mechanochemical synthesis of ceramic powders are simplicity and low cost, the main disadvantages are the low crystallinity and calcium-deficient nonstoichiometry (Ca/P molar ratio 1.50 – 1.64) of the HAp powders, as this results in their partial or total transformation into β-TCP during calcination (Bose et al., 2009). Hence, control over crystallinity degree of HAp nanostructures for specific applications is a challenging task. Based on experimental results, we conclude that the chemical composition of initial materials, milling time, milling media, and atmosphere are important parameters that affect the structural properties (crystallite size, lattice strain, crystallinity degree) and morphological features of HAp nanostructures during mechanochemical process. For example, mechanical activation of $Ca(OH)_2$ and P_2O_5 powder mixture lead to the formation of single phase HAp with low fraction of crystallinity (Figure 9). According to this mechanochemical reaction (7), nanocrystalline HAp with an average crystallite size of about 14 nm was produced after 2 h of milling in polymeric vial under air atmosphere. In addition the fraction of crystallinity was around 7 %.

$$10Ca(OH)_2 + 3P_2O_5 \rightarrow Ca_{10}(PO_4)_6(OH)_2 + 9H_2O \quad (7)$$

Figure 9b shows the morphology and particle size distribution of the nanocrystalline HAp produced after 2 h of milling. From the FE–SEM micrograph, it is clear that the powders displayed an agglomerate structure which consisted of several small particles with the average size of about 58 nm. In the field of science and technology of particles, agglomerate size is one of the key factors that influence the densification behaviors of nanoparticles. Large particle size along with hard agglomerates shows lower densification in calcium phosphate ceramics due to the formation of large interagglomerate/intraagglomerate pores (Banerjee et

http://dx.doi.org/10.5772/50160

al., 2007). The large interagglomerate/intraagglomerate pores increase the diffusion distance, resulting in lowering the densification rate. Thus, to compensate for this, higher sintering temperature becomes necessary.

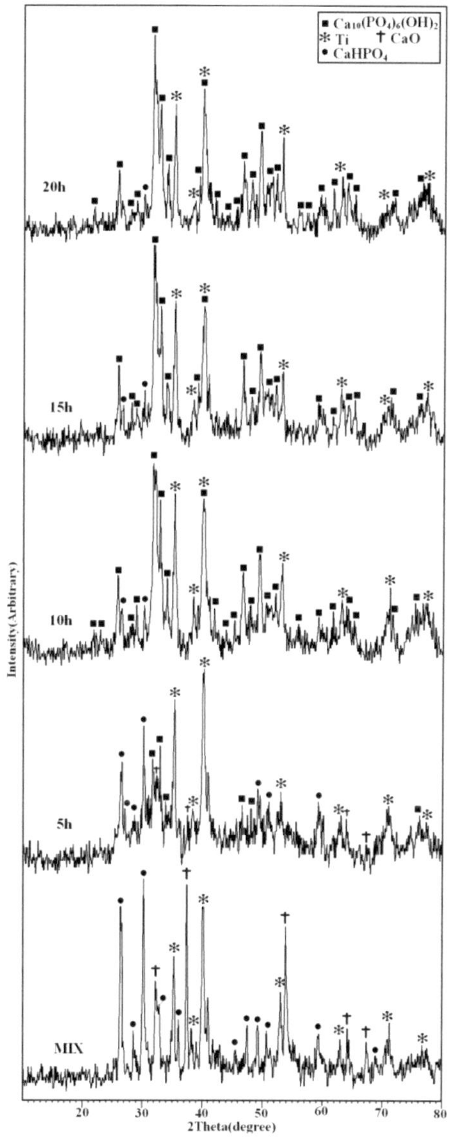

Figure 10. XRD patterns of the HAp-20%wt Ti nanocomposite after mechanochemical process for various time periods (Fahami et al., 2011).

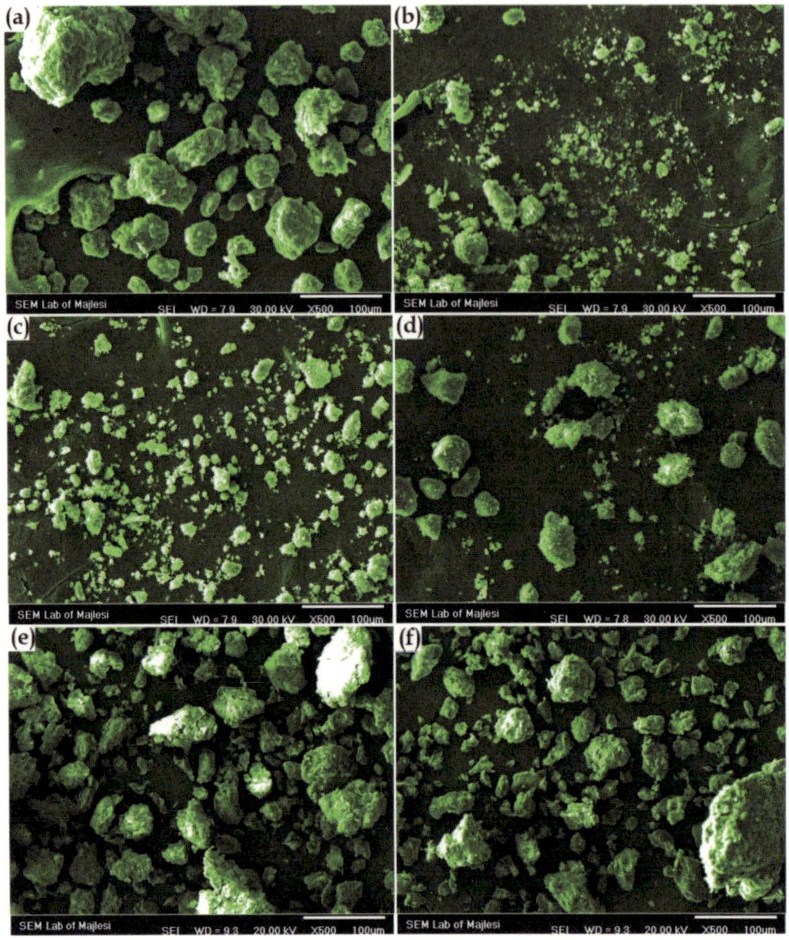

Figure 11. SEM micrographs of the HAp-20 wt.% Ti nanocomposite after different milling times (a) 5, (b) 10, (c) 15, (d) 20, (e) 40, and (f) 50 h.

3.1.4. Hydroxyapatite/titanium (HAp-Ti) nanocomposite

Apart from the displacement reactions to reduce oxides, chlorides, and sulfides to pure metals, mechanical alloying technique was also used to synthesize a large number of commercially important alloys, compounds, and nanocomposites using the mechanochemical reactions (Suryanarayana, 2001; De Castro & Mitchell, 2002; Balaz, 2008). An important characteristic of mechanosynthesized composites is that they have nanocrystalline structures which could improve the mechanical as well as biological properties (Silva et al., 2007). Nowadays, ceramic nanocomposites which play a crucial role in technology can be synthesized using surprisingly simple and inexpensive techniques such as a mechanochemical method which ordinarily include a two step process. Considering the above characteristics of the ceramic-based composites, the possibility of using one step mechanochemical process as a simple, efficient, and inexpensive method to prepare HAp-20wt.% Ti nanocomposite was investigated by our research group (Fahami et al., 2011). Furthermore, crystallite size, lattice strain, crystallinity degree, and morphological properties of products were determined due to the biological behaviour of HAp ceramics depends on structural and morphological features. For the preparation of HAp-20 wt.% Ti nanocomposite, anhydrous calcium hydrogen phosphate and calcium oxide mixture with Ca/P = 1.67 ratio was milled with the distinct amount of elemental titanium (20 wt.%) during 0, 5, 10, 15, and 20 h by a high energy planetary ball mill under highly purified argon gas atmosphere. The following reaction can be occurred at this condition (8):

$$6CaHPO_4 + 4CaO + Ti \rightarrow Ca_{10}(PO_4)_6(OH)_2 + Ti + 2H_2O \quad (8)$$

Figure 10 shows the XRD patterns of the samples after mechanochemical process for various time periods. At the initial mixture, only sharp characteristic peaks of $CaHPO_4$, CaO and Ti are observed. With increasing milling time to 5 h, the sharp peaks of starting materials degraded significantly, but the decreasing rate of each initial powder was differed. On the other hand, the appearance of weak peak between 31 and 32 confirms the formation of HAp phase. The main products of powder mixtures after 10 h of milling were HAp and Ti. The XRD patterns of the samples which are milled for 15 and 20 h indicate that increasing milling time to above 10 h does not accompany with any phase transformation. The determined amounts of crystallite size and lattice strain of the samples, after different milling time were presented in Table 1. According to Table 1, the crystallite size of HAp decrease with increasing milling time up to 20 h; whereas the change in crystallite size of Ti with increasing milling time is not linear. The calculated amount of crystallinity degree indicate that the increasing milling time dose not accompany by remarkable change in degree of crystallinity. Since the amorphous powders could find applications to promote osseointegration or as a coating to promote bone ingrowth into prosthetic implants (Sanosh et al., 2009), the resultant powders could be used to various biomedical applications.

Samples	Milling time (h)	Crystallite size (nm)		Lattice strain (%)		Crystallinity (%)
		HAp	Ti	HAp	Ti	
I	5	-	22.8	-	0.494	-
II	10	20.13	18.7	0.492	0.621	17
III	15	19.06	19.3	0.522	0.601	15
IV	20	13.13	22	1.209	0.530	13

Table 1. Comparison between structural features of the samples after different milling times.

The SEM micrographs of the samples after different milling times are presented in Figure 11. It can be seen that the particles of products can be attached together at specific surfaces and form elongated agglomerates which composed of many primary crystallites. The agglomerates with flaky-like structure formed after 10 h of milling. It seems that the existence of ductile Ti can be led to the more agglomeration during mechanochemical process. With increasing milling time to 20 h owing to sever mechanical deformation introduced into the powder, particle, and crystal refinement have occurred. Based on SEM observations, milling process reached steady state after 40 h of milling where the particles have become homogenized in size and shape. Figure 12 shows the SEM images of the HAp-20wt.% Ti nanocomposite after 40 and 50 h of milling and subsequent heat treatment at 700 C for 2h. According to SEM observations, the annealing of the milled samples at 700 C demonstrates the occurrence of grain growth.

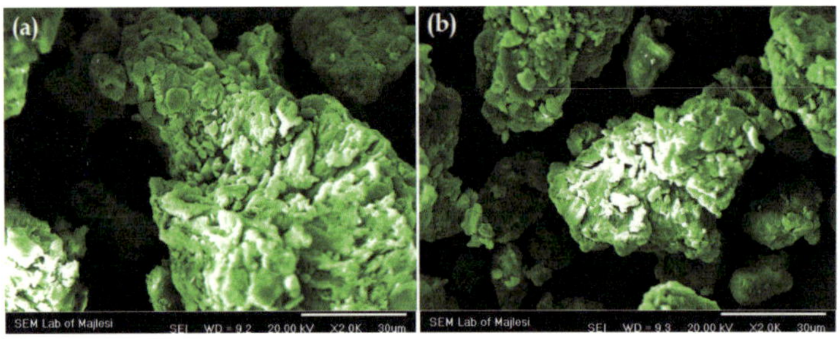

Figure 12. SEM micrographs of the HAp-20 wt.% Ti nanocomposite after different milling times (a) 40 and (b) 50 h and subsequent heat treatment at 700 C for 2h.

3.1.5. Hydroxyapatite/geikielite (HAp/$MgTiO_3$–MgO) nanocomposite

In the field of nanocomposites, an ideal reinforcing material for calcium phosphate-based composites has not yet been found. Nevertheless, different approaches have been extensively investigated in order to develop calcium phosphate-based composites. Despite a large number of studies on the synthesis of HAp and TCP composites (Viswanath & Ravishankar, 2006; Rao & Kannan, 2002; Nath et al., 2009; Jin. et al., 2010; Cao & Kuboyama, 2010; Hu et al., 2010), no systematic investigations on the preparation of HAp/$MgTiO_3$–MgO are performed. Therefore, a novel approach to synthesis of HAp/$MgTiO_3$–MgO nanocomposite has developed by our research group (Fahami et al., 2012). In this procedure, the starting reactant materials are $CaHPO_4$, CaO, titanium dioxide (TiO_2), and elemental magnesium (Mg). Synthesis of HAp/$MgTiO_3$–MgO composite nanopowders consists of: (i) mechanical activation of powder mixture, and (ii) subsequently thermal treatment at 700 C for 2 h. The obtained mixture was milled in a high energy planetary ball mill for 10 h according to the following reaction.

$$6CaHPO_4 + 4CaO + TiO_2 + 2Mg \rightarrow Ca_{10}(PO_4)_6(OH)_2 + TiO_2 + 2MgO + 2H_2\uparrow \quad (9)$$

$$MgO + TiO_2 \rightarrow MgTiO_3 \quad \Delta G_{298K} = -25.410\ kJ,\ \ \Delta H_{298K} = -26.209\ kJ \qquad (10)$$

Figure 13 shows the XRD profiles of the $CaHPO_4$, CaO, TiO_2, and Mg powder mixture after 10 h of milling and after thermal annealing at 700 C for 2 h. As can be seen, the product of mechanochemical process in presence of 20 wt.% (TiO_2, Mg) is HAp/MgO–TiO_2 composite. From Figure 13b, it was verified the existence of HAp and geikielite ($MgTiO_3$) phases together with minor MgO phase after the annealing at 700 ºC. This suggests that the thermal treatment at 700 ºC led to the formation of $MgTiO_3$ by the following reaction:

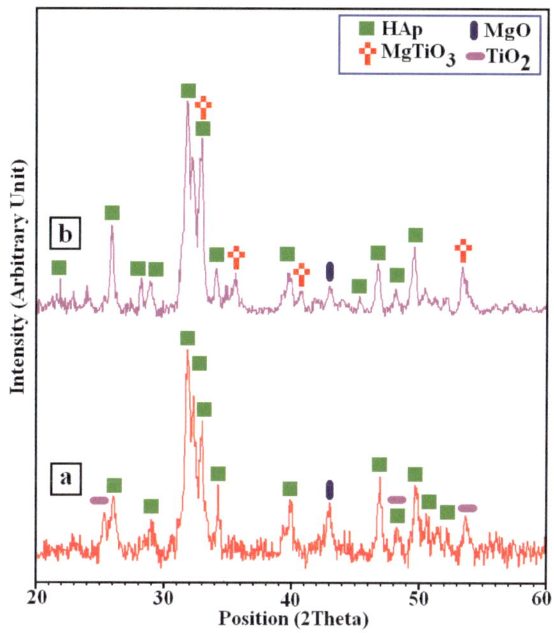

Figure 13. XRD patterns of the samples: (a) HAp/MgO–TiO_2 after 10 h of milling, and (b) HAp/$MgTiO_3$–MgO after 10 h of milling + heat treatment at 700 °C (Fahami et al., 2012).

Figure 14 shows the SEM micrographs and EDX results of the samples after milling and thermal treatment. As can be seen in Figure 15a, a very fine structure was formed after 10 h of milling. After thermal treatment at 700 ºC (Figure 14c), continuous evolution of the morphological features was appeared. The mean size of the powder particles increased after thermal treatment; however, only a slight change in particle size was observed in heat treated sample at 700 ºC compare to the milled powder. After milling and subsequent thermal treatment at 700 ºC, the products were composed of fine particles with a mean particle size of about 482 and 510 nm, respectively. Figures 14b and d represent the EDX results for HAp-based composites which are synthesized after 10 h of milling and subsequent heat treatment at 700 ºC. EDX data show that the main elements of the calcium phosphate-based composite nanopowders are calcium, phosphorus, oxygen, magnesium, and titanium. The EDX of HAp crystal, present in the HAp/MgO $-TiO_2$ composite exhibit a molar ratio Ca/P = 1.93, whereas in the HAp/$MgTiO_3$–MgO composite the molar ratio of calcium to phosphorus is greater (Ca/P = 2.87). These results suggest that the HAp crystals are closer to the expected value for the molar ratio of calcium to phosphorus ratio for the standard HAp (Ca/P = 1.67) (Cacciotti et al., 2009) and commercial HAp (2.38) (Silva et al., 2002), respectively. It is noteworthy to mention that chemically stable contaminants were not detected due to the excessive adhesion of powders to the vial and balls. Figure 15 demonstrates the TEM images of the HAp/$MgTiO_3$–MgO composite nanopowders produced after 10 h of milling and subsequent annealing at 700 ºC for 2h. As can be seen, the agglomerates with mean size of about 322 nm were developed after thermal treatment at 700 ºC. In this sample, the cluster-like shape particles were composed of fine spheroidal shape crystals with a mean size of about 55 nm. It should be mentioned that chemical interactions at the contacting surface of crystals resulted in cluster-like shape aggregates which were composed of fine spheroidal shape crystals. This phenomenon is referred to the nature of milling process which originates through repeated welding, fracturing and re-welding of fine powder particles (Suryanarayana, 2001; De Castro & Mitchell, 2002). It is found that MgO-doped HAp/TCP ceramics present high density and significantly enhance mechanical properties without any phase transformation of β-TCP to α-TCP up to 1300 ºC (Farzadi et al., 2011). Moreover, a patent (Sul, 2008) reported the biocompatibility and osteoconductivity of magnesium titanate oxide film implant for utilizing in several medical fields such as dentistry, orthopedic, maxillofacial, and plastic surgery. Therefore, the presence of MgO and $MgTiO_3$ phases along with HAp in outputs can be enhanced the biological and mechanical properties of HAp-based bioceramics.

3.2. Mechanochemical synthesize of fluorapatite nanostructures

The inorganic matrix of the bone is based on HAp doped with different quantities of cations, such as Na^+, K^+ and Mg^{2+}, and anions, such as $CO_3^{\ 2-}$, $SO_4^{\ 2-}$ and F^-. Among them, F^- plays a leading role because of its influence on the physical and biological characteristics of HAp (Nikcevic et al., 2004). In the recent years, fluoridated HAp (FHAp and FAp) has attracted much attention as a promising material to replace HAp in biomedical applications (Kim et al., 2004a; Fathi & Mohammadi Zahrani, 2009). It is found that the incorporation of fluoride ions into the HAp structure considerably increases the resistance of HAp to biodegradation and thermal decomposition (Fathi et al., 2009). In addition, fluoridated hydroxyapatite could provide better protein adsorption (Zeng et al., 1999) and comparable or better cell attachment than HAp (Kim et al., 2004b). This substitution also has positive effects on proliferation, morphology and differentiation of osteoblastic-like cells and promotes the bioactivity (Fathi et al., 2009). For all these reasons, synthesis of FHAp and FAp is of great value and has been widely investigated by multiple techniques, such as precipitation (Chen & Miao, 2005), sol–gel (Cheng et al., 2006), hydrolysis (Kurmaev et al., 2002), hydrothermal (Sun et al., 2012), and mechanochemical methods (Nikcevic et al., 2004).

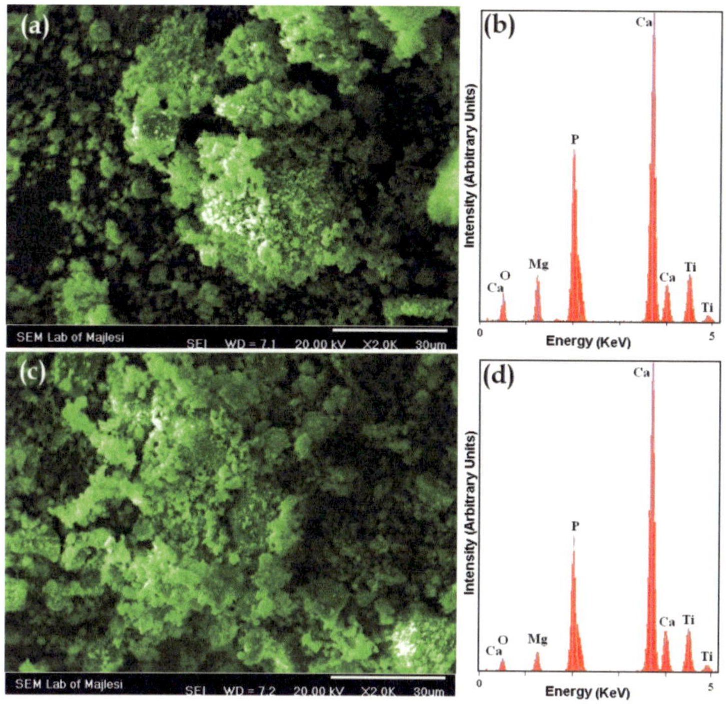

Figure 14. SEM micrographs and EDX results of the samples: (a-b) HAp/MgO–TiO_2 after 10 h of milling, (c-d) HAp/$MgTiO_3$–MgO after 10 h of milling + heat treatment at 700 °C for 2 h.

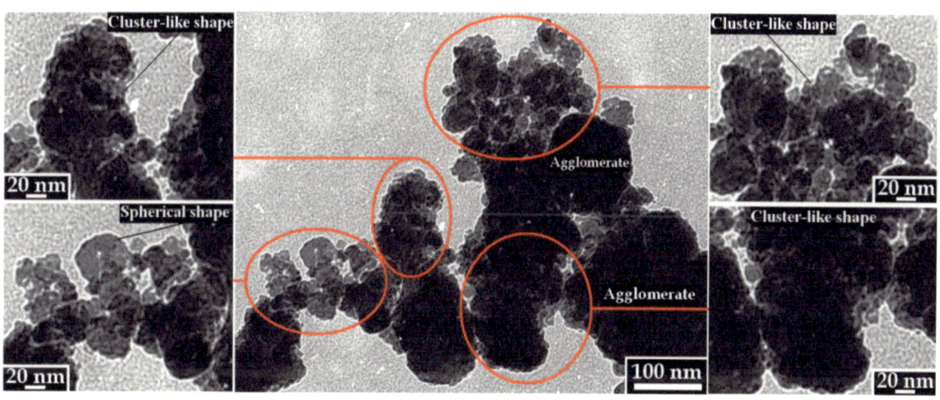

Figure 15. TEM images of the HAp/M_gTiO_3–MgO composite nanopowders after 10 h of milling + heat treatment at 700 °C for 2 h (Fahami et al., 2012).

Nikcevic et al. (Nikcevic et al., 2004) synthesized nanostructured fluorapatite/fluorhydroxyapatite and carbonated fluorapatite/fluorhydroxyapatite by mechanochemical process. Powder mixture of $Ca(OH)_2$-P_2O_5-CaF_2 were milled in planetary ball mill. A carbonated fluorhydroxyapatite, FHAp was formed after 5 h of milling and carbonated fluoroapatite was formed after 9 h of milling. Complete transformation of the carbonated form of FAp into the single phase of FAp occurred after 9 h milling and thermally treating. After that, Zhang et al. (Zhang et al., 2005) synthesized FHAp from the starting materials of $CaCO_3$, $CaHPO_4.2H_2O$, and CaF_2 via a mechanochemical–hydrothermal route. The mechanism study revealed that under such mechanochemical–hydrothermal conditions the formation reactions of FHAp were completed in two stages. The starting materials firstly reacted into a poorly crystallized calcium-deficient apatite and the complete incorporation of fluoride ions into apatite occurred in the second stage.

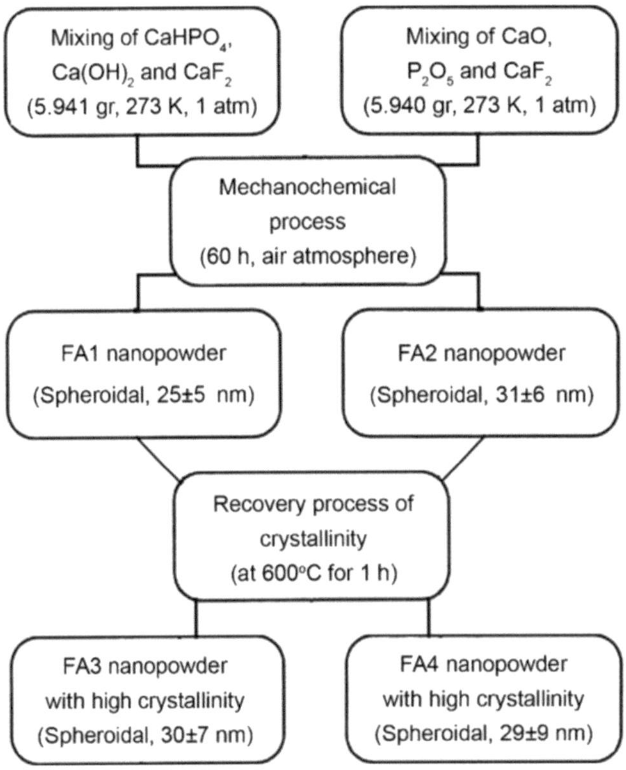

Figure 16. Flow sheet of FAp nanoparticles preparation (Ebrahimi-Kahrizsangi et al., 2011).

3.2.1. Single-crystal fluorapatite nanoparticles

Along with the development of mechanochemical processes, Mohammadi Zahrani & Fathi (Mohammadi Zahrani & Fathi, 2009) evaluated the effect of ball milling parameters on the synthesis of FAp nanopowder; also, the effect of fluoridation on bioresorbability and bioactivity of apatite was studied. Fluoridated hydroxyapatite nanopowders with 100% (FAp) were synthesized via mechanical alloying method. The results showed that the size and number of balls had no significant effect on the synthesizing time and grain size of FAp, while decreasing the rotation speed or ball to powder weight ratio increased synthesizing time and the grain size of FAp. In vitro test indicated that the bioactivity of FAp was less than HAp since the dissolution rate, precipitation amount and the size of precipitated bone-like apatite crystals on the surface of FAp samples was clearly lower than HAp (Nikcevic et al., 2004; Zhang et al., 2005). Recently, Ebrahimi-Kahrizsangi et al. (Ebrahimi-Kahrizsangi et al., 2011) synthesized FAp nanostructures from the starting materials of $CaHPO_4$, $Ca(OH)_2$, CaO, P_2O_5 and CaF_2 via mechanochemical process. The suitability of using the mechano-chemical process to prepare a high crystalline phase of FAp was studied. FAp nanopowders with different structural characteristics synthesized through novel dry mechanochemical processes, which are presented in Figure 16.

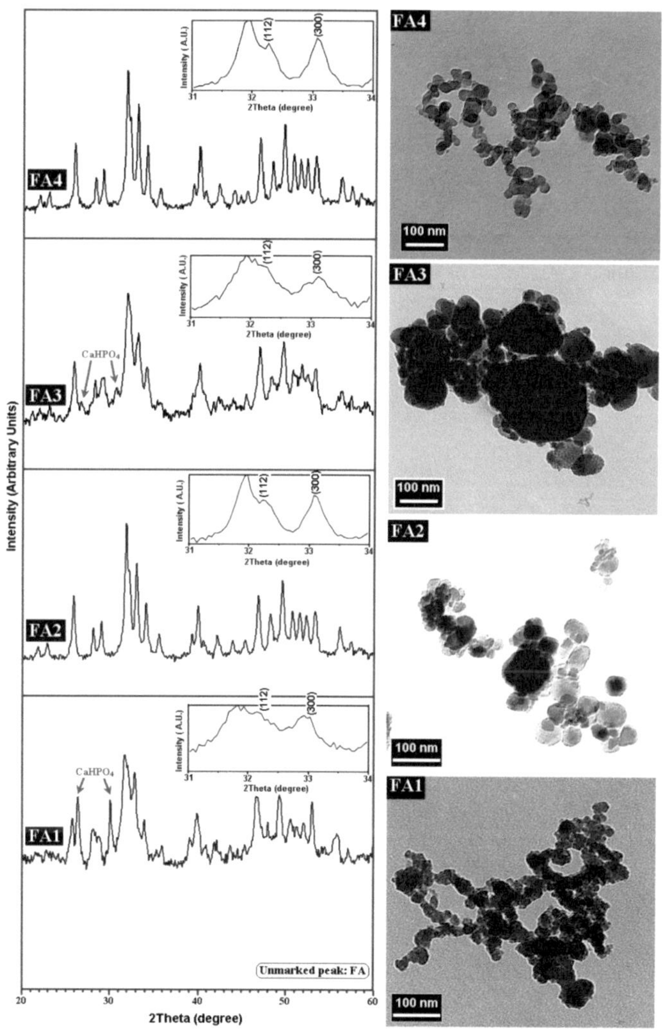

Figure 17. XRD patterns and morphological features of the FAp nanostructures (Ebrahimi-Kahrizsangi et al., 2011).

The purpose of the milling was twofold: first, to activate the following reactions via mechanochemical processes, and second, to produce the nanostructured FAp.

$$6CaHPO_4 + 3Ca(OH)_2 + CaF_2 \rightarrow Ca_{10}(PO_4)_6F_2 + 6H_2O \tag{11}$$

$$9CaO + 3P_2O_5 + CaF_2 \rightarrow Ca_{10}(PO_4)_6F_2 \tag{12}$$

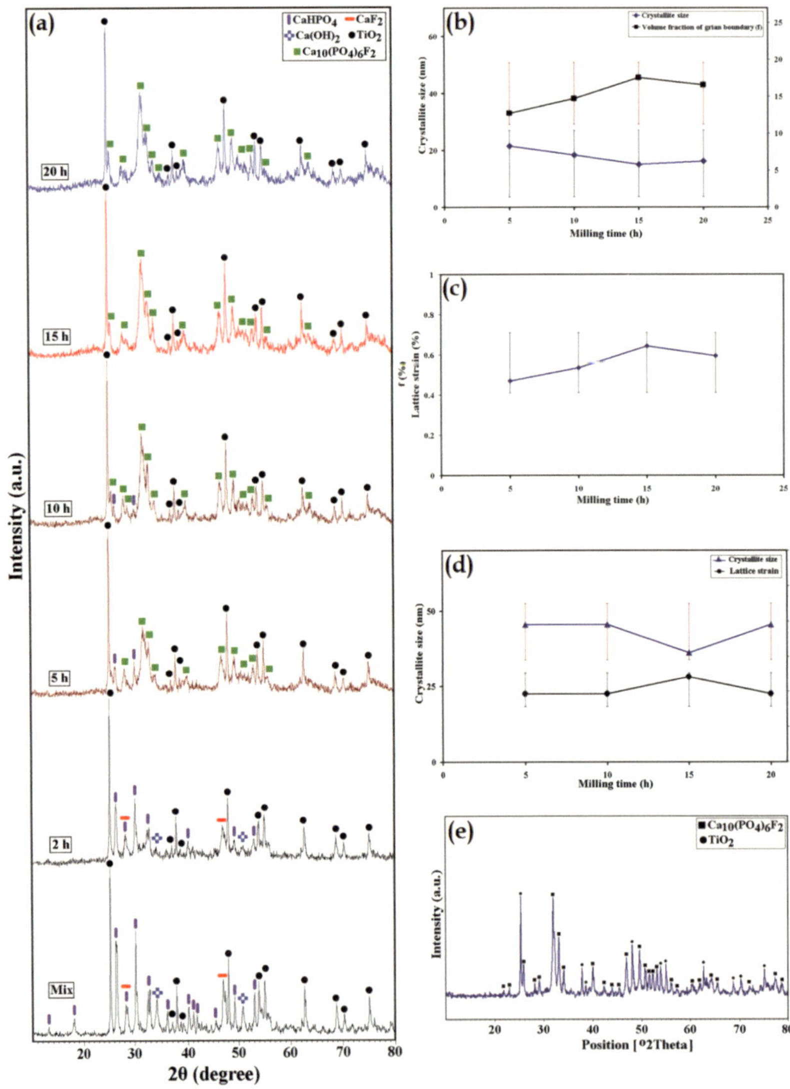

Figure 18. a) XRD patterns, and (b-d) structural features of $CaHPO_4$-$Ca(OH)_2$-CaF_2-TiO_2 powder mixture, mechanically alloyed for 2-20 h. (e) XRD profile of the FAp–TiO_2 nanocomposite after heat treatment at 650 C (Ebrahimi-Kahrizsangi et al., 2011).

Figure 17 illustrates XRD patterns and morphological features of the FAp nanostructures. XRD patterns of the FA1 and FA2 samples showing that the powders synthesized through the two different mechanochemical processes are mostly FAp. Complete agreement with the standard card of fluorapatite (JCPDS #15-0876) was not observed for FA1 due to the presence of additional peaks at $2\theta = 26.59°$ and $30.19°$. These additional peaks were attributed to $CaHPO_4$ from the starting materials. In FA2, complete agreement with JCPDS #15-0876 was observed, allowing FA2 to be used as the pure FAp phase sample when required. The results from the structural studies indicate that the maximum lattice disturbance in the apatite structure after the mechanochemical process was at the (0 0 2) plane. According to TEM images, the FA1 particles are spheroidal with an average diameter of 25 ± 5 nm. However, it should be noted that the particles do not possess high regularity in shape; in other words, their surfaces are not smooth. The particles show a high tendency towards agglomeration. In addition, the sample FA2 possesses a mostly spherical structure with an average diameter of 31 ± 6 nm. The TEM image of the FA3 sample confirms that the particles are spheroidal with an average diameter of 30 ± 7 nm. As shown in Figure 17, the TEM image of sample FA4 shows the particles to be more spherical than the unheated FA2 particles, and the average size of the particles is 29 ± 9 nm. Based on the obtained data, the maximum particle size measured with TEM is below the crystallite size calculated from the line broadening of the X-ray diffraction peak. Thus we concluded that, after 60 h of milling and subsequent thermal treatment at 600 C, this method gives rise to single-crystal FAp with average sizes of 30, 37, 37 and 38 nm for FA1, FA2, FA3 and FA4, respectively. In most reports concerning the synthesis of FAp, the particle shapes are plates (Rameshbabu et al., 2006) or polyhedral (Barinov et al., 2004; Fathi & Mohammadi Zahrani, 2009), but nanoparticles with spheroidal morphology were successfully prepared by current approach. Because the spherical geometry rather than irregular shape is important for achieving osseointegration (Hsu et al., 2007; Nayar et al., 2006), the products synthesized via mechanochemical processes are preferred for medical applications.

3.2.2. *Fluorapatite-titania (FAp-TiO_2) nanocomposite*

As a fact that the incorporation of bioinert ceramics into calcium phosphate-based materials has demonstrated significant improvement in mechanical properties without substantial compromise in biocompatibility, some attempts have been made to develop FHAp-based composites such as: FHAp–Al_2O_3 (Adolfsson et al., 1999), and FHAp–ZrO_2 (Kim et al., 2003; Ben Ayed & Bouaziz, 2008) composites. However, only a few studies have been devoted to the use of solid state reaction in order to prepare FAp nanocomposites (Bouslama et al., 2009). Therefore, synthesis of FAp–TiO_2 nanocomposite which can present advantages of both TiO_2 and FAp were carried out by Ebrahimi-Kahrizsangi et al. (Ebrahimi-Kahrizsangi et al., 2011). Based on XRD patterns and FT-IR spectroscopy, correlation between the structural features of the nanostructured FAp–TiO_2 and the process conditions was investigated. The starting reactant materials are $CaHPO_4$, $Ca(OH)_2$, CaF_2, and TiO_2. In the production of the nanocomposite, a distinct amount of titanium dioxide (20 wt.%) was mixed with CaH-PO_4, $Ca(OH)_2$ and CaF_2 according to reaction (13), and were milled in planetary ball mill for 2, 5, 10, 15, and 20 h under ambient air atmosphere. The aims of the milling were twofold: the first one was to activate the following reaction via one step mechanochemical process, and the secondly, was to produce the FAp–TiO_2 nanocomposite.

$$6CaHPO_4 + 3Ca(OH)_2 + CaF_2 + TiO_2 \rightarrow Ca_{10}(PO_4)_6F_2 + TiO_2 + 6H_2O \quad (13)$$

Figure 18a shows the XRD patterns of $CaHPO_4$–$Ca(OH)_2$–CaF_2–TiO_2 powder mixture, mechanically alloyed for 2-20 h. An XRD pattern of the mixture before milling is given in the same figure for comparison. For the powder mixture, milled for 2 h, all the sharp peaks corresponding to $Ca(OH)_2$, CaF_2 have diminished, and those corresponding to $CaHPO_4$ have been broadened, indicating that a significant refinement in crystallite and particle sizes of the starting powders had occurred together with a degree of amorphization at the initial stage of mechanical activation. Also, the X-ray pattern of the sample, milled for 2 h, shows the most intense peaks for TiO_2. Upon 5 h of mechanical activation, several new broadened peaks especially between $2\theta = 31 - 34$ appear to emerge, corresponding to FAp phase. This suggests that nanocrystalline FAp phase has been formed as a result of mechanical activation. According to XRD profile, the main products of mechanochemical process after 5 h of milling were FAp and TiO_2. Also, the two minor peaks observed in XRD patterns correspond to $CaHPO_4$. When the mechanical activation time is extended to 15 h, all the peaks corresponding to $CaHPO_4$ have disappeared and only those belonging to FAp and TiO_2 are detectable.

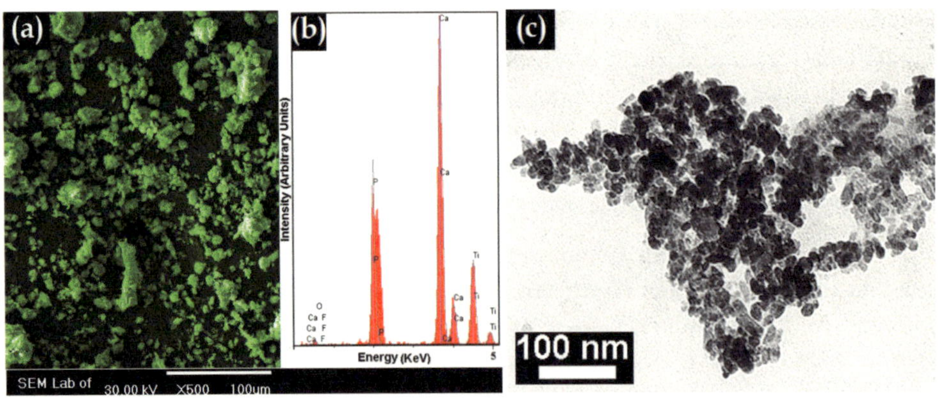

Figure 19. SEM micrograph and TEM image of FAp–TiO_2 nanocomposite after 20 h of milling (Ebrahimi-Kahrizsangi et al., 2011).

Figures 18b and c show the variations of the crystallite size, lattice strain, and the volume fraction of grain boundaries of FAp as a function of milling time. Mechanical activation up to 15 h leads to a rapid decrease in the crystallite size to less than 16 nm, and a large increase in the volume fraction of grain boundary to 17.55%. Upon 15 h of mechanical activation, the crystallite size and the volume fraction of grain boundary reach about 16 nm and 16.54%, respectively. The evaluation of the lattice strain indicates that the lattice strain significantly increased with mechanical activation until 15 h, and then decreased slightly with further milling up to 20 h. The determined amounts of the crystallite size and the lattice strain of TiO_2 as reinforcement are presented in Figure 18 d. The crystallite size of TiO_2 remains nearly constant with milling until 10 h, and then decreases severely with further mechanical activation up to 15 h. After 20 h of milling, the lattice strain decreases which leads to an increase in the crystallite size. According to data presented in Figure 18 with further mechanical activation up to 20 h for both matrix and reinforcement, the lattice strain can be decreased at higher milling intensities because of the enhanced dynamical recrystallisation. Because the calcium phosphate ceramics with higher crystallinity degree has lower activity towards bioresorption and lower solubility in physiological environment (Sanosh et al., 2009), thermal recovery of crystallinity was performed at 650 C for 2 h. Some increase in the peak intensity of the crystalline FAp phase was observed in the XRD pattern (Figure 18e). The crystalline phase of titanium dioxide (TiO_2) similarly appeared upon annealing. After heat treatment, the breadth of the fundamental diffraction peaks decrease as compared to the results of the milled powder which can be attributed to an increase in crystallite size and a decrease in lattice strain. The determined amounts of the structural features indicated that the crystallite size and lattice strain of FAp reached 43.3 nm and 0.30%, respectively.

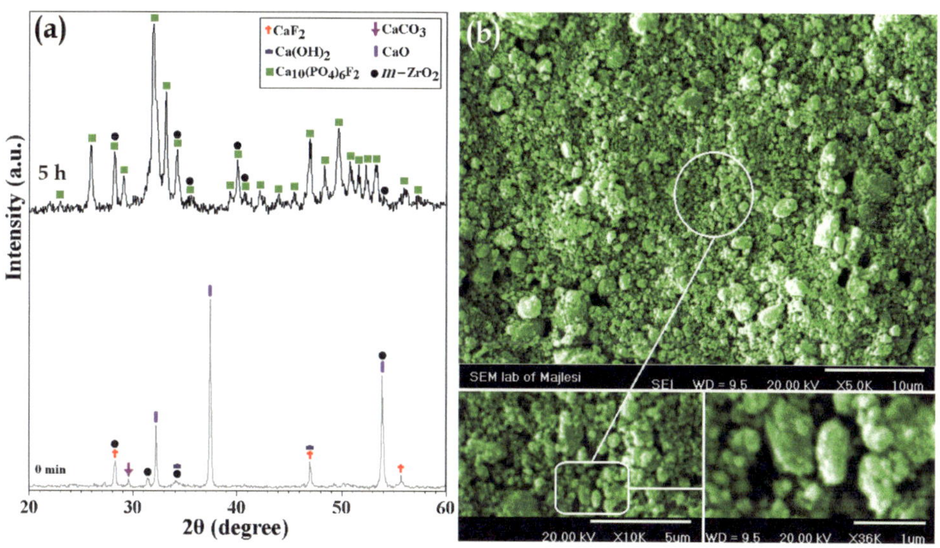

Figure 20. XRD patterns and morphological features of the FAp-ZrO_2 composite nanopowders after 5 h of milling.

The SEM micrograph, EDX result, and TEM image of the mechanosynthesized FAp–20 wt. %TiO_2 nanocomposite after 20 h of mechanical activation time are shown in Figure 19. After 20 h of milling, the mean size of powder particles decreased due to severe mechanical deformation introduced into the powder. It also emerged from SEM image, that the fine agglomerates consist of significantly finer agglomerates/particles that cannot be seen individually in micrographs because of their exceptionally small sizes. Therefore, the size and morphology of fine powders was determined by using transmission electron microscopy. The results of measurements of elemental composition by EDX confirm that very homogeneous distribution of components is formed during one step mechanochemical process particularly after 20 h of milling. The results of EDX analysis also reveal that no chemically stable contaminants are detected due to the excessive adhesion of powders to the milling media. From TEM images it is clearly seen that the synthesized powder after 20 h of milling has an appropriate homogeneity. Also, the particles of products are in average size of about 15 nm after 20 h of milling, respectively. Due to high surface energy, the nanostructured materials can improve the sinterability and, thus, improve mechanical properties (Suryanarayana, 2001). Of course, sintering behavior not only depends on particle size, but also on particle size distribution and morphology of the powder particles. Large particle size along with hard agglomerates shows lower densification in calcium phosphate ceramics. On the other hand, difference in shrinkage between the agglomerates is also responsible to produce small cracks in the sintered calcium phosphate-based ceramic (Banerjee et al., 2007). Therefore, preparation of agglomerate free or soft agglomerated nanostructured FAp–20 wt.%TiO_2 can be an important parameter to achieve good mechanical properties for dense nanostructure. The results of SEM and TEM images suggest that the formation of FAp–20 wt.%TiO_2 nanocomposite after 20 h of milling does not accompany hard agglomerates and, therefore, the synthesized powder can present appropriate mechanical properties.

3.2.3. *Fluorapatite-Zirconia (FAp-ZrO_2) nanocomposite*

Among the reinforcement materials of ceramic-based bionanocomposites, ZrO_2 as a bioinert reinforcement has been studied extensively because of its relatively higher mechanical strength and toughness (Rao & Kannan, 2002; Ben Ayed & Bouaziz, 2008; Evis, 2007). However, the addition of ZrO_2 results in lowering the decomposition temperature of microcrystalline HAp- and FHAp-based composites below the sintering temperature which causes an adverse influence on the mechanical properties (Kim et al., 2003). These phenomena are related to structural features of composite that are affected by the synthesis process. Generally, in order to prepare HAp/FHAp–ZrO_2 composites, calcium phosphate source chemicals and ZrO_2 powders are mixed, cold pressed and then sintered at high temperatures (Rao & Kannan, 2002). Under these conditions, the resulting product could have microcrystalline structure. Since the nanocrystalline structure compared to the microcrystalline structure is more important to achieve high thermal stability and mechanical properties, the FAp–ZrO_2 composites with nanostructural characteristics are preferred for medical applications. In fact, the composite nanopowders can improve the sinterability as well as mechanical properties due to high surface energy. For all these reasons, our research group was considered to synthesis of FAp–ZrO_2 nanocomposite with appropriate structural features via one step mechanochemical process. For this aim, commercially available calcium oxide (CaO), phosphorous pentoxide (P_2O_5), calcium fluoride (CaF_2), and monoclinic zirconia (m–ZrO_2) were used as starting reagents. The mechanochemical synthesis was performed in a planetary ball mill without using any process control agent (PCA).

Figure 20 shows the XRD patterns and SEM images of FAp-ZrO_2 composite nanopowders after mechanical activation for 5 h. An XRD pattern of the mixture before milling is also given in the same figure for comparison. As can be seen in this figure, only sharp characteristic peaks of CaO and CaF_2 could be detected at the initial mechanical activation of starting powder mixture. Phosphoric acid was formed immediately upon addition of P_2O_5 to the reaction mixture due to very high hydrophilic of P_2O_5. Therefore, characteristic peaks of P_2O_5 could not be observed. On the other hand, the CaO is not stable and will spontaneously react with H_2O and CO_2 from the air that leading to formation of $Ca(OH)_2$ and $CaCO_3$ in powder mixture. According to the XRD patterns, it can be seen that after 5 h of milling, all the peaks corresponding to CaO have vanished and only those belonging to FAp and *m*–ZrO_2 were visible. Figure 20b shows the morphological characteristics of the FAp-ZrO_2 composite nanopowders after 5 h of milling at different magnifications. According to this figure, a homogeneous microstructure was obtained after 5 h of milling which is important for the improvement in mechanical properties.

4. Conclusion and future directions

Hydroxyapatite- and fluorapatite-based nanocomposite powders have been developed and demonstrate huge potential for a variety of biomedical applications such as controlled drug release, bone cements, tooth paste additive, and dental implants. Different types of calcium phosphate-based nanocomposites can be synthesized by various approaches for instance wet chemical methods, hydrothermal processes, solid–state reaction, sol–gel method, and mechanochemical processes. Compared to various synthesis processes, mechanosynthesis method present a number of advantages such as high efficiency and enabling to synthesis a wide range of novel advanced materials (nanocomposites). It has been proved that the structural, mechanical, and biological properties of bioceramics can be significantly improved by using calcium phosphate-based nanocomposites as the advanced materials. For this reason, many attempts have been made to improve the mechanical properties as well as structural features of bioceramics through the incorporation of ceramic second phases. Mechanochemical synthesis of hydroxyapatite- and fluorapatite-based nanocomposite powders are of great interest and should be further explored. Although mechanochemical process have demonstrated their great potential for synthesis of hydroxyapatite- and fluorapatite-based nanocomposite powders, several challenges still remain. Mechanosynthesis of nanocomposites with precisely controlling their chemical composition, phases, biological characteristics, mechanical properties, and interfacial features is still a challenging task. Depending on the preparation circumstances (milling atmosphere, milling time, milling temperature, and type of milling media) and chemical composition of initial materials, the properties and performance of the nanocomposites can vary significantly; therefore, the ability to reproduce calcium phosphate-based nanocomposites with unique characteristics is very important for their wide use as biomaterials. With the increased interests and intensive research and development in the field of mechanochemistry, it is expected that, mechanosynthesized nanocomposites will have a promising future and will make a significant influence on the advanced materials industry.

Acknowledgements

The authors are grateful to research affairs of Islamic Azad University, Najafabad Branch and Iranian Nanotechnology Initiative Council (INIC) for supporting this research.

Author details

Bahman Nasiri–Tabrizi[1], Abbas Fahami[1], Reza Ebrahimi–Kahrizsangi[1] and Farzad Ebrahimi[2]

1 Materials Engineering Department, Najafabad Branch, Islamic Azad University, Najafabad, Isfahan, Iran

2 Department of Mechanical Engineering, Faculty of Engineering, Imam Khomeini International University, Qazvin, Iran

References

[1] Adamopoulos, O., & Papadopoulos, T. (2007). Nanostructured bioceramics for maxillofacial applications. *Journal of Materials Science: Materials in Medicine*, 18, 1587-1597.

[2] Adolfsson, E., Nygren, N., & Hermansson, L. (1999). Decomposition mechanisms in aluminum oxide-apatite systems. *Journal of the American Ceramic Society*, 82, 2909-2912.

[3] Avvakumov, E., Senna, M., & Kosova, N. (2002). *Soft mechanochemical synthesis: A basis for new chemical technologies*,, Kluwer Academic Publishers.

[4] Balamurugan, A., Kannan, S., & Rajeswari, S. (2002). Bioactive sol-gel hydroxyapatite surface for biomedical application-in vitro study. *Trends in Biomaterials & Artificial Organs*, 16, 18-20.

[5] Balaz, P. (2008). *Mechanochemistry in Nanoscience and Minerals Engineering*, Springer.

[6] Banerjee, A., Bandyopadhyay, A., & Bose, S. (2007). Hydroxyapatite nanopowders: synthesis, densification and cell materials interaction. *Materials Science and Engineering C*, 27, 729-735.

[7] Barinov, S. M., Shvorneva, L. I., Ferro, D., Fadeeva, I. V., & Tumanov, S. V. (2004). Solid solution formation at the sintering of hydroxyapatite-fluorapatite ceramics. *Science and Technology of Advanced Materials*, 5, 537-541.

[8] Ben, Ayed. F., & Bouaziz, J. (2008). Sintering of tricalcium phosphate-fluorapatite composites with zirconia. *Journal of the European Ceramic Society*, 28, 1995-2002.

[9] Bose, S., Xue, W., Banerjee, A., & Bandyopadhyay, A. (2009). Spherical and anisotropic hydroxyapatite nanocrystals Challa S. S. R. Kumar ed. *Nanomaterials for the Life Sciences 2, Nanostructured Oxides*, WILEY-VCH Verlag GmbH & Co, 407-447.

[10] Bouslama, N., Ben, Ayed. F., & Bouaziz, J. (2009). Sintering and mechanical properties of tricalcium phosphate-fluorapatite composites. *Ceramics International*, 35, 1909-1917.

[11] Cacciotti, I., Bianco, A., Lombardi, M., & Montanaro, L. (2009). Mg-substituted hydroxyapatite nanopowders: Synthesis, thermal stability and sintering behaviour. *Journal of the European Ceramic Society*, 29, 2969-2978.

[12] Cao, H., & Kuboyama, N. (2010). A biodegradable porous composite scaffold of PGA/β-TCP for bone tissue engineering. *Bone*, 46, 386-395.

[13] Chen-W, Ch., Riman, R. E., Ten, Huisen. K. S., & Brown, K. (2004). Mechanochemical-hydrothermal synthesis of hydroxyapatite from nonionic surfactant emulsion precursors. *Journal of Crystal Growth*, 270, 615-623.

[14] Cheng, K., Zhang, S., & Weng, W. (2006). Sol-gel preparation of fluoridated hydroxyapatitein Ca(3 $_2$-PO(OH)$_{3-x}$(OEt)$_x$-HPF$_6$ system. *Journal of Sol-Gel Science and Technology*, 38, 13-17.

[15] Chen, Y., & Miao, X. (2005). Thermal and chemical stability of fluorohydroxyapatite ceramics with different fluorine contents. *Biomaterials*, 26, 1205-1210.

[16] Choi-Y, W., Kim-E, H., Oh, S. Y., & Koh, Y. H. (2010). Synthesis of poly(ε-caprolactone)/hydroxyapatite nanocomposites using in-situ co-precipitation. *Materials Science and Engineering C*, 30, 777-780.

[17] De Castro, C. L., & Mitchell, B. S. (2002). Synthesis functionalization and surface treatment of nanoparticles. *Nanoparticles from mechanical attrition*, Stevenson Ranch, CA: American Scientific Publishers, 1-14.

[18] Ducheyne, P., & Qiu, Q. (1999). Bioactive ceramics: the effect of surface reactivity on bone formation and bone cell function. *Biomaterials*, 20, 2287-2303.

[19] Ebrahimi-Kahrizsangi, R., Nasiri-Tabrizi, B., & Chami, A. (2010). Synthesis and characterization of fluorapatite-titania (FAp-TiO_2) nanocomposite via mechanochemical process. *Solid State Sciences*, 12, 1645-1651.

[20] Ebrahimi-Kahrizsangi, R., Nasiri-Tabrizi, B., & Chami, A. (2011). Characterization of single-crystal fluorapatite nanoparticles synthesized via mechanochemical method. *Particuology*, 9, 537-544.

[21] El Briak-Ben, Abdeslam. H., Ginebra, M. P., Vert, M., & Boudeville, P. (2008). Wet or dry mechanochemical synthesis of calcium phosphates? Influence of the water content on DCPD-CaO reaction kinetics. *Acta Biomaterialia*, 4, 378-386.

[22] Enayati-Jazi, M., Solati-Hashjin, M., Nemati, A., & Bakhshi, F. (2012). Synthesis and characterization of hydroxyapatite/titania nanocomposites using in situ precipitation technique. *Superlattices and Microstructures*, 51, 877-885.

[23] Evis, Z. (2007). Reactions in hydroxylapatite-zirconia composites. *Ceramics International*, 33, 987-991.

[24] Fahami, A., Ebrahimi-Kahrizsangi, R., & Nasiri-Tabrizi, B. (2011). Mechanochemical synthesis of hydroxyapatite/titanium nanocomposite. *Solid State Sciences*, 13, 135-141.

[25] Fahami, A., Nasiri-Tabrizi, B., & Ebrahimi-Kahrizsangi, R. (2012). Synthesis of calcium phosphate-based composite nanopowders by mechanochemical process and subsequent thermal treatment. *Ceramics International*.

[26] Farzadi, A., Solati-Hashjin, M., Bakhshi, F., & Aminian, A. (2011). Synthesis and characterization of hydroxyapatite/β-tricalcium phosphate nanocomposites using microwave irradiation. *Ceramics International*, 37, 65-71.

[27] Fathi, M. H., & Mohammadi Zahrani, E. (2009). Fabrication and characterization of fluoridated hydroxyapatite nanopowders via mechanical alloying. *Journal of Alloys and Compounds*, 475, 408-414.

[28] Fathi, M. H., Mohammadi Zahrani, E., & Zomorodian, A. (2009). Novel fluorapatite/niobium composite coating for metallic human body implants. *Materials Letters*, 63, 1195-1198.

[29] Fini, M., Savarino, L., Nicoli, Aldini. N., Martini, L., Giavaresi, G., Rizzi, G., Martini, D., Ruggeri, A., Giunti, A., & Giardino, R. (2003). Biomechanical and histomorphometric investigations on two morphologically differing titanium surfaces with and without fluorohydroxyapatite coating: An experimental study in sheep tibiae. *Biomaterials*, 24, 3183-3192.

[30] Gergely, G., Wéber, F., Lukács, I., Tóth, A. L., Horváth, Z. E., Mihály, J., & Balázsi, C. (2010). Preparation and characterization of hydroxyapatite from eggshell. *Ceramics International*, 36, 803-806.

[31] Gonzalez, G., Sagarzazu, A., & Villalba, R. (2006). Mechanochemical transformation of mixtures of $Ca(OH)_2$ and $(NH_4)_2HPO_4$ or 2O5. *Materials Research Bulletin*, 41, 1902-1916.

[32] Gu, M., Huang, C., Xiao, S., & Liu, H. (2008). Improvements in mechanical properties of TiB_2 ceramics tool materials by dispersion of Al_2O_3 particles. *Materials Science and Engineering A*, 486, 167-170.

[33] Gu, Y. W., Loh, N. H., Khor, K. A., Tor, S. B., & Cheang, P. (2002). Spark plasma sintering of hydroxyapatite powders. *Biomaterials*, 23, 37-43.

[34] Honarmandi, P., Honarmandi, P., Shokuhfar, A., Nasiri-Tabrizi, B., & Ebrahimi-Kahrizsangi, R. (2010). Milling media effects on synthesis, morphology and structural

characteristics of single crystal hydroxyapatite nanoparticles. *Advances in Applied Ceramics*, 109, 117-122.

[35] Hsu, Y. H., Turner, I. G., & Miles, A. W. (2007). Fabrication and mechanical testing of porous calcium phosphate bioceramic granules. *Journal of Materials Science: Materials in Medicine*, 18, 1931-1937.

[36] Hu, H., Liu, X., & Ding, Ch. (2010). Preparation and in vitro evaluation of nanostructured TiO_2/TCP composite coating by plasma electrolytic oxidation. *Journal of Alloys and Compounds*, 498, 172-178.

[37] Jallot, E., Nedelec, J. M., Grimault, A. S., Chassot, E., Grandjean-Laquerriere, A., Laquerriere, P., & Laurent-Maquin, D. (2005). STEM and EDXS characterisation of physico-chemical reactions at the periphery of sol-gel derived Zn-substituted hydroxyapatites during interactions with biological fluids. *Colloids and Surfaces B: Biointerfaces*, 42, 205-210.

[38] Jin, H. H., Min, S. H., Song, Y. K., Park, H., Ch, , & Yoon, S. Y. (2010). Degradation behavior of poly(lactide-co-glycolide)/β-TCP composites prepared using microwave energy. *Polymer Degradation and Stability*, 95, 1856-1861.

[39] Kalita, S. J., Bhardwaj, A., & Bhatt, H. A. (2007). Nanocrystalline calcium phosphate ceramics in biomedical engineering. *Materials Science and Engineering C*, 27, 441-449.

[40] Kano, J., Zhang, Q., Saito, F., Baron, M., & Nzihou, A. (2006). Synthesis of hydroxyapatite with the mechanochemical treatment products of PVC and CaO Trans I Chem E, Part B. *Process Safety and Environmental Protection*, 84(B4), 309-312.

[41] Khaghani-Dehaghani, M. A., Ebrahimi-Kahrizsangi, R., Setoudeh, N., & Nasiri-Tabrizi, B. (2011). Mechanochemical synthesis of Al_2O_3-TiB_2 nanocomposite powder from Al-TiO_2-H_3BO_3 mixture. *Int. Journal of Refractory Metals and Hard Materials*, 29, 244-249.

[42] Kim, H. W., Kong, Y. M., Bae, Ch. J., Noh, Y. J., & Kim, H. E. (2004b). Sol-gel derived fluor-hydroxyapatite biocoatings on zirconia substrate. *Biomaterials*, 25, 2919-2926.

[43] Kim, H. W., Kong, Y. M., Koh, Y. H., & Kim, H. E. (2003). Pressureless sintering and mechanical and biological properties of fluor-hydroxyapatite composites with zirconia. *Journal of the American Ceramic Society*, 86, 2019-2026.

[44] Kim, H. W., Li, L. H., Koh, Y. H., Knowles, J. C., & Kim, H. E. (2004a). Sol-Gel preparation and properties of fluoride-substituted hydroxyapatite powders. *Journal of the American Ceramic Society*, 87, 1939-1944.

[45] Kivrak, N., & Tas, A. C. (1998). Synthesis of calcium hydroxyapatite-tricalcium phosphate composite bioceramic powders and their sintering behavior. *Journal of the American Ceramic Society*, 81, 2245-2252.

[46] Komlev, V. S., Barinov, S. M., Orlovskii, V. P., & Kurdyumov, S. G. (2001). Porous Ceramic Granules of Hydroxyapatite. *Refractories and Industrial Ceramics*, 42, 242-244.

[47] Kong, Y. M., Kim, S., & Kim, H. E. (1999). Reinforcement of hydroxyapatite bioceramic by addition of ZrO_2 coated with Al_2O_3 . *Journal of the American Ceramic Society*, 82, 2963-2968.

[48] Kurmaev, E. Z., Matsuya, S., Shin, S., Watanabe, M., Eguchi, R., Ishiwata, Y., Takeuchi, T., & Iwami, M. (2002). Observation of fluorapatite formation under hydrolysis of tetracalcium phosphate in the presence of KF by means of soft X-ray emission and absorption spectroscopy. *Journal of Materials Science: Materials in Medicine*, 13, 33-36.

[49] Landi, E., Tampieri, A., Celotti, G., & Sprio, S. (2000). Densification behavior and mechanisms of synthetic hydroxyapatites. *Journal of the European Ceramic Society*, 20, 2377-2387.

[50] Lee-H, H., Shin, U. S., Won-E, J., & Kim-W, H. (2011). Preparation of hydroxyapatite-carbon nanotube composite nanopowders. *Materials Letters*, 65, 208-211.

[51] Liu, F., Wang, F., Shimizu, T., Igarashi, K., & Zhao, L. (2006). Hydroxyapatite formation on oxide films containing Ca and P by hydrothermal treatment. *Ceramics International*, 32, 527-531.

[52] Marchi, J., Greil, P., Bressiani, J. C., Bressiani, A., & Müller, F. (2009). Influence of synthesis conditions on the characteristics of biphasic calcium phosphate powders. *International Journal of Applied Ceramic Technology*, 6, 60-71.

[53] Mayera, I., & Featherstone, J. D. B. (2000). Dissolution studies of Zn-containing carbonated hydroxyapatites. *Journal of Crystal Growth.*, 219, 98-101.

[54] Mobasherpour, I., Solati-Hashjin, M., & Kazemzadeh, A. (2007). Synthesis of nanocrystalline hydroxyapatite by using precipitation method. *Journal of Alloys and Compounds*, 430, 330-333.

[55] Mochales, C., El Briak-Ben, Abdeslam. H., Ginebra, M. P., Terol, A., Planell, J. A., & Boudeville, Ph. (2004). Dry mechanochemical synthesis of hydroxyapatites from DCPD and CaO: influence of instrumental parameters on the reaction kinetics. *Biomaterials*, 25, 1151-1158.

[56] Mishra, S. K., Das, S. K., & Pathak, L. C. (2006). Sintering behavior of self-propagating high temperature synthesized ZrB_2-Al_2O_3 composite powder. *Materials Science and Engineering A*, 426, 229-34.

[57] Mohammadi, Zahrani. E., & Fathi, M. H. (2009). The effect of high-energy ball milling parameters on the preparation and characterization of fluorapatite nanocrystalline powder. *Ceramics International*, 35, 2311-2323.

[58] Nakano, T., Tokumura, A., & Umakoshi, Y. (2002). Variation in crystallinity of hydroxyapatite and the related calcium phosphates by mechanical grinding and subsequent heat treatment. *Metallurgical and Materials Transactions A*, 33, 521-528.

[59] Nasiri-Tabrizi, B., Honarmandi, P., Ebrahimi-Kahrizsangi, R., & Honarmandi, P. (2009). Synthesis of nanosize single-crystal hydroxyapatite via mechanochemical method. *Materials Letters*, 63, 543-546.

[60] Nath, S., Tripathi, R., & Basu, B. (2009). Understanding phase stability, microstructure development and biocompatibility in calcium phosphate-titania composites, synthesized from hydroxyapatite and titanium powder mixture. *Materials Science and Engineering C*, 29, 97-107.

[61] Nayar, S., Sinha, M. K., Basu, D., & Sinha, A. (2006). Synthesis and sintering of biomimetic hydroxyapatite nanoparticles for biomedical applications. *Journal of Materials Science: Materials in Medicine*, 17, 1063-1068.

[62] Nikcevic, I., Jokanovic, V., Mitric, M., Nedic, Z., Makovec, D., & Uskokovic, D. (2004). Mechanochemical synthesis of nanostructured fluorapatite/fluorhydroxyapatite and carbonated fluorapatite/fluorhydroxyapatite. *Journal of Solid State Chemistry*, 177, 2565-2574.

[63] Otsuka, M., Matsuda, Y., Hsu, J., Fox, J. L., & Higuchi, W. I. (1994). Mechanochemical synthesis of bioactive material: effect of environmental conditions on the phase transformation of calcium phosphates during grinding. *Bio-Medical Materials and Engineering*, 4, 357-362.

[64] Palmer, C. A., & Anderson, J. J. B. (2001). Position of the American dietetic association: The impact of fluoride on health. *Journal of the American Dietetic Association*, 101, 126-132.

[65] Pan, H., Tao, J., Yu, X., Fu, L., Zhang, J., Zeng, X., Xu, G., & Tang, R. (2008). Anisotropic demineralization and oriented assembly of hydroxyapatite crystals in enamel: smart structures of biominerals. *Journal of Physical Chemistry B*, 112, 7162-7165.

[66] Pushpakanth, S., Srinivasan, B., Sreedhar, B., & Sastry, T. P. (2008). An in situ approach to prepare nanorods of titania hydroxyapatite (TiO_2-HAp) nanocomposite by microwave hydrothermal technique. *Materials Chemistry and Physics*, 107, 492-498.

[67] Rajkumar, M., Meenakshisundaram, N., & Rajendran, V. (2011). Development of nanocomposites based on hydroxyapatite/sodium alginate: Synthesis and characterisation. *Materials Characterization*, 62, 469-479.

[68] Rameshbabu, N., Sampath, Kumar. T. S., & Prasad, Rao. K. (2006). Synthesis of nanocrystalline fluorinated hydroxyapatite by microwave processing and its in vitro dissolution study. *Bulletin of Material Science*, 29, 611-615.

[69] Ramesh, S., Tan, C. Y., Tolouei, R., Amiriyan, M., Purbolaksono, J., Sopyan, I., & Teng, W. D. (2012). Sintering behavior of hydroxyapatite prepared from different routes. *Materials and Design*, 34, 148-154.

[70] Rao, R. R., & Kannan, T. S. (2002). Synthesis and sintering of hydroxyapatite-zirconia composites. *Materials Science and Engineering C*, 20, 187-193.

[71] Ren, F., Leng, Y., Xin, R., & Ge, X. (2010). Synthesis, characterization and ab initio simulation of magnesium-substituted hydroxyapatite. *Acta Biomaterialia*, 6, 2787-2796.

[72] Rhee, S. H. (2002). Synthesis of hydroxyapatite via mechanochemical treatment. *Biomaterials*, 23, 1147-1152.

[73] Sanosh, K. P., Chu, M. C., Balakrishnan, A., Lee, Y. J., Kim, T. N., & Cho, S. J. (2009). Synthesis of nano hydroxyapatite powder that simulate teeth particle morphology and composition. *Current Applied Physics*, 9, 1459-1462.

[74] Schneider, O. D., Stepuk, A., Mohn, Dirk., Luechinger, Norman. A., Feldman, K., & Stark, W. J. (2010). Light-curable polymer/calcium phosphate nanocomposite glue for bone defect treatment. *Acta Biomaterialia*, 6, 2704-2710.

[75] Silva, C. C., Graca, M. P. F., Valente, M. A., & Sombra, A. S. B. (2007). Crystallite size study of nanocrystalline hydroxyapatite and ceramic system with titanium oxide obtained by dry ball milling. *Journal of Materials Science*, 42, 3851-3855.

[76] Silva, C. C., Pinheiro, A. G., Figueiro, S. D., Goes, J. C., Sasaki, J. M., Miranda, M. A. R., & Sombra, A. S. B. (2002). Piezoelectric properties of collagen-nanocrystalline hydroxyapatite composites. *Journal of Materials Science*, 37, 2061-2070.

[77] Silva, C. C., Pinheiro, A. G., Miranda, M. A. R., Goes, J. C., & Sombra, A. S. B. (2003). Structural properties of hydroxyapatite obtained by mechanosynthesis. *Solid State Sciences*, 5, 553-558.

[78] Silva, C. C., Valente, M. A., Graça, M. P. F., & Sombra, A. S. B. (2004). Preparation and optical characterization of hydroxyapatite and ceramic systems with titanium and zirconium formed by dry high-energy mechanical alloying. *Solid State Sciences*, 6, 1365-1374.

[79] Suchanek, W. L., Byrappa, K., Shuk, P., Riman, R. E., Janas, V. F., & Ten, Huisen. K. S. (2004). Preparation of magnesium-substituted hydroxyapatite powders by the mechanochemical-hydrothermal method. *Biomaterials*, 25, 4647-4657.

[80] Sul, Y. T. (2008). *Osseoinductive magnesium-titanate implant and method of manufacturing the same*, United State Patent Patent No.: US 7,452,566 B2.

[81] Sun, Y., Yang, H., & Tao, D. (2012). Preparation and characterization of Eu^{3+}-doped fluorapatite nanoparticles by a hydrothermal method. *Ceramics International*.

[82] Suryanarayana, C. (2001). Mechanical alloying and milling. *Progress in Materials Science*, 46, 1-184.

[83] Takacs, L. (2002). Self-sustaining reactions induced by ball milling. *Progress in Materials Science*, 47, 355-414.

[84] Takacs, L., Balaz, P., & Torosyan, A. R. (2006). Ball milling-induced reduction of MoS_2 with Al. *Journal of Materials Science*, 41, 7033-7039.

[85] Tian, T., Jiang, D., Zhang, J., & Lin, Q. (2008). Synthesis of Si-substituted hydroxyapatite by a wet mechanochemical method. *Materials Science and Engineering C*, 28, 57-63.

[86] Toriyama, M., Lavaglioli, F., , A., Krajewski, A., Celotti, G., & Piancastelli, A. (1996). Synthesis of hydroxyapatite-based powders by mechano-chemical method and their sintering. *Journal of the European Ceramic Society*, 16, 429-436.

[87] Viswanath, B., & Ravishankar, N. (2006). Interfacial reactions in hydroxyapatite/alumina nanocomposites. *Scripta Materialia*, 55, 863-866.

[88] Wu-C, S., Hsu-C, H., Wu-N, Y., & Ho-F, W. (2011). Hydroxyapatite synthesized from oyster shell powders by ball milling and heat treatment. *Materials Characterization*, 62, 1180-1187.

[89] Xia, Z. P., Shen, Y. Q., Shen, J. J., & Li, Z. Q. (2008). Mechanosynthesis of molybdenum carbides by ball milling at room temperature. *Journal of Alloys and Compounds*, 453, 185-90.

[90] Yeong, K. C. B., Wang, J., & Ng, S. C. (2001). Mechanochemical synthesis of nanocrystalline hydroxyapatite from CaO and $CaHPO_4$. *Biomaterials*, 22, 2705-2712.

[91] Zeng, H., Chittur, K. K., & Lacefield, W. R. (1999). Analysis of bovine serum albumin adsorption on calcium phosphate and titanium surfaces. *Biomaterials*, 20, 377-384.

[92] Zhang, H., Zhu, Q., & Xie, Z. (2005). Mechanochemical-hydrothermal synthesis and characterization of fluoridated hydroxyapatite. *Materials Research Bulletin*, 40, 1326-1334.

[93] Zhou, H., & Lee, J. (2011). Nanoscale hydroxyapatite particles for bone tissue engineering. *Acta Biomaterialia*, 7, 2769-2781.

[94] Hench, Larry L. (1998). Bioceramics. *Journal of the American Ceramic Society*, 81(7), 1705-1728.

Chapter 12

Application of Nanocomposites for Supercapacitors: Characteristics and Properties

Dongfang Yang

Additional information is available at the end of the chapter

http://dx.doi.org/10.5772/50409

1. Introduction

Supercapacitors, ultracapacitors or electrochemical capacitors (ECs), are energy storage devices that store energy as charge on the electrode surface or sub-surface layer, rather than in the bulk material as in batteries, therefore, they can provide high power due to their ability to release energy more easily from surface or sub-surface layer than from the bulk. Since charging-discharging occurred on the surface, which does not induce drastic structural changes upon electroactive materials, supercapacitors possess excellent cycling ability. Due to those unique features, supercapacitors are regarded as one of the most promising energy storage devices. There are two types of supercapacitors: electrochemical double layer capacitors (EDLCs) and pseudocapacitors. In EDLCs, the energy is stored electrostatically at the electrode–electrolyte interface in the double layer, while in pseudocapacitors charge storage occurs via fast redox reactions on the electrode surface. There are three major types of electrode materials for supercapacitors: carbon-based materials, metal oxides/hydroxides and conducting polymers. Carbon-based materials such as activated carbon, mesoporous carbon, carbon nanotubes, graphene and carbon fibres are used as electrode active materials in EDLCs, while conducting polymers such as polyaniline, polypyrrole and polythiophene or metal oxides such as MnO_2, V_2O_5, and RuO_2 are used for pseudocapacitors. EDLCs depends only on the surface area of the carbon-based materials to storage charge, therefore, often exhibit very higher power output and better cycling ability. However, EDLCs have lower energy density values than pseudocapacitors since pseudocapacitors involve redox active materials to store charge both on the surface as well as in sub-surface layer.

Although carbon-based materials, metal oxides/hydroxides and conducting polymers are the most common electroactive materials for supercapacitor, each type of material has its own unique advantages and disadvantages, for example, carbon-based materials can provide high

INTECH
open science | open minds

© 2012 Yang; licensee InTech. This is a paper distributed under the terms of the Creative Commons Attribution License (http://creativecommons.org/licenses/by/3.0), which permits unrestricted use, distribution, and reproduction in any medium, provided the original work is properly cited.

power density and long life cycle but its small specific capacitance (mainly double layer capacitance) limits its application for high energy density devices. Metal oxides/hydroxides possess pseudocapacitance in additional to double layer capacitance and have wide charge/discharge potential range; however, they have relatively small surface area and poor cycle life. Conducting polymers have the advantages of high capacitance, good conductivity, low cost and ease of fabrication but they have relatively low mechanical stability and cycle life. Coupling the unique advantages of these nano-scale dissimilar capacitive materials to form nanocomposite electroactive materials is an important approach to control, develop and optimize the structures and properties of electrode material for enhancing their performance for supercapacitors. The properties of nanocomposite electrodes depend not only upon the individual components used but also on the morphology and the interfacial characteristics. Recently, considerable efforts have been placed to develop all kinds of nanocomposite capacitive materials, such as mixed metal oxides, conducting polymers mixed with metal oxides, carbon nanotubes mixed with conducting polymers, or metal oxides, and graphene mixed with metal oxides or conducting polymers. Design and fabrication of nanocomposite electroactive materials for supercapacitors applications needs the consideration of many factors, such as material selection, synthesis methods, fabrication process parameters, interfacial characteristics, electrical conductivity, nanocrystallite size, and surface area, etc. Although significant progress has been made to develop nanocomposite electroactive materials for supercapacitor applications, there are still a lot of challenges to be overcome. This chapter will summarize the most recent development of this new area of research including the synthesis methods currently used for preparing nanocomposite electroactive materials, types of nanocomposite electroactive materials investigated,structural and electrochemical characterization of nanocomposites, unique capacitive properties of nanocomposite materials, and performance enhancement of nanocomposite electroactive materials and its mechanism.

2. Fabrication and characterization of nanocomposite active electrode materials

2.1. Fabrication methods

To prepare mixed metal oxide nanocomposites, various synthesis methods including solid state reactions (i.e. thermal decomposition of mechanical mixtures of metal salts), mechanical mixing of metal oxides (i.e. ball milling), and chemical co-precipitation and electrochemical anodic deposition from solutions containing metal salts, have been used. For example (examples in section 2.1 will be described in more details in section 3 and the references will also be given in section 3), Mn-Pb and Mn-Ni mixed oxide nanocomposites were prepared by reduction of $KMnO_4$ with Pb(II), and Ni(II) salts to form amorphous mixed oxide precipitant. Mn-V-W oxide, and Mn-V-Fe oxide were then directly deposited by anodic deposition on conductive substrates from aqueous solution consisting of mixed metal salts. Directly anodizing Ti–V alloys in ethylene glycol with HF electrolyte was used to synthesis mixed

V_2O_5–TiO_2 nanotube arrays. Hydrothermal process was also used to prepare SnO_2–Al_2O_3 mixed oxide nanocomposites involving urea as the hydrolytic agent in an autoclave.

Carbon nanotubes (CNTs)-metal oxide nanocomposites were prepared by either mechanically mixing CNTs with metal oxides in a mortar, or depositing metal oxides directly on CNTs by metal-organic chemical vapour deposition (CVD), wet-chemical precipitation, or electrochemically deposition. For example, IrO_2 nanotubes were deposited on multiwall CNTs using metal-organic CVD with the iridium source of $(C_6H_7)(C_8H_{12})Ir$ at 350∘C to form IrO_2-CNTS nanocomposite. The CNTs themselves were initially grown on stainless steel plate using thermal CVD. The MnO_2-CNTs nanocomposites were synthesized by direct current anodic deposition of MnO_2 from the $MnSO_4$ solution over electrophoretically deposited CNTs on the Ni substrate. RuO_2-CNTs was formed by impregnating CNTs with a ruthenium nitrosylnitrate solution and then followed by heat treatment to form composite electrode.

Nanocomposite of a conducting polymer with metal oxide, CNTs or graphene (GN) were mainly synthesized by in situ polymerization in solutions containing monomers of the conducting polymer and suspension of CNTs, metal oxide nanoparticles or GN nanosheets. For example, CNTs– polyaniline (PANI) nanocomposite was prepared from a solution consisting of CNTs and aniline monomer. With addition of an oxidant solution containing $(NH_4)_2S_2O_8$, polymerization of aniline on the surface of CNTs occurred to form CNTs–PANI nanocomposite. MoO_3-Poly 3,4-ethylenedioxythiophene (PEDOT) nanocomposites was synthesized by adding 3,4-ethylenedioxythiophene monomer into a lithium molybdenum nanoparticle suspension, and subsequently, Iron (III) chloride ($FeCl_3$) was added to the suspension as the oxidizing agent under microwave hydrothermal conditions for polymerization to occur. GN-PEDOT nanocomposite was chemically synthesized by oxidative polymerization of ethylene dioxythiophene using ammonium peroxydisulfate [$(NH_4)_2S_2O_8$)] and $FeCl_3$ as oxidizing agents in a solution containing sodium polystyrene sulfonate Na salt, HCl, EDOT monomer and GN. G–PANI nanocomposite was chemically synthesized by oxidative polymerization of aniline monomers using ammonium peroxydisulfate [$(NH_4)_2S_2O_8$)] in solution containing GN.

GN-metal oxide nanocomposites were prepared by chemical precipitation of metal oxide in the presence of GN nanosheets in the solution. For example, GN-CeO_2 nanocomposite was prepared by adding KOH solution dropwise into a $Ce(NO_3)_3$ aqueous solution in the presence of 3D GN material, followed by filtering, and drying. The GN-SnO_2/CNTs nanocomposite was synthesized by ultrasonicating the mixture of chemically functionalized GN and SnO_2–CNTs in water. Normally, GN sheets were synthesized via exfoliation of graphite oxide in hydrogen environment at low temperature while graphite oxide (GO) was prepared normally by Hummers method.

2.2. Structure, electrical, chemical composition and surface area characterization

X-ray diffraction (XRD), scanning electron microscopy/energy-dispersive analysis (SEM/EDX), high-resolution transmission electron microscopy (HRTEM), infrared spectra (IR) and the Brunauer–Emmett–Teller (BET) specific surface areas were the most common analytical techniques to characterize the morphologies, structures, chemical composition and surface area of nanocomposite electroactive materials. XRD analysis was carried out for the nanocomposite samples containing metal oxides to examine the crystallinity and crystal phases of the oxide materials. IR spectra were used for the identification of the characteristic bands of a polymer for nanocomposites consisting of conducting polymers. SEM was used for morphological analysis of nanocomposites, while EDX was used to determine their chemical composition. The electrical conductivity of nanocomposites was obtained normally using a four-probe technique. To do the measurement, the nanocomposite samples were ground into fine powders and then were pressed as pellets. The weight loss of nanocomposite material and the heat flow associated with the thermal decomposition during synthesis or heat treatment were studied by thermogravimetric analysis (TGA) and differential thermal analysis (DTA).

2.3. Electrochemical characterizations

Cyclic voltammetry (CV) was usually conducted to characterize the nanocomposite electrode in a three-electrode cell in either aqueous electrolytes or organic electrolytes using an electrochemistry workstation. The working electrode was metal plate or mesh (e.g. nickel, aluminium, stainless steel) coated with a mixture of nanocomposite and conductive carbon such as acetylene black with a binder such as PTFE or polyvinylidenedifluoride (PVDF). The reference electrodes were either saturated calomel electrode (SCE), Ag/AgCl or others. The counter electrode was typically platinum foil. The specific current and specific capacitance of nanocomposite was determined by the CV current value, scan rate and the weight of nanocomposites.

Galvanostatic charge-discharge cycling was performed with two electrode system having identical electrodes made of same nanocomposite electroactive material. Constant current densities ranging from 0.5 to 10 mA/cm^2 were typically employed for charging/discharging the cell in the voltage range 0-1 V for aqueous electrolytes or 0-2.7 V for organic electrolytes. The discharge capacitance (C) is estimated from the slope (dV/dt) of the linear portion of the discharge curve using the expression.

$$C = I\, dV / dt \tag{1}$$

The weight of the active material of the two electrodes is same in a symmetric supercapacitor. The specific capacitance (Cs) of the single electrode can thus be expressed as:

$$C_s = 2C / m \tag{2}$$

where m is the active material mass of the single electrode. The energy density (E_d) of the capacitor can be expressed as,

$$E_d = ½ \left(C_s V^2_{max}\right) \tag{3}$$

The coulomb efficiency 'η' was evaluated using the following relation,

http://dx.doi.org/10.5772/50409

$$h = t_d / t_c \times 100\% \quad (4)$$

where t_c and t_d are the time of charge and discharge respectively.

Experiments of electrochemical impedance spectra (EIS) were also performed with a two electrode system having identical electrodes made of same nanocomposite active electrode materials at open circuit potential (OCP) over the frequency range 10 kHz–10mHz with a potential amplitude normally of 5 mV. The impedance spectra usually show a single semi-circle in the high frequency region and nearly vertical line in the low frequency region for a supercapacitor, which indicates that the electrode process is controlled by electrochemical reaction at high frequencies and by mass transfer at low frequencies. The intercept of the semi-circle with real axis (Z_{real}) at high frequencies is the measure of internal resistance (R_s) which may be due to (i) ionic resistance of the solution or electrolyte, (ii) intrinsic resistance of the active electrode materials and (iii) interfacial resistance between the electrode and current collector. The origin of the semi-circle at higher frequency range is due to ionic charge transfer resistance (R_{ct}) at the electrode–electrolyte interface. The diameter of the semi-circle along the real axis (Z_{dia}) gives the charge transfer resistance R_{ct}.

3. Performance of various types of nanocomposite active electrode materials

3.1. Mixed pseudocapacitive metal oxide nanocomposites

Metal oxides like such as RuO_2, MnO_2, Co_3O_4, NiO, SnO_2, Fe_3O_4, and V_2O_5 have been employed as electroactive materials for pseudocapacitors. Those metal oxides typically have several redox states or structures and contribute to the charge storage in pseudocapacitors via fast redox reactions. The remarkable performance of RuO_2 in supercapacitors (exhibits the highest specific capacitance values of 720 F g^{-1}) has stimulated many interests in investigating metal oxide system for supercapacitor applications. The commercial use of RuO_2, however, is limited owing to its high cost and toxic nature. Other simple metal oxides usually have some limitations such as poor electrical conduction, insufficient electrochemical cycling stability, limited voltage operating window and low specific capacitance. Those limitations need to be addressed in order for commercial applications of supercapacitors based on metal oxides. Mixed binary or ternary metal oxides systems, such as Ni–Mn oxide, Mn–Co oxide, Mn–Fe oxide, Ni–Ti oxide, Sn–Al oxide, Mn–Ni–Co oxide, Co–Ni–Cu oxide and Mn–Ni–Cu oxide have shown improved properties as electroactive materials for pseudocapacitors and have shed new lights in this area of research. The following section summarizes the recent development in seeking electroactive mixed metal oxide nanocomposites for pseudocapacitors.

3.1.1. Mixed manganese oxides

The natural abundance and low cost of Mn oxides, along with their satisfactory energy-storage performance in mild electrolytes and environmental compatibility, has made them the most promising new electroactive material for the pseudocapacitor applications. However, Mn oxides has limitations such as low surface areas, poor electrical conductivity and relatively small specific capacitance value. To improve the electrochemical performance of Mn oxides for pseudocapacitors, many efforts have been devoted to incorporate various transition metals into Mn oxides to form mixed metal oxide nanocomposites with controlled micro-/nanostructures in order to improve their electrochemical characteristics. The understanding of their synergistic effect and the eventual design of an integrated material architecture in which each component's properties can be optimized and a fast ion and electron transfer will be guaranteed still remains a great challenge.

Jiang et al. [1] designed and synthesized MnO_2 nanoflakes-$Ni(OH)_2$ nanowires composites that can be used in both neutral and alkaline electrolytes and have very high cycling stability. The nanocomposites with 70.4 wt.% MnO_2 content exhibited specific capacitance of 355 F g^{-1} with excellent cycling stability (97.1% retention after 3000 cycles) in 1 M Na_2SO_4 neutral aqueous solution. In 1M KOH aqueous alkaline solution, the MnO_2-$Ni(OH)_2$ nanocomposite with 35.5 wt. % MnO_2 content possessed a specific capacitance of 487.4 F g^{-1} also with excellent cycling stability. Such excellent capacitive behaviours are attributed by the authors to the unique MnO_2–$Ni(OH)_2$ core–shell nanostructures as depicting in Figure 1(b). The interconnected MnO_2 nanoflakes were well-dispersed on the surface of $Ni(OH)_2$ nanowires that creates highly porous surface morphology. This integrated structure can provide high surface area and more active sites for the redox reactions. The specific capacitance and Coulombic efficiency as function of cycle number at a current density of 10 A g^{-1} for up to 3000 cycles is also shown in Figure 1(a) for the MnO_2–$Ni(OH)_2$ nanocomposite. After long cycling, the $Ni(OH)_2$–MnO_2 nanocomposites are overall preserved with little structural deformation, as shown in Figure 1(c) and (d). Oxides of Pb, Fe, Mo, and Co were also incorporated into MnO_2 to form mixed metal oxide nanocomposites. Kim et al.[2] synthesized mixed oxides of Mn with Pb or Ni by reduction of $KMnO_4$ with either lead(II) acetate-manganese acetate or nickel(II) acetate-manganese acetate reducing solutions. Characterization of the nanocomposite electrodes were carried out using cyclic voltammetry, galvanostatic charge-discharge, XRD, BET analysis, and TGA. The results showed that by introducing Ni and Pb into MnO_2, the surface area of the mixed oxide increased due to the formation of micropores. The specific capacitance increased from 166 F g^{-1} (for MnO_2) to 210 and 185 F g^{-1} for Mn-Ni and Mn-Pb mixed oxides, respectively. Kim et al. [2] also found that annealing of the nanocomposites can affect their capacitance: transition from amorphous to a crystalline structure occurred at high temperature (400 ºC) reduces the specific capacitance. Binary Mn–Fe oxide was electroplated on graphite substrates by Lee et al. [3] at a constant applied potential of 0.8V vs. SCE in a mixed plating solution of $Mn(CH_3COO)_2$ and $FeCl_3$. The electrochemical behaviours of the as-deposited and the annealed mixed oxide nanocomposites were characterized by cyclic voltammetry in 2M KC1 solution. Lee et al. found that as-deposited Mn–Fe binary oxide has porous structure and is amorphous. After annealing at 100°C to remove the adsorbed water, the partially hydrous mixed oxide has optimized ionic and

http://dx.doi.org/10.5772/50409

electronic conductivity and gives rise to the best pseudocapacitive performance. However, if the annealing temperature is increased to higher, the mixed oxide loses it porosity and slowly crystalizes which leads to the decrease in specific capacitance. A series of Mn and Mo mixed oxides (i.e. Mn-Mo-X (X= W, Fe, Co)) were investigated by Ye et al.[4] and they found that the specific capacitance of Mo doped Mn oxides are higher than that of pure Mn oxide. The Mn-Mo-Fe oxide reach a high specific capacitance value of 278 F g^{-1} in aqueous 0.1 M Na_2SO_4 electrolyte at a scan rate of 20 mV s^{-1} and has a rectangular-shaped voltammogram. The improvement in capacitance of Mn oxides doped with molybdenum was attributed by the authors to the formation of nanostructure and the existence of low crystallinity. The above results show that mixed metal oxides with amorphous structure have better specific capacitance than that of crystalline structure. Incorporation of various transition metals into Mn oxides creates more porous structures, therefore increase their specific capacitance.

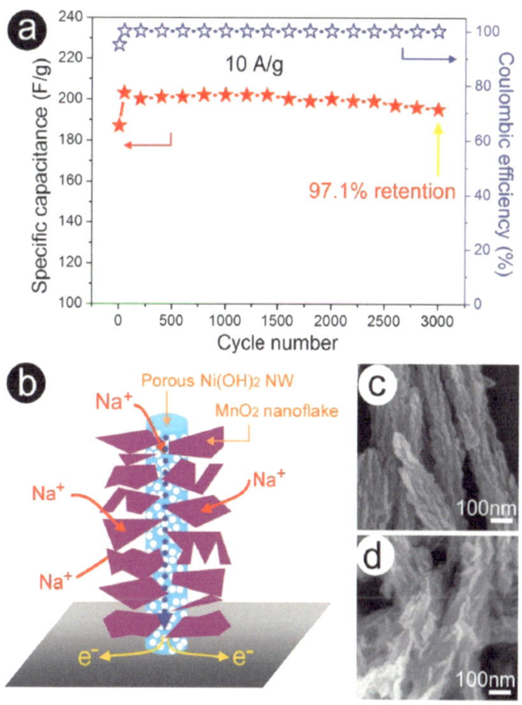

Figure 1. a) Specific capacitance as a function of cycle number at 10 A g^{-1}, (b) schematic of the charge storage advantage of the $Ni(OH)_2$–MnO_2 core–shell nanowires, (c) and (d) SEM images of the $Ni(OH)_2$–MnO_2 core–shell nanowires before and after 3000 cycles (from ref. 1)

Advance thin film physical vapour deposition methods were also used to prepare mixed metal oxide nanocomposite for supercapacitor electroactive materials research. Thin films of manganese oxide doped with various percentages of cobalt oxide were grown by pulsed laser deposition (PLD) on silicon wafers and stainless steel substrates at our laboratory [5]. Before investigated Co-doped manganese oxide film, our team [6] developed different PLD processing parameters (i.e. temperature, oxygen pressure) to produce various chemical compositions and phases of manganese oxides such as pure crystalline phases of Mn_2O_3 and Mn_3O_4 as well as amorphous phase of MnO_x. He then evaluated the pseudo-capacitance behaviours of these different phases of manganese oxides and found that the crystalline Mn_2O_3 phase has the highest specific current and capacitance, while the values for crystalline Mn_3O_4 films are the lowest. The specific current and capacitance values of the amorphous MnO_x films are in between Mn_2O_3 and Mn_3O_4. The specific capacitance of Mn_2O_3 films of 120 nm thick reaches 210 F g^{-1} at 1 mV s^{-1} scan rate with excellent stability and cyclic durability. He then doped amorphous MnO_x and crystalline Mn_2O_3 phases with Co_3O_4 and characterized the mixed Co-Mn oxide films with X-ray diffraction and CVs. The CVs recorded at a 20 mV s^{-1} scan rate for un-doped and Co-doped amorphous MnO_x films are shown in Figure 2(a), and their specific capacitance determined from the CV curves at scan rates of 5, 10, 20 and 50 mV s^{-1} are shown in Figure 2(b). The CVs in Figure 2 shows that the Co-doped amorphous MnO_x films have larger specific currents and capacitances than the un-doped amorphous MnO_x film. Low cobalt doping (3.0 atm.%) had the greatest increase in capacitance, followed by 9.3 atm.% cobalt doping. The 22.6 atm.% cobalt doping had the least increase in specific capacitance. The operating potential window (between H_2 evolution and O_2 evolution due to decomposition of water) was shifted about 100 mV toward more negative potentials for all the Co-Mn mixed oxide films. At a 5 mV/s scan rate, the 3.0 atm.% Co-doped MnOx film reached 99 F g^{-1}, which is more than double that the 47 F g^{-1} observed for the un doped MnO_x film. This result indicates that Co doping significantly improves the pseudo-capacitance performance of amorphous manganese oxide. However, Co-doped crystalline Mn_2O_3 films did not show an improvement in specific current and capacitance compared with un-doped Mn_2O_3 crystalline films. High Co doping level (20.7 atm.% doped) in the crystalline Mn_2O_3 films actually decreased both the specific current and capacitance values. These findings demonstrate that elemental doping is an effective way to alter the performance of pseudo-capacitive metal oxides. Our work also demonstrated that thin film deposition techniques such as PLD are very promising techniques for screening high performance mixed oxide active materials for supercapacitor applications.

3.1.2. Other mixed metal oxides

In addition to mixed manganese oxides, many other mixed metal oxides have also been investigated as electroactive materials for supercapacitor applications. Co_3O_4-$Ni(OH)_2$ nanocomposites were synthesized by electrochemical deposition on the Ti substrate in a solution of $Ni(NO_3)_2$, $Co(NO_3)_2$ and NH_4Cl, then follows by heat treatment at 200°C [7]. The Co_3O_4-$Ni(OH)_2$ electrodes exhibited high specific capacitance value of 1144 F g^{-1} at 5 mV s^{-1} and long-term cycliability. The excellent capacitive behaviours of Co_3O_4-$Ni(OH)_2$ nanocomposite was attributed by the authors to the porous network structures that favour electron and ions transportation as well as faradic redox reactions of both couples of Co^{2+}/Co^{3+} and Ni^{2+}/Ni^{3+}. Y. Yang et al. [8] prepared mixed V_2O_5–TiO_2 nanotube arrays by anodizing Ti–V alloys with different V compositions using an ethylene glycol with 0.2 M HF as the electrolyte at a comparably high anodization voltage. Well-defined nanotube structures were grown for alloys with vanadium content up to 18 at%. The mixed V_2O_5–TiO_2 nanotube arrays were found to exhibit greatly enhanced capacitive properties compared with pure TiO_2 nanotubes. The specific capacitance of the mixed V_2O_5–TiO_2 nanotubes can reach up to 220 F g^{-1} with an energy density of 19.56 Wh kg^{-1} and was found to be very stable in repeated cycles. Another interesting mixed oxide is SnO_2–Al_2O_3 mixed oxide [9], which shows much greater electrochemical capacitance than pure SnO_2 and was electrochemically and chemically stable even after cycling1000 times.

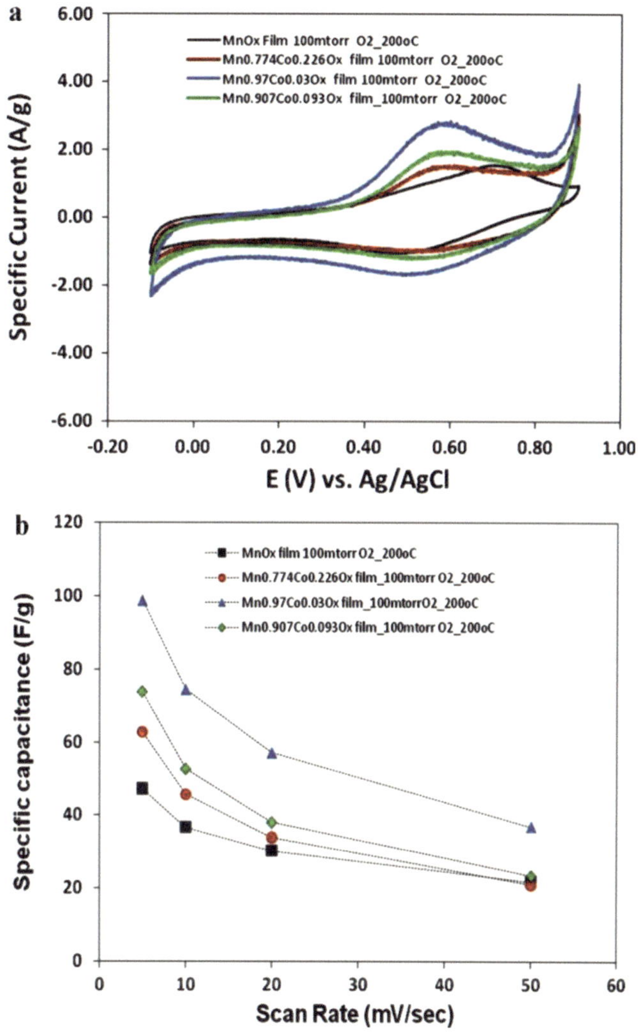

Figure 2. Cyclic voltammetry (a) and specific capacitance (b) of amorphous MnO_x film and various Co-doped amorphous MnO_x films deposited by PLD at 200∘C in 100 mTorr of O_2 (from ref. 6).

Spinel nickel cobaltite (doped or un-doped, such as $NiCo_2O_4$ and $NiMn_xCo_{2-x}O_{4-y}$ ($x \leq 1.0$)) possesses much better electronic conductivity than that of NiO and Co_3O_4. They are low-cost and have multiple oxidation states, and therefore, are also exploited for supercapacitor applications. C. Wang et al. [10] prepared nanostructured $NiCo_2O_4$ spinel platelet like particles with narrow size distribution of 5–10 nm by co-precipitation process. The $NiCo_2O_4$ has excellent conductivity and showed a high-specific capacitance of 671 F g^{-1} under a mass loading of 0.6 mg cm^{-1} at a current density of 1 A g^{-1}. Chang et al. [11] also prepared $NiCo_2O_4$ and $NiMn_xCo_{2-x}O_{4-y}$ ($x \leq 1.0$) using a precipitation route. They found that the spinel structural of $NiCo_2O_4$ is retained with a quarter of the Co ions replaced with Mn. The presence of Mn significantly suppresses crystallite growth upon thermal treatment, and greatly enhances the specific capacitance of the spinel. At the scan rate of 4 mV s^{-1}, the specific capacitance is found to increase from 30 F g^{-1} for Mn content x = 0 to 110 F g^{-1} for x = 0.5. The $NiMn_{0.5}Co_{1.5}O_4$ powder has been found by the authors to be much smaller surface area than the $NiCo_2O_4$ powder. Therefore, the remarkable capacitance enhancement exhibited by the $NiMn_{0.5}Co_{1.5}O_4$ electrode is not due to microstructural variations of the oxide powders. The capacitance enhancement is attributed by the authors to the facile charge-transfer characteristic of the Mn ions, which enables a greater amount of charge transferred between the oxide and the aqueous electrolyte species over the same potential window.

3.2. Carbon nanotubes based nanocomposites

Carbon nanotubes (CNTs) have superior material properties such as high chemical stability, aspect ratio, mechanical strength and activated surface area as well as outstanding electrical properties, which make them good electroactive material candidates for supercapacitors. The electrodes made from CNTs exhibit a unique pore structure for change storage; however, there are limitations for further increasing the effective surface area of the CNTs, as well as relatively high materials cost which limit the commercial application of CNTs based supercapacitors. To improve the performance of CNTs, they are composited with conductive polymers and metal oxides. This section will summarize the recent development of CNTs based nanocomposites for supercapacitor applications.

3.2.1. Carbon nanotubes and polymer nanocomposites

Techniques that can be used to synthesize CNTs include Arc discharge, chemical vapour deposition, and laser ablation. Kay et al. [12] synthesized single-walled CNTs by dc arc discharge of a graphite rod under helium gas using Ni, Co, and FeS as catalysts. Then they prepared single-walled CNTs-polypyrrole (PPY) nanocomposite using in situ chemical polymerization of pyrrole monomer in solution with single-walled CNTs suspension. Figure 3 shows the FE-SEM images of as-grown single-walled CNTs, pure PPY, and single-walled CNT-PPY nanocomposite powder formed by the in situ chemical polymerization. The as-grown single-walled CNTs are randomly entangled and cross-linked, and some carbon nanoparticles are also observed, as shown in Figure 3(a). In Figure 3(b), the image for pure PPY synthesized without single-walled CNTs present in solution shows a typical granular morphology with granule size of about 0.2-0.3 mm. Figure 3(c) demonstrates that the individual carbon nanotube bundles are uniformly coated with PPY which indicates that in situ chemical polymerization of pyrrole can effectively coated all the CNTs. The electrode prepared using single-walled CNTs-PPY nanocomposite as active materials show very high specific capacitance: a maximum specific capacitance of 265 F g^{-1} from the single-walled CNT-PPY nanocomposite electrode containing 15 wt. % of the conducting agent was obtained. Figure 4 shows the specific capacitances of the as-grown single-walled CNTs, pure PPY, and single-walled CNTs-PPY nanocomposite electrodes as a function of discharge current density. In comparison to the pure PPY and as-grown single-walled CNTs electrodes, the single-walled CNTs-PPY nanocomposite electrode shows very high specific capacitance. The improvement in the specific capacitance of the CNTs-PPY nanocomposite was attributed by the authors to the increase in active surface area of pseudocapacitive PPY by CNTs. CNTs was also composited with polyaniline (PANI) by Deng et al. [13]. In their experiments, the CNTs–PANI was prepared using direct polymerization of aniline monomer with oxidant agent, $(NH_4)_2S_2O_8$, in acidic solution containing CNTs suspension, similar to the CNTs-PPY nanocomposite prepared by Kay et al. The CNTs–PANI nanocomposite achieved a specific capacitance of 183 F g^{-1}, almost 4 times higher than pure CNTs (47 F g^{-1}).

Figure 3. The FE-SEM images of (a) as-grown single-walled CNTs, (b) pure PPY, and (c) single-walled CNTs-PPY powder (from ref. 12)

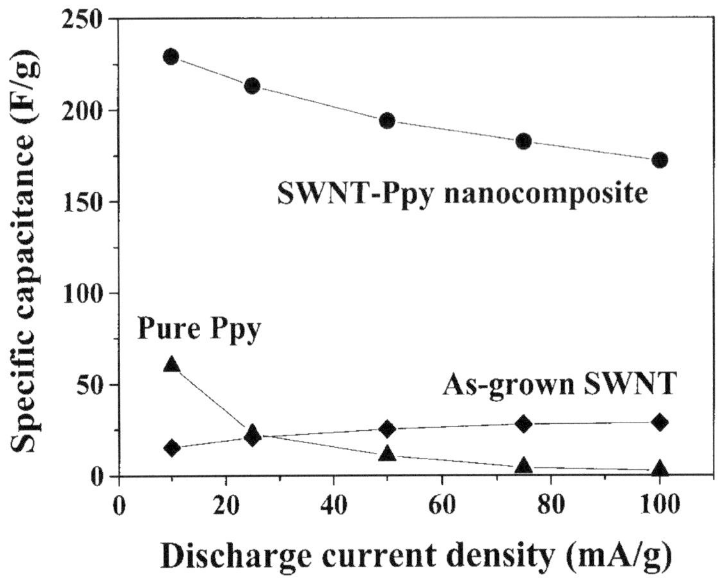

Figure 4. The specific capacitances of the as-grown single-walled CNTs, pure PPY, and single-walled CNTs-PPY, single-walled CNTs PPY nanocomposite electrodes as a function of ischarge current density at a charging voltage of 0.9 V for 10 min (from ref.

3.2.2. Carbon nanotubes and ruthenium oxide nanocomposites

Although RuO_2 shows remarkable performance as supercapacitors electrode active materials, its high cost has limited its commercial applications. To fully utilize the expensive RuO_2 in an electrode, it is necessary to disperse it over high surface area materials such as CNTs. Y-T. Kim et al. [14, 15] discovered a new way to uniformly disperse RuO_2 over the whole surface area of CNTs: firstly, they oxidized the multi-walled CNTs in a concentrated H_2SO_4–HNO_3 mixture to introduce the surface carboxyl groups, and then they prepared multi-walled CNTs-RuO_2 nanocomposites with a conventional sol–gel method. The surface carboxyl groups formed on CNTs allow RuO_2 to disperse more effectively since bond formation between RuO_2 and carboxyl group protects against agglomeration of RuO_2 as illustrated in Figure 5. Figure 5 schematically shows that for purified multi-walled CNTs, RuO_2 can be spontaneously reduced to metallic Ru on the CNTs surface and subsequently covered with $RuO_x(OH)_y$ via the reaction with NaOH to form core-shell structures. For oxidized CNTs, the positive charged Ru precursor ions have limited contact with the CNTs surface to be reduced due to negatively charged carboxyl groups. Surface carboxyl groups act not only as protectors against spontaneous reduction of Ru ions but also as anchorage centres for Ru which enhance the dispersity of RuO_2 and hinder their agglomeration into large particles. TEM images in figure 5 show the dramatic difference in particle size of RuO_2 nanoparticle and dispersity between RuO_2 –pure CNTs and RuO_2–oxidized CNTs.

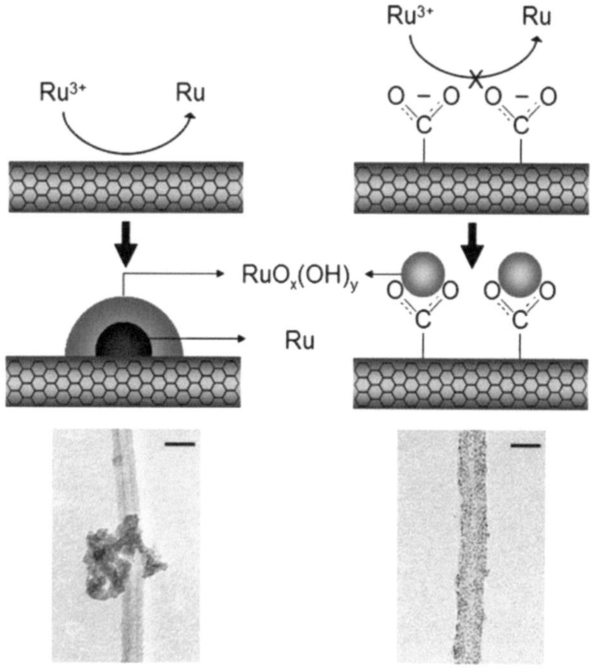

Figure 5. Schematic diagram of the different formation mechanisms of RuO_2 on purified multi-walled CNTs and oxidized multi-walled CNTs in the preparation process and their actual TEM images scale bar (20 nm). (from ref. 15)

Another way to effectively disperse RuO_2 over CNTs was developed by Hsieh et al. [16] who grew vertically aligned multi-walled CNTs films directly by CVD on the Ti current collector using thin nickel layers as the catalyst. Hydrous ruthenium dioxide was then directly deposited onto the surface of CNTs electrodes by electrochemical CV deposition from an aqueous acidic solution of ruthenium trichloride ($RuCl_3.nH_2O$). The SEM morphology of the composites shows that the surfaces of the multi-walled CNT/Ti electrodes were coated uniformly with hydrous ruthenium dioxide, which increased the utilization of the electroactive RuO_2 material. Electrochemical measurements showed that the $RuO_2.nH_2O$-CNTs nanocomposites have high capacitance of 1652 F g^{-1} and decay rate of 3.45% at 10 mV s^{-1} in a 1.0 M H_2SO_4 aqueous electrolyte within the potential range from -0.1 to 1.0 V. Figure 6 shows the comparison of specific capacitance of $RuO_2.nH_2O$-CNTs nanocomposite electrode with $RuO_2.nH_2O$ and multi-walled CNTs electrodes at a scan rate of 10 mV/s. The results in the figure 6 clearly show that the capacitance of $RuO_2.nH_2O$-CNTs was much higher (4.7 times) than those of pure materials. Chemical impregnation of ruthenium nitrosylnitrate solution $(Ru(NO)(NO_3)_x(OH)_y$ on the CVD-grown multi-walled CNTs following by calcination at 350∘C process was used by Lee et al. [17] to form RuO_2-CNTs nanocomposites. The specific capacitance of the nanocomposite was found to be as high as 628 F g^{-1}. The authors believed that nanoporous three-dimensional structure of RuO_2-CNTs nanocomposite facilitated the electron and ion transfer. Byung Chul Kim et al. [18] used the electrochemical potentiodynamic deposition method to prepare RuO_2-CNTS and Ru/Co oxides-CNTs nanocomposites from $RuCl_3$, and 0.1M $CoCl_2$ + 0.05M $RuCl_3$ solutions, respectively. All the composites showed considerable increase in capacitance values. The Ru/Co mixed oxides-CNTs showed superior performance (570 F g^{-1}) at high scan rates (500 mV s^{-1}) when compared to the RuO_2 electrode (475 F g^{-1}). This increase in capacitance at high scan rates is attributed by the authors to the enhanced electronic conduction of Co in the composites.

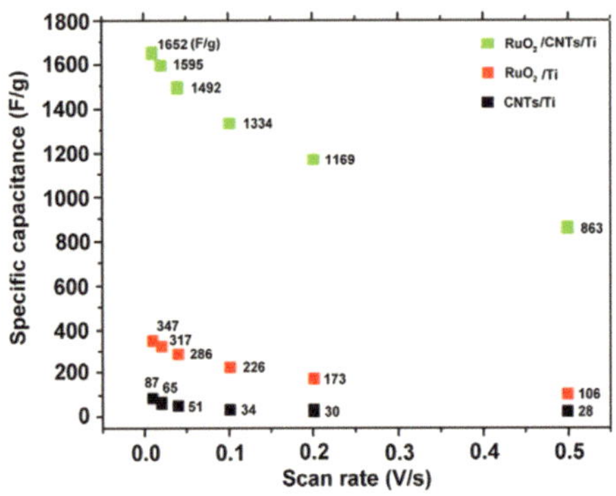

Figure 6. Specific capacitance of $RuO_2.nH_2O$-CNT, $RuO_2.nH_2O$, and multi-walled CNTs after 80 scan cycles with scan rates from 10 to 500 mV/s (from ref. 16).

3.2.3. Carbon nanotubes and other metal oxide nanocomposites

Besides RuO_2, there are many other metal oxides that were composited with CNTs to form electroactive materials for supercapacitors. Chen et al. [19] used thermal CVD to grow multi-walled CNTs on a stainless steel plate and then on top of CNTs, IrO_2 nanotubes were deposited using metal-organic CVD with the iridium source of $(C_6H_7)(C_8H_{12})Ir$ at 350∘C. The IrO_2 square nanotube crystals were grown on the upper section of the CNTs thin film. Figure 7 shows the morphologies of multi-walled CNTs, IrO_2 nanotubes, and IrO_2 nanotubes – multi-walled CNTs nanocomposite. The cross-sectional view of multi-walled CNTs, figure 7(b), shows that there are upper and lower sections in the CNTs thin film. The upper section, ~2 μm in thickness, consists of entangled carbon nanotubes without distinct orientation. The lower section, approximately 4 μm thick, is composed of largely parallel nanotubes, aligned in the vertical direction. These nanotubes act as the templates for the IrO_2 nanotube growth. Figures 7(c) and (d) show a top view and cross-sectional view of IrO_2 nanotubes grown on stainless steel substrate. Figure 7(e) and (f) show IrO_2 nanotubes grown over CNTs. Figure 7(e) indicates that the grown IrO_2 nanotubes had a high density along the wires of multi-walled CNTs in the upper section. In comparison to multi-walled CNTs, the nanostructured IrO_2–CNTs increases the capacitance by a factor of six, from 15 to 69 F g^{-1}, and reduces the resistance from 90 to 60 Ω. Such a hierarchical structure provides a high surface area for electrical charge storage, and a double-layer capacitance in conjunction with pseudocapacitance. Electrochemical anodic deposition of MnO_x•nH_2O films from $MnSO_4$•$5H_2O$ solution on CNTs coated Ni substrates was used by Lee et al. [20] to form amorphous MnO_x-CNTs nanocomposite electrode. The CNTs were electrophoretically deposited on the Ni substrate by applying a dc voltage of 20V before the deposition of MnOx•nH2O. The MnO_x-CNTs nanocomposite electrodes have shown much better energy storage capabilities than MnO_x deposited directly on the Ni substrate: the specific capacitances were 415 as obtained from CV measurements with a scan rate of 5 mV/s and it preserved 79% of its original capacitance value after 1000 cycles. The authors attributed the improvement to the low resistance and large surface area of the nanocomposite electrodes. Other metal oxide CNTs nanocomposites being investigated include NiO-CNTs[21], V_2O_5-CNTs [22] and SnO_2–V_2O_5–CNTs [23]. The V_2O_5-CNTs composite was prepared by electrochemical deposition of V_2O_5 on vertically aligned multi-walled CNTs and it can reach a specific capacitance of 713.3 F g^{-1} at 10 mV s^{-1}. The SnO_2–V_2O_5–CNTs was prepared by simply mixing CNTs and SnO_2–V_2O_5 mixed oxide powder in a mortar prior and fixing on the surface of a graphite electrode that was impregnated with paraffin. At a scan rate of 100 mV s^{-1}, the SnO_2–V_2O_5–CNTs electrode provides 121.4 F g^{-1} specific capacitance.

Other types of high surface area carbon materials such as activated carbons, carbon fibres and carbon aerogels were also composited with metal oxide to form high performance electroactive materials. Those examples include vapour-grown carbon fibre (VGCF) - RuO_2 xH_2O nanocomposite prepared by a thermal decomposition [24], $RuO_2.xH_2O$-mesoporous carbon nanocomposites prepared using impregnation [25], ZnO–activated carbon nanocomposite electrode by simply mixing [26] and MoO_3-graphite prepared by ball milling [27].

3.3. Pseudocapacitive polymer and metal oxide nanocomposites

Electronically conducting polymers derived from monomers such as pyrrole, aniline, and thiophene have unique properties, such as good environmental stability, electroactivity, and unusual doping/de-doping chemistry, therefore, they are suitable for active electrode material usage in supercapacitors. When used as electrode materials, these polymers have advantage over carbon-based materials since they have both electrochemical double layer capacitance and pseudocapacitance which arises mainly from the fast and reversible oxidation and reduction processes related to the π-conjugated polymer chain. However, conducting polymers have problems of typical volumetric shrinkage during ejection of ions (doped ions) and low conductance at de-doped state which would result in high ohmic polarization of supercapacitors. In order to solve this problem, conducting polymers were mixed with metal oxides to form nanocomposites and the synergistic effect of the polymer–metal oxide nanocomposites has been exploited. Such nanocomposites were found to have the advantages of polymers such as flexibility, toughness and coatability and metal oxides such as hardness, and durability. They also possess some synergetic properties which are different from that of parent materials. This section will summarize the recent development in this serial of materials.

Figure 7. SEM image of CNTs top view (a), CNTs cross-sectional view (b), IrO_2 nanotubes top view (c), IrO_2 cross-sectional view (d), IrO_2–CNTs top view (e), IrO_2–CNTs cross-sectional view (f). The inset of (b) and (d) are magnified images. (from ref. 19).

3.3.1. Polyaniline (PANI) and metal oxide nanocomposites

Polyaniline (PANI) is one of the most important conducting polymers because of its ease of synthesis at low cost, good processability, environmental stability and easily tuneable conducting properties. The synthesis and studies of composites of PANI and metal oxides such as MnO_2, SnO_2 and $MnWO_4$ have been carried out. In a PANI-metal oxide nanocomposite, PANI not only serves as an electroactive material for energy storage but also as a good coating layer to restrain metal oxides from dissolution in acidic electrolytes. Chen et al. [28] synthesized a very high performance PANI-MnO_2 nanocomposite using the following procedure: first, the hydroxylated MnO_2 nanoparticles were surface modified with silane coupling agent, ND42, then the obtained surface modified MnO_2 nanoparticles (ND-MnO_2) were washed and dried. Electro-co-polymerization of aniline and ND-MnO_2 nanoparticles was conducted on a carbon cloth in an electrolyte solution containing ND-MnO_2, aniline, H_2SO_4 and Na_3PO_3. The co-polymerization was preceded through successive cyclic voltammetric scans. The whole synthesis process is illustrated in Figure 8. Electro-co-polymerization method was also used to prepare unmodified PANI–MnO_2 nanocomposite and pure

PANI. The SEM images of PANI–MnO_2, PANI-ND-MnO_2 films reveal that the addition of MnO_2 nanoparticles promotes the one dimensional growth of PANI, which substantially reduces the size of the nanorods and increases the surface area/internal space of the composite films (Figure 9). PANI-ND-MnO_2 composite film has an average specific capacitance of ~80 F g^{-1} and a very stable coulombic efficiency of ~98% over 1000 cycles. It also exhibit high intrinsic electrical conductivity and good kinetic reversibility. The excellent properties were attributed by the authors to the improved interaction between MnO_2 and PANI and the increased effective surface area in PANI-ND-MnO_2 film, due to the surface modification of MnO_2 nanoparticles with the silane coupling reagent. Significantly high specific capacitor was achieved with PANI-SnO_2 nanocomposites prepared by Hue et al. [29] using a chemical method in which SnO_2 nanoparticles and aniline were dispersed in sodium dodecylbenzenesulfonate solution and then, ammonium persulfate was added to the above mixture to start polymerization. The PANI-SnO_2 nanocomposite thus prepared had a high specific capacitance of 305.3 F g^{-1} with a specific energy density of 42.4Wh kg^{-1} and a coulombic efficiency of 96%. The energy storage density of the composite was about three times as compared with pure SnO_2.

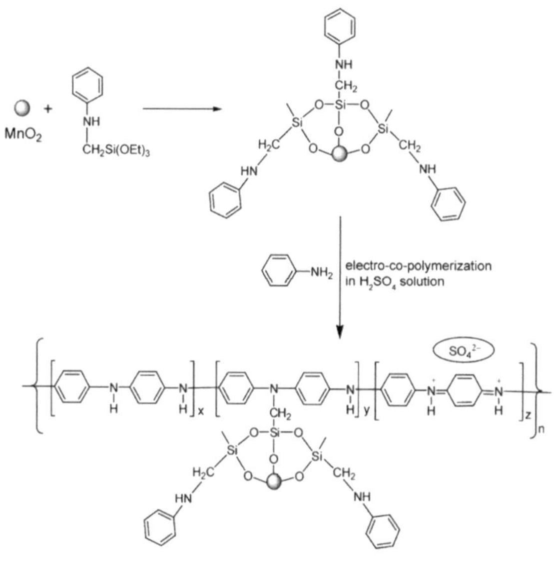

Figure 8. A schematic diagram illustrates the reaction pathway for the synthesis of PANI-ND-MnO_2 nanocomposite film (from ref. 28).

Wang et al. [30] developed an innovative way to synthesize PANI-MnO_2 nanocomposites. This so-called "interfacial synthesis" utilized the interfacial region between an organic phase and an aqueous phase to synthesize the composite. The organic phase was prepared by dissolving aniline monomers into inorganic Trichloromethane ($CHCl_3$) solution, while the aqueous phase was obtained by dissolving potassium permanganate in distilled water. When the aqueous solution was added into the organic solution, an interface was formed immediately between the two phases and the reaction occurred. During the reaction, aniline was diffused from the organic solution to the interface and was chemically oxidized into polyaniline. At the same time, MnO_4^- was reduced to manganese oxide precipitate. Finally, the PANI-MnO_2 nanocomposite was formed and remained in the aqueous solution. For comparison, conventional chemical co-precipitation of MnO_2-PANI composite was performed to make the conventional PANI-MnO_2 composite. Both synthesis processes was schematically illustrated in Figure 10. The interfacial synthesized MnO_2-PANI composite shows larger specific surface area (124 m^2 g^{-1}) and more uniform pore-size distribution than the composite prepared by chemical co-precipitation as shown in Figure 11. It exhibits a higher specific capacitance of 262 F g^{-1} (about twice amount of conventionally prepared MnO_2-PANI composite) with better cycling stability. The authors attributed the observed enhanced electrochemical properties of the interfacial synthesized MnO_2-PANI composite electrode to its unique hollow microstructure with well-defined mesoporosity and the coexistence of conducting PANI. Other interesting PANI metal oxide nanocomposites include PANI-$MnWO_4$ nanocomposites, which was prepared in situ polymerization of aniline monomer in solution containing $MnWO_4$ nanoparticles [31]. The composite has shown good electrochemical properties: with 50% of $MnWO_4$ loading, the PANI-$MnWO_4$ nanocomposites shows high specific capacitance of 475 F g^{-1}, much higher than that of the physical mixture of PANI and $MnWO_4$ (346 F g^{-1}).

3.3.2. Other polymer and metal oxide nanocomposites

Beside Polyaniline (PANI), other conducting polymers that have been used to composite with metal oxides to form electroactive materials including polypyrrole (PPY), polythiophene and their derivatives such as Poly 3,4-ethylenedioxythiophene (PEDOT). PEDOT is a stable and environmentally friendly polymer and has controllable electrical conductivity. However, PEDOT suffers from problems such as volumetric swelling and shrinkage during the insertion and ejection of ions. PEDOT was comprised with pseudocapacitive metal oxides such as $MnFe_2O_4$, $CoFe_2O_4$ and MoO_3 to improve its property. The synergistic effect of composite formation plays a significant role to increase the capacitance value. Sen et al. [32] prepared PEDOT–$NiFe_2O_4$ nanocomposites by chemical polymerization of EDOT monomer in solution containing nickel ferrite nanoparticles ($NiFe_2O_4$). Pure PEDOT polymer in both n-hexane medium and aqueous medium was also synthesized by similar procedure in the absence of $NiFe_2O_4$ nanoparticles. Electrochemical CVs and impedance spectroscopy were used to characterize the PEDOT–$NiFe_2O_4$ nanocomposite as well as pure PEDOT synthesized in organic medium and aqueous medium, and $NiFe_2O_4$ nanoparticles prepared by sol–gel procedure. Figure 12 shows typical Nyquist impedance spectra of the four compounds over a frequency range of 10 kHz–10 mHz with a potential amplitude of 5mV. The impedance results show that introduction of $NiFe_2O_4$ nanoparticles into the PEDOT not only helps to reduce the intrinsic resistance (the intercept of the semi-circle with real axis (Z′) at high frequencies is the measure of internal resistance) through the development more mesoporous structures but also increase the kinetics of electron transfer through redox process leading to the enhancement of pseudocapacitance in the composite materials (pseudocapacitance values were also determined from the impedance by fittings the spectra with Randles equivalent circuit). The PEDOT– $NiFe_2O_4$ nanocomposite shows high specific capacitance (251 F g^{-1}) in comparison to $NiFe_2O_4$ (127 F g^{-1})

and PEDOT (156 F g^{-1}) where morphology of the pore structure was believed to play a significant role over the total surface area. PEDOT was also composited with MoO_3 by Murugan et al. [33] using chemical polymerization of EDOT monomer with $FeCl_3$ as oxidizing agent in MoO_3 suspension. The nanocomposite also has much higher specific capacitance (300 F g^{-1}) compared to that of pristine MoO_3 (40 mF g^{-1}). The improved electrochemical performance was attributed by the authors to the intercalation of electronically conducting PEDOT between MoO_3 layers and an increase in surface area.

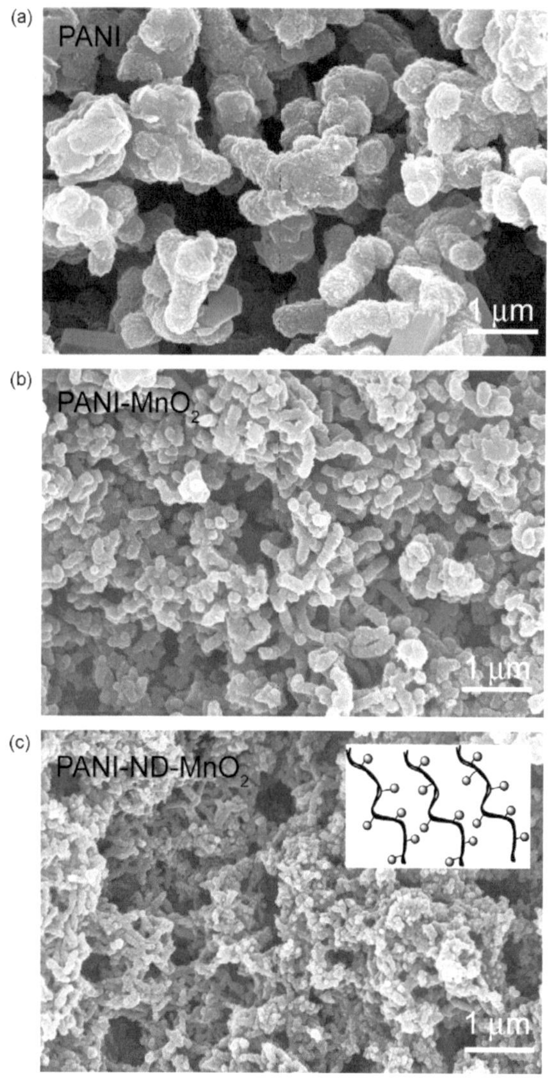

Figure 9. SEM images of (a) PANI, (b)PANI–MnO_2, and (c)PANI-ND-MnO_2 composite(from ref. 28).

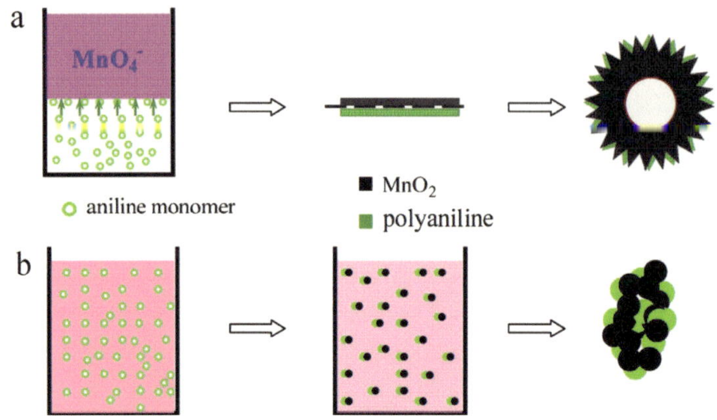

Figure 10. Schematic illustration of the formation mechanisms of MnO_2-PANI composites: (a) interfacial synthesis and (b) chemical co-precipitation (from ref. 30).

Polypyrrole (PPY) is also a promising conducting polymer material due to its highly reversible redox reaction. Although the electrical conductivity of intrinsic PPY is low, doping of surfactants can enhance effectively the electrical conductivity of PPY. p-Toluenesulfonic acid (p-TSA) was used as a dopant by Dong et al. [34] to prepare MnO_2-PPY/TSA nanocomposite for supercapacitor applications. TSA and pyrrole were dispersed ultrasonically in deionised water to form a homogeneous solution. With the addition of $KMnO_4$ or FeCl3•6H2O oxidant, redox reactions occurred and MnO_2-PPY/TSA nanocomposite was produced. Micrographs and BET isotherm measurements showed that the particle and the pore size of the MnO_2-PPY/TSA nanocomposite are much smaller than those of the MnO_2-PPY. Electrochemical measurements showed that the MnO_2-PPY/TSA nanocomposite electrode exhibited a higher specific capacitance of ~376 F g^{-1} at 3 mA cm^{-2} and better cycling stability in 0.5M Na_2SO_4 solution than the MnO_2-PPY. Another polymer metal oxide composite that shows promising supercapactive properties is MnO_2-poly(aniline-co-o-anisidine) [35], which has specific capacitance of the 262 F g^{-1} in 1M Na_2SO_4 at a current density of 1A g^{-1}. All the above results presented by various authors have demonstrated that the development of novel metal oxide-conducting polymers nanocomposite holds great potential applications in high-performance electrochemical capacitors.

3.4. Graphene based nanocomposites

Graphene (GN) is a two-dimensional monolayer of sp^2-bonded carbon atoms. It has attracted increasing attention in recent years, due to its extraordinarily high electrical and thermal conductivities, great mechanical strength, large specific surface area, and potentially low manufacturing cost. The excellent properties of high specific surface area (2675 m^2 g^{-1}) and high electrical conductivity have made it a suitable material for supercapacitor applications. Use of thermally exfoliated GN nanosheets as supercapacitor electrode materials has been reported to give a maximum specific capacitance of 117 F g^{-1} in aqueous H_2SO_4 electrolyte. For supercapacitors made of chemically modified GN, a specific capacitance of 135 F g^{-1} in

http://dx.doi.org/10.5772/50409

aqueous KOH electrolyte has been reported. However, when drying GN during electrode preparation process, the irreversible agglomeration and restacking of GN due to van der Waals interactions to form graphite becomes a major problem for GN based supercapacitors. The agglomeration adversely affects supercapacitor performance by preventing electrolyte from penetrating into the layers. This problem can be avoided by the introduction of spacers into the GN layers. CNTs, metal oxides and conducting polymers can be used as the spacers. Spacers can ensure high electrochemical utilization of GN layers; in addition, electroactive spacers also contribute to the total capacitance. In this section, recent developments on GN-based nanocomposite materials for supercapacitor applications will be reviewed.

Figure 11. SEM micro-images of MnO2-PANI nanocomposites synthesized by (a) interfacial synthesis and (b) chemical co-precipitation (from ref. 30).

3.4.1. Graphene and metal oxide nancomposites

Metal oxides such as CeO_2, RuO_2, V_2O_5 and SnO_2 were used to composite with GN to form advance nanocomposite for supercapacitor applications. Synergistic effect contributed from GN and metal oxide due to improved conductivity of metal oxide and better utilization of GN is expected to contribute to enhance the pseudocapacitance. Jaidev et al. [36] prepared $RuO_2{\bullet}xH_2O$-GN nanocomposite by hydrothermal treatment of GN nanosheets, synthesized via exfoliation of graphite oxide in hydrogen environment, with ruthenium chloride in a Teflon-lined autoclave. A symmetrical supercapacitor was fabricated using electrodes prepared by mixing the as-prepared RuO_2 xH_2O-GN, activated carbon and Nafion (binder) on conducting carbon fabric. The hybrid nanocomposite shows a maximum specific capacitance of 154 F g^{-1} and energy density of about 11Wh kg^{-1} at a specific discharge current of 1 A g^{-1} (20 wt.% Ru loading). The composite also shows a maximum power density of 5 kW kg^{-1} and coulombic efficiency of 97% for a specific discharge current of 10 A g^{-1}. CeO_2 was also deposited onto the 3D GN by chemical precipitation of 3D GN materials contained $Ce(NO_3)_3$ solution with KOH [37]. CeO_2-GN nanocomposite gave high specific capacitance (208 F g^{-1} or 652 mF cm^{-2}) and long cycle life although the specific surface area of the composite decreases as compared with pure GN. Bonso et al. synthesized GN-V_2O_5 nanocomposite by mixing V_2O_5 sol with the GN/ethanol dispersion and stirred for many days [38]. The thus-prepared GN-V_2O_5 composite electrode achieved specific capacitance value of 226 F g^{-1} in 1 M LiTFSI in acetonitrile. In contrast, the specific capacitance of just V_2O_5 was 70 F g^{-1} and just GN was 42.5 F g^{-1}, demonstrating the synergistic effect of combining the two materials.

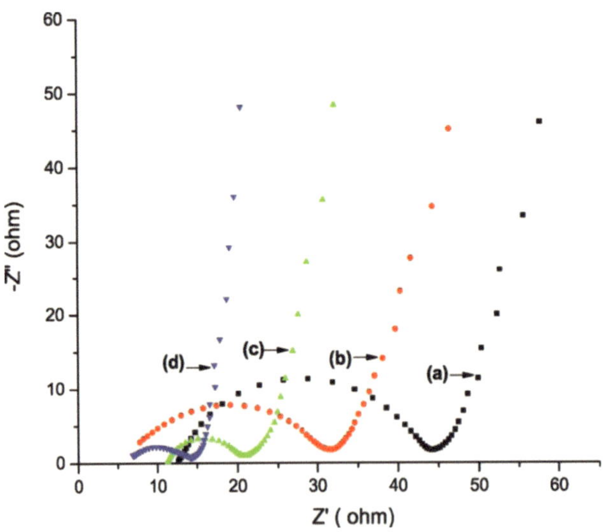

Figure 12. Typical Nyquist impedance plot at open circuit potential (OCP) over a frequency range of 100 kHz–10 mHz with a potential amplitude of 5mV for (a) PEDOT-Aq, (b) PEDOT-Org, (c) nano- $NiFe_2O_4$ and (d) PEDOT– $NiFe_2O_4$ composite electrodes (from ref. 32)

http://dx.doi.org/10.5772/50409

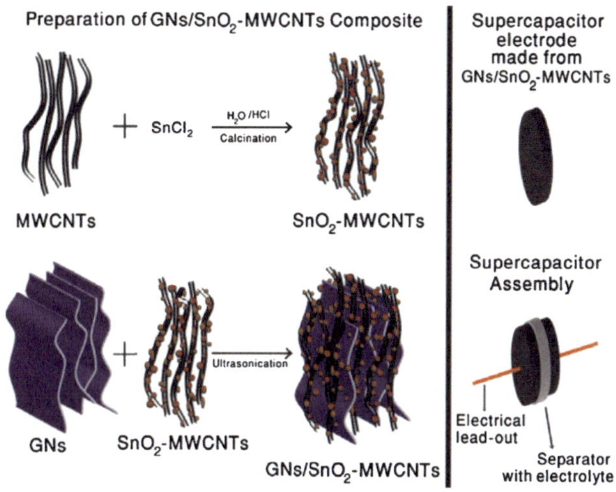

Figure 13. Schematic of preparation of supercapacitor electrode material (from ref. 39).

Transition metal oxide nanoparticles loaded CNTs has been demonstrated as an excellent electroactive material for supercapacitor applications. It is expected that a nanocomposite obtained by dispersion of metal oxide nanoparticles loaded multi-walled CNTs (MWCNTs) into GN can be a good electroactive materials for supercapacitors particularly to increase their cycling stability due to more open structures. Rakhi et al. [39] has prepared the GN-SnO_2/CNTs nanocomposite by ultrasonically mixing of chemically functionalized GN and SnO_2–CNTs. The SnO_2/CNTs was first prepared by chemical precipitation of SnO_2 from $SnCl_2$ solution containing functionalized multi-walled CNTs. The SnO_2/CNTs precipitate was filtrated, washed, dried and calcined. It was then mixed with functionalized GN by ultrasonication to obtain a homogeneous GN-SnO_2/CNTs suspension. Finally, the solid was filtered, washed and dried in a vacuum. To produce chemically functionalized graphene, GN was dispersed in concentrated nitric and sulphuric acid mixture. The process for preparation of GN-SnO_2/CNTs composite is illustrated in Figure 13. The TEM images of multiwall CNTs and SnO_2/CNTs are shown in Figure 14 (a) and (b) respectively. Multi-walled CNTs have an average inner diameter of 10 nm, an outer diameter of 30 nm and an average length in the range of 10–30 μm. Figure 14(b) suggests an uniform distribution of SnO_2 nanoparticles over the surface of multi-walled CNTs. High resolution TEM image of SnO_2/CNTs (inset of Figure 14(b)) reveals that the SnO_2 nanoparticles are highly crystalline in nature with an average particle size of 4–6 nm. TEM images of large area GN and GN-SnO_2/CNTs composite are shown in Figure 14(c) and (d) respectively. SnO_2/CNTs are seen to occupy the surface of GN. Symmetric supercapacitor devices were fabricated by the authors using GN and GN-SnO_2/CNTs composite electrodes. The latter gave remarkable results with a maximum specific capacitance of 224 F g–1, power density of 17.6 kW kg^{-1} and an energy density of 31 Wh kg^{-1}. The results demonstrated that dispersion of metal oxide loaded multi-walled CNTs improved the capacitance properties of GN. The fabricated supercapacitor device exhibited excellent cycle life with ~81% of the initial specific capacitance retained after 6000 cycles.

3.4.2. Graphene and polymer nancomposites

Conducting polymers such as PANI and PEDOT were composited with GN to improve the electrochemical performance of GN for supercapacitor applications. GN-PANI nanocomposite was chemically synthesized by oxidative polymerization of aniline monomer using ammonium peroxydisulfate [$(NH_4)_2S_2O_8$)] as the oxidizing agent in the GN and aniline mixing solution [40]. The presence of GN in polyaniline shows the penetrating network like structure in GN–PANI nanocomposite film, whereas the GN platelets are making the network structure with polyaniline. The high specific capacitance and good cyclic stability have been achieved using 1:2 aniline to GN ratio by weight. The result of Gómeza et al. [40] has proved that the presence of GN in network of polyaniline changes the composite structure. The supercapacitor fabricated using GN–PANI shows the specific capacitance of 300–500 F g^{-1} at a current density of 0.1A g^{-1}.

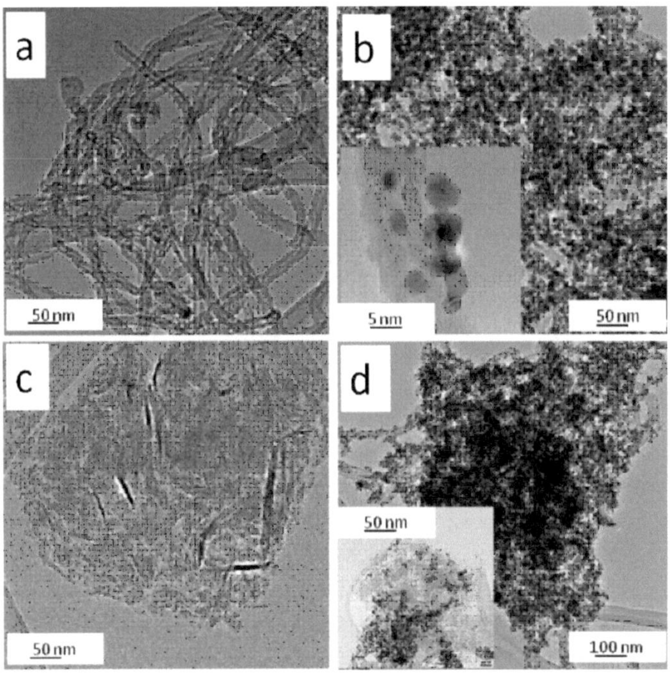

Figure 14. TEM images of (a) MWCNTs, (b) SnO_2–MWCNTs (inset shows HRTEM image), (c) GNs and (d) GNs/SnO2–MWCNTs composite (from ref. 39)

GN-PANI composite film with layered structure was obtained via filtration of an aqueous dispersion consisting of positively charged PANI nanofibres and negative charged chemically converted GN sheets that form a stable composite dispersion via electrostatic interaction with the assistance of ultrasonication [41]. The conductivity of GN-PANI film was one order higher than that of pure PANI nanofibres film. The symmetric supercapacitor device using GN-PANI films exhibited a high capacitance of 210 F g^{-1} at 0.3 A g^{-1}, and this capacitance can be maintained for about 94% (197 F g^{-1}) as the discharging current density was increased from 0.3 to 3 A g^{-1}. Due to the synergic effect of both components, the performance of GN-PANI based capacitor is much higher than those of the supercapacitors based on pure chemically converted GN or PANI-nanofibre films. The GN-PANI film has a layered structure as shown in its cross-section scanning electron micrograph (SEM) of Figure 15 (a), which is probably caused by the flow assembly effect of GN sheets during filtration. The magnified SEM image (Figure 15(b)) reveals that PANI nanofibres are sandwiched between chemically converted GN layers. The interspaces between the chemically converted GN layers are in the range of 10-200 nm. This morphology endows GN-PANI film with additional specific surface area comparing with that of the compact GN film prepared under the same conditions (Figure 15c). Filtrating of the dispersion PANI-nanofibres also produced a porous film as shown in Figure 15d, however, the mechanical property of this film is poor and it usually breaks into small pieces after drying.

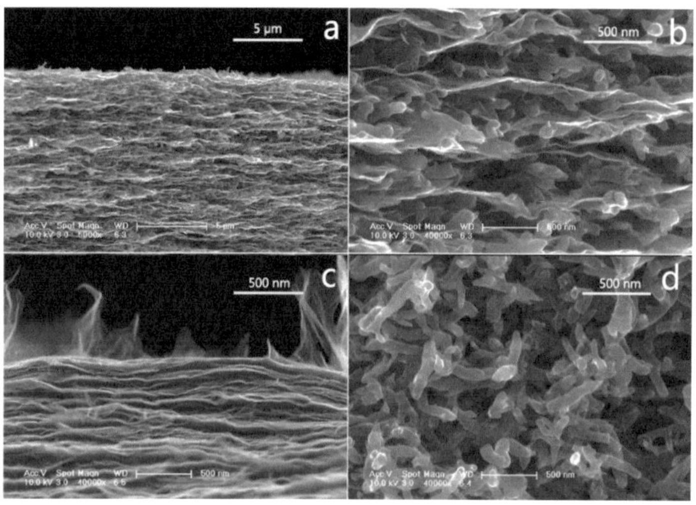

Figure 15. Cross-section SEM images of GN-PANI (a, b), pure chemically converted GN (c), and PANI nanofibre (d) films prepared by vacuum filtration (from ref. 41)

GN was also composited with PEDOT by F. Alvi et al. [42], by chemically oxidative polymerization of ethylene dioxythiophene (EDOT) using ammonium peroxydisulfate $[(NH_4)_2S_2O_8)]$ and $FeCl_3$ as oxidizing agents. In the solution of EDOT monomer, GN was added into it at EDOT to GN ratio of 1:1. The GN-PEDOT nanocomposite dramatically improves the electrochemical performance comparing to PEDOT based supercapacitors. The GN-PEDOT has also provided faster electrochemical reaction with an average capacity of 350 F g^{-1}. All the above results been demonstrated that the improvement in supercapacitor performance of GN based electroactive material can be achieved by compositing with metal oxides or conducting polymers. GN in those nanocomposites can act as nanoscale supports for dispersing metal oxides or conducting polymers to increase their surface area. GN can also provide the electronic conductive channels for metal oxides and conducting polymers. In addition, GN nanosheets can restrict the mechanical deformation of the polymers during the redox process due to its unique structural and mechanical properties. Graphene-based nanocomposites are expected to have great future for their application in supercapacitors.

4. Conclusion and future directions

Nanocomposite electroactive materials that have been developed so far have demonstrated huge potential for supercapacitor applications. Different types of nanocomposite electroactive materials, such as mixed metal oxides, polymers mixed with metal oxides, carbon nanotubes mixed with polymers, or metal oxides, and graphene mixed with metal oxides or polymers, can be fabricated by various processes such as solid state reactions, mechanical mixing, chemical co-precipitation, electrochemical anodic deposition, sol-gel, in situ polymerization and other wet-chemical synthesis. It has been shown that significant improvement in term of specific surface area, electrical and ionic conductivities; specific capacitance, cyclic stability, and energy and power density, of supercapacitors can be achieved by using nanocomposite electroactive materials. This can be attributed to the complementary and synergy behaviours of the consisting material components, the unique interface characteristics and the significant increase in surface areas, as well as nano-scale dimensional effects. Electrochemical double layer capacitors using carbon based electroactive materials and pseudocapacitors using metal oxide or conducting polymers as electroactive materials are the two types of most common supercapacitor structures. Compared with the electrochemical double layer capacitors, pseudocapacitors present a number of advantages such as high energy density and low materials cost, but suffer from poor cyclic stability and lower power density. Asymmetric supercapacitors using one electrode of pseudocapacitive materials and the other carbon based double capacitive materials are of great interest and are the current research focus. Nanocomposite pseudocapactive materials have great potential for asymmetric supercapacitor applications. The key issue is to fully utilize nanocomposites' excellent intrinsically properties, especially their high surface area and high conductivity, and to improve the synergistic effect of different electroactive components.

Although nanocomposite films have demonstrated their great potential for supercapacitor applications, several challenges still remain. Synthesis of nanocomposite electroactive materials with precisely controlling their chemical composition ratio, micro/nanostructures, phases, surface area and interfacial characteristics is still challenging. Depending on the preparation technique and process parameters, the property and behaviours of the nanocomposite electroactive materials can vary significantly; therefore, the ability to reproducibly synthesize nanocomposite materials with consistent properties is very important for their wide uses in supercapacitors. Degradation of the nanocomposite electroactive materials stemming from aggregation of the nano-scale components due to the relatively strong forces between them, micro/nanostructure changes due to charging-discharging cycling and materials contaminations due to impurity introduced for original reactants or during synthesis processes etc. has to be resolved before their large-scale adoption by the industry. Most importantly, the costs of materials and their synthesis processes have to be reduced significantly. With the increase in interest and intensive research and development, it is expected that, nanocomposite electroactive materials will have a promising future and will bring a huge change to the energy storage industries.

Acknowledgements

The author would like to thank Transport Canada and National Research Council of Canada (NRC) for supporting the publication of this chapter. The author is also indebted to his NRC colleagues: Dr. Sylvain Pelletier and Ms. Nathalie Legros for their initiation of supercapacitor research at NRC, and Dr. Alexis Laforgue, Dr. Lucie Robitaille, Dr. Yves Grincourt, Dr. Lei Zhang, and Dr. Jiujun Zhang for their collaboration in the supercapacitor research project. He would also like to thanks his research team members: Mr. Brian Gibson, Mr. Marco Zeman, Mr. Robinet Romain, Mr. Benjamin Tailpied, and Ms. Gaëlle LEDUC for their dedication to supercapacitor research. Thank is also given to Ms. Catherine Yang for her editorial review on this chapter.

Author details

Dongfang Yang*

Address all correspondence to: Email: dongfang.yang@nrc.gc.ca

National Research Council Canada, 800 Collip Circle, London, Ontario,, Canada

References

[1] Jiang, H., Li, C., Sun, T., & Ma, J. (2012). High-performance supercapacitor material based on $Ni(OH)_2$ nanowire-MnO_2 nanoflakes core–shell nanostructures. *Chem. Commun.*, 48, 2606-2608.

[2] Kim, H., & Popov, N. (2003). Synthesis and Characterization of MnO2Based Mixed Oxides as Supercapacitors. *Journal of the Electrochemical Society*, 150(3), D56-D62.

[3] Lee, M. T., Chang, J. K., Hsieh, Y. T., & Tsai, W. T. (2008). Annealed Mn-Fe binary oxides for supercapacitor applications. *Journal of Power Sources*, 185-1550.

[4] Ye, Z. G., Zhou, X. L., Meng, H. M., Hua, X. Z., Dong, Y. H., & Zou, A. H. (2011). The Electrochemical Characterization of Electrochemically Synthesized MnO2 based Mixed Oxides for Supercapacitor Applications. *Advanced Materials Research*, 287(290), 1290-1298.

[5] Yang. (2012). Pulsed laser deposition of cobalt-doped manganese oxide thin films for supercapacitor applications. *Journal of Power Sources*, 198-416.

[6] Yang, D. (2011). Pulsed laser deposition of manganese oxide thin films for supercapacitor Applications. *Journal of Power Sources*, 196-8843.

[7] Zhong, J. H., Wang, A. L., Li, G. R., Wang, J. W., Ou, Y. N., & Tong, Y. X. (2012). $Co_3O_4/Ni(OH)_2$ composite mesoporous nanosheet networks as a promising electrode for supercapacitor applications. *J. Mater. Chem.*, 22-5656.

[8] Yang, Y., Kim, D., Yang, M., & Schmuki, P. Vertically aligned mixed 2O5-TiO2 nanotube arrays for supercapacitor applications. 2011, *Chem. Commun.*, 47, 7746-7748.

[9] Jayalakshmi, M., Venugopal, N., Phani, Raja. K., & Mohan, Rao. (2006). Nano SnO2 - Al_2O_3 mixed oxide and SnO_2-Al_2O_3-carbon composite oxides as new and novel electrodes for supercapacitor applications. *Journal of Power Sources*, 158, 1538-1543.

[10] Wang, C., Zhang, X., Zhang, D., & Yao, C. (2012). Facile and low-cost fabrication of nanostructured NiCo2O4 spinel with high specific capacitance and excellent cycle stability. *Electrochimica Acta*, 63-220.

[11] Chang, S. K., Lee, K. T., Zainal, Z., Tan, K. B., Yusof, N. A., Yusoff, W., Lee, J. F., & Wu, N. L. (2012). Structural and electrochemical properties of manganese substituted nickel cobaltite for supercapacitor application. *Electrochimica Acta*, 67.

[12] An, K. H., Jeon, K. K., Heo, J. K., Lim, S. C., Bae, D. J., & Lee, Y. (2002). High-Capacitance Supercapacitor Using a Nanocomposite Electrode of Single-Walled Carbon Nanotube and Polypyrrole. *Journal of The Electrochemical Society*, 149(8), A1058-A1062.

[13] Deng, M., Yang, B., & Hu, Y. (2005). Polyaniline deposition to enhance the specific capacitance of carbon nanotubes for supercapacitors. *Journal of Materials Science Letters*.

[14] Tadai, K., & Mitani, T. (2005). Highly dispersed ruthenium oxide nanoparticles on carboxylated carbon nanotubes for supercapacitor electrode materials. *J. Mater. Chem.*, 15-4914.

[15] Kim, Y., & Mitani, T. (2006). Oxidation treatment of carbon nanotubes: An essential process in nanocomposite with RuO_2 for supercapacitor electrode materials. *Appl. Phys. Lett.*, 89, 033107.

[16] Hsieh, T. F., Chuang, C. C., Chen, W. J., Huang, J. H., Chen, W. T., & Shu, C. M. (2012). Hydrous ruthenium dioxide/multi-walled carbon-nanotube/titanium electrodes for supercapacitors. Oxidation treatment of carbon nanotubes:. *An essential process in nanocomposite with RuO2 for supercapacitor electrode materials*, 50, 1740-1747.

[17] Lee, J. K., Pathan, H. M., Jung, K. D., & Joo, O. S. (2006). Electrochemical capacitance of nanocomposite films formed by loading carbon nanotubes with ruthenium oxide. *Journal of Power Sources*, 159-1527.

[18] Kim, B. C., Wallace, G. G., Yoon, Y. I., Ko, J. M., & Too, C. (2009). Capacitive properties of RuO_2 and Ru-Co mixed oxide deposited on single-walled carbon nanotubes for high-performance supercapacitors. *Synthetic Metals*, 159-1389.

[19] Chen, Y. M., Cai, J. H., Huang, Y. S., Lee, K. Y., & Tsai, D. (2011). Preparation and characterization of iridium dioxide-carbon nanotube nanocomposites for supercapacitors. *Nanotechnology*, 22(115706), 7.

[20] Lee, C. Y., Tsai, H. M., Chuang, H. J., Li, S., Lin Yi, P., & Tseng, T. (2005). Characteristics and Electrochemical Performance of Supercapacitors with Manganese Oxide-Carbon Nanotube Nanocomposite Electrodes. *Journal of The Electrochemical Society*, 152(4), 716A-A720.

[21] Lee, J. Y., Liang, K., An, K. H., & Lee, Y. (2005). Nickel oxide/carbon nanotubes nanocomposite for electrochemical capacitance. *Synthetic Metals*, 150-153.

[22] Liu, M. Y., Hsieh, T. F., Hsieh, C. J., Chuang, C. C., & Shu, C. M. (2011). Performance of vanadium oxide on multi-walled carbon nanotubes/titanium electrode for supercapacitor application. *Advanced Materials Research*, 311-313, 414-418.

[23] Jayalakshmi, M., Mohan, Rao. M., Venugopal, N., & Kim, K. B. (2007). Hydrothermal synthesis of SnO_2-2O5 mixed oxide and electrochemical screening of carbon nanotubes (CNT), V_2O_5, V_2O_5-CNT, and SnO_2-V_2O_5-CNT electrodes for supercapacitor applications. *Journal of Power Sources*, 166, 578-583.

[24] Lee, B. J., Sivakkumar, S. R., Ko, J. M., Kim, J. H., Jo, S. M., & Kim, D. (2007). Carbon nanofibre/hydrous RuO_2 nanocomposite electrodes for supercapacitors. *Journal of Power Sources*, 168-546.

[25] Cormier, Z. R., Andreas, H. A., & Zhang, P. (115). Temperature-Dependent Structure and Electrochemical Behaviour of RuO_2/Carbon Nanocomposites. *J. Phys. Chem*, 115, 19117-19128.

[26] Selvakumar, M., Bhat, D. K., Aggarwal, A. M., Iyer, S. P., & Sravani, B. (2010). *Physica*, 405-2286.

[27] Tao, T., Chen, Q., Hu, H., & Chen, Y. (2012). MoO_3 nanoparticles distributed uniformly in carbon matrix for supercapacitor applications. *Materials Letters*, 66-102.

[28] Chen, L., Sun, L., Luan, F., Liang, Y., Li, Y., Liu, X., & , X. (2010). Synthesis and pseudocapacitive studies of composite films of polyaniline and manganese oxide nanoparticles. *Journal of Power Sources*, 195-3742.

[29] Hu, Z., Xie, Y., Wang, Y., Mo, L., Yang, Y., & Zhang, Z. (2010). Synthesis and pseudocapacitive studies of composite films of polyaniline and manganese oxide nanoparticles. *Journal of Power Sources*, 195-3742.

[30] Wang, J., Yang, Y., & Huang, Z. (2012). . Interfacial synthesis of mesoporous MnO2/polyaniline hollow spheres and their application in electrochemical capacitors. *Journal of Power Sources*, 204-236.

[31] Saranya, S., & Selvan, R. K. (2012). Synthesis and characterization of polyaniline/$MnWO_4$ nanocomposites as electrodes for pseudocapacitors. *Applied Surface Science*, 258-4881.

[32] Sen, P. (2010). Electrochemical performances of poly 3 4ethylenedioxythiophene)-$NiFe_2O_4$ nanocomposite as electrode for supercapacitor. *Electrochimica Acta*, 55, 4677-4684.

[33] Murugan, A. V., Viswanath, A. K., & Gopinath, C. S. (2006). Highly efficient organic-inorganic poly (3 4-ethylenedioxythiophene)-molybdenum trioxide nanocomposite electrodes for electrochemical supercapacitor. *J. Appl. Phys.*, 100(074319), 1-5.

[34] Dong, Z. H., Wei, Y. L., Shi, W., & Zhang, G. (2011). Characterisation of doped polypyrrole/manganese oxide nanocomposite for supercapacitor electrodes. *Materials Chemistry and Physics*, 131-529.

[35] Yang, X., Wang, G., Wang, R., & Li, X. (2010). A novel layered manganese oxide/poly (aniline-co-o-anisidine) nanocomposite and its application for electrochemical supercapacitor. *Electrochimica Acta*, 55-5414.

[36] Jaidev, Jafri. R. I. (2011). Ramaprabhu S. Hydrothermal synthesis of $RuO_2.xH_2O$/graphene hybrid nanocomposite for supercapacitor application. International Conference on Nanoscience,. *Technology and Societal Implications*, 8-10.

[37] Wang, Y., Guo, C. X., Liu, J., Chen, T., Yang, H., & Li, C. (2011). CeO_2 nanoparticles/graphene nanocomposite-based high performance supercapacitor. *Dalton Trans.*, 40(6388).

[38] Bonso, J. S., Rahy, A., Perera, S. D., Nour, N., Seitz, O., Chabal, Y. J., Balkus Jr, K. J., Ferraris, J. P., & Yang, D. (2012). Exfoliated graphite nanoplatelets-2O5 nanotube composite electrodes for supercapacitors. *Journal of Power Sources*, 203, 227-232.

[39] Rakhi, R. B., & Alshareef, H. N. (2011). Enhancement of the energy storage properties of supercapacitors using graphene nanosheets dispersed with metal oxide-loaded carbon nanotubes. *Journal of Power Sources*, 196-8858.

[40] Gómeza, H., Ramb, M. K., Alvia, F., Villalba, P., & Stefanakos, E. (2011). Graphene-conducting polymer nanocomposite as novel electrode for Supercapacitors. *Journal of Power Sources*, 196-4102.

[41] Wu, Q., Xu, Y., Yao, Z., & Liu, A. (2010). Supercapacitors Based on Flexible Graphene/Polyaniline Nanofiber Composite Films. ACS. *Nano*, 4(4), 1963-1970.

[42] Stefanakos, E., Goswami, Y., & Kumar, A. (2011). Graphene-polyethylenedioxythiophene conducting polymer nanocomposite based supercapacitor. *Electrochimica Acta*, 56-9406.

Chapter 13

Conducting Polymer Nanocomposites for Anticorrosive and Antistatic Applications

Hema Bhandari, S. Anoop Kumar and S. K. Dhawan

Additional information is available at the end of the chapter

http://dx.doi.org/10.5772/50470

1. Introduction

Intrinsically conducting polymers (ICPs) have been considered for use in various applications. One of the most important applications of these materials which are attracting considerable attention in the most recent times is in corrosion protection of oxidizable metals [1]. The effective use of conducting polymers for corrosion protection of metals can be carried out by different methods; like formulation of polymers with paints, by electro-deposition of conducting polymers onto metal surface and by direct addition of polymers in the corrosive solution as corrosion inhibitors. Coatings on the surface of metals by polymeric materials have been widely used in industries for the protection of these materials against corrosion [2-13]. Some specific conducting polymers like polyaniline and its derivatives, have been found to display interesting corrosion protection properties. In the past decade, the use of polyaniline as anticorrosion coatings had been explored as the potential candidates to replace the chromium-containing materials, which have adverse health and environmental concerns [14-17]. A polymer behaves as a barrier when it exists in the electronically and ionically insulating state. An important feature of the polymer coating in its conductive state is the ability to store large quantity of charge at the interface formed with a passive layer on a metal. This charge can be effectively used to oxidize base metal to form a passive layer. Thus, the conducting polymer film was also capable of maintaining a stationary potential of the protected metal in the passive range [18]. Application of conducting polymers like polyaniline to corrosion protection of metals is, however, subject to some limitations. First, charge stored in the polymer layer (used to oxidize base metal and to produce passive layer) can be irreversibly consumed during the system's redox reactions. Consequently, protective properties of the polymer coating may be lost with time. Also, porosity and anion exchange properties of conducting polymers could be disadvantageous, particularly when it comes to

INTECH
open science | open minds

© 2012 Bhandari et al.; licensee InTech. This is a paper distributed under the terms of the Creative Commons Attribution License (http://creativecommons.org/licenses/by/3.0), which permits unrestricted use, distribution, and reproduction in any medium, provided the original work is properly cited.

pitting corrosion caused by small aggressive anions (*e.g.*, chlorides) [19]. An interesting alternative is to consider conducting polymer based composite systems. Composite materials play an important role due to their light weight and improved corrosion resistance. These materials usually comprise of a polymer matrix in which fibres and/or small filler particles are thoroughly dispersed. Silicon dioxide particles, for example, comprise one of the common fillers in composite materials such as plastics and films. Conducting polyaniline/inorganic nanocomposites have also attracted more and more attention. A number of different metals and metal oxide particles have so far been encapsulated into the shell of conducting PANI to produce a host of composites materials. These composite materials have shown better mechanical, physical and chemical properties, due to combining the qualities of conducting PANI and inorganic particles [20-22]. Among various inorganic particles, SiO_2 nanoparticles have attractive attention due to their excellent reinforcing properties for polymer materials [23]. However, SiO_2 is an insulator and a lot of works have been done to expand the applications of insulator SiO_2 as fillers and improve the processability of polyaniline [24-28]. Corrosion protective coatings on the mild steel surface by electrochemical deposition of conducting polymers have been extensively studied [29-33]. These types of coatings lack long lasting ability of metals in a corrosive medium. In order to improve the adherence ability and efficiency of conducting polymer coating on the metal surface, the use of liquids paints have also been performed by many researchers [34-38].

Present chapter is based on the preparation of conducting polymer nanocomposites coating onto the mild steel surface by using powder coating techniques. Powder coatings are often used as an alternation to liquids paints finishing or traditional liquid finishing. The key benefits of powder coating techniques are cost effective, environmentally friendly, excellence of finish and performance. This shows single coat finishes with no primer or any other solvent required and high film thickness can be achieved with single coat. Preparation of PANI/SiO_2 nanocomposites was carried out using in-situ polymerization and evaluation of corrosion protection effect for the polyaniline/SiO_2 nanocomposite materials on the mild steel surface. Corrosion protection performance of these composites was compared with that of the polyaniline by performing a series of electrochemical measurements of corrosion potential, polarization resistance, and corrosion current in 1.0 M HCl solution.

In order to improve anticorrosion performance of mid steel in 3.5 % NaCl aqueous medium, preparation of highly hydrophobic polyaniline-SiO_2 nano-composite coating have also been developed. There are several ways to develop hydrophobic surfaces i.e. by electrodeposition method [39-41], solvent casting of polymers [42], layer-by-layer deposition [43-45], chemical vapor deposition [46, 47], dip-coating and/or self-assembly [48-51] and chemical grafting [52-55]. Most of those methods cannot be easily developed and they need very strict conditions of preparation, and low adhesion coatings are often obtained. However, development of highly hydrophobic conducting polymer nanocomposite coatings via powder coating method was found to be highly adhesive, long lasting and more convenient [56, 57].

2. Corrosion study

The corrosion inhibition performance study was carried out at room temperature in aqueous solution of 1.0 M HCl/ 3.5 % NaCl by using Tafel extrapolation and chrono-amperometry methods. Experiments were carried in a conventional three electrode cell assembly using Autolab Potentiostat/ Galvanostat, PGSTAT100 (Nova Software). In three electrode cell assembly, pure iron of dimension 1 cm x 1 cm is taken as working electrode embedded in araldite epoxy, Pt as counter electrode and saturated calomel electrode (SCE) as reference electrode. The cleaning of the working iron electrode was carried out by 1/0, 2/0, 3/0 and 4/0 grade emery papers. The electrodes were then thoroughly cleaned with acetone and trichloroethylene to remove any impurities on the surface.

2.1. Tafel extrapolation method

Tafel extrapolation method involves the measurement of over potentials for various current densities. Figure 1 shows a potential vs. log absolute current plot for an applied potential scan. The linear Tafel segments to the anodic and cathodic curves (-0.2 to + 0.2 V versus corrosion potential) were extrapolated to corrosion potential to obtain the corrosion current densities. The slope gives the Tafel slopes (b_a and b_c) and the intercept corresponds to i_{corr}. The corrosion current density [i_{corr} (A/cm^2)] was calculated with the Stern-Geary equation [58] (Eq.1);

$$i_{corr} = \frac{b_a . b_c}{2.3(R_P)(b_a + b_c)} \tag{1}$$

Corrosion rate (C.R) in mm/year can also be calculated by using following relationship [59] (Eq.2);

$$C.R = 3.268 \times 10^3 \frac{i_{corr}}{\rho} \frac{MW}{z} \tag{2}$$

where MW is the molecular weight of the specimen (g/mole), ρ is density of the specimen (g/m^3) and z is the number of electrons transferred in corrosion reactions.

The corrosion protection efficiency (% P.E.) was determined from the measured i_{corr} (corrosion current densities with blank mild steel electrode (i^0_{corr}) without coatings and corrosion current densities with a mild steel electrode coated with polymer coated (i^c_{corr}) values by using the following relationship;

$$P.E.(\%) = \frac{i^0_{corr} - i^c_{corr}}{i_{0_{corr}}} \times 100 \tag{3}$$

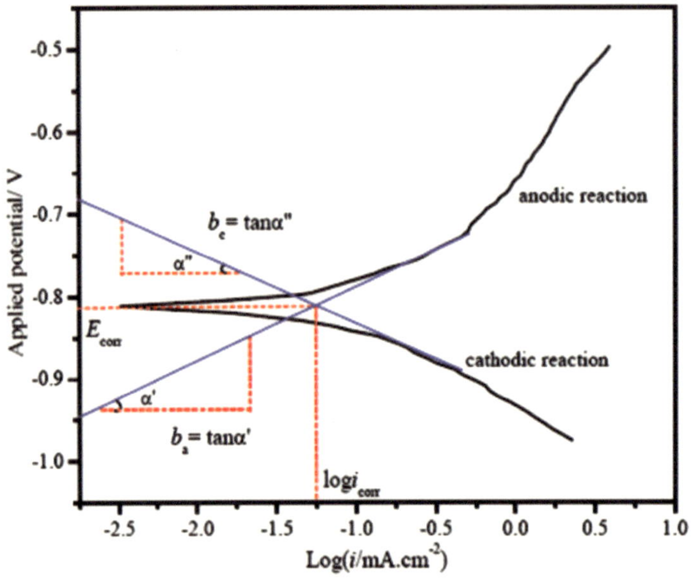

Figure 1. Schematic polarization curve showing Tafel extrapolation.

2.2. Weight loss method

The weight loss methods have also been performed for corrosion study. Polymer coated mild steel specimens of dimension 4 x 3.5 cm^2 have been tested for same span of time by immersing the samples in 1.0 M HCl or aqueous 3.5 % NaCl solution for 60 days. The uncoated and polymers coated mild steel specimens were weighed in an electronic balance with an accuracy of 0.1 mg. before immersion in saline medium. After the 60 days of immersion the mild steel specimens were withdrawn from the tested solution, washed thoroughly with distilled water followed by acetone and dried with air, then weighed again. The performance of the coating was examined visually and through calculation of the weight loss. Weight loss (W.L) is expressed as the loss in the weight per unit area or per unit area per unit time (g cm^{-2} h^{-1}) as follows:

$$W.L = \frac{w_0 - w_1}{a.t} \tag{4}$$

where, w_0 = initial weight of the sample before immersion (g); w_1 = weight of the sample after immersion (mg); a= surface area (cm^2) of specimen; t = end time (h) of each experiment. If we introduce the density of metal; d (g/cm^3), the loss in the thickness of metal per unit time can be calculated. Corrosion rate (C.R) in mm/year can also be calculated by weight loss method as follows:

$$C.R(mm / year) = \frac{(w_0 - w_1) \times 87.6}{a.t.d} \quad (5)$$

2.3. Surface study

Surface studies comprise the analysis of surface of metals before and after corrosion in order to estimate the rate as well as mechanism of corrosion. Techniques like scanning electron microscopy and electron probe micro analysis are used to study the structure, chemical composition of corrosion product formed onto metal surface. Other techniques like atomic force microscopy and ellipsometry are used to study the surface of metals with and without corrosion.

3. Synthesis of the SiO_2 nanoparticles

The syntheses of mono disperse uniform- sized SiO_2 nanoparticles were carried out by using ammonia as catalyst and ethanol as solvent. Hydrolysis method of tetra-ethylorthosilicates (TEOS) was used for synthesizing SiO_2 nanoparticles. Aqueous ammonia (0.1M) was added to a solution containing ethanol (1.0 M) and 20 ml of deionized water which was stirred for 1 h then 0.05M TEOS was added and again stirred for 1 h at room temperature. Appearance of white turbid suspension indicating the formation of silicon dioxide, this suspension was retrieved by centrifugation and further calcination at 823 K for 6 hours.

4. Preparation of PANI/SiO_2 composites:

4.1. Chemical oxidative polymerization

PANI/SiO_2 composites were prepared by *in situ* chemical oxidative polymerization of aniline using APS as an oxidant. Weight ratio of aniline and SiO_2 was taken as 1:1 for preparation of PANI/SiO_2 nanocomposites. Aniline was adsorbed on SiO_2 particles and 0.2 M phosphoric acid/0.2 M perfluro octanoic acid (PFOA) was added in this solution. Polymerization was initiated by drop wise addition of ammonium persulphate solution (APS) (0.1 M, $(NH)_4S_2O_8$ in distilled water). The polymerization was carried out at a temperature of 0-3 C for a period of 4-6 h. The synthesized polymer composite was isolated from reaction mixture by filtration and washed with distilled water to remove oxidant and oligomers and followed by drying in the vacuum oven at 60°C.

4.2. Electro-chemical polymerization

The electrochemical polymerization of 0.1 M aniline and aniline in the presence of SiO_2 in 0.2 M H_3PO_4 /0.2 M PFOA was carried out between −0.20 to 1.5 V on platinum electrode vs. Ag/AgCl reference electrode. The polymer film growth was studied by sweeping the potential between −0.20 to 1.5V on Pt electrode at a scan rate of 20 mV/s. Prior to polymerization, the solution was deoxygenated by passing argon gas through the reaction solution for 30 min. Peak potential values of the corresponding polymer and PANI-SiO_2 composites were recorded in 0.2 M H_3PO_4 /0.2 M PFOA medium.

5. Preparation of PANI/SiO_2 composites coated mild steel

Mild steel coated with PANI/SiO_2 nanocomposites electrode of dimension 1 cm x 1 cm were employed to carry out the corrosion studies. Surface treatments were applied on the samples including the cleaning of the electrode was carried out by 1/0, 2/0, 3/0 and 4/0 grade emery papers. The electrodes were then thoroughly cleaned with acetone and trichloroethylene to remove any impurities on the surface. The powder polymer was mixed with epoxy formulation in various proportions ranging from 1.0 % to 6.0 wt. %. The polymer-epoxy powder coating was applied to a thickness of 45 ± 3 μm using an electrostatic spray gun. After obtaining uniform coverage of the powder the powder-coated panels were placed in air drying oven for curing at 140°C for 20 min. The adhesion of the coating was tested by tape test as per ASTM D3359-02 and found to pass the test.

6. Characterization of PANI-SiO_2 nanocomposites

6.1. Electrochemical behaviour

Figure 2 shows the electrochemical growth behaviour of aniline and aniline-SiO_2 in 0.2 M H_3PO_4. The polymer film growth was studied by sweeping the potential between -0.20 and 1.5 V on Pt electrode at a scan rate of 20 mV/s. Peak potential values of the corresponding PANI and PSC were recorded in H_3PO_4 medium. First anodic peak (oxidation peak) corresponds to the oxidation of monomer. During the first reverse sweep, a reduction peak appears which shows that the formation of oligomers and polymer on electrode surface as shown in Figure 2.

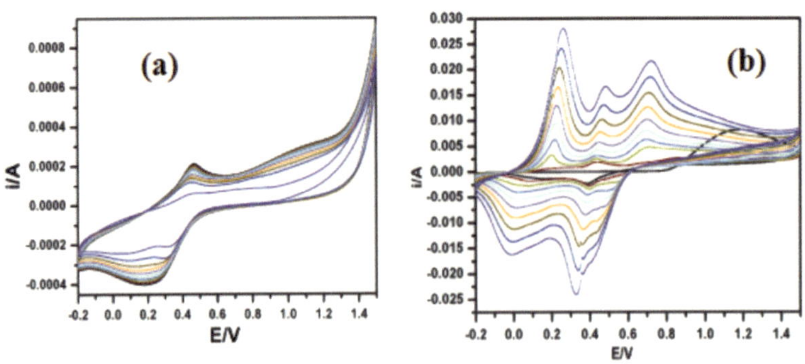

Figure 2. Electrochemical growth behaviour of (a) 0.1M aniline in 0.2M H_3PO_4 medium and (b) aniline- SiO_2 in 0.2M H_3PO_4 medium in potential range between -0.2 V to 1.5 V vs. Ag/AgCl at scan rate of 20 mV/s.

After the first scan, well defined oxidation and reduction peaks of polymers between 0.2 and 0.6 V vs. Ag/AgCl appeared. The current values of each oxidation and reduction peaks are greater than that of a previous cycle which indicate the built up of an electro active polymeric material on the electrode surface.

Figure 3 shows the cyclic voltammogram of PANI-SiO_2 composite and the inset figure shows the cyclic voltammogram of PANI in H_3PO_4 medium. We have observed quite interesting observation when we recorded the cyclic voltammogram of aniline in phosphoric acid medium and when SiO_2 nanoparticles were incorporated in the monomer matrix. On recording the cyclic voltammogram of aniline in H_3PO_4 medium, it was observed that anodic peak potential is observed at 0.456 V.

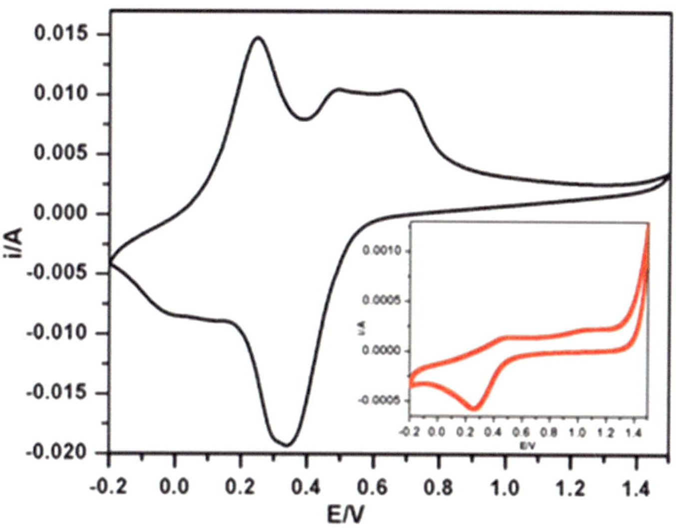

Figure 3. Cyclic voltammogram of PSC and inset shows the cyclic voltammogram of PANI in H_3PO_4 medium at a scan rate of 20 mV/s.

However, in the presence of SiO_2 matrix, these peaks appeared at 0.247 V. The reason for getting these deviations in H_3PO_4 medium is that phosphoric acid is a weak protonic acid whose pKa1 value is 2.21 which results in shifting of peak potential value to 0.456 V. Conventionally, in strong acidic medium like 1.0 M HCl medium, these values of peak potential for PANI are observed at 0.1 V. However, when the CV was recorded for aniline in the presence of SiO_2 and H_3PO_4 medium, there is a possibility that protons from phosphoric acid medium might have protonated SiO_2 resulting in generation of well-defined peaks as observed in the growth behavior of CV resulting in observing peak potential values at 0.247 V (EaI). These experiments were repeated by us number of times and each time this type of cyclic voltammogram was observed which has led us to draw the above conclusion. Protons from phosphoric acid might have led to the formation of protonated silica thereby shifting of peak potential values which might have enhanced the electropolymerization of aniline.

6.2. FTIR spectra

Figure 4 shows the FTIR spectra of SiO_2, PANI and PANI-SiO_2. PANI showed the main characteristics bands at 1565 and 1475 cm^{-1} attributed to the stretching mode of C=N and C=C, the bands at 1292 and 1245 cm^{-1} indicating the C–N stretching mode of benzenoid ring and the band at 1117 - 1109 cm^{-1} is assigned to a plane bending vibration of C–H mode which is found during protonation [60]. The FTIR spectra of SiO_2 indicated that the characteristic peak at 1081 cm^{-1} and 807 cm^{-1} are assigned to the stretching and bending vibration of Si–O–Si respectively. By comparing the peaks of PANI and PSC, it was observed that some peaks of PSC were shifted due to the presence of SiO_2 particles in polymer matrix. For example, the peaks at 1565 cm^{-1}, 1475 cm^{-1}, 1292 cm^{-1}, and 1245 cm^{-1} shifted to higher wavenumbers, and the bending vibration of Si–O–Si peak at 1056 cm^{-1} shifted to the lower wavenumbers. These changes also indicate that an interaction exists between PANI molecule and SiO_2 particles. These peaks were also observed in PSC indicating the interaction of SiO_2 particles in polyaniline chain.

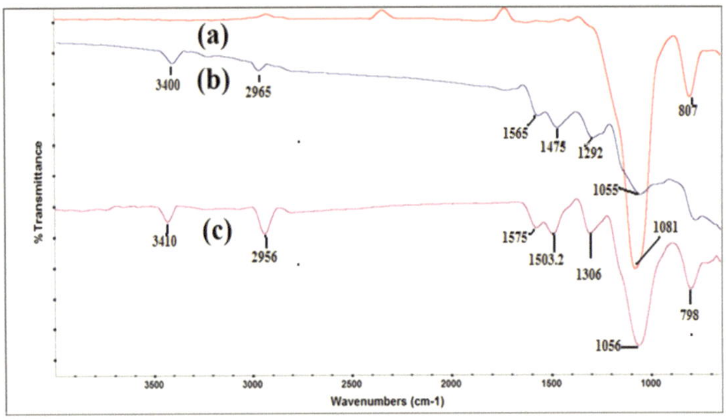

Figure 4. FTIR spectra of (a) SiO_2 (b) PANI and PSC.

6.3. Thermogravimetric analysis

Figure 5 shows the thermo-gravimetric curves (TG) of pure SiO_2, PANI and their composites. The materials were heated from 25 to 800°C under a constant heating rate of 10°C/min and in the inert atmosphere of nitrogen gas (60 ml/min). The SiO_2 particle has excellent thermal stability up to 800°C and weight loss was only 0.15 %. The TGA curve of PSC indicated, first weight loss at 110°C may be attributed to the loss of water and other volatiles species. The weight loss in the second step at about 280°C involves the loss of phosphate ions as well as onset of degradation of polyaniline backbone. The increasing SiO_2 content slightly affects the decomposition temperature (DT) which increases from 280°C (PANI) to 295°C (PSC). The third weight loss step between 300 to 550°C can be ascribed to the complete degradation of dopant as well as polymeric backbone. The composites show little weight loss between the 500-800°C and the residue remaining in this region gives an approximate estimate of filler content. Therefore, the final weight of SiO_2 incorporated in polymer was found to 21 %. The results indicate that actually incorporated SiO_2 fraction is less than the ratio of aniline: SiO_2 taken in the initial reaction mass.

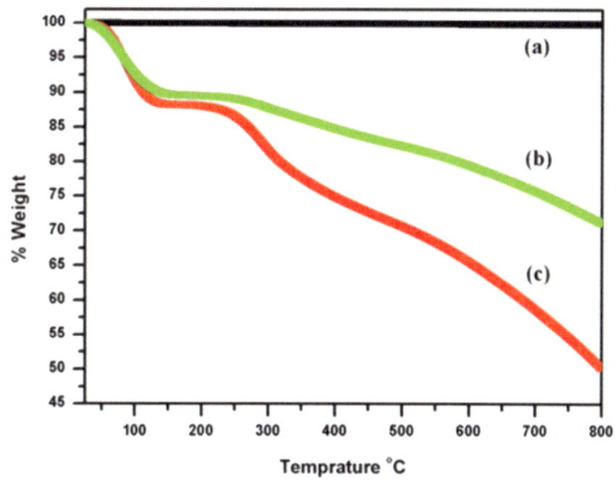

Figure 5. Thermal gravimetric analysis of SiO_2, PANI and (c) PSC doped with H_3PO_4.

The TGA data clarify that these composites are thermally stability up to 295°C, which envisages them as a good candidate for melt blending with conventional thermoplastics like polyethylene, polypropylene, polystyrene etc.

6.4. UV-Visible spectra

Figure 6 shows the UV absorption spectra of polyaniline and its composite with SiO_2. We have measured the UV absorption spectra of polymer using dimethyl sulfoxide (DMSO) as a solvent from 250 to1100 nm.

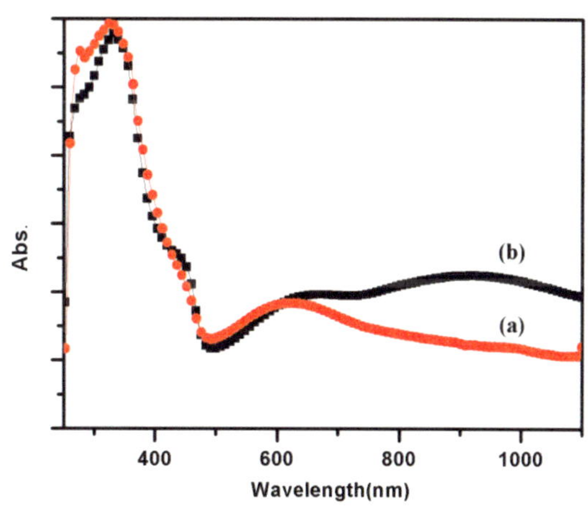

Figure 6. UV–Visible spectra of (a) PANI & (b) PSC in DMSO.

The UV-visible absorption data indicates that λ_{max} values in case of polyaniline doped with o-phosphoric acid medium in DMSO solvent lies at 326nm, 431nm and 619 nm whereas in PSC composite these values lies at 336 nm, 447 nm and 654 nm. In case of polyaniline, π-π^* transition [61, 62] occurs at 326 nm whereas in case of PSC composite, this transition value lies at 336 nm.This indicates that the addition of SiO_2 particles absorbed in aniline matrix and on polymerization in o-phosphoric acid medium may have caused some interactions with polymer matrix resulting in shifting of bands from 326 nm to 336 nm. This is the reason of shifting of polaronic bands which also shows a shift from 431 nm to 447 nm and 619 nm to 654 nm.

7. Anticorrosive properties of PANI and PSC coated mild steel in 1.0 M HCl medium.

7.1. Chronoamperometry method

Figure 7 shows the chronoamperometric response of uncoated, epoxy coated, PANI and PSC coated mild steel sample in 1.0 M HCl. After the samples reached a stable OCP (open circuit potential), a potential in the range of 1.2 V vs SCE was applied and current was recorded as a function of time. It was observed that the current of PANI and PSC coated mild steel sample remained at a very small value as compared with the uncoated mild steel electrode indicating the good protective properties by these polymers coating. Moreover, it has been observed that the current density value of PSC coated mild steel was lower than that of PANI-coated mild steel sample. The decrease in current density with increasing amount of PSC material in epoxy resin. Hence, chronoamperometric test results showed that mild steel coated with PANI/SiO_2 composites shows the higher corrosion protection performance as compared to PANI coated mild steel samples. This statement was further confirmed by other corrosion test methods like Tafel extrapolation and weight loss methods.

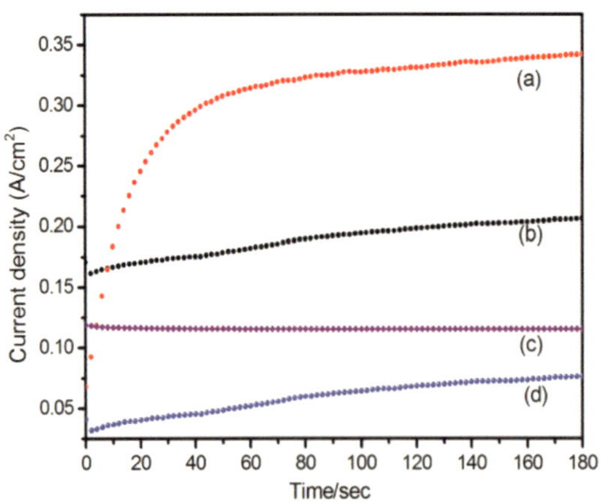

Figure 7. Chronoamperometric response of (a) uncoated (b) epoxy coated (c) PANI and (d) PSC coated mild steel sample in 1.0 M HCl.

http://dx.doi.org/10.5772/50470

7.2. Tafel extrapolation method

Tafel polarization behaviour of mild steel in 1.0 M HCl with uncoated, epoxy coated, PANI and PSC coated mild steel are shown in the Figure 8& Figure 9.

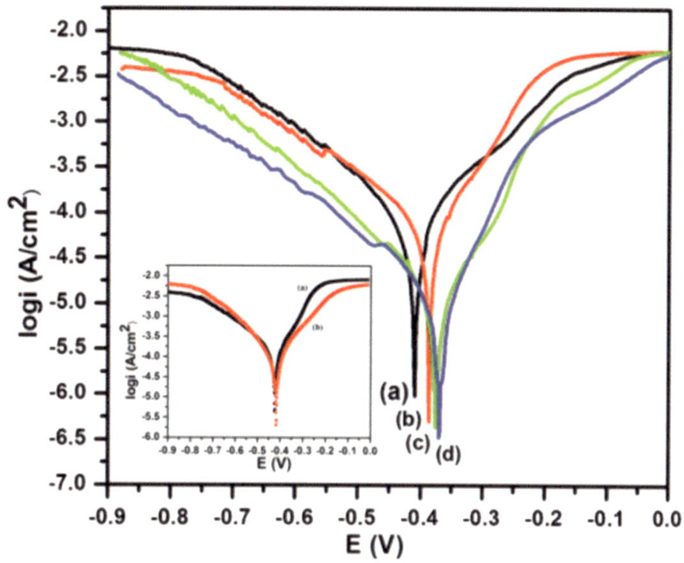

Figure 8. Tafel curves of PANI coated mild steel electrode with different loading level of PANI in epoxy resin (a) 1.5% (b) 3.0% (c) 4.5% (d) 6.0% whereas the inset shows (a) blank mild steel electrode and (b) epoxy coated mild steel electrode in 1.0 M HCl.

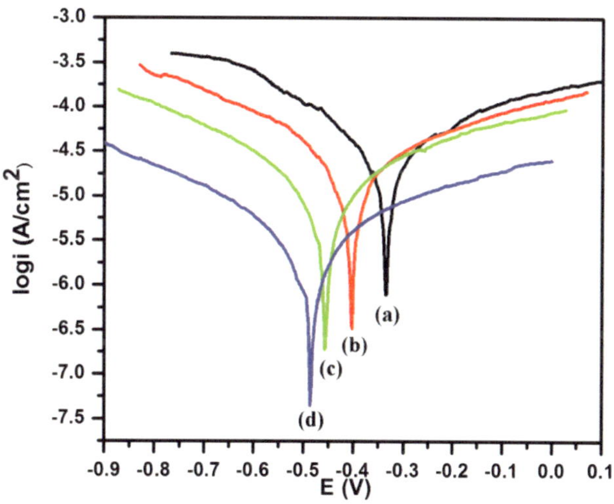

Figure 9. Tafel curves of PSC coated mild steel electrode with different loading level of PSC in epoxy resin (a) 1.5% (b) 3% (c) 4.5% (d) 6.0%.

The corrosion kinetic parameters derived from these curves are given in the Table 1. As shown in Table 1, PSC coated mild steel sample showed a remarkable current shift from 132 μA to 0.09 μA versus Ag/AgCl in the corrosion current (i_{corr}), relative to the value of the uncoated mild steel.

The significant reduction in the corrosion current density (i_{corr}) in polymer coated mild steel indicated the effective corrosion protection performance of these polymers. The corrosion current values (i_{corr}) were found to be decreased from 132 μA/cm^2 for uncoated mild steel sample to 107.6 μA/cm^2 for epoxy coated mild steel sample to 0.09 μA /cm^2 for PSC coated mild steel samples.

The corrosion current values (i_{corr}) decreased with increasing the concentration of PSC in epoxy resin. i_{corr} value decreased from 15.4 μA/cm^2 at 1.5 wt. % to 0.09 μA/cm^2 at 6.0 wt. % loading of conducting material in epoxy resin as shown in Figure 9. While PANI coated mild steel showed the i_{corr} in the range of 10.9 μA/cm^2 at 6.0 % loading as shown in Figure 8.

Sample name	Loading level of polymer (%)	I_{corr} (μA/cm^2)	Corrosion rate (mm/year)	Protection efficiency (%)
Blank mild steel	-	132.0	1.54	--
Epoxy coated mild steel	--	107.6	1.26	18.48
PANI	1.5	98.8	1.15	25.15
	3.0	75.4	0.88	42.88
	4.5	20.5	0.23	84.47
	6	10.9	0.13	91.74
PANI-SiO_2 Composite (PSC)	1.5	15.4	0.18	88.33
	3.0	9.09	0.11	93.11
	4.5	5.12	0.05	96.12
	6	0.09	0.0011	99.93

Table 1. Tafel parameters for corrosion of mild steel in 1.0 M HCl with different loading level of PANI & PSC in epoxy resin.

The corrosion protection efficiency calculated from Tafel parameter revealed that PANI coated mild steel showed 25% protection efficiency at 1.5 wt.% loading while PSC coated mild steel showed 88 % P.E at the same loading level. Up to 99.93 % protection efficiency have been achieved by using 6.0 wt.% loading of PSC in epoxy resin.

7.3. SEM studies of uncoated and coated mild steel observed by weight loss method

The scanning electron micrographs of SiO_2 particles, PANI and PSC are shown in Figure 10. SiO_2 particles showed spherical shaped morphology and PANI showed globular morphology. Figure 10 b shows the TEM image of of SiO_2 particles, which indicates the dimension of

SiO_2 particles, was found to be 90-100 nm. Morphology of PSC indicates the incorporation of SiO_2 particles in PANI matrix. SEM image of PSC revealed that the entrapment of SiO_2 particles in the globular space of PANI matrix during in situ polymerisation of polyaniline.

Figure 10. SEM micrographs powder sample of (a) SiO_2 (c) PANI (d) PSC and (b) TEM image of SiO_2 particles.

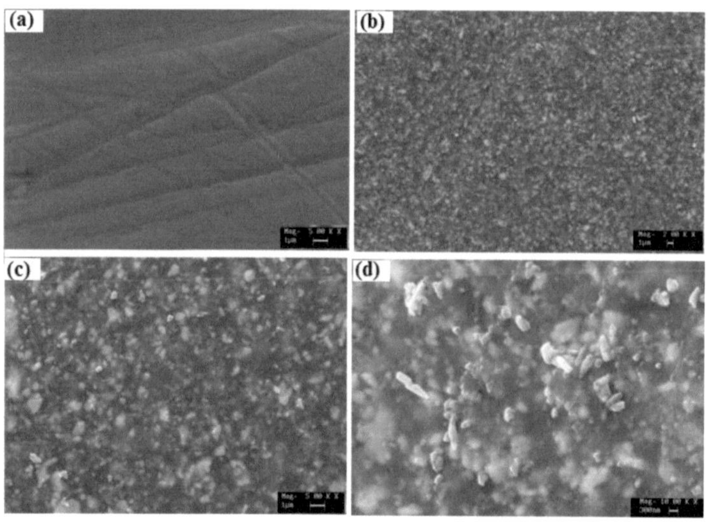

Figure 11. SEM micrographs of (a) blank mild steel electrode (b) blank epoxy resin coated electrode (c) PANI coated (d) PSC coated electrode before immersion in 1.0M HCl.

SEM images of uncoated, epoxy coated and polymer coated samples before and after the immersion test of 60 days have been shown in Figure 11 & Figure 12. These images clearly show the formation of large pits on the surface of mild steel after immersion. These pits and cracks were developed during the corrosion of mild steel in acidic medium. In the case PANI coated sample, few pits still appeared on mild steel surface. While, PSC coated mild steel samples did not show any cracks and pits on the metal surface. No detachment of coating from mild steel substrate was also observed after the immersion of these samples in 1.0 M HCl medium for 60 days of immersion indicating strong adherence of PSC composite to the mild steel substrate and it is resistant to corrosion in aqueous 1.0 M HCl solution as shown in Figure 13.

When epoxy coated mild steel sample was immersed in the acidic medium for 60 days, detachment of coating from mild steel substrate have been observed. The pits were also appeared on the metal surface as shown in Figure 12 b.

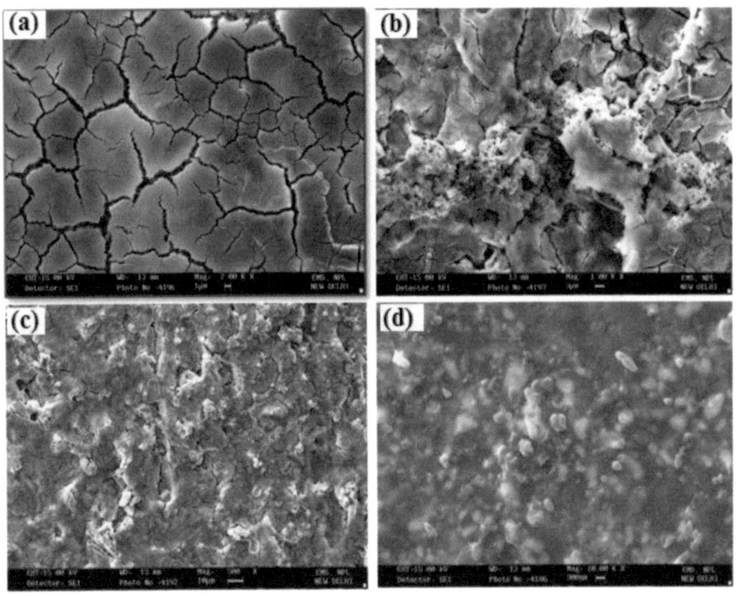

Figure 12. SEM micrographs of (a) blank mild steel electrode (b) blank epoxy resin coated electrode (c) PANI coated (d) PSC coated electrode after immersion in 1.0 M HCl for 60 days.

It was found that the PSC content has a great influence on the anticorrosive performance of the coating. The corrosion protection effect of PSC coated mild steel sample improved slowly when PSC content in epoxy formulation increases from 1.5 to 3.0 wt. % Afterward, an excellent corrosion protection effect appears at 6.0 wt. % loading of PSC content in epoxy resin.

Corrosion rates (C.R) in mm/year have also been calculated by weight loss method and the values have been given in Table 2. It was observed that the corrosion rate was highest for uncoated mild steel in HCl medium. After 60 days of immersion, the C.R value of uncoated mild steel was found to 7.25 mm/year. Epoxy and PANI coated samples showed C.R. value of 6.37 mm/year and 1.90 mm/year respectively. While in the case of PSC coated sample in 6.0 % loading, C.R. value reduced to 0.73 mm/year.

Sample name	Loading level of polymer (%)	Initial Weight (mg)	Weight After immersion in HCl for 60 days (mg)	Weight loss (mg)	Weight loss (%)	C.R (mm/ year)	Protection Efficiency (%)
Blank mild steel	-	30354.6	17256.30	13098.30	43.15	7.25	-
Blank epoxy	0	31113.4	19598.61	11514.79	37.00	6.37	12.10
PANI	1.5	31717.2	21518.09	10199.11	32.15	5.64	22.13
	3.0	31601.2	23698.65	7902.55	25.00	4.37	39.67
	4.5	31289.1	27021.37	4267.73	13.63	2.36	67.42
	6	30454.0	27018.20	3435.80	11.28	1.90	73.77
PANI-SiO2 Composite	1.5	32487.2	29606.00	881.20	8.86	1.59	78.00
	3.0	32891.5	30397.90	2493.60	7.58	1.38	80.96
	4.5	31856.7	30031.30	1825.40	5.73	1.01	86.06
	6	31773.7	30454.20	1319.50	4.15	0.73	89.93

Table 2. Weight loss parameter of uncoated and coated mild steel samples after immersion test in 1.0 M HCl for 60 days.

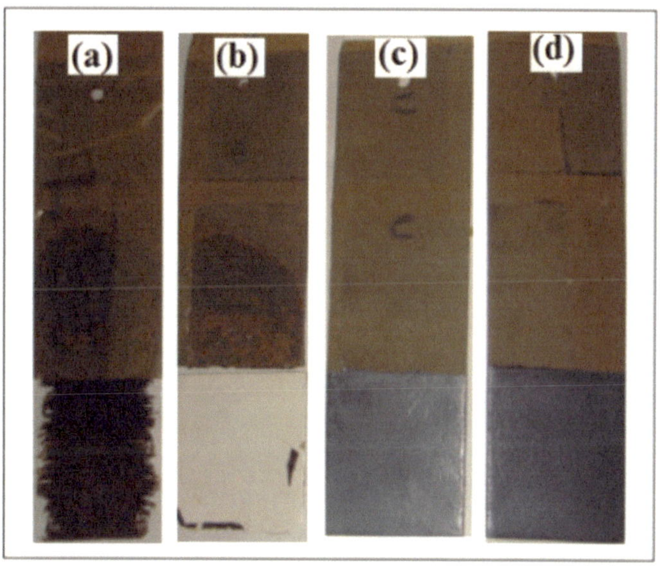

Figure 13. Photographs of (a) blank mild steel electrode (b) blank epoxy resin coated electrode (c) PANI coated (d) PSC coated electrode after immersion in 1.0 M HCl for 60 days.

8. Characterization of hydrophobic PANI - SiO_2 Nanocomposites (HPSC)

8.1. Electrochemical behaviour

Figure 14 shows the electrochemical growth behavior of aniline and aniline-SiO_2 in 0.2 M PFOA solution. Electrochemical polymeriztion was carried out at 0.9 V on platinum electrode vs Ag/AgCl reference electrode. The polymer film growth was studied by sweeping the potential between -0.20 and 0.9 V on Pt electrode at a scan rate of 20 mV/s.

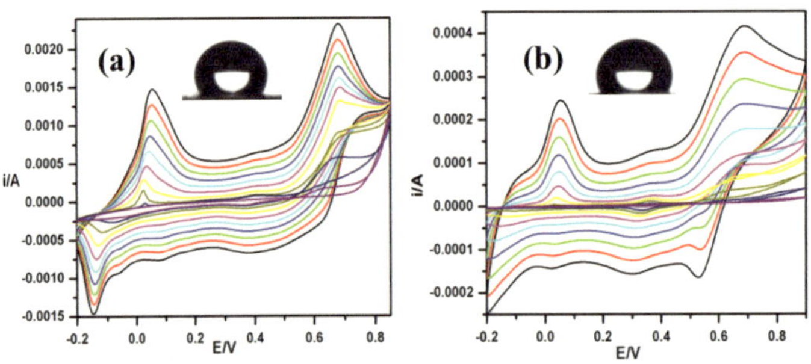

Figure 14. Electrochemical growth behaviour of (a) 0.1M aniline in 0.2M PFOA medium and (b) aniline- SiO_2 in 0.2M PFOA medium in potential range between -0.2 V to 0.9 V vs. Ag/AgCl at scan rate of 20 mV/s.

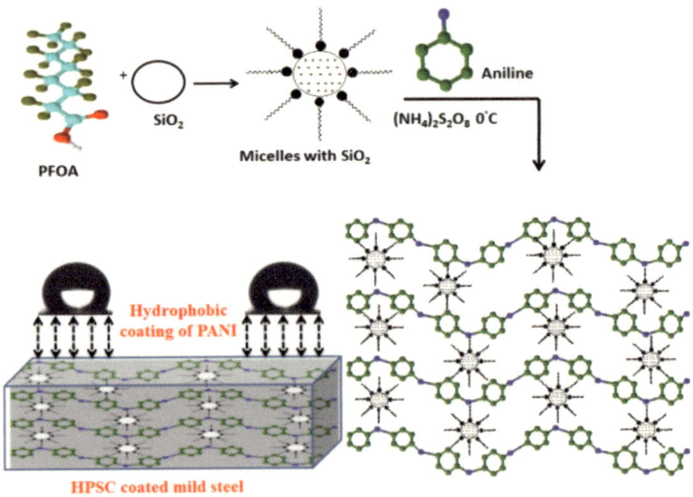

Figure 15. Schematic representation of formation of HPSC coating onto mild steel surface.

Peak potential values of the corresponding PANI and HPSC were recorded in PFOA medium. First anodic peak (oxidation peak) corresponds to the oxidation of monomer. The intensity of this peak gradually decreases with subsequent scans. During the first reverse sweep, a reduction peak appears which shows that the formation of oligomers and polymer on electrode surface as shown in Figure 14. Figure 15 shows the Schematic representation of formation of HPSC coating onto mild steel surface. After the first scan, well defined oxidation and reduction peaks of polymers between 0.2 and 0.6 V vs. Ag/AgCl appeared. The current values of each oxidation and reduction peaks are greater than that of a previous cycle which indicate the built up of an electroactive polymeric material on the electrode surface. Moreover, it was observed that current value of PANI film was found to be higher than that of HPSC film which revealed higher conductivity of PANI as compare to HPSC coating on electrode surface.

Cyclic voltammogram of HPSC and PANI in PFOA medium indicates that the first peak potential value of PANI in PFOA medium lies at 0.15 V. Incorporation of SiO_2 particle in PANI, the first peak potential value shifted from 0.15 V to 0.041 V vs Ag/AgCl as shown in Figure 16.

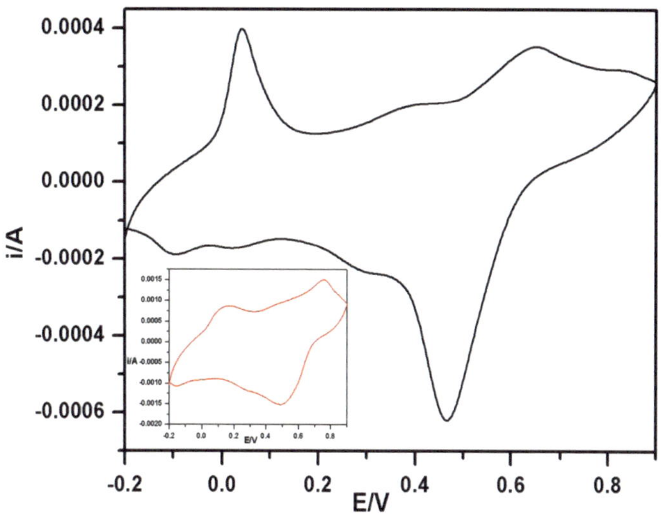

Figure 16. Cyclic voltammogram of HPSC and inset figure shows the cyclic voltammogram of PANI in PFOA medium at a scan rate of 20 mV/s.

This implies that the polymerization of aniline leads to larger peak potential shift as compared to the aniline in presence of SiO_2 which indicates the presence of SiO_2 particles in the polymer chain induce some change in configurations along the polymer backbone which is responsible for the negative shift in the oxidation potential.

8.2. FTIR spectra

The Figure 17 shows the FTIR spectra of SiO_2, PANI and HPSC. PANI showed the main characteristics bands at 1554 and 1438 -1440 cm^{-1} attributed to the stretching mode of C=N and C=C, the bands at 1250 cm^{-1} indicating the C–N stretching mode of benzenoid ring.

The FTIR spectra of SiO_2 indicated that the characteristic peak at 1081 cm^{-1} and 807 cm^{-1} are assigned to the stretching and bending vibration of Si–O–Si respectively. These peaks were also observed in HPSC indicating the interaction of SiO_2 particles in polyaniline chain. HPSC and PANI showed a characteristic strong peak at 1738 cm^{-1} due to C=O stretching mode and peak at 1365 cm^{-1}due to C-F stretching mode [61], which indicates the interaction of PFOA dopant in the polymer chain.

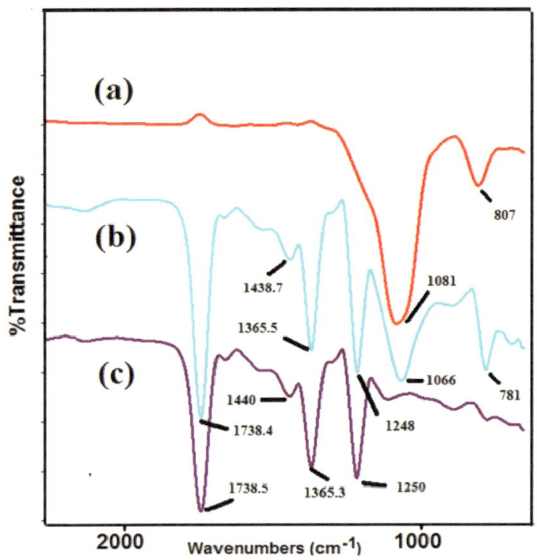

Figure 17. FTIR spectra of (a) SiO_2 (b) PANI and (c) HPSC.

8.3. Wettability of HPSC coating

The surface wettability was measured by static contact angle measurements with water (γ=72.8 mN/m) to determine surface hydrophobicity. The drop volume used for the measurements was 2.0 μL. The PANI-SiO_2 nanocomposite (HPSC) coated electrodes exhibited hydrophobic properties with static water contact angle of about 115° as shown in Figure 14.

9. Anticorrosive properties of HPSC coated mild steel in 3.5 % NaCl solution

9.1. Tafel Extrapoaltion method

Tafel polarization behaviour of mild steel in 3.5 % NaCl solution with uncoated, epoxy coated, PANI and HPSC coated mild steel are shown in the Figure 18& Figure 19. The corrosion kinetic parameters derived from these curves are given in the Table 3. As shown in Table 3. HPSC coated mild steel sample showed a remarkable current density shift from 106.5

μA/cm² to 4.36 μA /cm² versus Ag/AgCl in the corrosion current (I_{corr}), relative to the value of the uncoated mild steel.

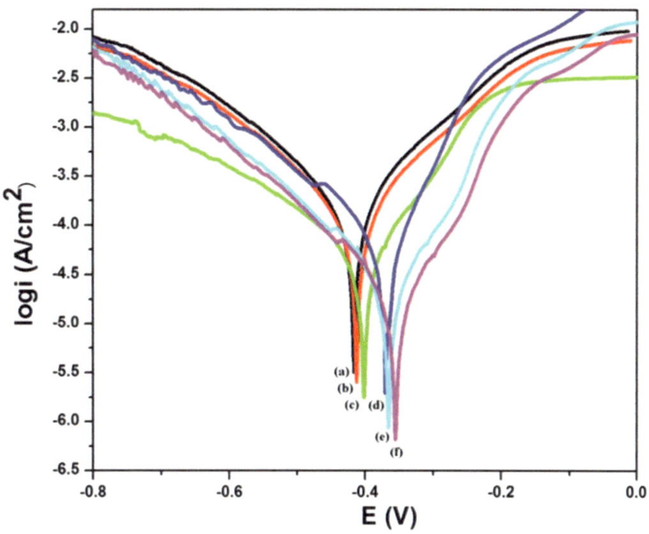

Figure 18. Tafel curves of uncoated and polymer coated mild steel electrode in 3.5 % NaCl solution (a) blank electrode (b) epoxy coated mild steel (c) PANI coated mild steel at 1.5 wt. % loading (d) 3.0 wt. % loading (e) 4.5 wt.% loading and (f) 6.0 wt. % loading of PANI in epoxy resin.

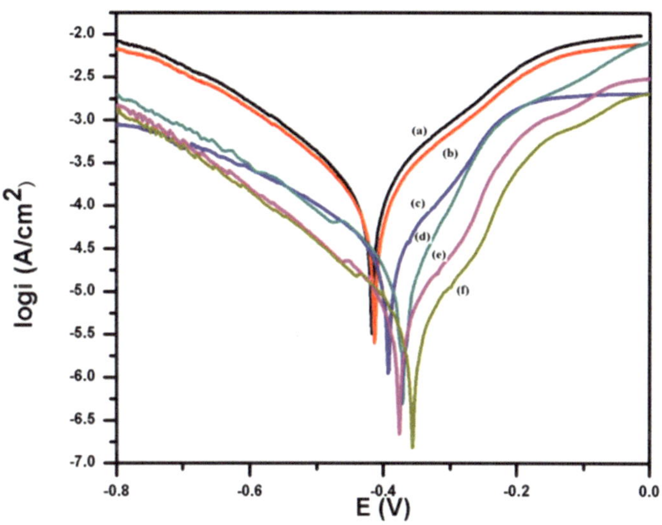

Figure 19. Tafel curves of (a) blank mild steel (b) epoxy coated mild steel (c) HPSC coated mild steel at 1.5 wt. % loading (d) 3.0 wt. % loading (e) 4.5 wt.% loading (f) 6.0 wt. % in 3.5 % NaCl solution.

The significant reduction in the corrosion current density (i_{corr}) in polymer coated mild steel indicated the effective corrosion protection performance of these polymers. The corrosion current values (i_{corr}) found to be decreased from 106.5 μA/cm² for uncoated mild steel sample to 98 μA/cm² for epoxy coated mild steel sample to 4.36 μA /cm² for HPSC coated mild steel samples. The corrosion current values (i_{corr}) decreased with increasing the concentration of HPSC in epoxy resin, i_{corr} value decreased from 32.6 μA/cm² at 1.5 wt. % to 4.36 μA/cm² at 6.0 wt. % loading of conducting material in epoxy resin.

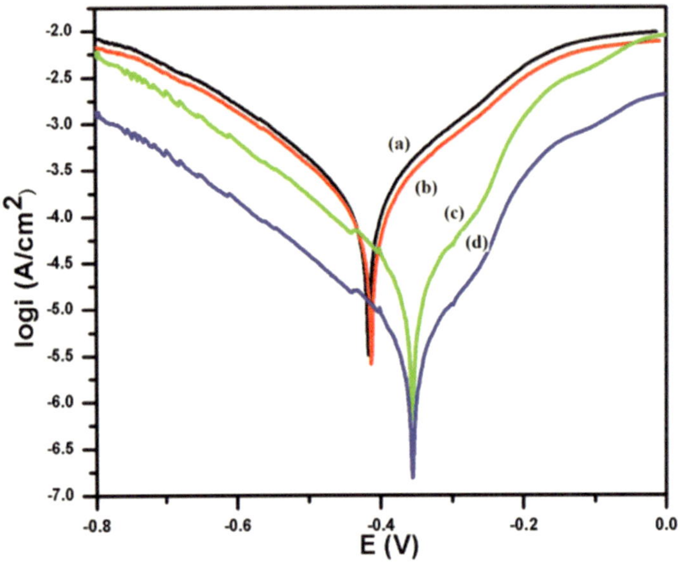

Figure 20. Tafel curves of (a) blank electrode (b) epoxy coated (c) PANI coated mild steel at 6.0 wt. % loading and (d) HPSC coated mild steel at 6.0 wt. % loading in 3.5 % NaCl solution.

Sample name	Loading level of polymer (%)	I_{corr} (μA/cm²)	Corrosion rate (mm/year)	Protection efficiency (%)
Blank mild steel	-	106.5	1.51	0
Epoxy coated mild steel	0	98.0	1.39	7.9
PANI	1.5	64.5	0.91	39.43
	3.0	61.6	0.87	42.16
	4.5	30.2	0.42	71.6
	6	20.4	0.28	80.84
HPSC	1.5	32.6	0.45	69.39
	3.0	12.8	0.17	87.98
	4.5	6.16	0.08	94.21
	6	4.36	0.06	96.0

Table 3. Tafel parameters for corrosion of mild steel in 3.5% NaCl with different loading level of PANI & HPSC in epoxy resin.

While PANI coated mild steel showed the i_{corr} in the range of 20.4 μA/cm² at 6.0 % loading. The corrosion protection efficiency calculated from Tafel parameter revealed that PANI coated mild steel showed the protection efficiency 39.4 % at 1.5 wt.% loading while HPSC coated mild steel showed 69.4 % P.E at the same loading level.

Up to 96 % protection efficiency have been achieved by using 6.0 wt. % HPSC in epoxy resin. While in case of PANI, only 80.84% protection has been achieved at 6.0 wt. % loading, as shown in Figure 20.

9.2. Weight loss method

Table 4 shows the values of the weight loss from mild steel samples during the immersion test. The results revealed that HPSC coated samples were more protectable to mild steel than that of only PANI coated samples in same immersion time. After the immersion of coated and uncoated samples in 3.5% NaCl solution for 60 days, it was observed that uncoated and epoxy coated samples showed the maximum weight loss of 34.18 % and 29.11 % respectively

Sample name	Loading level of polymer (%)	Initial Weight (before immersion) (mg)	Final Weight After immersion (mg)	Weight loss (mg)	Weight loss (%)	C.R (mm/ year)	P.E (%)
Blank mild steel	-	32632.1	21478.45	11153.7	34.18	4.18	0
Blank epoxy	0	30918.7	21918.27	9000.4	29.11	3.56	14.83
PANI	1.5	32798.2	24913.5	7884.7	24.04	2.94	29.67
	3.0	30416.2	23502.6	6913.6	22.73	2.78	33.49
	4.5	32678.1	26642.5	6035.6	18.47	2.26	45.93
	6	31567.2	28369.4	3197.8	10.13	1.24	70.33
HPSC	1.5	31494.2	25541.8	5952.4	18.9	2.32	44.50
	3.0	30566.5	25935.7	4630.8	15.15	1.86	55.50
	4.5	32929.7	29926.5	3003.2	9.12	1.12	73.20
	6	30804.7	30102.4	702.3	2.28	0.28	93.30

Table 4. Weight loss parameter of uncoated and coated mild steel samples after immersion test in 3.5 % NaCl for 60 days.

PANI coated mild steel showed the weight loss up to 10.13 % at 6.0 wt. % loading while HPSC coated samples at the same loading level showed negligible weight loss (i.e < 3 %) after 60 days of immersion in 3.5 % NaCl medium. Corrosion rate (C.R) in mm/year have also been calculated by weight loss method. It was observed that the corrosion rate was highest for uncoated mild steel in 3.5 % NaCl medium. After 60 days of immersion, the C.R value of uncoated mild steel was found to 4.18 mm/year. Epoxy and PANI coated samples showed C.R. value of 3.56 mm/year and 1.24 mm/year respectively. While in the case of HPSC coated sample in 6 % loading, C.R. value reduced to 0.28 mm/year.

The appearance of the uncoated and HPSC coated mild steel samples after exposure to salt spray fog for 35 days is shown in Figure 21.

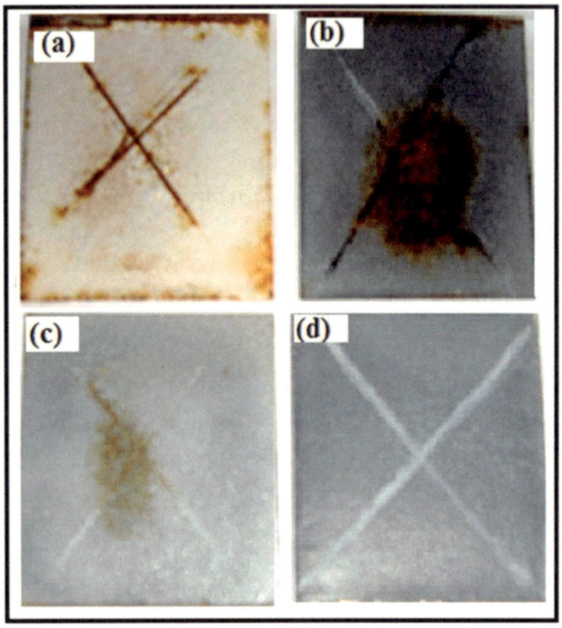

Figure 21. Photograph of (a) epoxy coated (b) PANI (at 6 wt.% loading) coated (c) HPSC (at 1.5 wt. % loading) and (d) HPSC (6.0 wt. % loading mild steel after 35 days of exposure to salt spray test.

It was observed that epoxy coated and PANI coated (at 6.0 wt. % loading) mild steel have more corrosion extended area from the scribes as shown in Figure 21 (a) and 21(b) while HPSC coated mid steel (at 1.0 wt. % loading) showed less corrosion extended area as compared to epoxy and PANI coated mid steel as shown in Figure 21(c). However, HPSC containing coating sample (at 6.0 wt.% loading) were found to be free from rust and blister as shown in Figure 21 (d). Moreover, there was no spreading of rust along the scribed areas.

9.3. SEM studies of uncoated and coated mild steel before and after immersion test

It was observed that SiO_2 particles showed spherical shaped morphology as shown in Figure 10 a. The scanning electron micrographs of PANI and HPSC are shown in Figure 22. PANI doped with PFOA showed uniform net like morphology as shown in Figure 22(a). Morphology of HPSC was entirely different with incorporation of SiO_2 particles in PANI matrix during polymerisation. SEM image of HPSC revealed that the entrapment of SiO_2 particles in the globular space of PANI matrix during in-situ polymerisation of polyaniline as shown in Figure 22 (b).

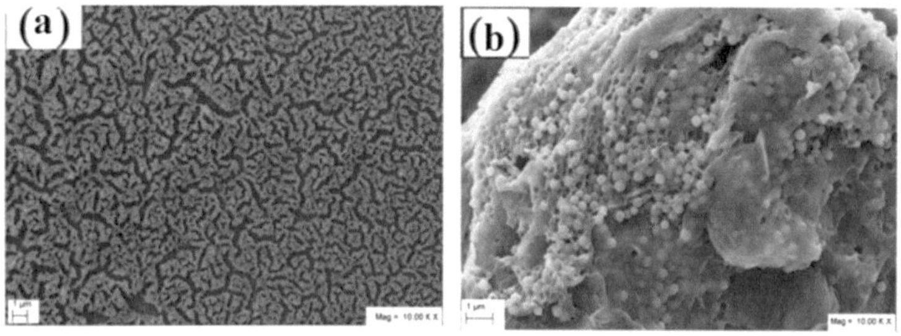

Figure 22. SEM micrographs of powder sample of (a) PFOA doped PANI, and (b) HPSC.

SEM images of uncoated, epoxy coated and polymer coated samples before and after the immersion test of 60 days have been shown in Figure 23 and 24 respectively. These images clearly show the formation of large pits on the surface of mild steel after immersion. These pits and cracks were developed during the corrosion of mild steel in NaCl medium.

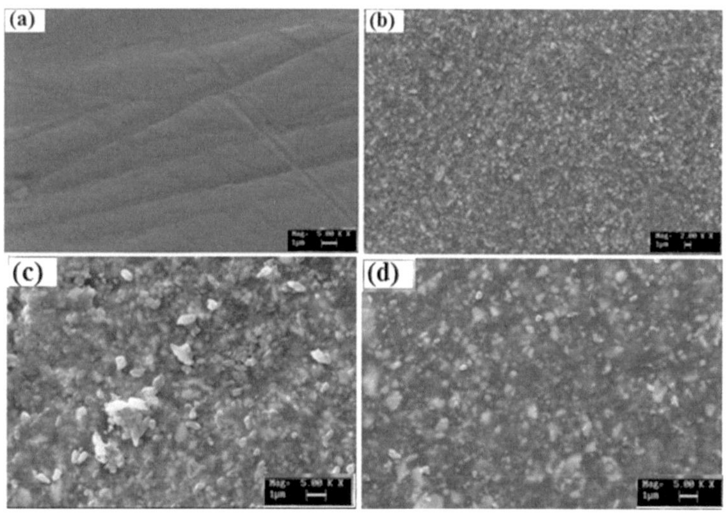

Figure 23. SEM micrographs of mid steel electrode (a) blank (b) epoxy coated (c) PANI coated (d) HPSC coated before immersion.

When epoxy coated mild steel sample was immersed in the acidic medium for 60 days, detachment of coating from mild steel substrate have been observed. The pits were also appeared on the metal surface as shown in Figure 24. In the case PANI coated sample, few pits still appeared on mild steel surface. While, HPSC coated mild steel samples did not show any cracks and pits on the metal surface. No detachment of coating from mild steel substrate was also observed after the immersion of these samples in 3.5 % NaCl medium for 60 days of immersion indicating strong adherence of HPSC composite to the mild steel substrate and it is resistant to corrosion in aqueous 3.5 % NaCl medium.

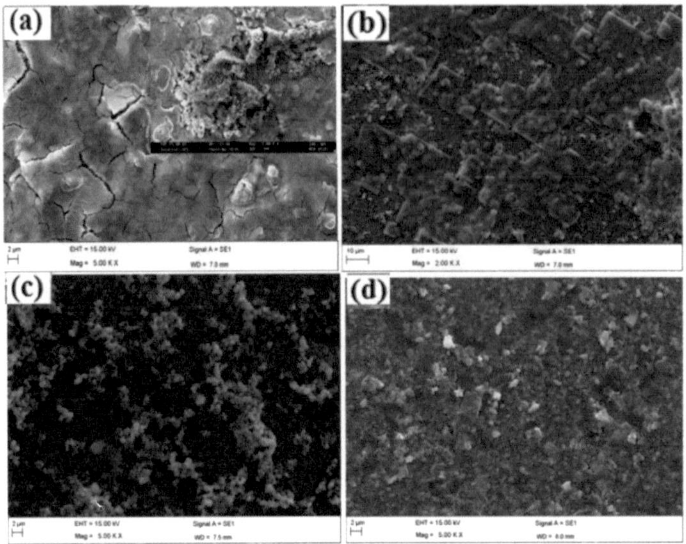

Figure 24. SEM micrographs of mid steel electrode (a) blank (b) epoxy coated (c) PANI coated (d) HPSC coated after immersion.

It was found that the HPSC content has a great influence on the anticorrosive performance of the coating. The corrosion protection effect of HPSC coated mild steel sample improved slowly when HPSC content in epoxy formulation increases from 1.5 to 3.0 wt. % Afterward, an excellent corrosion protection effect appears at 6.0 wt. % loading of HPSC content in epoxy resin.

9.4. Mechanism of corrosion protection of PANI-SiO_2 nanocomposites

The corrosion studies show that the PANI-SiO_2 nanocomposites containing coating showed better corrosion protection as compared to PANI coating which may be due to the redox property and uniform distribution of PANI in the coating containing PANI-SiO_2 nano-composites. Earlier studies [63-64] have shown that the redox property of PANI coating on metal surface plays an important role to protect the metal by passivating the pin holes. Corrosion protection of metals occurs via reduction of PANI–Emeraldine salt (PANI–ES) to PANI–Leucosalt (PANI–LS) with the concomitant release of phosphate dopant [19]. Phosphate ions help to form passive film on mild steel at the defect. PANI–LS is assumed to undergo a subsequent re-oxidation by dissolved oxygen to PANI–ES. Due to this cyclic reaction, the coating containing PANI is able to offer higher corrosion protection. However, in case of PANI-SiO_2 nano composites containing coating, these composites have a dual protection mechanism; forming a passive layer and simultaneously acting as a physical barrier to avoid chloride ion penetration. Moreover, it acts as a barrier between metal surface and corrosive environment. Entrance of water and corrosive ions on the metal surface causes the defects in the paint coating and therefore the protective property of the coating is decreased. Due to uniform distribution of PANI, the possibility of forming uniform passive layer on the mild steel surface is more since PANI has been shown to protect the mild steel surface by passive film formation. Furthermore, powder coating technique also plays an important role for achieving high quality, durable and good anticorrosive coatings.

Corrosion protection property of these coating may also be attributed to the PSC/HPSC content in epoxy resin which can react with epoxy to form highly adherent, dense and non porous polymer film on the mild steel surface. On the other hand, presence of SiO_2 nanoparticles entrapped in PANI chain provide the reinforcement to PANI chain which reduce the degradation of polymer chain in corrosive condition.

10. Antistatic performance of the conducting polymers nanocomposites based on nanotubes of poly (aniline-co-1-amino-2-naphthol-4-sulphonic acid)/LDPE composites

10.1. Introduction

Electrostatic charge dissipation has become an important issue within the electronic components such as data storage devices, chips carriers and computer internals. Antistatic protection is also required for parts where relative motion between dissimilar materials occurs [65] like weaving machine arms, airplane tyres etc. Conventional polymers commonly being used for packaging of various electronic equipments but due to their inherent electrical insulating nature, these polymers failed to dissipate the static or electrostatic charge. The generation of static electricity on the materials leads to a variety of problems in manufacturing and consumer use. Moreover, electronic components are susceptible to damage from electrostatic discharge. Thus the challenge is to convert inherently insulating thermoplastic to a product that would provide an effective antistatic material. Various attempts have been made to achieve the antistatic polymers with retained mechanical properties such as addition of antistatic agents [66], conducting additives [67] and fillers like carbon powder [68] and carbon nanotubes [69] etc. The electrical conductivity of the polymeric material depends on the amount, type and shape of the conducting filler [70]. According to electronic industries association (EIA) standards, in ESD protected environments, the optimal surface conductivity should be in the range of 10^{-6} to 10^{-10} S/cm. However the functioning of antistatic agents is critically dependent on the relative humidity [71] whereas the metal and carbon filled materials suffer from the problems like bleeding and poor dispersion [72]. Moreover, it has been observed that carbon black loaded static controlling materials usually contain 15-20 % carbon black. The addition of carbon black at higher concentrations showed a negative effect on the proccessability of the compound and mechanical properties such as increase in melt viscosity and decrease in impact resistance. Use of conducting blends and composites with conventional polymers as an electrostatic charge dissipative material is one of the promising application of conducting polymers which combines the mechanical properties of conventional polymers and electrical properties of conducting polymers. Polyaniline is one of the most promising intrinsically conducting polymer (ICPs) because of its good environmental stability and high electrical conductivity, which can be reversibly controlled by a change in the oxidation state and protonation of the imine nitrogen groups. Blending of polyaniline with conventional polymers like polypropylene, ABS, LDPE etc. can also be used to improve the proccessability of polyaniline creating new materials with specific properties for the desired application at low cost that can also be used for different applications like electromagnetic shielding and corrosion prevention where conductivity, proccessability and mechanical properties of the materials are of the primary importance. Hence desired properties of conducting polymers can be enhanced by mixing it with a polymer that has good mechanical properties and the unique combination of electrical and mechanical properties of conducting copolymers blends with insulating polymers seems to have great potential for their use in many applications [73-75].

11. Antistatic measurements

Antistatic or electrostatic charge dissipative performances of blends of conducting polymers were measured by Static decay meter, John Chubb Instrument and Static Charge Meter. The detailed method is given below.

11.1. Static decay meter

Static decay meter is very useful device for measurement of static decay time of the conducting polymer blends in the form of injection moulded sheets and blown film. The samples of conducting copolymer blends (LDPE/conducting copolymer) were cut in to the 15 x 15 cm^2 blown film and was used for measurement of static decay time on Static Decay Meter by measuring the time on applying a positive voltage of 5000 V and recording the decay time on going down to 500 V. Similarly the static decay time was measured by applying a negative voltage of 5000 V. Here, these measurements were carried out on Static Decay Meter; model 406D, Electro-tech System, Inc., USA. The model 406D Static Decay Meter is designed to test the static dissipative characteristics of material by measuring the time required for charge test sample to discharge to a known, predetermined cut-off level. Three manually selected cut-off threshold at 50%, 10% and 1% of full charge are provided and samples are charged by an adjustable 0 to ±5kV high voltage power supply.

11.2. John Chubb Instrument

John Chubb Instrument (JCI 155 v5) charge decay test unit is a compact instrument for easy and direct measurement of the ability of materials to dissipate static electricity and to assess whether significant voltage will arise from practical amount of charge transferred to surface [76]. The JCI 176 Charge Measuring Sample Support (connected with JCI 155 v5) provides a convenient unit to support film and layer materials (and also powder and liquids). The samples of conducting copolymer blends (LDPE/conducting copolymers) i.e. 45 x 54 mm^2 blown film were used for measurement of static decay time on John Chubb Instrument (Model JCI 155 v5) by measuring the time on applying the positive as well as negative high corona voltage of 5000 V on the surface of material to be tested and recorded the decay time at 10 % cut-off. A fast response electrostatic field meter observes the voltage received on the surface of sample and measurements were to observe how quickly the voltage falls as the charge is dissipated from the film. The basic arrangement for measuring the corona charge transferred to the test sample during corona charge decay measurements is shown in Figure 24. Charge is measured as a combination of two components-'conduction charge' and 'induction charge'. The 'conduction' component is that which couples directly to the sample mounting plates within the time of application of corona charging and the time for the plate carrying the corona discharge points to move away.

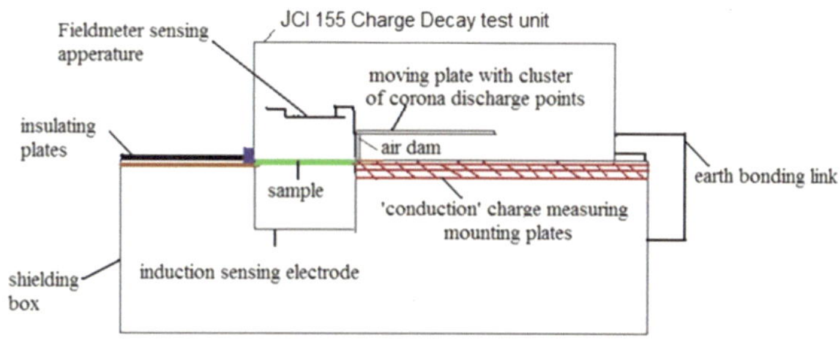

Figure 25. Schematic arrangement of JCI 155 v5 on JCI 176 charge measuring sample support.

The 'inducting' relates to the charge that has been deposited but has not coupled out directly to the mounting plates and the total charge transferred to the sample can be measured as:

$$Q_{total} = Q_{(conduction)} + f * Q_{(induction)} \quad (6)$$

where the factor 'f' is actually close to 2.2. This factor can be determined experimentally. The film and layer polymeric samples are easily mounted in the JCI 176 between the two hinged flat metal plates. The aperture in the sample mounted plates, to which the conduction charge is measured, are 5 mm larger all round than the 45 x 54 mm^2 test aperture of the JCI 155. The JCI 155 Charge Decay Test unit sits on top of the JCI 176 Charge Measuring Sample Support into the recess between the boundary edges. The measurements are recorded in the form of graphs (ESD/ JCI-graphs) which show the decay of surface voltage with respect to decay time.

12. Synthesis of poly(aniline-co-1-amino-2-naphthol-4-sulphonic acid)

Copolymers of 1-amino-2-naphthol-4-sulphonic acid (ANSA) and aniline of varying composition (i.e. by varying the co-monomer feed compositions in the initial feed) were synthesised by chemical oxidative polymerization in the presence of PTSA. Polymerization was initiated by the drop wise addition of ammonium persulphate solution (0.1 M APS in distilled water). The polymerization was carried out at a temperature of 0 C for a period of 4-6 h. Their copolymers were synthesised by varying the molar ratio of co-monomers in the initial feed. The synthesized copolymers were isolated from reaction mixture by filtration and washed with distilled water to remove oxidant and oligomers.

PTSA doped copolymers of aniline and ANSA (poly(AN-co-ANSA) in 80:20 molar ratio and 50:50 molar ratio is abbreviated as PANSA2-PTS and PANSA5-PTS respectively whereas PTSA doped polyaniline is abbreviated as PANI-PTS.

13. Preparation of LDPE-Conducting Copolymer Film

Composites of copolymers with LDPE were prepared by melt blending method. Required amount of LDPE and copolymers were loaded in internal mixer for 20-30 minutes at around 60 rpm. Blending of copolymers with LDPE was carried out in twin-screw extruder at the temperature range from 140-150°C by melt mixing method. The blown film of the copolymer/LDPE composite was made by Haake Blown Film instrument at the temperature range of 160°C where speed of screw was maintained at 40 rpm. PTSA doped copolymers of aniline and ANSA (poly(AN-co-ANSA) in 80:20 and 50:50 molar ratio blended with LDPE is abbreviated as PANSA2-PTS/LDPE and PANSA5-PTS/LDPE respectively.

14. Characterization

14.1. Characterization of PTSA doped PANI and copolymers of AN and ANSA

ANSA is a tri-functional monomer having three functional groups (i.e. $-NH_2$, -OH and $-SO_3H$) along with two fused benzene rings.

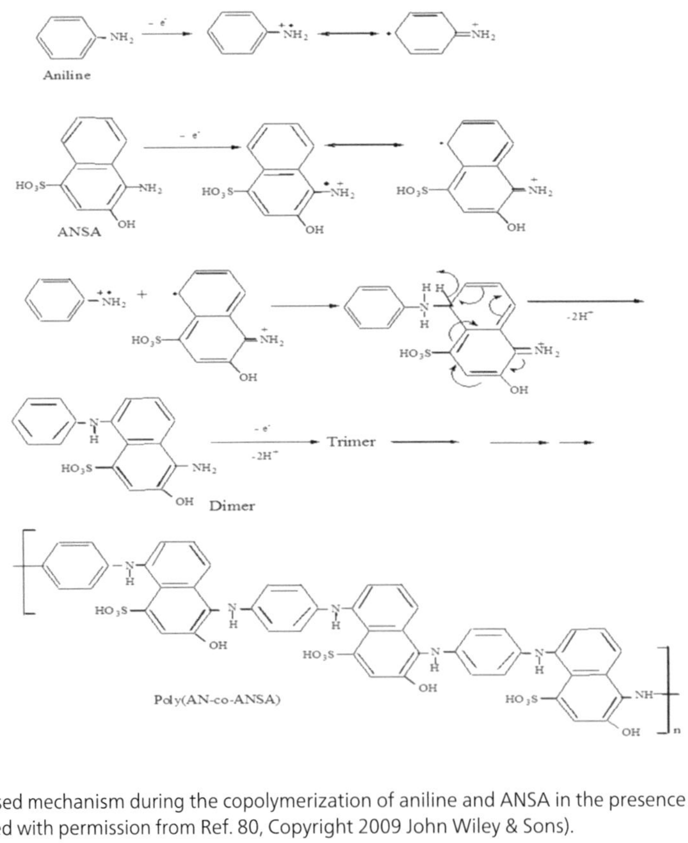

Figure 26. Proposed mechanism during the copolymerization of aniline and ANSA in the presence of p-toluene sulphonate (Reproduced with permission from Ref. 80, Copyright 2009 John Wiley & Sons).

This monomer can be copolymerized with aniline to give different materials and it has been observed that the participation of functional groups (-NH_2 and –OH) in the polymerization depends upon the reaction conditions. It is proposed that polymerization of ANSA in the presence of PTSA occurred selectively through –NH_2 group (figure 26) as confirmed by structural characterization (FTIR and NMR spectroscopy) [77].

14.1.1. Morphological Characterization

Figure 27 shows SEM micrographs of PTSA copolymers of ANSA and AN. PANI-PTS show a globular sponge like structure (Figure 27a) and morphology changed with varying copolymer composition. PTSA doped copolymers of aniline and ANSA exhibit hollow tube like morphology. The use of 1-amino-2-naphthol-4-sulphonic acid as a comonomer as well as nature of external dopant played an important role for achieving the tubular morphology. In case of PTSA doped copolymers of aniline and ANSA in ratio of 80:20 (PANSA2-PTS), the globular morphology of the resultant copolymer tend to change to the tube forms (Figure 27b).

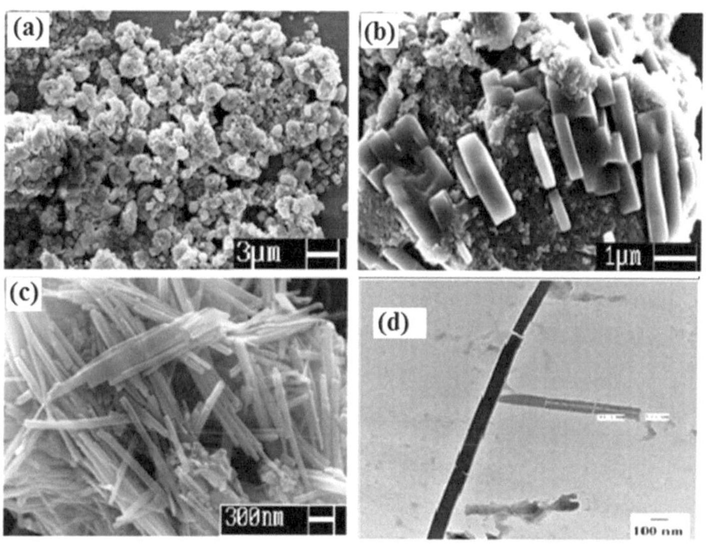

Figure 27. SEM image of (a) PANI-PTS (b) PANSA2-PTS; (c) PANSA5-PTS and (d) TEM image of PANSA5-PTS (Reproduced with permission from Ref. 77, Copyright 2009 John Wiley & Sons).

However, well defined tubes were formed when molar ratio of aniline/ANSA was 50:50 in the presence of PTSA (Figure 27c). The difference in the morphology between polyaniline and its copolymers with ANSA may be related to the different reactivities of the two monomers, nature of reaction media and reaction route.

TEM image of PTSA doped copolymer of aniline and ANSA in 50:50 molar ratios (Figure 26d) shows that these tubes are hollow with outer diameter of 80-90 nm.

14.1.2. Conductivity

Room temperature conductivity values of PTSA doped samples are summarised in Table 5, which reveals that the room temperature conductivity of PANI-PTS was found to be better than PTSA doped copolymers.

Sample Designation	Room temperature conductivity (S/cm)	Thermal stability (°C)
PANI-PTS	1.72	200
PANSA2-PTS	4.48×10^{-1}	195
PANSA5-PTS	1.98×10^{-2}	188

Table 5. Room temperature conductivity and thermal stability.

On increasing the molar ratio of ANSA in copolymer, conductivity tends to decrease accordingly due to the presence of three functional groups in ANSA unit which exerted a strong steric effect on the doping process hence induces additional deformation along the polymer backbone.

14.2. Characterization of LDPE/Conducting Polymer composite film

14.2.1. Thermal Properties

These copolymers (PANSA2-PTS/PANSA5-PTS) can be melt blended with conventional polymers like LDPE. Figure 28 shows the TG traces of blown films of PTSA doped copolymer/LDPE composites. The degradation temperature of pure LDPE blown film was around 400°C. Thermal stability of the blown film of copolymer/LDPE blends (0.5-1.0 wt % loading) was also found to be same as LDPE.

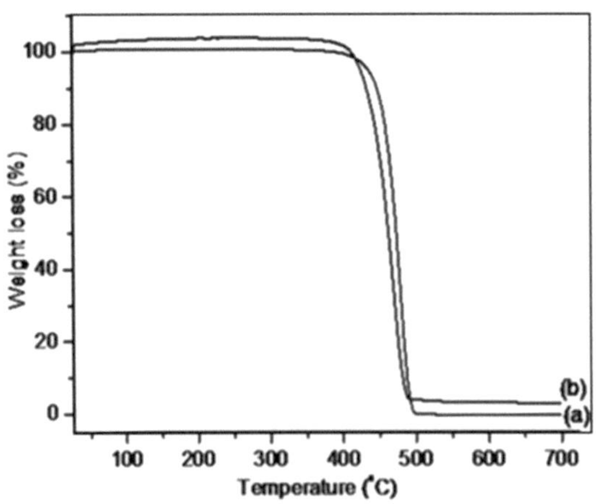

Figure 28. TG traces of (a) LDPE/PANSA2 -PTS and (b) LDPE/PANSA5-PTS films at 1.0 wt. % loading.

http://dx.doi.org/10.5772/50470

14.2.2. Mechanical properties

The mechanical properties of PTSA doped poly (AN-co-ANSA)/LDPE film was measured and the results are summarised in the Table 6. In case of pure LDPE film, the tensile modulus and yield stress were 141 MPa and 16.3 MPa respectively. However, the inclusion of poly (AN-co-ANSA) in LDPE led to decrease in both tensile modulus and yield stress depending upon the molar ratio of ANSA in the copolymer chain as well as type of dopant.

Sample Designation	Loading of copolymers In LDPE (wt. %)	Tensile modulus (MPa)	Yield Stress (MPa)	Ultimate elongation (%)	Conductivity of LDPE/ copolymers film (S/cm)
LDPE	1.0	141	16.3	187	"/10^{-12}
LDPE/PANSA2-PTS	1.0	129	12.1	176	1.28×10^{-6}
	0.5	134	13.1	180	2.22×10^{-9}
LDPE/PANSA5-PTS nanocomposites	1.0	120	10.3	166	8.18×10^{-7}
	0.5	131	12.3	171	4.13×10^{-9}

Table 6. Mechanical and electrical properties of LDPE films in the absence/presence of conducting copolymers.

In the case of film prepared by composites of LDPE/conducting copolymer (99/1 w/w or 99.5/0.5 w/w), tensile modulus, yield stress and % elongation decreased (Table 6). Tensile modulus decreased from 141 MPa (LDPE) to 120 MPa (LDPE/PANSA5-PTS) at a concentration of 1.0 % w/w. Yield stress also decreased for 16.3 MPa (LDPE) 10.3 MPa (LDPE/PANSA5-PTS). Similarly, the ultimate elongation also decreased in the same manner.

At a loading of 0.5 % (w/w) of PTSA doped copolymers in LDPE, tensile modulus also decreased from 134 MPa (for PANSA2-PTS/LDPE) to 120 MPa (for PANSA5/LDPE). Similarly, 0.5 wt. % loading of PTSA doped copolymers with LDPE, yield stress reduced from 13.1 MPa in case of PANSA2-PTS/LDPE to 12.3 MPa for PANSA5-PTS/LDPE film and the ultimate elongation was also found to be 180 % and 171 % for PANSA2-PTS/LDPE and PANSA5-PTS/LDPE composite films respectively. Hence, the mechanical strength of PTSA doped poly(AN-co-ANSA)/LDPE blended films was found to be better in case 0.5 wt.% loading of copolymers than that of 1.0 wt. % loading. Moreover, it has also been observed that mechanical properties of PTSA doped copolymers-LDPE film were different from that of self doped copolymers-LDPE films. Mechanical strength of the poly(AN-co-ANSA)/LDPE composites decreased with increasing the molar ratio of ANSA in the copolymer (Table 6).

14.2.3. Electrical Properties

Room temperature conductivity values of PTSA doped copolymers/LDPE composite film are summarised in Table 6. The room temperature conductivity of copolymers of aniline with ANSA decreased from 4.48×10^{-1} to 1.98×10^{-2} S/cm depending on the molar ratio of

ANSA in the copolymer feed and type of dopant. The conductivity values copolymers were found to be 4.48 x 10^{-1} S/cm and 1.98 x10^{-2} S/cm for PANSA2-PTS and PANSA5-PTS respectively (Table 5). On blending with LDPE at 1.0 wt %, conductivity value decreased from 1.28 x 10^{-6} S/cm to 8.18 x 10^{-7} S/cm respectively. When the loading level of copolymers with LDPE reduced from 1.0 wt % to 0.5 wt %, the conductivity of the resultant composites decreased. 0.5 % (w/w) loading of LDPE films based on PANSA2-PTS and PANSA5-PTS had conductivity value in the order of 2.22 x 10^{-9} S/cm and 4.13 x 10^{-9} S/cm respectively.

14.2.4. Morphological Characterization

Figure 29 show the SEM images of LDPE films in the presence of PTSA doped copolymers at 0.5 wt. % loading. When these copolymers were blended with LDPE, the copolymer domains were found to disperse in the LDPE matrix as evident by the appearance of tubes and needle like granules in the LDPE matrix (Figure 29). In addition, the formation of the conducting path is evident and agrees with the results relating to electrical conductivity of the composites. In copolymer composites (matrix and dispersed phase), the level of interaction between the two components and mode of dispersion in the matrix, influence the electrical and mechanical properties of the composites [79]. The SEM micrographs of the LDPE/copolymer film showed two different phases i.e. conducting copolymer and non conducting matrix (LPDE). Interconnection of conducting phase in the non-conducting matrix creates a conducting path along the LDPE matrix.

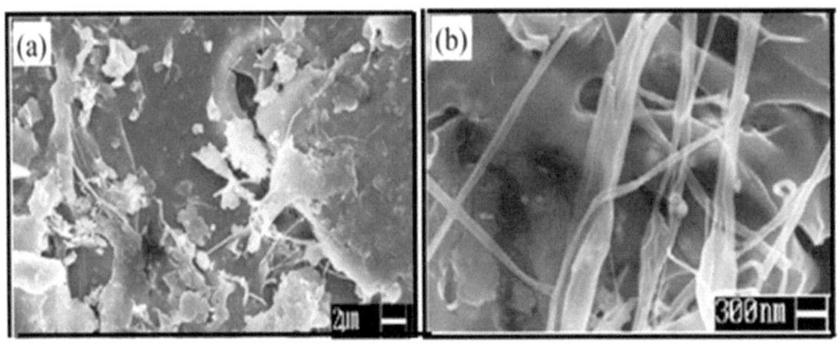

Figure 29. SEM images of (a) PANSA2-PTS/LDPE, (b) PANSA5-PTS/LDPE composite films at 0.5% (w/w) loading (Reproduced with permission from Ref. 77, Copyright 2009 John Wiley & Sons).

Moreover, it has also been observed that the conduction mechanism and transportation of charge carrier in the blends depend on the loading level and mode of dispersion of the conducting materials. PANSA5-PTS nanotubes at 0.5 wt. % loading with LDPE, the surface conductivity was found to be in the order of 10^{-9} S/cm, which is suitable for their use in ESD protection applications.

Hence, it may be presumed that when the sufficient amount of conducting material is loaded in the polymer matrix, the conducting particles get closer and form linkage which makes an easy path for conduction of charge carrier throughout the blend which shows sufficient loading and good dispersion of conducting material in the polymer matrix (i.e. LDPE). While in the case of very low loading of conducting material in the polymer matrix, the gap between conducting particles in the polymer matrix is large with the result that no conduction path in the blend.

Hence, the conductivity of films based on blends depends on the morphology of conducting material. The nanotubular or fibre like morphology of conducting materials which form a network in the whole blend facilitate the conduction of charge carrier through the continuous structure of the chain of conducting material in the insulating matrix at very low loading of conducting material in the insulating matrix.

15. Antistatic Behaviour of LDPE/Copolymer Film

The results of static decay time on application of positive/negative voltage of 5000 V and recording the decay time on going down to 500 V are summarised in Table 7. It was observed that blank LDPE film shows a static decay time of 120.1 sec. It decreased upon addition of copolymer and was found to be dependent on the amount of copolymer. LDPE film having 1.0 % (w/w) and 0.5 % (w/w) of PANSA2-PTS showed a decay time of 0.1 sec. and 1.4 sec. respectively at 10 % cut-off. However, the PANSA5-PTS/LDPE film showed a static decay time of 0.8 sec. at a loading of 0.5 wt. % and 0.2 sec. at 1.0 wt.% loading [80]. Any material which showed a static decay time less than 2.0 sec passes the criteria for its use as antistatic material. Based on the above observations, we can say that LDPE film prepared by blending of conducting copolymer based on AN and ANSA at 1.0 % w/w loading, can be used as an effective antistatic film. Similar measurements were recorded with copolymer composite film with a cut-off value of 50 % and the results are summarised in Table 7.

Sample Designation	Loading of copolymers in LDPE (wt. %)	Static decay time (at 10 % Cut off) (Sec.)		Static decay time (at 50 % Cut off) (Sec.)	
		Positive voltage	Negative voltage	Positive voltage	Negative voltage
Blank LDPE	--	120.1	110.8	94.9	93.1
LDPE/PANSA2-PTS	1.0	0.1	0.1	0.01	0.01
	0.5	1.4	1.5	0.2	0.3
LDPE/PANSA5-PTS nanocomposites	1.0	0.2	0.1	0.01	0.01
	0.5	0.8	0.9	0.1	0.1

Table 7. Antistatic behaviour of LDPE/copolymer composite films.

Static decay measurements were also performed on John Chubb Instrument (JCI 155 v5) charge decay test unit by measuring the time on applying the positive as well as negative high corona voltage of 5000 V on the surface of material to be tested and recorded the decay time at 10 % cut off. A fast response electrostatic field meter observes the voltage received on the surface of sample and measurements were to observe how quickly the voltage falls as the charge is dissipated from the film. Graphs obtained from these experiments have been shown in the Figure 30, which show the decay of surface voltage and decay time.

The surface voltage and surface charge received by the materials depend on nature of materials. When positive or negative high corona voltage (i.e. 5000 V) was applied to the surface of the material, only a limited amount of voltage was received by the blend depending on the nature of materials. When high corona voltage was applied on the surface of insulating material, only some voltage was drained away and greater amount of voltage were retained on its surface. This surface voltage decays at particular time. Moreover, the surface charge received by the blends was also calculated during the experiment.

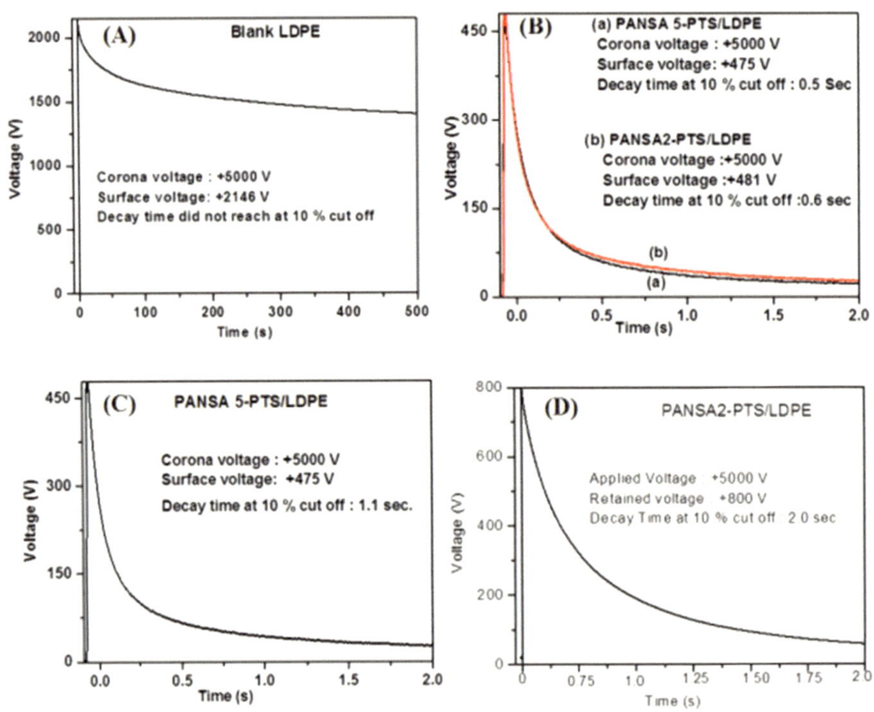

Figure 30. ESD-graphs of LDPE film in the presence of AN-ANSA copolymers (A) blank LDPE film, (B) Copolymer/LDPE nanocomposites film at 1.0 wt.% loading, (C) PANSA5-PTS at 0.5 wt. % loading and (D) PANSA2-PTS at 0.5 wt. % loading (Reproduced with permission from Ref. 77, Copyright 2009 John Wiley & Sons).

Hence the charge retention capability of conducting materials was found to be very low thus they quickly dissipate this surface charge. The static decay time of blank LDPE film was found to be very high on applying the positive and negative corona voltage of 5000 V. The peak at 2146 V indicate that the LDPE film has received 2146 V at the surface corresponding to 55.79 nC of static charge, which get dissipated very slowly and was not found to be able to dissipate it up to 10 % cut off as shown in the Figure 30A. Due to insulating nature of the material, lot of charges were found to be retained on the surface of LDPE film. Blending of 1.0 wt. % of conducting copolymer with LDPE, decreases the charge retention capability by reducing the decay time. In case of PANSA5-PTS/ LDPE film at 1.0 wt. % loading of PANSA 5-PTS in LDPE, the peak started at 475 V on applying the voltage of +5000 V, which indicates that the voltage received at the surface is only 475 V corresponding to 9.51 nC of charges which dissipated quickly, around 0.5 sec., at 10% cut off as shown in the Figure 30B (curve a).

Similar trend has been found for negative polarity charging at the same corona voltage. In the case of film samples prepared by blending of 0.5 wt. % PANSA5-PTS with LDPE, 475 V of voltage and 10.45 nC of charges were received by its surface which was dissipated up to 10 % cut off in 1.1 sec (Figure 30C). on the other hand, LDPE + 0.5 wt. % PANSA2-PTS showed + 800 V of surface voltage received by the composite on applying the voltage of +5000 V which showed the large decay time (2.0 sec at 10 % cut off). Similar behaviour was observed at negative polarity charging. Hence the ESD protection performance of the conducting blends not only depends on the loading level of conducting materials but also depend on the morphology and dispersion of conducting materials in the polymer matrix. Nanocomposites based on LDPE/PANSA5-PTS film showed better ESD performance as compared to LDPE/PANSA2-PTS film.

16. Conclusions

PANI/SiO_2 nanocomposites were prepared by chemical oxidation polymerization of aniline and SiO_2 by using ammonium persulfate (APS) as an oxidant in the presence of phosphoric acid/PFOA medium. FTIR, UV-Visible, cyclic voltammetry and SEM techniques confirmed the interaction of PANI with SiO_2 particles. The excellent corrosion protection performance by PSC coated mild steel could be due to the strong adherence of polymer film which uniformly covers the entire electrode surface as has shown by the surface morphology. The corrosion current densities were lowered several orders of magnitude with these coatings. The coating had good protective efficiency which increased with increasing the loading of PSC to the maximum of 99 % at 6.0 wt.% loading and reduced to about 89.93 % after 60 days of immersion in highly corrosive environment confirming the improved coating performance. Weight loss method also revealed that PSC coated samples showed very low weight loss as well as negligible corrosion rate as compared to PANI coated samples at same immersion time, which indicates the better protection and adhesion of PSC onto the mild steel surface as compared to PANI in strong acidic condition.

In order to improve anticorrosion performance of iron in 3.5 % NaCl aqueous medium, preparation of highly hydrophobic polyaniline-SiO_2 nano-composites (HPSC) have also been carried out by chemical oxidation polymerization. Water repellent property of the PSC has been developed by using fluorinated dopant i.e. perfluoro-octanoic acid (PFOA). HPSC coating were evaluated for protection of mild steel from corrosion in 3.5 % NaCl aqueous solution. Suitable coating with HPSC was formed on mild steel using epoxy resin by powder coating technique which showed the contact angle in the range of 115°. Corrosion protection efficiency of mild steel coated HPSC in 3.5 % NaCl aqueous solution has been evaluated using Tafel Extrapolation method, surface morphology, salt spray test and weight loss methods. The results reveals that the HPSC coating showed the significant reduction in the corrosion current density reflects the better protection of mild steel in marine environment. The coating had good protective efficiency which increased with increasing the loading of HPSC to the maximum of 96 % at 6.0 wt.% loading and reduced to about 93.3 % after 60 days of immersion in 3.5 % NaCl solution confirming the improved coating performance.

Presence of SiO_2 nanoparticles entrapped in PANI chain which was evident my surface morphology of composite coating, provide the reinforcement to PANI chain which reduce the degradation of polymer chain in corrosive environment. PSC/HPSC coating protect metal by dual mechanism by forming passivating layer as well as act as a physical barrier. Furthermore the role of powder coating technique for achieving high quality, durable and good anticorrosive coatings have also been explained. These studies revealed that the polyaniline-SiO_2 nanocomposites has excellent corrosion protection properties and it can be considered as a potential material for corrosion protection of mild steel in corrosive medium like.1.0 M HCl as well as 3.5 % NaCl solution.

In order to carry out the effective use of conducting polymer for antistatic application, nanocomposites based on poly(aniline -co- 1-amino-2-naphthol-4-sulphonic acid) (PANSA5-PTS) with low density polyethylene (LDPE) have been developed. The copolymer nanotubes of aniline and ANSA were synthesised in tosyl medium in 50: 50 molar ratio. Formation of nanotubes of copolymers was confirmed by morphological characterization using SEM and TEM. Dimension of nanotubes of PANSA5-PTS was found to be 80-90 nm. Blending of copolymers with LDPE was carried out in twin screw extruder by melt blending method by loading 0.5 wt. % and 1.0 wt. % of the conducting copolymer in LDPE matrix. The conductivity of the blown film of poly (AN-co-ANSA) /LDPE composites was found to be in the range of 1.28×10^{-6} to 4.13×10^{-9} S/cm. Thermo gravimetric traces of copolymers reveals that these copolymers were thermally stable from 180°C to 195°C. Such copolymers were successfully melt blended with LDPE and conducting film was prepared using film blending technique. Antistatic performance of PANSA5-PTS/LDPE nanocomposite have compared with PANSA2-PTS/LDPE composites to show the influence of nanotubes in composites. Static charge measurements carried out on the films shows that no charge is present on the surface. Copolymer/LDPE composites films (1.0 % w/w) showed static decay time in the order of 0.1 to 0.2 sec. at 10 % cut-off on recording the decay time from 5000 V to 500 V. When the loading level of copolymers in LDPE was reduced to 0.5 wt. %, only the nanocomposites based on PANSA5-PTS showed better good performance to ESD protection. Better antistatic behavior shown by these copolymers at very low loading in LDPE was investigated by their nanotubular morphology. Blending of 0.5 and 1.0 wt. % of PTSA doped copolymers with selective composition of ANSA and aniline with LDPE has a great potential to be used as effective antistatic films. The loading level, morphology of the conducting material, and its proper dispersion with insulating matrix affect the properties like surface conductivity, mechanical properties, and its performance to application for electrostatic charge dissipation.

Author details

Hema Bhandari, S. Anoop Kumar and S. K. Dhawan

*Address all correspondence to: skdhawan@mail.nplindia.ernet.in

CSIR–National Physical Laboratory, India

References

[1] Zhua, H., Zhonga, L., Xiaoa, S., & Gan, F. (2004). *Electrochim. Acta*, 49, 5161.

[2] Jones, D. A. (1992). *Principles and Prevention of Corrosion, Macmillan Publishing, Chap. 1*, New York.

[3] Lacroix, J. C., Camalet, J. L., Aeiyach, S., Chane-Ching, K. I., Petitjean, J., Chauveau, E., & Lacaze, P. C. (2000). *J. Electroanal. Chem.*, 481, 76.

[4] Kinlen, P. J., Menon, V., & Ding, Y. (1999). *J. Electrochem. Soc.*, 146, 3690.

[5] Kinlen, P. J., Ding, Y., & Silverman, D. C. (2002). *Corrosion*, 58, 490.

[6] de Souza, S., da Silva, J. E. P., de Torrosi, S. I. C., Temperani, M. L. A., & Torresi, R. M. (2001). *Electrochem. Solid State Lett.*, 4, B27.

[7] Samui, A. B., Patankar, A. S., Rangarajan, J., & Deb, P. C. (2003). *Prog. Org. Coat.*, 47.

[8] Dominis, A. J., Spinks, G. M., & Wallace, G. G. (2003). *Prog. Org. Coat.*, 48, 43.

[9] Sathiyanarayanan, S., Muthukrishnan, S., & Venkatachari, G. (2006). *Prog. Org. Coat.*, 55, 5.

[10] Plesu, N., Ilia, G., Pascariu, A., & Vlase, G. (2006). *Synth. Met.*, 156, 230.

[11] Su, S. J., & Kuramuto, N. (2001). *Synth. Met.*, 114, 147.

[12] Gurunathan, K., Amalnerker, D. P., & Trivedi, D. C. (2003). *Mater. Lett.*, 57, 1642.

[13] Sathiyanarayanan, S., Muthukrishnan, S., Venkatachari, G., & Trivedi, D. C. (2005). *Prog. Org. Coat.*, 53, 297.

[14] Deberry, D. W. (1985). *J. Electrochem. Soc.*, 132, 1027.

[15] Wessling, B. (1991). *Synth. Met*, 41, 907.

[16] Elsenbaumer, R. L., Lu, W. K., & Wessling, B. (1994). Seoul, Korea. *Int. Conf. Synth. Met.*, Abstract No. APL(POL)1.

[17] Wrobleski, D. A., Benicewicz, B. C., Thompson, K. G., & Byran, C. (1994). *J. Polym. Prepr. (Am. Chem. Soc., Div. Polym. Chem.)*, 35, 265.

[18] Spinks, G. M., Dominis, A. J., Wallace, G. G., & Tallman, D. E. (2002). *J. Solid State Electrochem*, 6, 85.

[19] Sathiyanarayanan, S., Azim, S. S., & Venkatachari, G. (2007). *Electrochimica Acta*, 52, 2068.

[20] Majumdar, G., Goswami, M., Sarma, T. K., Paul, A., & Chattopadhyay, A. (2005). *Langmuir*, 21, 1663.

[21] Chowdhury, D., Paul, A., & Chattopadhyay, A. (2005). *Langmuir*, 21, 4123.

[22] Feng, X. M., Yang, G., Xu, Q., Hou, W. H., & Zhu, J. J. (2006). *Macromol Rapid Commun*, 27, 31.

[23] Hasan, M., Zhou, Y., Mahfuz, S., & Jeelani, S. (2006). *Materials Science and Engineering: A*, 429, 181.

[24] Li, X., Dai, N., Wang, G., & Song, X. (2008). *J Appl Polym Sci*, 107, 403.

[25] Xia, H. S., & Wang, Q. (2003). *J Appl Polym Sci*, 87, 1811.

[26] Zengina, H., & Erkan, B. (2010). *Polym. Adv. Technol.*, 21, 216.

[27] Stejskal, J., Kratochvı'l, P., Armes, S. P., Lascelles, S. F., Riede, A., Helmstedt, M., Prokes, J., & Krivka, I. (1996). *Macromolecules*, 29, 6814.

[28] Al-Dulaimi, A. A., Hashim, S., & Khan, M. I. (2011). *Sains Malaysiana*, 40, 757.

[29] Beck, F., Michaelis, R., Scholoten, F., & Zinger, B. (1994). *Electrochimica Acta*, 39, 229.

[30] Camalet, J. L., Lacroix, J. C., Aeiyach, S., Chane-Ching, K., & Lacaze, P. C. (1998). *Synth. Met.*, 93, 133.

[31] Kilmartin, P. A., Trier, L., & Wright, G. A. (2002). *Synthetic Metals*, 131, 99.

[32] Meneguzzi, A. A. P., Ferreira, C. A., Pham, M. C., Delamar, M., & Lacaze, P. C. (1999). *Electrochim. Acta*, 44, 2149.

[33] Bhandari, H., Choudhary, V., & Dhawan, S. K. (2010). *Thin Solid Film*, 519, 1031.

[34] Kinlen, P. J., Menon, V., & Ding, Y. J. (1999). *Electrochem. Soc.*, 146, 3690.

[35] Talo, A., Passiniemi, P., Forse´n, O., & Yla¨saari, S. (1997). *Synth. Met.*, 85, 1333.

[36] Wessling, B., & Posdorfer, J. (1999). *Electrochim. Acta.*, 44, 2139.

[37] Iribarren, J. I., Armelin, E., Liesa, F., Casanovas, J., & Aleman, C. (2006). *Material and corrosion*, 57, 683.

[38] Mc Andrew, T. P., Miller, S. A., Gilleinski, A. G., & Robeson, L. M. (1996). *Polym. Mater. Sci. Eng.*, 74, 204.

[39] Mc Hale, G., Shirtcliffe, N. J., Aqil, S., Perry, C. C., & Newton, M. I. (2004). *Phys. Rev.Lett.*, 93, 36102.

[40] Shirtcliffe, N. J., Mc Hale, G., Newton, M. L., Chabrol, G., & Perry, C. C. (2004). *Adv.Mater.*, 16, 1929.

[41] Wu, X. F., & Shi, G. Q. (2006). *J. Phys. Chem. B.*, 110, 11247.

[42] Jiang, L., Zhao, Y., & Zhai, J. (2004). *Angew. Chem. Int. Ed.*, 43, 4338.

[43] Jiang, W. H., Wang, G. J., He, Y. N., An, Y. L., Wang, X. G., Song, Y. L., & Jiang, L. (2005). Chem. J. Chin. Univ. (Chinese) , , 26, 1360.

[44] Han, J. T., Zheng, Y., Cho, J. H., Xu, X., & Cho, K. J. (2005). *Phys. Chem. B.*, 109, 20773.

[45] Soeno, T., Inokuchi, K., & Shiratori, S. (2004). *Appl. Surf. Sci.*, 237, 543.

[46] Li, H., Wang, X., Song, Y., Liu, Y., Li, Q. L., & Zhu, D. (2001). *Angew. Chem.*, 113, 1793.

[47] Wu, Y., Sugimura, H., Inoue, Y., & Takai, O. (2002). *Chem. Vap. Deposition*, 8, 47.

[48] Wang, S. T., Feng, L., & Jiang, L. (2006). *Adv. Mater.*, 18, 767.

[49] Qu, M. N., Zhang, B. W., Song, S. Y., Chen, L., Zhang, J. Y., & Cao, X. P. (2007). *Adv. Funct. Mater.*, 17, 593.

[50] Pan, Q. M., Jin, H. Z., & Wang, H. B. (2007). *Nanotechnology*, 18, 355605.

[51] Yabu, H., Takebayashi, M., Tanake, M., & Shimomura, M. (2005). *Langmuir*, 21, 3235.

[52] Li, X., Reindhout, D., & Crego-Calama, M. (2007). *Chem. Soc. Rev*, 36, 1350.

[53] Callies, M., & Quere, D. (2005). *Soft Matter*, 1, 55.

[54] Ma, M. L., & Hille, R. M. (2006). *Curr. Opin. Colloid Interface Sci.*, 11, 193.

[55] Blossey, R. (2003). *Nat. Mater.*, 2, 301.

[56] Anoop, Kumar. S., Bhandari, H., Sharma, C., Khatoon, F., & Dhawan, S. K. (2012). *Polymer International.*

[57] Anoop, Kumar. S., Bhandari, H., Sharma, C., Khatoon, F., & Dhawan, S. K. (2012). *Polymer (Paper under process).*

[58] Stern, M., & Geary, A. (1957). *J. Electrochem Soc.*, 104, 56.

[59] Spinks, G. M., Dominis, A. J., Wallace, G. G., & Tallman, D. E. (2002). *J. Solid State Electrochem.*, 6, 85.

[60] Silverstein, R. M., & Webster, F. X. (2002). *Spectrometric identification of organic compounds* (sixth edition), John Wiley and Son, Wiley, India, 165.

[61] Kim, H., Foster, C., Chiang, J., & Heegar, A. J. (1989). *Synth. Metals.*, 29, 285.

[62] Pron, A., & Rannou, P. (2002). Progr. Polym. Sci. ; , 27, 135.

[63] Huerta-Vilca, D., Moraes, S. R., & Motheo, A. J. (2003). *J. Braz. Chem. Soc.*, 14, 52.

[64] Mc Andrew, T. P., Miller, S. A., Gilleinski, A. G., & Robeson, L. M. (1996). *Polym. Mater. Sci. Eng.*, 74, 204.

[65] Kusy, R. P. (1986). in Metal filled polymers, edited by S.K. Battacharya (Dekker New York , 1.

[66] Rosner, R. B. (2001). *IEEE Transaction on device and material reliability*, 1.

[67] Edenbaum, J., & Reinhold, V. N. *Plastics Additives and Modifiers Handbook*, New York, 615.

[68] Narkis, M., Lidor, G., Vaxman, A., & Zuri, L. (1999). *J. Electrostatics*, 47, 201.

[69] Li, C., Liang, T., Lu, W., Tang, C., Hu, X., Cao, M., & Liang, J. (2004). *Composite Sci. Tech.*, 64, 2089.

[70] Narkis, M., Ram, A., & Stein, A. Z. (1980). *J. Appl. Polym. Sci.*, 25, 1515.

[71] Kobayashi, T., Wood, A., Takemura, A., & Ono, A. (2006). *Barbara Jnl. Electrostat.*, 64, 377.

[72] Yukishige, H., Koshima, Y., Tanisho, H., & Kohara, T. (1996). *US patent 5571859*.

[73] Angelopoulos, M. (2001). Conducting polymers in microelectronics. *IBM Journal of Research and Development*, 45(1), 57-76.

[74] Laska, J., Żak, K., & Prón, A. (1997). Conducting blends of polyaniline with conventional polymers. *Synthetic Metals*, 84(1), 117-118.

[75] Mitzakoff, S., & De Paoli, M. A. (1999). European Polymer Journal ., 35(10), 1791-1798.

[76] Chubb, J. N. (2002). *J. Electrostatics*, 54, 233.

[77] Bhandari, H., Bansal, V., Choudhary, V., & Dhawan, S. K. (2009). *Polymer International*, 58, 489.

[78] Chen, S. A., & Hwang, G. W. (1996). *Macromolecules*, 29, 3950.

[79] Bhandari, H. (2011). Synthesis, Characterization and Evaluation of Conducting Copolymers for Corrosion Inhibition and Antistatic Applications. *PhD thesis*, Indian Institute of Technology, Delhi, India.

[80] Bhandari, H., Singh, S., Choudhary, V., & Dhawan, S. K. (2011). *Polymers for Advanced Technologies*, 22(9), 1319-1328.

[81] Strumpler, R., & Glatz-Reichenbach, J. (1999). Conducting polymer composites. J. Electroceramics ., 3(4), 329-346.

Chapter 14

Electroconductive Nanocomposite Scaffolds: A New Strategy Into Tissue Engineering and Regenerative Medicine

Masoud Mozafari, Mehrnoush Mehraien,
Daryoosh Vashaee and Lobat Tayebi

Additional information is available at the end of the chapter

http://dx.doi.org/10.5772/51058

1. Introduction

Nanocomposites are a combination of a matrix and a filler, where at least one dimension of the system is on the nanoscale being less than or equal to 100 nm. Much work has focused on the construction of nanocomposites due to the structural enhancements in physico-chemical properties, and functionality for any given system [1-6]. The physico-chemical enhancements result from the interaction between the elements being near the molecular scale. Nanocomposite materials have also received interest for tissue engineering scaffolds by being able to replicate the extracellular matrix found *in vivo*. Currently, researchers have created composite materials for scaffold formation which incorporate two or more materials. Some of these materials consist of minerals for bone tissue engineering including calcium, hydroxyapatite, phosphate, or combinations of different polymers, such as poly (lactic acid), poly (ε-caprolactone), collagen and chitosan, and many other different combinations [7-9]. Other work has focused on doping the polymer scaffolds with specific growth hormones or adhesion sequences to influence how cells attach to the scaffold and cause the scaffold to become a drug delivery vehicle for different kind of tissue engineering applications [10]. Among different materials used in preparation of nanocomposits, conducting polymers are one of the effective materials that can be employed to facilitate communication with neural system for regenerative purposes.

However, the major obstacle concerning the electrically conducting polymers has been the difficulty associated with the processing of them [11]. To overcome this problem, most researchers have electrospun conducting polymers by blending them with other spinnable

INTECH
open science | open minds

© 2012 Mozafari et al.; licensee InTech. This is a paper distributed under the terms of the Creative Commons Attribution License (http://creativecommons.org/licenses/by/3.0), which permits unrestricted use, distribution, and reproduction in any medium, provided the original work is properly cited.

polymers, compromising the conductivity of the nanocomposite fibres [12-16]. Blending of conducting polymers with other polymers positively affects the properties of the resultant nanocomposite fibres. In addition, sometimes for making benefid from condicting polymers and the specific properties of them we can have just a small thib coating of the polymer on the surface of nanocomposite.

The term of "Tissue Inducible Biomaterials" has been recently applied based on the principles of biology and engineering to design nanocomposite scaffolds that restore, maintain or improve the general function of damaged tissues. To gain tissue induction activity and assist tissue regeneration, the nanocomposite scaffolds need to be designed based on nanostructural properties, surface modifications or incorporation of molecules into them. Among different approaches and materials for the preparation of scaffolds, get benefit from conducting polymers seems to be more interesting and promising. Electroconductive polymers exhibit excellent electrical properties and have been explored in the past few decades for a number of applications. In particular, due to the ease of synthesis, cytocompatibility, and good conductivity, some kind of conducting polymers have been extensively studied for biological and medical applications. Different forms of conducting polymers such as polypyrrole (PPy), polythiophene (PT), polyaniline (PANI), poly(3,4-ethylenedioxythiophene) (PEDOT) etc. are used in our daily life due to their uniqe properties which can be applied in different applications. These materials have a conjugated π electron system with "metal-like" electrical conductivity. Due to the rich chemistry of conducting polymers, they have attracted the attention of many researchers and leading to the publication of thousands of papers. The most important property of conducting polymers is their electrical conductivity, so the first approach is to study their electrical-related biological behaviors. Neurons are well known for the membrane-potential-wave style signal transduction. Hence, early studies were focused on the electrical stimulation to the neuron cells using conducting polymers as electrodes. The results showed that the electric conducting polymers can be used as biological electrodes and the neuron growth can be enhanced under an electrical field.

Using conducting polymers in nanocomposite scaffold design is relatively new in tissue engineering applications [17]. It has been demonstrated that these conducting nanocomposites are able to accept and modulate the growth of different cell types [18] including endothelial cells, [19] nerve cells [20] and chromaffin cells [21]. It has been demonstrated that using conducting nanocomposite scaffolds are most promissing in nerve tissue engineering. These electroconductive polymers have been recognized as potential nanocomposite scaffold materials to electrically stimulate tissues for therapeutic purposes in tissue engineering scaffolds. Based on the literature search within the last decade, the present chapter summarized the strategy of electroconductive nanocomposite scaffolds for tissue engineering and regenerative medicine purposes.

http://dx.doi.org/10.5772/51058

2. Conductive polymers

2.1. General approaches and considerations

Conducting polymers are a special class of materials with electronic and ionic conductivity [22]. The structures of the widely used conducting polymers are depicted in Fig. 1 [23]. These polymers have immense applications in the fields of drug delivery, neuroprosthetic devices, cardiovascular applications, bioactuators, biosensors, the food industry and etc.

One of the first electrically conducting polymers, polypyrrole (PPy) was introduced in the 1960s, but little was understood about this polymer at that time [24]. In 1977, a research team reported a 10 million-fold increase in the conductivity of polyacetylene doped with iodine as the first inherently conducting polymer [25,26]. Unlike polyacetylene, polyphenylenes, are known to be thermally stable as a result of their aromaticity [27]. Polyheterocycles, such as PPy, polythiophene (PT), polyaniline (PANI), and poly(3,4-ethylenedioxythiophene) (PEDOT), developed in the 1980s, have since emerged as another class of aromatic conducting polymers that exhibit good stabilities, conductivities, and ease of synthesis [28]. Table 1 shows a list of different conducting polymers and their conductivities [29].

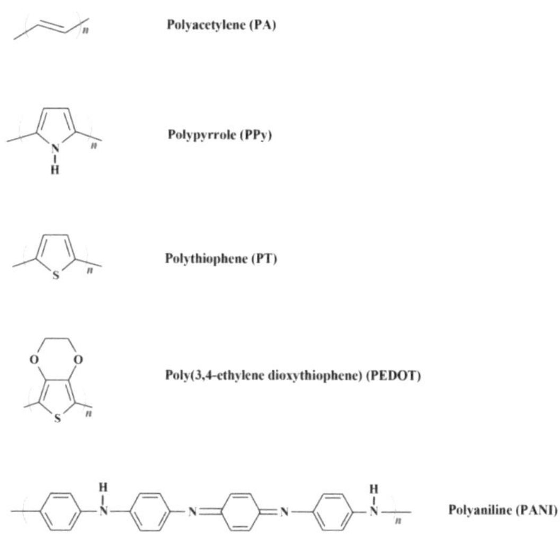

Figure 1. Chemical structures of various conducting polymers

Conducting polymers have an inherently unstable backbone, resulting from the formation of alternate single and double bonds along with the monomer units during polymerization. The delocalized π bonding electrons, produced across the conjugated backbone, provide an electrical pathway for mobile charge carriers which are introduced through doping. Consequently, the electronic properties, as well as many other physicochemical properties, are determined by the structure of the polymer backbone and the nature and the concentration of the dopant ion [30].

Conducting polymer	Maximum Conductivity (Siemens/cm)	Type of doping
Polyacetlene (PA)	200-1000	n,p
Polyparaphenylene (PPP)	500	n,p
Polyparaphenylene sulfide (PPS)	3-300	p
Polyparavinylene (PPv)	1-1000	p
Polypyrrole (PPy)	40-200	p
Polythiophene (PT)	10-100	p
Polyisothionaphthene (PITN)	1-50	p
Polyaniline (PANI)	5	n,p

Table 1. Some of the common conducting polymers and their conductivity [29].

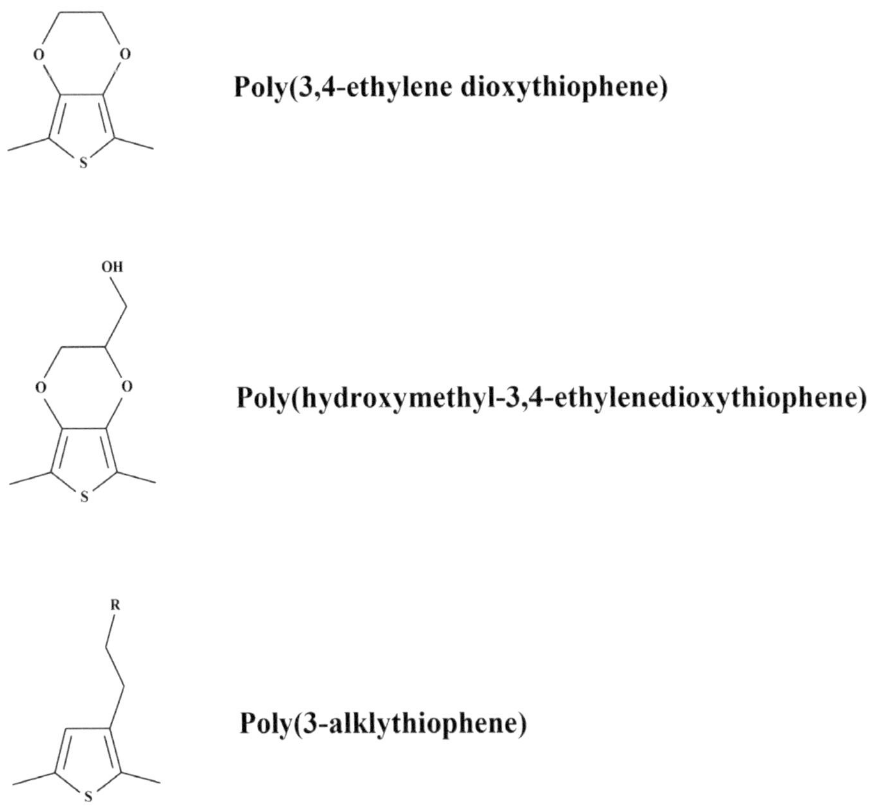

Figure 2. Typical monomer structures used to fabricate Poly(3,4-ethylene dioxythiophene), Poly(hydroxymethyl- 3,4-ethylenedioxythiophene) and Poly(3-alklythiophene) [30]

Conjugated aliphatics, including polyacetylene, and benzene derivatives such as PANI, have been largely ruled out for biomedical applications due to their oxidative degradation in air and the cytotoxic nature of their by-products. Although recent research has shown that the emeraldine salt of PANI (EPANI) can be successfully fabricated in a biocompatible form [31,32], modern biomedical conducting polymers are typically composed of heterocyclic aromatics, such as derivatives of thiophene and pyrrole [33,34]. Specifically, PEDOT and PPy have been widely studied for their superior environmental and electrochemical stability [35-37]. Fig. 2 shows the chemical structure of various thiophene derivatives including EDOT, EDOT-MeOH and 3-alkylthiophene [30].

2.2. Surface modification of conducting polymers

For biomedical approaches, sometimes we need to modify the outer surface of the materials to induce special features. Conducting polymers can be also modified to enhance the functionality of nanocomposites. The surface modifications of conducting polymers have some concerns including:

- Enhancement of the charge transport of carriers between the implant and tissue
- Mediating the large difference in mechanical modulus
- Improvement of biodegradation
- Decreasing the impedance to enhance the sensitivity of the recording site
- Cell response enhancement
- Bioactivity enhancement

Surface modification and functionalization of conducting polymers with different biomolecules or dopants has allowed us to modify them with biological sensing elements, and to turn on and off different signalling pathways required for cellular processes. In this way, conducting polymers can show significant enhancement in cell proliferation and differentiation. Thus, conducting polymers provides an excellent opportunity for fabrication of highly selective, biocompatible, specific and stable nanocomposite scaffolds for tissue engineering of different organs [39,40].

2.3. General use of conducting polymers

A range of applications for conducting polymers are currently being considered, such as the development of tissue-engineered organs [41], controlled drug release [42], repaire of nerve chanels [43], and the stimulation of nerve regeneration [44]. In addition, electrically active tissues (such as brain, heart and skeletal muscle) provide opportunities to couple electronic devices and computers with human or animal tissues to create therapeutic body–machine interfaces [45]. The conducting and semiconducting properties of this class of polymers make them important for a wide range of applications. The important properties of various conducting polymers and their potential applications are discussed in Table 2 [23].

Conducting polymer	properties	applications
polypyrrole (PPy)	Highly conductive Opaque Brittle Amorphous structure	biosensors drug delivery bioactuators Nerve tissue engineering Cardiac tissue engineering Bone tissue engineering
polythiophenes (PT)	Good electrical conductivity Good optical property	Biosensors Food industry
polyaniline (PANI)	A semiflexible rod polymer Requires simple doping/dedoping chemistry Exists as bulk films or dispersions High conductivity up to 100 S/cm	Biosensors Drug delivery Bioactuators Nerve tissue engineering Cardiac tissue engineering
poly(3,4-ethylenedioxythiophene) (PEDOT)	High temperature stability Transparent conductor Moderate band gap Low redoxpotential conductivity up to 210 S/cm	Biosensors Antioxidants Drug delivery neural prosthetics

Table 2. Properties and applications of some common conducting polymer [23]

2.4. Conductivity mechanism

Generally, polymers with loosely held electrons in their backbones can be called conducting polymers. Each atom on the backbone has connection with a π bond, which is much weaker than the σ bonds in the backbone. These atoms have allways a conjugated backbone with a high degree of π-orbital overlap [46]. It is known that the neutral polymer chain can be oxidized or reduced to become either positively or negatively charged through doping process [47]. It is also known that conducting polymers could not be perfectly conductive without using dopants, and doping of π-conjugated polymers results in high conductivity [24]. The doping process is influenced by different factors such as polaron length, chain length, charge transfer to adjacent molecules and conjugation length [46]. There have been different dopants for the addition of H^+ (protonation) to the polymers. For example, strong inorganic hydrochloric acid (HCl), organic and aromatic acids containing different aromatic substitution have been used as dopants for PANI. It is also reported that the surface energies of the doped conducting polymers vary greatly, depending on the choice of the dopants and doping level. Recently, PPy doped with nonbiologically active dopants (tosylate) and it has been

http://dx.doi.org/10.5772/51058

characterized for biological interactions as they can trigger cellular responses in biological applications. However, the incorporation of more biologically active dopants can significantly modify PPy-based nanocomposites for biomedical applications [48].

One of the most important challenges of nanocomposite scaffolds based on conducting polymers is their inherent inability to degrade in the body, which may induce chronic inflammation [49]. Hence, belending of conducting polymers with biodegradable polymers seems to solve the problem. PPy and PANI are the most importan conducting polymers for tissue engineering, and they are important in terms of their biocompatibility and cell signaling especially for nerve tissue engineering [24].

2.5. Polypyrrole

PPy is among one of the first conducting polymers that studied a lot for its effect on the behaviour of cells. This material has been reported to support cell adhesion and growth of different cells [50]. This conducting synthetic polymer has numerous applications in tissue engineering and drug delivery. Recently, Moroder *et al.* [51] studied the properties of polycaprolactone fumarate–polypyrrole (PCLF–PPy) nanocomposite scaffolds under physiological conditions for application as conductive nerve conduits. In their study, PC12 cells cultured on PCLF–PPy nanocomposite scaffolds were stimulated with regimens of 10 μA of either a constant or a 20 Hz frequency current passed through the scaffolds for 1 h per day. The surface resistivity of the scaffolds was 2 kΩ and the nanocomposite scaffolds were electrically stable during the application of electrical stimulation. As can be seen in Fig. 3, *in vitro* studies showed significant increases in the percentage of neurite bearing cells, number of neurites per cell and neurite length in the presence of electrical stimulation compared with no electrical stimulation. They concluded that the electrically conductive PCLF–PPy nanocomposite scaffolds possed the material properties necessary for application in nerve tissue engineering.

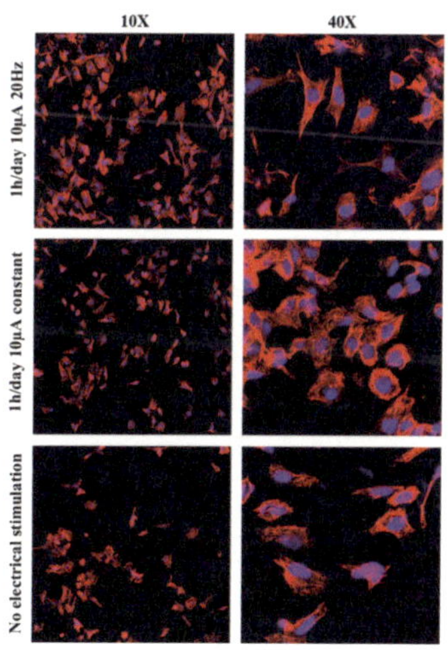

Figure 3. Fluorescence microscopy of PC12 cells at 10× and 40× magnification after undergoing different electrical stimuli treatment regimens for 48h

2.6. Polyaniline

PANI is an oxidative polymeric product of aniline under acidic conditions and is commonly known as aniline black [52]. The exploration of PANI for tissue-engineering applications has progressed more slowly than the development of PPy for similar applications. However, recently there has been more evidence of the ability of PANI and PANI variants to support cell growth [53]. Recently, Fryczkowski *et al.* [54] synthesized three-dimensional nanocomposite fibres of poly(3-hydroxybutyric acid) (PHB) and dodecylbenzene sulfonic acid (DBSA) doped polyaniline in chlorophorm/trifluoroethanol mixture, using electrospinning method. The morphology, electro-active properties and supermolecular structure of nanofibres webs have been analyzed and discussed. Obtained nanofibres are potentially applicable as nanocomposite scaffolds for tissue engineering. According to their results, there were limitations in composition of blended system and the PHB:PANI:solvent ratio needed to be optimized in order to obtain reasonable spinnability of compositions, and even small amount of PANI caused changes in super molecular structure of PHB/PANI nanofibres.

2.7. Poly (3, 4-ethylenedioxythiophene)

Although PPy and PANI have been the most extensively conductive polymers for tissue engineering and regenerative medicine, recently the potential of polythiophene conductive polymer for tissue engineering have been approved. This polymer has received significant attention due to a wide range of promising electronic and electrochemical applications [55,56]. PEDOT can be considered as the most successful polythiophene due to its specific characteristics [57-65]. PEDOT can also be considered as the most stable conducting polymer currently available because of not only high conductivity but also unusual environmental and electrochemical stabilities in the oxidized state [57-60]. Recently, Bolin *et al.* [66] reported electronically conductive and electrochemically active 3D-nanocomposite scaffolds based on electrospun poly(ethylene terephthalate) (PET) nanocomposite fibers. They employed vapour phase polymerization to achieve a uniform and conformal coating of PEDOT doped with tosylate on the nano-fibers. They observed that the PEDOT coatings had a large impact on the wettability, turning the hydrophobic PET fibers super-hydrophilic. According to Fig. 4, the SH-SY5Y cells adhered well and showed healthy morphology. These electrically active nanocomposite scaffolds were used to induce Ca^{2+} signalling in SH-SY5Y neuroblastoma cells. Their reported nanocomposite fibers represented a class of 3D host environments that combined excellent adhesion and proliferation for neuronal cells with the possibility to regulate their signalling.

2.8. Piezoelectric polymeric nanocomposites

Recent studies on the application of conductive materials showed that piezoelectric polymeric materials can also be considered for tissue engineering applications. Piezoelectric polymeric materials can generate surface charges by even small mechanical deformations [67]. Poly(vinylidenefluoride) (PVDF) is a synthetic, semicrystalline polymer with piezoelectric properties that can be potentially used for biomedical application due to their unique molecular structure [107]. An electrical charged porous nanocomposite could be a promising approach for a number of tissue engineering applications. Reported data on piezoelectric polymeric nanocomposites showed that after electrical stimulation, cellular interaction and tissue growth might be improved [68].

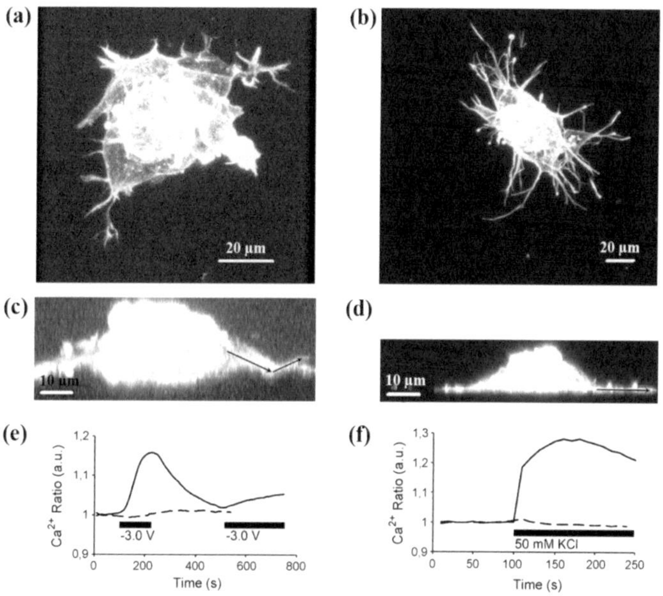

Figure 4. Confocal micrograph top view *Y*-axis projection of Tritc-phalloidin stained cluster of SH-SY5Y cells growing on (a) VPP-PEDOT coated nano-fiber (b) cell culture treated glass. Confocal micrograph side view Z-projection of Tritc-phalloidin stained cluster of SH-SY5Y cells growing on (c) VPP-PEDOT coated nano-fiber surface and (d) cell culture treated glass. Arrows indicate the direction of neurites. (e) Solid line shows intracellular Ca^{2+} flux in FURA-2-AM loaded SH-SY5Y cells cultured on nano-fiber surface. A potential of -3.0V is applied at 100 s. The potential is turned off at 250 s and turned on again at 500 s. Dashed line shows cell treated with 50μM nifedipine in order to block the VOCCs and stimulated with -3.0V at 100 s until 380 s. (f) Solid line shows intracellular Ca^{2+} flux in FURA-2-AM loaded SH-SY5Y cells cultured in cell culture dish 50mM KCl was added at 100 s. Dashed line shows cell treated with 50μM nifedipine in order to block the VOCCs and stimulated in the same way [66].

3. Applications of conducting polymers

3.1. Applications of conducting polymers: general view

Conductive polymers exhibit attractive properties such as ease of synthesis and processing [69]. The unique properties of this type of materials have recently given a wide range of applications in the biological field. Research on conductive polymers for biomedical applications expanded extrimly in the 1980s, and they were shown via electrical stimulation, to modulate cellular activities (e.g. cell adhesion, migration, DNA synthesis and protein secretion) [70-73]. Since then many studies have been done on nerve, bone, muscle, and cardiac cells. The unique characteristics of conducting polymers have been shown to be useful in many biomedical applications, specially tissue engineering nanocomposite scaffolds and drug delivery devices [74]. In comparison to other conductive materials for biological applications, conducting polymers are inexpensive, easy to synthesize, and versatile. In addition, conducting polymers permit control over the level and duration of electrical stimulation for tissue engineering applications.

3.2. Use and modification of conducting polymers for drug delivery

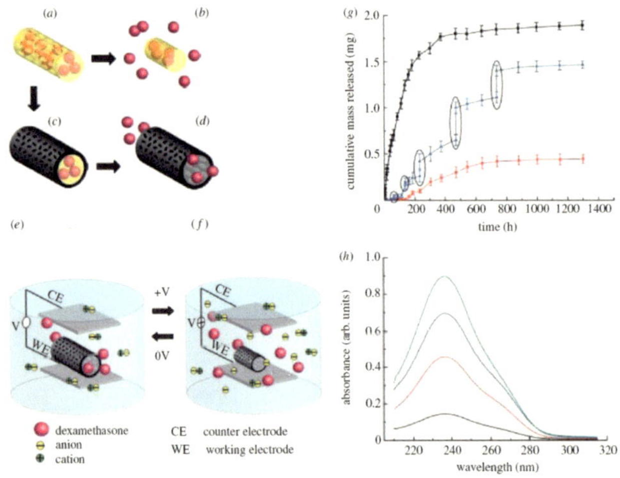

Figure 5. a) dexamethasone-loaded electrospun PLGA, (b) hydrolytic degradation of PLGA fibres leading to release of the drug and (c) and (d) electrochemical deposition of PEDOT around the dexamethasone-loaded PLGA fibre slows down the release of dexamethasone. (e) PEDOT nanotubes in a neutral electrical condition. (f) External electrical stimulation controls the release of dexamethasone from the PEDOT nanotubes. By applying a positive voltage, electrons are injected into the chains and positive charges in the polymer chains are compensated. (g) Cumulative mass release of dexamethasone from: PLGA nanoscale fibres (black squares), PEDOT-coated PLGA nanoscale fibres (red circles) without electrical stimulation and PEDOT-coated PLGA nanoscale fibres with electrical stimulation of 1 V applied at the five specific times indicated by the circled data points (blue triangles). (h) UV absorption of dexamethasone-loaded PEDOT nanotubes after 16 h (black), 87 h (red), 160 h (blue) and 730 h (green). [80]

Developing novel drug-delivery systems will open up new applications that were previously unsuited to traditional delivery systems. The use of conducting polymers in drug delivery is an excellent approach due to their biocompatibility and their possibility of using them in *in vivo* applications for real time monitoring of drugs in biological environments [75]. Controlled drug release can also be facilitated using a change in conductive polymer redox state to increase permeation of drugs such as dexamethasone [76]. Electrical stimulation of conductive polymers has been used to release a number of therapeutic proteins and drugs like nerve growth factor [77], dexamethasone [78] and heparin [79]. Another study demonstrated the use of PEDOT nanotubes polymerized on top of electrospun poly(lactic-co-glycolic acid) (PLGA) nanocomposite fibres for the potential release of the drug dexamethasone. Here, dexamethasone was incorporated within the PLGA nanocomposite fibres and then PEDOT was polymerized around the dexamethasone-loaded PLGA nanocomposite. As the PLGA fibres degraded, dexamethasone molecules remained inside the PEDOT nanotubes. These PEDOT nanotubes favoured controlled drug release upon electrical stimulation. Fig. 5 demonstrates the incorporation and release mechanism of dexamethasone from PEDOT nanotubes due to electrical stimulation. This drug-delivery system had the potential of immense interest for the treatment of cancer and tissue engineering and regenerative medicin [80].

3.3. Use and modification of conducting polymers for bioactuators

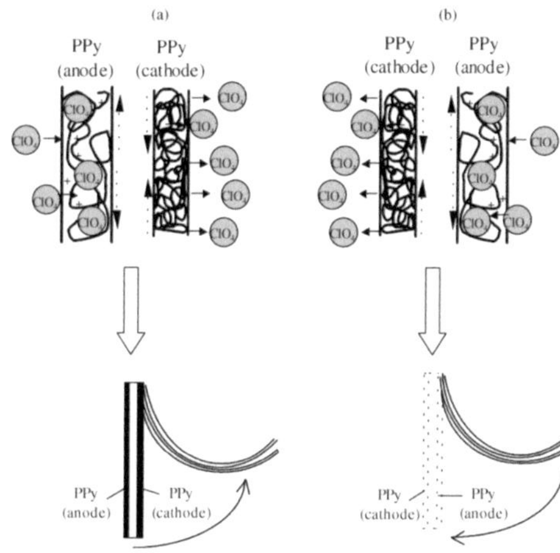

Figure 6. The triple layer device (polypyrrole(ClO_4^-)/non-conducting and adherent polymer/polypyrrole(ClO_4^-)) and its macroscopic movement produced as a consequence of volume change in the polypyrrole films. (a) A current flows and the left polypyrrole film acting as the anode is swelled by the entry of the hydrated counter ions (ClO_4^-). Simultaneously, the right film acting as the cathode contracts and shrinks because of the expulsion of the counter ions. These volume changes and the constant length of the non-conducting film promote the movement of the triple layer towards the polypyrrole film that is being contracted. (b) By changing the direction of the current, the movement takes place in the opposite direction. The muscle works in $LiClO_4$ aqueous solution [83]

Bioactuators are devices that are used to create mechanical force, which in turn can be used as artificial muscles. The phenomenon of change in the volume of the conducting polymers scaffold upon electrical stimulation has been employed in the construction of bioactuators. In artificial muscle applications, two layers of conducting polymers are placed in a triple layer arrangement, where the middle layer comprises a non-conductive material [81]. When current is applied across the two conducting polymers films, one of the films is oxidized and the other is reduced. The oxidized film expands owing to the inflow of dopant ions, whereas the reduced film expels the dopant ions and in the process shrinks, as depicted in Fig. 6 [81]. Conducting actuators have many features that make them ideal candidates for artificial muscles, including that they:

- can be electrically controlled,
- have a large strain which is favourable for linear, volumetric or bending actuators,
- possess high strength,
- require low voltage for actuation (1 V or less),

- can be positioned continuously between minimum and maximum values,
- work at room/body temperature,
- can be readily microfabricated and are light weight, and
- can operate in body fluids [82].

3.4. Use and modification of conducting polymers for tissue engineering applications

The essential properties of conductive polymers desired for tissue engineering and regenerative medicine are conductivity, reversible oxidation, redox stability, biocompatibility, hydrophobicity, three-dimensional geometry and surface topography. Conductive polymers are widely used in tissue engineering due to their ability to subject cells to an electrical stimulation. Studies have addressed cell compatibility when a current or voltage is applied to PPy. An advantage offered by conducting polymers is that the electrochemical synthesis allows direct deposition of a polymer on the surface while simultaneously trapping the protein molecules [84].

In a recent study the release of NGF from PPy nanocomposites by using biotin as a co-dopant during the electrical polymerization was investigated [85]. In this research, NGF was biotinylated and immobilized to streptavidin entrapped within PPy nanocomposites doped with both biotin and dodecylbenzenesulfonate. The release of heparin from hydrogels immobilized onto PPy nanocomposites could also be triggered by electrical stimulation [86]. PVA hydrogels were covalently immobilized onto PPy via grafting of aldehyde groups to PPy and chemical reaction of these with hydroxyl groups from the hydrogel as shown in Fig. 7.

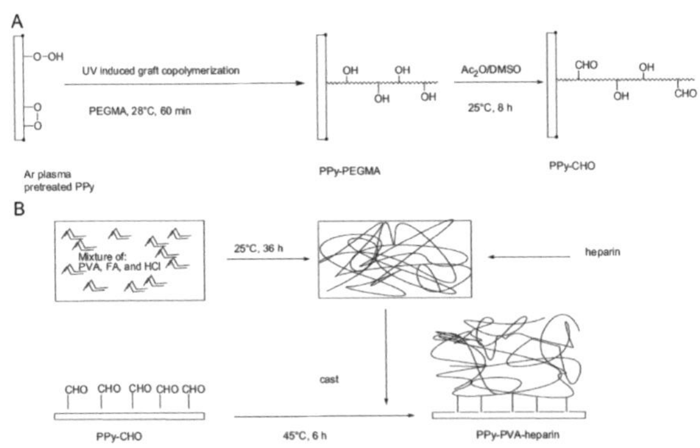

Figure 7. Controlled release of heparin from poly(vinyl alcohol) (PVA) hydrogels immobilized on PPy. (A) Post-polymerization of PPy to incorporate aldehyde groups. (B) Covalent immobilization of PVA hydrogels containing heparin on PPy substrates. Controlled release of heparin was obtained by electrical stimulation of PPy [148].

Electrically conducting polymers have attracted much interest for the construction of nerve guidance channels. The use of conducting polymers can help locally deliver electrical stimulus. It can also provide a physical template for cell growth and tissue repair and allow precise external control over the level and duration of stimulation [87,88]. The importance of conducting polymeric nanocomposites is based on the hypothesis that such composites can be used to host the growth of cells, so that electrical stimulation can be applied directly to the cells through the composite, proved to be beneficial in many regenerative medicine strategies, including neural and cardiac tissue engineering [89].

Recently, Li *et al.* [90] blended PANI with a natural protein, gelatin, and prepared nanocomposite fibrous scaffolds to investigate the potential application of such a blend as conductive scaffold for tissue engineering applications. As can be seen in Fig. 8, SEM analysis of the scaffolds containing less than 3% PANI in total weight, revealed uniform fibers with no evidence for phase egregation, as also confirmed by DSC.

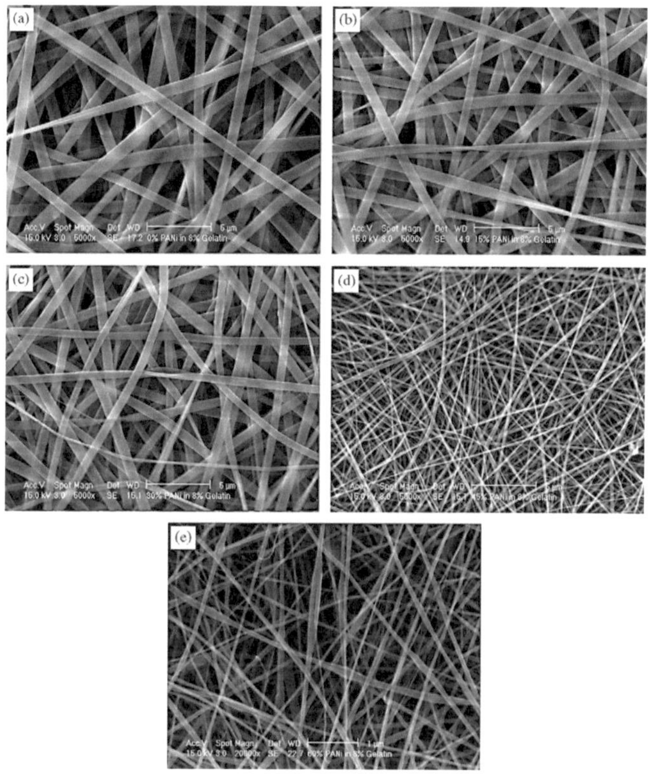

Figure 8. SEM micrographs of gelatin fibers (a) and PANi-gelatin blend fibers with ratios of (b) 15:85; (c) 30:70; (d) 45:55; and (e) 60:40. Original magnifications are 5000× for (a–d) and 20000× for (e). Figure shows the electrospun fibers were homogeneous while 60:40 fibers were electrospun with beads [90]

To test the usefulness of PANI/gelatin blends as a fibrous matrix for supporting cell growth, H9c2 rat cardiac myoblast cells were cultured on fiber-coated glass cover slips. Cell cultures were evaluated in terms of cell proliferation and morphology. According to Fig. 9, the results indicated that all PANI/gelatin blend fibers supported H9c2 cell attachment and proliferation to a similar degree as the control tissue culture-treated plastic (TCP) and smooth glass substrates.

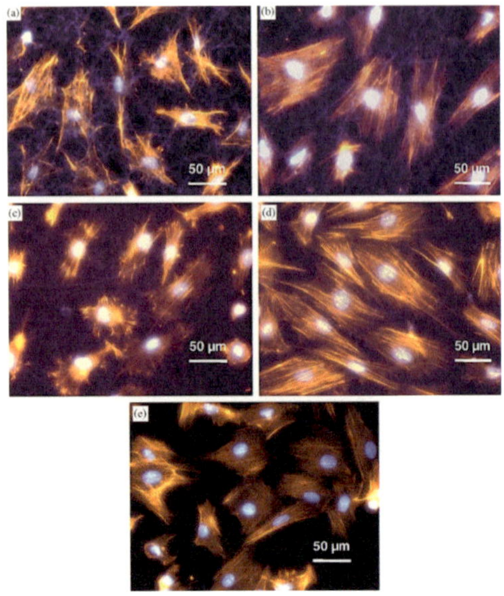

Figure 9. Morphology of H9c2 myoblast cells at 20 h of post-seeding on: (a) gelatin fiber; (b) 15:85 PANI/gelatin blend fiber; (c) 30:70 PANI/gelatin blend fiber; (d) 45:55 PANI/gelatin blend fibers; and (e) glass matrices. Staining for nuclei-bisbenzimide and actin cytoskeleton-phalloidin, fibersautofluorescence, original magnification 400× [90].

Depending on the concentrations of PANI, the cells initially displayed different morphologies on the fibrous substrates, but after 1week all cultures reached confluence of similar densities and morphology. Taken together they suggested that PANI/gelatin blend nanocomposite fibers could provide a novel conductive material well suited as biocompatible scaffolds for tissue engineering.

4. Conclusion

Tissue engineering is a new concept which is a growing area of research, in which cells are seeded on nanocomposite scaffolds and then implanted in defected part of body. Appropriate stimuli (chemical, biological, mechanical and electrical) can be applied and over a relatively short time new tissue can be formed to help restore function in the patient. The ideal scaffolds should have an appropriate surface chemistry and microstructures to facilitate cellular attachment, proliferation and differentiation. In addition, the scaffolds should possess adequate mechanical strength and biodegradation rate without any undesirable by-products. Among different materials, conducting polymers are one of the materials that can be employed to facilitate communication with neural system for regenerative purposes. In this chapter the recent methods of the synthesis of nanocomposite scaffolds using different conducting polymers was reviewed. The ability of conductive scaffolds to accept and modulate the growth of a few different cell types including endothelial, nerve, and chromaffin cells have shown a bright future in the field of tissue engineering and regenerative medicine.

http://dx.doi.org/10.5772/51058

Acknowledgements

This review chapter book is partially based upon work supported by Air Force Office of Scientific Research (AFOSR) High Temperature Materials program under grant no. FA9550-10-1–0010 and the National Science Foundation (NSF) under grant no. 0933763.

Author details

Masoud Mozafari[1], Mehrnoush Mehraien[1], Daryoosh Vashaee[2] and Lobat Tayebi[1*]

*Address all correspondence to: lobat.tayebi@okstate.edu

1 Helmerich Advanced Technology Research Center, School of Material Science and Engineering, Oklahoma State University, USA

2 Helmerich Advanced Technology Research Center, School of Electrical and Computer Engineering, Oklahoma State University, USA

References

[1] Mozafari, M., Moztarzadeh, F., Rabiee, M., Azami, M., Tahriri, M., Moztarzadeh, Z., & Nezafati, N. (2010). Development of Macroporous Nanocomposite Scaffolds of Gelatin/Bioactive Glass Prepared Through Layer Solvent Casting Combined with Lamination Technique for Bone Tissue Engineering. *Ceramics International*, 36, 2431-2439.

[2] Mozafari, M., Rabiee, M., Azami, M., & Maleknia, S. (2010). Biomimetic Formation of Apatite on the Surface of Porous Gelatin/Bioactive Glass Nanocomposite Scaffolds. *Applied Surface Science*, 257, 1740-1749.

[3] Mozafari, M., Moztarzadeh, F., Rabiee, M., Azami, M., Tahriri, M., & Moztarzadeh, Z. (2010). Development of 3D Bioactive Nanocomposite Scaffolds Made from Gelatin and Nano Bioactive Glass for Biomedical Applications. *Advanced Composites Letters*, 19, 91-96.

[4] Hamlekhan, A., Mozafari, M., Nezafati, N., Azami, M., & Hadipour, H. (2010). A Proposed Fabrication Method of Novel PCL-GEL-HAp Nanocomposite Scaffolds for Bone Tissue Engineering Applications. *Advanced Composites Letters*, 19, 123-130.

[5] Poursamar, S. A., Azami, M., & Mozafari, M. (2011). Controllable Synthesis and Characterization of Porous Polyvinyl Alcohol/Hydroxyapatite Nanocomposite Scaffolds via an in Situ Colloidal Technique. *Colloids and Surfaces B: Biointerfaces*, 84, 310-316.

[6] Hamlekhan, A., Moztarzadeh, F., Mozafari, M., Azami, M., & Nezafati, N. (2011). Preparation of Laminated Poly(ε-caprolactone)-Gelatin-Hydroxyapatite Nanocomposite Scaffold Bioengineered via Compound Techniques for Bone Substitution. *Biomatter*, 1, 1-11.

[7] Ghafari Nazari, A., & Mozafari, M. (2012). Simulation of Structural Features on Mechanochemical Synthesis of $Al_2O_3TiB_2$ Nanocomposite by Optimized Artificial Neural Network. *Advanced Powder Technology*, 23, 220-227.

[8] Hamlehkhan, A., Mozafari, M., Nezafati, N., Azami, M., & Samadikuchaksaraei, A. (2012). Novel Bioactive Poly(ε-caprolactone)-Gelatin-Hydroxyapatite Nanocomposite Scaffolds for Bone Regeneration. *Key Engineering Materials*, 493-494, 909-915.

[9] Baghbani, F., Moztarzadeh, F., Gafari Nazari, A., Razavi Kamran, A. H., Tondnevis, F., Nezafati, N., Gholipourmalekabadi, M., & Mozafari, M. (2012). Biological Response of Biphasic Hydroxyapatite/Tricalcium Phosphate Scaffolds Intended for Low Load-Bearing Orthopaedic Applications. *Advanced Composites Letters*, 21, 16-24.

[10] Jalali, N., Moztarzadeh, F., Mozafari, M., Asgari, S., Motevalian, M., & Naghavi Alhosseini, S. (2012). Surface Modification of Poly(lactide-co-glycolide) Nanoparticles by d-α-tocopheryl Polyethylene Glycol 1000 Succinate as Potential Carrier for the Delivery of Drugs to the Brain. *Colloids and Surfaces A: Physicochemical and Engineering Aspects*, 392, 335-342.

[11] Pomfret, S. J., Adams, P. N., Comfort, N. P., et al. (2000). Electrical and Mechanical Properties of Polyaniline Fibres Produced by a One-Step Wet Spinning. *Polymer*, 41, 2265-2269.

[12] Yu, Q. Z., Shi, M. M., Deng, M., et al. (2008). Morphology and Conductivity of Polyaniline Sub-Micron Fibers Prepared by Electrospinning. *Mater Sci Eng B*, 150, 70-76.

[13] Veluru, J. B., Satheesh, K. K., Trivedi, D. C., et al. (2007). Electrical Properties of Electrospun Fibers of PANI-PMMA Composites. *J Eng Fibers Fabrics*, 2, 25-31.

[14] Bishop, A., & Gouma, P. (2005). Leucoemeraldine Based Polyaniline-Poly-Vinylpyrrolidone Electrospun Composites and Bio-Composites: A Preliminary Study of Sensing Behavior. *Rev Adv Mater Sci*, 10, 209-214.

[15] Desai, K., Lee, J. S., & Sung, C. (2004). Nanocharacterization of Electrospun Nanofibers of Polyaniline/Poly Methyl Methacrylate Blends Using SEM, TEM and AFM. *Microsc Microanal*, 10, 556-557.

[16] Ju, Y. W., Park, J. H., Jung, H. R., et al. (2007). Electrochemical Properties of Polypyrrole/Sulfonated SEBS Composite Nanofibers Prepared by Electrospinning. *Electrochim Acta*, 52, 4841-4847.

[17] Schmidt, E., Shastri, V. R., Vacanti, J. P., & Langer, R. (1997). *Proc. Natl. Acad. Sci. Usa*, 94, 8948.

[18] De Giglio, E., Sabbatini, L., & Zambonin, P. G. J. (1999). *Biomater. Sci. Polym. Ed*, 10, 845.

[19] Garner, B., Georgevich, A., Hodgson, A. J., Liu, L., & Wallace, G. G. J. (1999). *Biomed. Mater. Res*, 44, 121.

[20] Valentini, R. F., Vargo, T. G., Gardellajr, J. A., & Aebischer, P. (1992). *Biomaterials*, 13, 193.

[21] Kotwal, A., & Schmidt, C. E. (2001). *Biomaterials*, 22, 1055.

[22] Xu, L. B., Chen, W., Mulchandani, A., & Yan, Y. (2005). Reversible Conversion of Conducting Polymer Films from Superhydrophobic to Superhydrophilic. . Angew. Chem. Int. Ed doi:10.1002/anie.200500868 , 44, 6009-6012.

[23] Rajeswari, R., Subramanian, S., Jayarama, R. V., Shayanti, M., & Seeram, R. (2010). Applications of Conducting Polymers and Their Issues in Biomedical Engineering. J. R. Soc. Interface published online 7 July, doi: 10.1098/rsif.2010.0120.focus.

[24] Street, G. B. (1986). Polypyrrole: from Powders to Plastics. In: Skotheim T. A., editor. *Handbook of conducting polymers*, vol. I. New York: Marcel Dekker, 265-291.

[25] Shirakawa, H., Louis, E. J., MacDiarmid, A. G., Chiang, C. K., & Heeger, A. J. (1977). Synthesis of Electrically Conducting Organic Polymers: Halogen Derivatives of Polyacetylene, (CH)x. *J Chem Soc Chem Commun*, 578-580.

[26] Heeger, A. J. (2001). Semiconducting and Metallic Polymers: the Fourth Generation of Polymeric Materials (Nobel Lecture). *Angew Chem Int Ed*, 40, 2591-2611.

[27] Feast, W. J. (1986). Synthesis of Conducting Polymers. In: Skotheim T. A., editor. *Handbook of conducting polymers*, vol. I. New York: Marcel Dekker, 1-43.

[28] Hong, S. Y., & Marnick, D. S. (1992). Understanding the Conformational Stability and Electronic Structures of Modified Polymers Based on Polythiophene. *Macromolecules*, 4652-4657.

[29] Guimard, N. K., Gomez, N., & Schmidt, C. E. (2007). Conducting Polymers in Biomedical Engineering. *Prog. Polym. Sci.*, 32, 876-921.

[30] Green, R. A., Baek, S., Poole-Warren, L. A., & Martens, P. J. (2010). Conducting Polymer-Hydrogels for Medical Electrode Applications. *Sci. Technol. Adv. Mater*, 11 014107 (13pp), doi: 10.1088/1468-6996/11/1/014107.

[31] Kamalesh, S., Tan, P., Wang, J., Lee, T., Kang, E. T., & Wang, C. H. (2000). *J. Biomed. Mater. Res.*, 52, 467.

[32] Wang, H. J., Ji, L. W., Li, D. F., & Wang, J. Y. (2008). *J. Phys. Chem. B*, 112, 2671.

[33] Shreyas, S. R., & Jessica, W. (2009). *Front. Neuroeng.*, 2, 6.

[34] Cogan, S. F. (2008). *Annu. Rev. Biomed. Eng.*, 10, 275.

[35] Li, G. C., & Pickup, P. G. (2000). *PCCP Phys. Chem. Chem. Phys.*, 2, 1255.

[36] Tourillon, G., & Garnier, F. (1983). *J. Electrochem. Soc.*, 130, 2042.

[37] Cui, X., & Martin, D. C. (2003). *Sensors Actuators B*, 89, 92.

[38] Yamato, H., Ohwa, M., & Wernet, W. J. (1995). *Electroanal. Chem.*, 397, 163.

[39] Ghasemi-Mobarakeh, L., Prabhakaran, M. P., Morshed, M., Nasr Esfahani, M. H., Baharvand, H., Kiani, S., & Al-Deyab, S. S. (2011). Ramakrishna S. Application of Conductive Polymers, Scaffolds and Electrical Stimulation for Nerve Tissue Engineering. *J Tissue Eng Regen Med*, 5, 17-35.

[40] Naghavi Alhosseini, S., Moztarzadeh, F., Mozafari, M., Asgari, S., Dodel, M., Samadikuchaksaraei, A., Kargozar, S., & Jalali, N. (2011). Synthesis and Characterization of Electrospun Polyvinyl Alcohol Nanofibrous Scaffolds Modified by Blending with Chitosan for Neural Tissue Engineering. *International Journal of Nanomedicine*, 6, 1-10.

[41] Otero, T. F., & Sansinena, J. M. (1998). Soft and Wet Conducting Polymers for Artificial Muscles. *Adv. Mater.*, 10, 491-494.

[42] Abidian, M. R., Kim, D. H., & Martin, D. C. (1998). Conducting Polymer Nanotubes for Controlled Drug Release. *Adv. Mater.*, 18, 405-409, doi:10.1002/adma.200501726.

[43] Abidian, M. R., Ludwig, K. A., Marzullo, T. C., Martin, D. C., & Kipke, D. R. (2009). Interfacing Conducting Polymer Nanotubes with the Central Nervous System: Chronic Neural Recording Using poly(3,4-ethylenedioxythiophene) Nanotubes. *Adv. Mater.*, 21, 3764-3770, doi:10.1002/adma.200900887.

[44] Schmidt, C.E., Shastri, V. R., Vacanti, J. P., & Langer, R. (1997). Stimulation of Neurite Outgrowth Using an Electrically Conducting Polymer. *Proc. Natl Acad. Sci. USA*, 94, 8948-8953.

[45] Warren, L. F., Walker, J. A., Anderson, D. P., Rhodes, C. G., & Buckley, L. J. (1989). A study of Conducting Polymer Morphology. *J. Electrochem. Soc.*, 136, 2286-2295.

[46] Breads, J. L., & Silbey, R. (1991). Conjugated Polymers. Kluwer Academic: Amsterdam, The Netherlands.

[47] Wong, J. Y., Langer, R., & Ingberi, D. E. (1994). Electrically Conducting Polymers Can Noninvasively Control the Shape and Growth of Mammalian Cells. *Proc Natl Acad Sci USA*, 91, 3201-3204.

[48] Sanchvi, A. B., Miller, K. P. H., Belcher, A. M., et al. (2005). Biomaterials Functionalization Using a Novel Peptide That Selectively Binds to a Conducting Polymer. *Nat Mater*, 4, 496-502.

[49] Huang, L., Hu, J., Lang, L., et al. (2007). Synthesis and Characterization of Electroactive and Biodegradable ABA Block Copolymer of Polylactide and Aniline Pentamer. *Biomaterials*, 28, 1741-1751.

[50] Wong, J. Y., Langer, R., & Ingber, D. E. (1994). Electrically Conducting Polymers Can Noninvasively Control the Shape and Growth of Mammalian Cells. *Proc Natl Acad Sci USA*, 91, 3201-3204.

[51] Moroder, P., Runge, M. B., Wang, H., Ruesink, T., Lu, L., Spinner, R. J., Windebank, A. J., & Yaszemski, M. J. (2011). Material Properties and Electrical Stimulation Regimens of Polycaprolactone Fumarate-Polypyrrole Scaffolds as Potential Conductive Nerve Conduits. *Acta Biomaterialia*, 7, 944-953.

[52] Nalwa, H. S. (1997). Handbook of Organic Conductive Molecules and Polymers. Wiley: New York.

[53] Mattioli-Belmonte, M., Giavaresi, G., Biagini, G., et al. (2003). Tailoring Biomaterial Compatibility: in Vivo Tissue Response Versus in Vitro Cell Behavior. *Int J Artif Organs*, 26, 1077-1085.

[54] Fryczkowski, R., & Kowalczyk, T. (2009). Nanofibres from Polyaniline/Polyhydroxybutyrate Blends. *Synthetic Metals*, 159, 2266-2268.

[55] Crispin, X., Marciniak, S., Osikowicz, W., Zotti, G., van Der Gon, A. W. D., Louwt, F., et al. (2003). *J Polym Sci Pol Phys*, 41, 2561.

[56] Sarac, A. S., Sonmez, G., & Cebeci, F. C. (2003). *J Appl Electrochem*, 33, 295.

[57] Breiby, D. W., Samuelsen, E. J., Groenendaal, L. B., & Struth, B. (2003). *J Polym Sci Pol Phys*, 41, 945.

[58] Jonsson, S. K. M., Birgersson, J., Crispin, X., Grezynsky, G., Osikowicz, W., van der Gon, A. W. D., et al. (2003). *Synthetic Met*, 139, 1.

[59] Ocampo, C., Oliver, R., Armelin, E., Alema´n, C., & Estrany, F. (2006). *J Polym Res*, 13, 193.

[60] Liesa, F., Ocampo, C., Alema´n, C., Armelin, E., Oliver, R., & Estrany, F. (2006). *J Appl Polym Sci*, 102, 1592.

[61] Marsella, M. J., & Reid, R. (1999). *J. Macromolecules*, 32, 5982.

[62] Otero, T. F., & Cortes, M. T. (2003). *Adv Mater*, 15, 279.

[63] Yu, H. H., Xu, B., & Swager, T. M. (2003). *J Am Chem Soc*, 125, 1142.

[64] Casanovas, J., Zanuy, D., & Alema´n, C. (2006). *Angew Chem Int Edit*, 45, 1103.

[65] Kros, A., Van Hovell, S. W. F. M., Sommerdijk, N. A. J. M., & Nolte, R. J. M. (2001). *Adv Mater*, 13, 1555.

[66] Bolin, M. H., Svennersten, K., Wang, X., Chronakis, I. S., Richter-Dahlfors, A., Jager, E. W. H., & Berggren, M. (2009). Nano-Fiber Scaffold Electrodes Based on PEDOT for Cell Stimulation. *Sensors and Actuators B*, 142, 451-456.

[67] Valentini, R. F., Vargo, T. G., Gardella, J. A., et al. (1992). Electrically Charged Polymeric Substrates Enhance Nerve Fibre Outgrowth in Vivo. *Biomaterials*, 13, 183-190.

[68] Weber, N., Lee, Y. S., Shanmugasundaram, S., Jaffe, M., & Arinzeh, T. L. (2010). Characterization and in Vitro Cytocompatibility of Piezoelectric Electrospun Scaffolds. *Acta Biomaterialia*, 6, 3550-3556.

[69] Heeger, A. J. (2002). Semiconducting and Metallic Polymers: the Fourth Generation of Polymeric Materials. *Synth Met*, 125, 23-42.

[70] Foulds, N. C., & Lowe, C. R. (1986). Enzyme Entrapment in Electrically Conducting Polymers. *J Chem Soc Faraday Trans*, 82, 1259-1264.

[71] Umana, M., & Waller, J. (1986). Protein Modified Electrodes: the Glucose/Oxidase/ Polypyrrole System. *Anal Chem*, 58, 2979-2983.

[72] Venugopal, J., Molamma, P., Choon, A. T., Deepika, G., Giri Dev, V. R., & Ramakrishna, S. (2009). Continuous Nanostructures for the Controlled Release of Drugs. *Curr. Pharm. Des.*, 15, 1799-1808.

[73] Adeloju, S. B., & Wallace, G. G. (1996). Conducting Polymers and the Bioanalytical Sciences: New Tools for Biomolecular Communication. *A review. Analyst*, 121, 699-703.

[74] Harwood, G. W. J., & Pouton, C. W. (1996). Amperometric Enzyme Biosensors for the Analysis of Drug and Metabolites. *Adv. Drug. Deliv. Rev.*, 18, 163-191, doi: 10.1016/0169-409X(95)00093-M.

[75] Stassen, I., Sloboda, T., & Hambitzer, G. (1995). Membrane with Controllable Permeability for Drugs. *Synth. Met.*, 71, 2243-2244, doi:10.1016/0379-6779(94)03241-W.

[76] Pernaut, J. M., & Reynolds, J. R. (2000). Use of Conducting Electroactive Polymers for Drug Delivery and Sensing of Bioactive Molecules. A Redox Chemistry Approach. *J. Phys. Chem. B*, 104, 4080-4090, doi:10.1021/jp994274o).

[77] Hodgson, A. J., John, M. J., Campbell, T., Georgevich, A., Woodhouse, S., & Aoki, T. (1996). Integration of Biocomponents with Synthetic Structures: Use of Conducting Polymer Polyelectrolyte Composites. *Proc. SPIE. Int. Soc. Opt. Eng.*, 2716, 164-176, doi: 10.1117/12.232137.

[78] Wadhwa, R., Lagenaur, C. F., & Cui, X. T. (2006). Electrochemically Controlled Release of Dexamethasone from Conducting Polymer Polypyrrole Coated Electrode. *J. Control. Rel.*, 110, 531-541, doi: 10.1016/j.jconrel.2005.10.027.

[79] Li, Y., Neoh, K. G., Cen, L., & Kang, E. T. (2005a). Controlled Release of Heparin from Polypyrrole-Poly(Vinyl Alcohol) Assembly by Electrical Stimulation. *J. Biomed. Mater. Res. A*, 73A, 171-181, doi:10.1002/jbm.a.30286.

[80] Abidian, M. R., Kim, D. H., & Martin, D. C. (2006). Conducting-Polymer Nanotubes for Controlled Drug Release. *Adv. Mater.*, 18, 405-409.

[81] Otero, T. F., & Cortes, M. T. (2003). A Sensing Muscle. *Sens. Actuat. B*, 96, 152-156, doi: 10.1016/S0925-4005(03)00518-5.

[82] Smela, E. (2003). Conjugated Polymer Actuators for Biomedical Applications. *Adv. Mater.*, 15, 481-494, doi: 10.1002/ adma.200390113.

[83] Otero, T. F., & Cortés, M. T. (2003). A Sensing Muscle. *Sensors and Actuators B*, 96, 152-156.

[84] Bartlett, P. N., & Whitaker, R. G. (1988). Modified Electrode Surface in Amperometric Biosensors. *Med. Biol. Eng. Comput.*, 28, 10-17, doi:10.1007/BF02442675.

[85] George, P. M., LaVan, D. A., Burdick, J. A., Chen, C. Y., Liang, E., & Langer, R. (2006). Electrically Controlled Drug Delivery from Biotindoped Conductive Polypyrrole. *Adv Mater*, 18, 577-581.

[86] Li, Y., Neoh, K. G., & Kang, E. T. (2005). Controlled Release of Heparin from Polypyrrole-Poly(Vinyl Alcohol) Assembly by Electrical Stimulation. *J Biomed Mater Res A*, 73A, 171-181.

[87] Chronakis, I. S., Grapenson, S., & Jakob, A. (2006). Conductive Polypyrrole Nanofibers via Electrospinning: Electrical and Morphological Properties. *Polymer*, 47, 1597-1603.

[88] Zhang, Q., Yan, Y., Li, S., et al. (2010). The Synthesis and Characterization of a Novel Biodegradable and Electroactive Polyphosphazene for Nerve Regeneration. *Mater Sci Eng C*, 30, 160-166.

[89] Bettinger, C. J., Bruggeman, J. P., Misra, A., et al. (2009). Biocompatibility of Biodegradable Semiconducting Melanin Films for Nerve Tissue Engineering. *Biomaterials*, 30, 3050-3057.

[90] Li, M., Guo, Y., Wei, Y., MacDiarmid, A. G., & Lelkes, P. I. (2006). Electrospinning Polyaniline-Contained Gelatin Nanofibers for Tissue Engineering Applications. *Biomaterials*, 27, 2705-2715.

[91] Buijtenhuijs, P., Buttafoco, L., Poot, A. A., Daamen, W. F., van Kuppevelt, T. H., Dijkstra, P. J., et al. (2004). Tissue Engineering of Blood Vessels: Characterization of Smooth-Muscle Cells for Culturing on Collagen-and-Elastinbased Scaffolds. *Biotechnol Appl Biochem*, 39, (Pt 2) 141-149.

[92] Lu, Q., Ganesan, K., Simionescu, D. T., & Vyavahare, N. R. (2004). Novel Porous Aortic Elastin and Collagen Scaffolds for Tissue Engineering. *Biomaterials*, 25(22), 5227-5237.

[93] Tan, K. H., Chua, C. K., Leong, K. F., Naing, M. W., & Cheah, C. M. (2005). Fabrication and Characterization of Three-Dimensional Poly(ether-ether-ketone)-Hydroxyapatite Biocomposite Scaffolds Using Laser Sintering. *Proc Inst Mech Eng [H]*, 219(3), 183-194.

[94] Di Martino, A., Sittinger, M., & Risbud, M. V. (2005). Chitosan: a Versatile Biopolymer for Orthopaedic Tissue-Engineering. *Biomaterials*, [Epub ahead of print].

[95] Kim, K., Yu, M., Zong, X., Chiu, J., Fang, D., Seo, Y. S., et al. (2003). Control of Degradation Rate and Hydrophilicity in Electrospun Non-Woven Poly(D,L-lactide) Nanofiber Scaffolds for Biomedical Applications. *Biomaterials*, 24, 4977-4985.

[96] Zong, X., Bien, H., Chung, C. Y., Yin, L., Fang, D., Hsiao, B. S., et al. (2005). Electrospun Fine-Textured Scaffolds for Heart Tissue Constructs. *Biomaterials*, 26(26), 5330-5338.

[97] Metzke, M., O'Connor, N., Maiti, S., Nelson, E., & Guan, Z. (2005). Saccharidepeptide Hybrid Copolymers as Biomaterials. *Angew Chem Int Ed Engl*, 44(40), 6529-6533.

[98] Boland, E. D., Matthews, J. A., Pawlowski, K. J., Simpson, D. G., Wnek, G. E., & Bowlin, G. L. (2004). Electrospinning Collagen and Elastin: Preliminary Vascular Tissue Engineering. *Front Biosci*, 9, 1422-1432.

[99] Li, M., Mondrinos, M. J., Gandhi, M. R., Ko, F. K., Weiss, A. S., & Lelkes, P. I. (2005). Electrospun Protein Fibers as Matrices for Tissue Engineering. *Biomaterials*, 26(30), 5999-6008.

[100] Rho, K. S., Jeong, L., Lee, G., Seo, B. M., Park, Y. J., Hong, S. D., et al. (2005). Electrospinning of Collagen Nanofibers: Effects on the Behavior of Normal Human Keratinocytes and Early Stage Wound Healing. *Biomaterials*, [Epub ahead of print].

[101] Riboldi, S. A., Sampaolesi, M., Neuenschwander, P., Cossu, G., & Mantero, S. (2005). Electrospun Degradable Polyesterurethane Membranes: Potential Scaffolds for Skeletal Muscle Tissue Engineering. *Biomaterials*, 26(22), 4606-4615.

[102] Ma, Z., Kotaki, M., Inai, R., & Ramakrishna, S. (2005). Potential of Nanofiber Matrix as Tissue-Engineering Scaffolds. *Tissue Eng*, 11(1-2), 101-109.

[103] Yang, F., Murugan, R., Wang, S., & Ramakrishna, S. (2005). Electrospinning of Nano/Micro Scale Poly(L-lactic acid) Aligned Fibers and Their Potential in Neural Tissue Engineering. *Biomaterials*, 26(15), 2603-2610.

[104] Khil, M. S., Bhattarai, S. R., Kim, H. Y., Kim, S. Z., & Lee, K. H. (2005). Novel Fabricated Matrix via Electrospinning for Tissue Engineering. *J Biomed Mater Res B Appl Biomater*, 72(1), 117-124.

[105] Khil, M. S., Cha, D. I., Kim, H. Y., Kim, I. S., & Bhattarai, N. (2003). Electrospun Nanofibrous Polyurethane Membrane as Wound Dressing. *J Biomed Mater Res B Appl Biomater*, 67(2), 675-679.

[106] Zeng, J., Yang, L., Liang, Q., Zhang, X., Guan, H., Xu, X., et al. (2005). Influence of the Drug Compatibility with Polymer Solution on the Release Kinetics of Electrospun Fiber Formulation. *J Control Release*, 105(1-2), 43-51.

[107] Buttafoco, L., Kolkman, N. G., Poot, A. A., Dijkstra, P. J., Vermes, I., & Feijen, J. (2005). Electrospinning Collagen and Elastin for Tissue Engineering Small Diameter Blood Vessels. *J Control Release*, 101(1-3), 322-324.

[108] MacDiarmid, A. G. (2001). Nobel lecture: Synthetic Metals: a Novel Role for Organic Polymers. *Rev Mod Phys*, 73, 701-712.

[109] Pedrotty, D. M., Koh, J., Davis, B. H., Taylor, D. A., Wolf, P., & Niklason, L. E. (2005). Engineering Skeletal Myoblasts: Roles of Three-Dimensional Culture and Electrical Stimulation. *Am J Physiol Heart Circ Physiol*, 288(4), H1620-1626.

[110] Azioune, A., Slimane, A. B., Hamou, L. A., Pleuvy, A., Chehimi, M. M., Perruchot, C., et al. (2004). Synthesis and Characterization of Active Esterfunctionalized Polypyrrole-Silica Nanoparticles: Application to the Covalent Attachment of Proteins. *Langmuir*, 20(8), 3350-3356.

[111] Arslan, A., Kiralp, S., Toppare, L., & Yagci, Y. (2005). Immobilization of Tyrosinase in Polysiloxane/Polypyrrole Copolymer Matrices. *Int J Biol Macromol*, 35(3-4), 163-167.

[112] Kim, D. H., Abidian, M., & Martin, D. C. (2004). Conducting Polymers Grown in Hydrogel Scaffolds Coated on Neural Prosthetic Devices. *J Biomed Mater Res A*, 71(4), 577-585.

[113] Kotwal, A., & Schmidt, C. E. (2001). Electrical Stimulation Alters Protein Adsorption and Nerve Cell Interactions with Electrically Conducting Biomaterials. *Biomaterials*, 22(10), 1055-1064.

[114] Sanghvi, A. B., Miller, K. P., Belcher, A. M., & Schmidt, C. E. (2005). Biomaterials Functionalization Using a Novel Peptide That Selectively Binds to a Conducting Polymer. *Nat Mater*, 4(6), 496-502.

[115] Lakard, S., Herlem, G., Valles-Villareal, N., Michel, G., Propper, A., Gharbi, T., et al. (2005). Culture of Neural Cells on Polymers Coated Surfaces for Biosensor Applications. *Biosens Bioelectron*, 20(10), 1946-1954.

[116] George, P. M., Lyckman, A. W., La Van , D. A., Hegde, A., Leung, Y., Avasare, R., et al. (2005). Fabrication and Biocompatibility of Polypyrrole Implants Suitable for Neural Prosthetics. *Biomaterials*, 26(17), 3511-3519.

[117] Wan, Y., Wu, H., & Wen, D. (2004). Porous-Conductive Chitosan Scaffolds for Tissue Engineering. 1. Preparation and Characterization. *Macromol Biosci*, 4(9), 882-890.

[118] Jiang, X., Marois, Y., Traore, A., Tessier, D., Dao, L. H., Guidoin, R., et al. (2002). Tissue Reaction to Polypyrrole-Coated Polyester Fabrics: an in Vivo Study in Rats. *Tissue Eng*, 8(4), 635-647.

[119] Bidez, P. R., Li, S., MacDiarmid, A. G., Venancio, E. C., Wei, Y., & Lelkes, P. I. (2006). Polyaniline, an Electroactive Polymer with Potential Applications in Tissue Engineering. *J Biomater Sci Polym*, 17(1-2), 199-212.

[120] Kamalesh, S., Tan, P., Wang, J., Lee, T., Kang, E. T., & Wang, C. H. (2000). Biocompatibility of Electroactive Polymers in Tissues. *J Biomed Mat Res*, 52(3), 467-478.

[121] Ahmad, N., & MacDiarmid, A. G. (1996). Inhibition of Corrosion of Steels with the Exploitation of Conducting Polymers. *Synth Met*, 78, 103-110.

[122] Yang, Y., Westerweele, E., Zhang, C., Smith, P., & Heeger, A. J. (1995). Enhanced Performance of Polymer Light-Emitting Diodes Using High-Surface Area Polyaniline Network Electrodes. *J Appl Phys*, 77, 694-698.

[123] MacDiarmid, A. G., Yang, L. S., Huang, W. S., & Humphrey, B. D. (1987). Polyaniline: Electrochemistry and Application to Rechargeable Batteries. *Synth Met*, 18, 393-398.

[124] Karyakin, A. A., Bobrova, O. A., Lukachova, L. V., & Karyakina, E. E. (1996). Potentiometric Biosensors Based on Polyaniline Semiconductor Films. *Sensors Actuators, B: Chem.*, B33(1-3), 34-38.

[125] Wei, Y., Lelkes, P. I., MacDiarmid, A. G., Guterman, E., Cheng, S., Palouian, K., et al. (2004). Electroactive Polymers and Nanostructured Materials for Neural Tissue Engineering. In: Qi-Feng Z., Cheng S. Z. D., editors. *Contemporary Topics in Advanced Polymer Science and Technology*, Beijing, China: Peking University Press, 430-436.

Chapter 15

Photonics of Heterogeneous Dielectric Nanostructures

Vladimir Dzyuba, Yurii Kulchin and
Valentin Milichko

Additional information is available at the end of the chapter

http://dx.doi.org/10.5772/50212

1. Introduction

Over last 20 years great scientific attention has been paid to nanostructures and nanocomposites based on nanoparticles of semiconductor materials (1eV<E_{gap}<3eV) since they exhibit a wide range of nonlinear properties and can be used in various applied fields [1-5]. However, the history of active investigation of dielectric nanostructures' properties started only recently [6-8]. These structures are the heterogeneous medium formed by liquid or solid dielectric matrices (e.g., polymer glasses and oils) and nanoparticles of dielectrics (Al_2O_3, SiO_2, MgO, etc.).

As some experiments have shown [6,8,9], such structures have nonlinear optical properties, whose dependence on the intensity and optical radiation wavelength is not typical of previously known nonlinear optical media [10-14]. The anomalous nonlinear optical properties are manifested in the fact that, firstly, the nanostructures' optical response on the radiation is of a non-thermal nature and occurs at radiation intensities below 1kW/cm^2 [9]. Second, despite the wide band gap of nanostructures' components (E_{gap}>3eV), this response takes place in the visible and infrared region of light spectrum, and reaches a maximum then decreases to zero under increasing intensity [8,9,15-17].

The dielectric nanostructures show other unexpected nonlinear optical properties. The nonlinear interaction of high-intensity radiation of different frequencies results in the generation of harmonics in conventional dielectric media. In the case of propagation of the low-intensity radiation of different frequencies in the dielectric nanostructures, the nonlinear interaction is manifested in the dependence of the light beam intensity on the intensity of another collinearly propagating light beam [18]. The two-frequency interaction observed in nano-

INTECH
open science | open minds

© 2012 Dzyuba et al.; licensee InTech. This is a paper distributed under the terms of the Creative Commons Attribution License (http://creativecommons.org/licenses/by/3.0), which permits unrestricted use, distribution, and reproduction in any medium, provided the original work is properly cited.

structures does not prevent the generation of harmonics, but this process requires radiation intensities four orders of magnitude higher.

The optical nonlinearity of dielectric nanostructures allows one to believe that they will be used to develop and create new optoelectronic [19-24] and fibre-optic devices to control [11], process and transmit the information [25]. Of no less interest are the prospects of using such nanostructures in new optical materials with controlled optical properties, in particular, photonic crystals [26] and the media generating optical solitons at low intensities. In addition, several international research groups have proposed using these structures to create the elements of electrical circuits, since the nonlinear properties appear in the range of THz and GHz radiation and under the influence of an electric potential [27-30].

The study of nonlinear optical properties of dielectric nanostructures containing nanoscale objects of different chemical natures, shapes and sizeshas shown that the existence of a low-threshold optical response is due to a number of conditions. The first is the presence of defect levels in the band gap of nanoparticles' charge carriers and this is manifested in the form of absorption bands in the nanoparticles' transmission spectrum [8,9,31-33]. Second, the radiation forming the nonlinear response of the nanostructure must have a frequency lying within the absorption band [8,9]. Third, the size and shape of the nanoparticles have to lead to the formation of a wide range of exciton states due to the quantum size effect [10,34-38]. Fourth, the matrix permittivity must be less than that of nanoparticle material, since the chemical nature of the matrix material significantly affects the formation of long-lived exciton states [9,31,39-42]. Fifth, the value of electric dipole moments induced by electrons phototransition should be substantially larger than dipole moments in the bulk material. It allows observing the optical nonlinearity of nanostructures with a low concentration of nanoparticles under low-intensity optical fields.

The theoretical description of the observed effects [43-45] is based on the fact that the occurrence of nontypical optical nonlinearity requires the existence of defect levels and the broad band of exciton states in the energy band gap of charge carriers. The radiation causes the electron transitions from the defect to the exciton levels, thereby creating the photo-induced population difference. This process is accompanied by the appearance of the nanoparticle electric dipole moment, herewith its module depends nonlinearly on the intensity and light wavelength. The theory conclusions and theoretical modelling of transmission spectrum and the behaviour of the nonlinear refractive index are very similar to the experimental results [9].

It follows from the theory that the nature of the nonlinearity is determined not only by the behaviour of the photo-induced dipole moment module in an external field, but also by the nanoparticle's orientation along the vector E. However, this orientation has a minor contribution to the nonlinearity, so the observed nonlinear optical response can take place in the case of unpolarized light and solid nanostructures, which is in agreement with the experiment.

This chapter is an original quantitative study of the nonlinear refraction and absorption of continuous low-intensity laser radiation in different heterogeneous dielectric nanostructures and compares these data with theoretical ones. In addition, the theory of nonlinear light transmission by dielectric nanostructures is discussed.

http://dx.doi.org/10.5772/50212

2. Experimental Chapter

2.1. Dielectric Nanocomposite Preparation and Spectral Features

To study the changes of optical characteristics of the heterogeneous dielectric nanostructuresunder continuous low-intensity radiation we used the dielectric Al_2O_3, SiO_2, TiO_2 and ZnO nanoparticles with 7,2; 8; 3,4 and 3,3eV band gaps of the bulk samples respectively [46]. The nanoparticles were purchased from Sigma Aldrich Company and investigated by AFM microscopy (Figures 2,3). The Al_2O_3 nanoparticles were used from work [9]. The averaged dimensions of nanoparticles are 45nm in diameter and 6nm in height for Al_2O_3, 20nm in diameter and 10nm in height for SiO_2, 15nm in diameter and 5nm in height for TiO_2 and 50nm in diameter and 40nm in height for ZnO. As a matrix for nanoparticles, dielectric immersion and transformer oils were used. The immersion oil consists of weak polar molecules and is based on cedar resin witha negative temperature gradient of refractive index $|dn/dT|=4*10^{-4}$. The transformer oil is a PDMS (polydimethylsiloxane) liquid having an optical transparency up to 200nm, chemical inertness, a high heat resistance ($|dn/dT|<10^{-7}$) and high stability of dielectric characteristics.

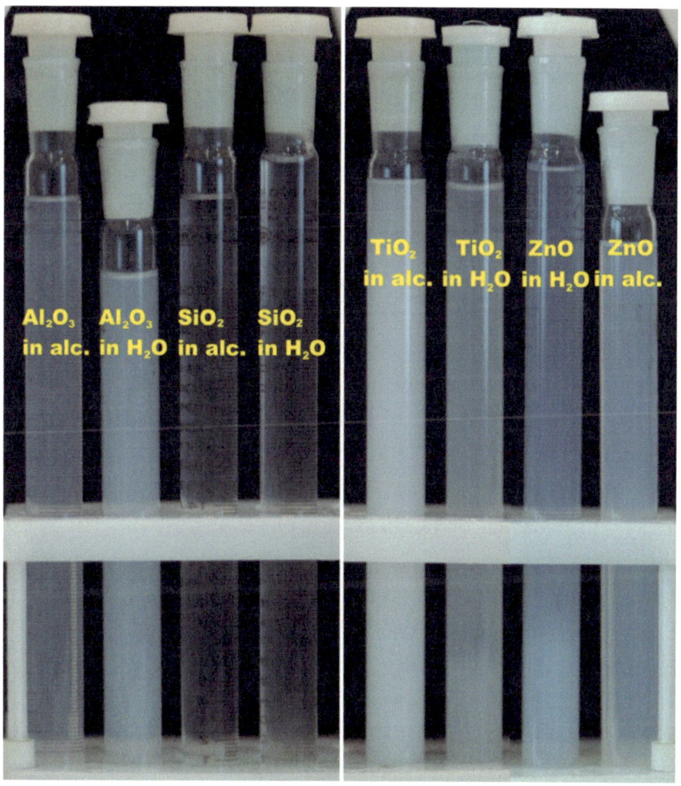

Figure 1. The nanoparticles suspension in isopropyl alcohol and distilled water.

The nanopowders were dissolved in isopropyl alcohol to precipitate the particles sticking together (Figure 1). After precipitation the upper isopropyl alcohol layers containing separate nanoparticles with small deviation in size were added into the oil (Al_2O_3 into immersion oil; SiO_2, TiO_2 and ZnO into PDMS) pre-heated to 40°C. Slow heating of the mixture to 60°C resulted in the appearance of convection currents which, in turn, form the uniform nanoparticles' distribution over the entire suspension volume and lead to alcohol evaporation. Then heterogeneous dielectric nanostructures (hence forth called the HDN) with nanoparticle volume concentration >1% were placed in a quartz cuvette 5 mm thick and 18.7 mm in length.

The nonlinear response of the medium on the radiation of certain frequencies can take place if this medium has nonlinear spectral characteristics that are related directly to the energy spectrum structure of charge carriers. When dealing with a nanoparticle we can expect the energy spectrum of its charge carriers to depend on the form and degree of nanoparticle surface development. Besides, the energy spectrum will depend on the matrix material and, to a greater extent, on its permittivity ε. Therefore, we studied the transmission spectra of Al_2O_3 (permittivity of the bulk sample is $\varepsilon_{stat}=10$), SiO_2 ($\varepsilon_{stat}=4,5$), TiO_2 ($\varepsilon_{stat}>86$) and ZnO ($\varepsilon_{stat}=8,8$) nanoparticles suspended in isopropyl alcohol ($\varepsilon_{stat}=24$), distilled water ($\varepsilon_{stat}=80$) and oil ($\varepsilon_{stat}=2,5$).

As it follows from the transmission spectra figures (Figure 4), the nanoparticles of broad-band dielectrics (Al_2O_3 and SiO_2) suspended in oil have a non-symmetric broad absorption band that is formed by exciton states with high density. The asymmetry of the absorption band is explained due to the broadening of the exciton levels. This band is not observable either in the bulk sample or in the nanoparticles' array suspended in other media.This can be explained by the fact that the electronic structure of nanoparticles embedded in a matrix depends strongly on the ratio between the permittivity of matrix ε_1 and nanoparticles ε_2 [39,40]. Given $\varepsilon_1/\varepsilon_2>1$, polarization interaction leads to attraction of positive charges to the inner surface of the nanoparticles and to the destruction of defect states by virtue of interaction the nanoparticle electrons from these levels with high-polarized matrix molecules. If $\varepsilon_1/\varepsilon_2<1$, polarization interaction causes repulsion of charges from the nanoparticle's surface into the interior, thus preserving these states. So, the propagation of visible radiation ($2,1eV<E<3,1eV$) into the HDN based on oil and Al_2O_3 nanoparticles results in electron transition from defect to exciton levels (Figure 5). In the case of water and alcohol matrixes there are no free electrons on the defect level, therefore, electron transition under ultraviolet radiation ($E>4eV$) is available. The same situation is exists for the HDN based on oil and SiO_2 nanoparticles: the exciton generation occurs under visible and ultraviolet radiation ($2eV<E<6eV$), however, less probable electron transitions from valence band to exciton levels can take place provided $E>2eV$ (Figure 5).

The TiO_2 and ZnO nanoparticles' array transmission spectra (Figure 6) inform that the nanoparticles of narrow-band dielectrics suspended in oil have a blurred edge of fundamental absorption that is formed by exciton states without any absorption band within (400-700)nm. The matrix permittivity effects only the position of fundamental absorption.

http://dx.doi.org/10.5772/50212

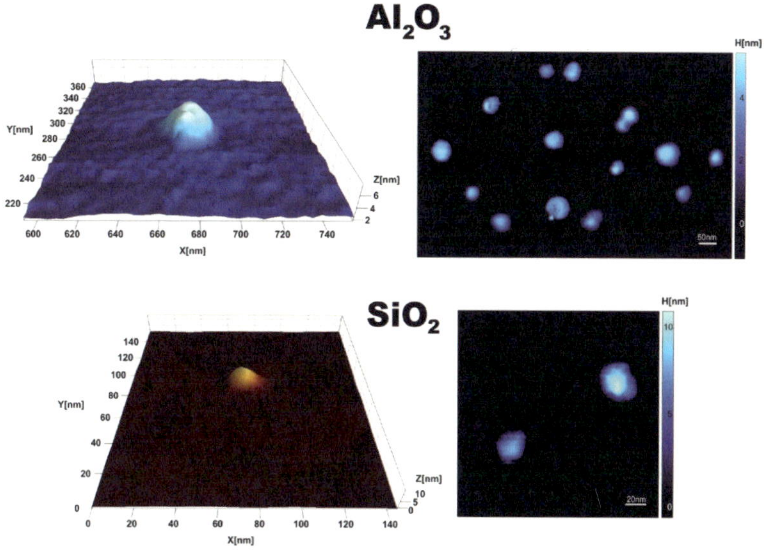

Figure 2. AFM images of Al_2O_3 and SiO_2 nanoparticles precipitated on a mica place. Defined dimensions are 45nm in diameter and 6nm in height for Al_2O_3, 20nm in diameter and 10nm in height for SiO_2.

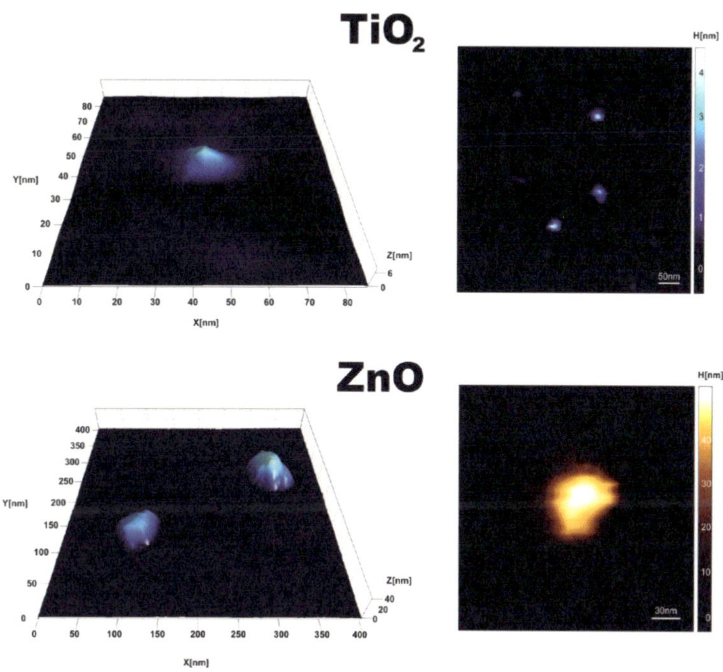

Figure 3. AFM images of $TiO_2$and ZnOnanoparticles precipitated on a mica place. Defined dimensions are15nm in diameter and 5nm in height for TiO_2, 50nm in diameter and 40nm in height for ZnO.

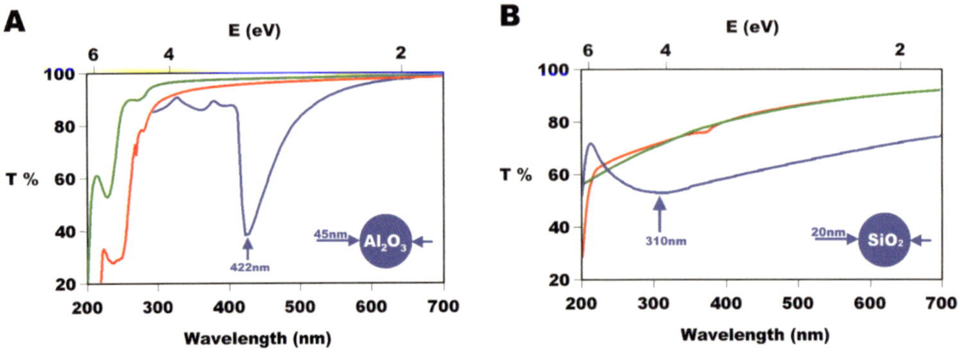

Figure 4. Transmission spectra of Al_2O_3 (A) and SiO_2 (B) nanoparticles' array dispersed in H_2O (green curve), isopropyl alcohol (red curve) and oil (blue curve). The curves were obtained by division of the T spectrum of nanoparticle suspension by that of the matrix.

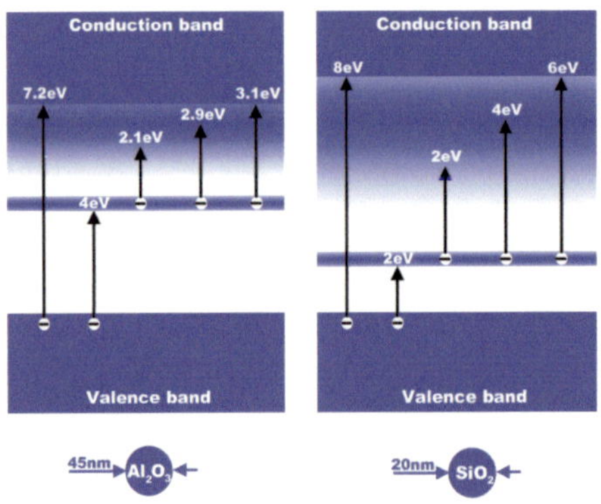

Figure 5. The energy band gap structure of Al_2O_3 and SiO_2 nanoparticles' array.

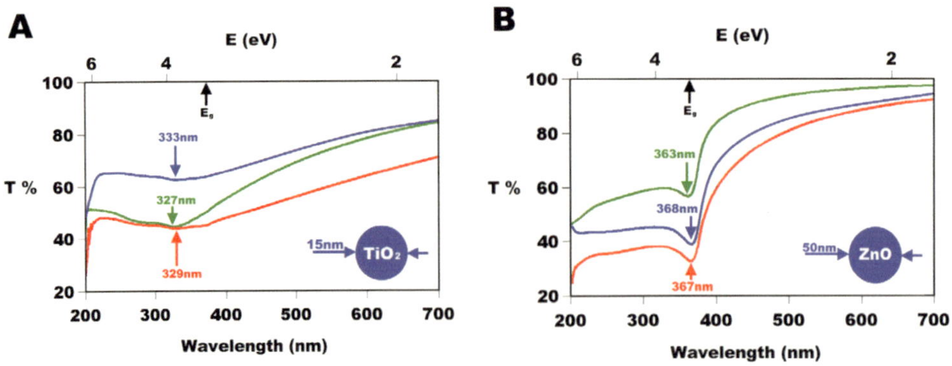

Figure 6. Transmission spectra of TiO_2 (A) and ZnO (B) nanoparticles' array dispersed in H_2O (green curve), isopropyl alcohol (red curve) and oil (blue curve). The curves were obtained by division of the T spectrum of nanoparticle suspension by that of the matrix.

http://dx.doi.org/10.5772/50212

2.2. Z-scan Experiment

The experimental study of the dependence of the HDN nonlinear optical response on the laser radiation of variable intensity was performed using the standard z-scan technique with open and closed apertures [47,48] (Figure 7). Since the nonlinear optical response of the HDN appears under low-intensity optical fields we used the semiconductor lasers providing focal intensities of up to 500 W/cm^2 for green and violet radiation. Such low intensities are at least four orders of magnitude lower than the pulsed mode intensity required for appearance of the nonlinear response in previously known environments [10-14].

In the z-scan experiments, the HDN based on Al_2O_3, SiO_2, TiO_2 and ZnO nanoparticles can be considered as thin experimental samples, since the interaction length of radiation with the samples is equal to L=5mm, being less than minimum Rayleigh range [47] $Z_0=(\pi\omega_0^2)/\lambda$=8.7mm.

The values the changes in refractive index Δn (I,λ) and absorption coefficient Δα(I, λ) one can calculate by the following:

$$\Delta n(I,\lambda)=\frac{\lambda\Delta T_{pv}\ln{}^{I_0}/_{I}}{0.812\pi(1-S)^{0.27}L\left(1-{}^{I}/_{I_0}\right)} \qquad (1)$$

$$\Delta a(I,\lambda)=\frac{2\sqrt{2}\Delta T}{L} \qquad (2)$$

that has been derived from the number of expressions:

$$\Delta T_{pv}\approx 0.406(1-S)^{0.27}\,|\,\Delta\Phi_0\,| \qquad (3)$$

$$\Delta\Phi_0=\frac{2\pi}{\lambda}n_2 I_0 L_{eff} \qquad (4)$$

where λ is a radiation wavelength and I_0 and Iare the input and output intensities, respectively. S is a fraction radiation transmitted by the aperture in the absence of the sample (S=0.04 and 0.06 provided the HDN based on Al_2O_3 in green and violet optical field, respectively; S=0.22 and 0.35 provided the HDN based on SiO_2, TiO_2, ZnO in green and violet optical field, respectively), $L_{eff}=L^*(1-e^{-\alpha L})$, L and α are the sample length and absorption coefficient, respectively and ΔT is a normalized change in integral transmitted intensity. The ratio of I/I_0 was determined due to the transmittance characteristics of the HDN (Figure 8). Since $I/I_0=(P_{out}S_0)/(P_{in}S)$, where P_{in} and P_{out} are input and output radiation powers, S_0 and S are the beam squares into the sample, and near the outer surface, the ratio P_{out}/P_{in} was defined from Figure 8 and S/S_0 negligibly exceeds the unite.

Linear behaviour of the HDN transmittance under increasing optical field is not reflected in the real behaviour that was most detailed in the study by z-scan with open aperture and will be discussed below. However, the z-scans have shown the changes of absorption ΔT<5% that is absolutely imperceptible in Figure 8.

It is necessary to clarify that the z-scan of low-intensity continuous radiation gives information about the total impact of all physical processes, excited by radiation in the matrix and nanoparticles, on the optical properties of the HDN.Thus, we have to divide the matrix effect from the nonlinear response caused by presence of nanoparticles.

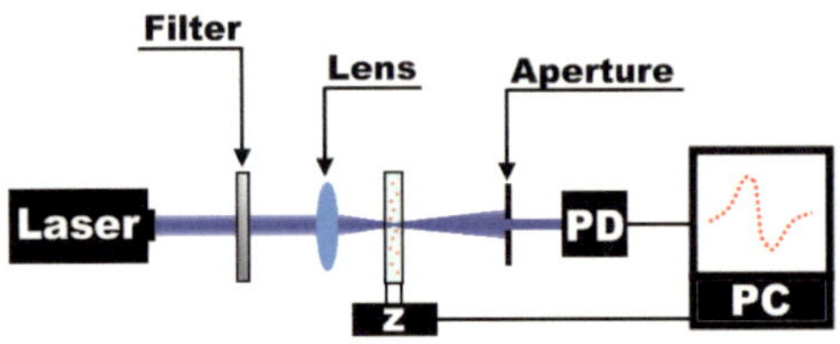

Figure 7. The experimental setup used for z-scan experiments. Setup includes: semiconductor sources of coherent continuous radiation (wavelengths of 532 and 442 nm with maximum power 22 and 35mW and beam diameters 1,1 and 0,95 mm, respectively); the PD-power photodetector; 75mm (focus diameters ω_0=71mkm for the green radiation and 92mkm for the violet) and 50mm (ω_0=46 mkm for the green radiation and 90 mkm for the violet) lens; Z is z shift and PC is a computer.

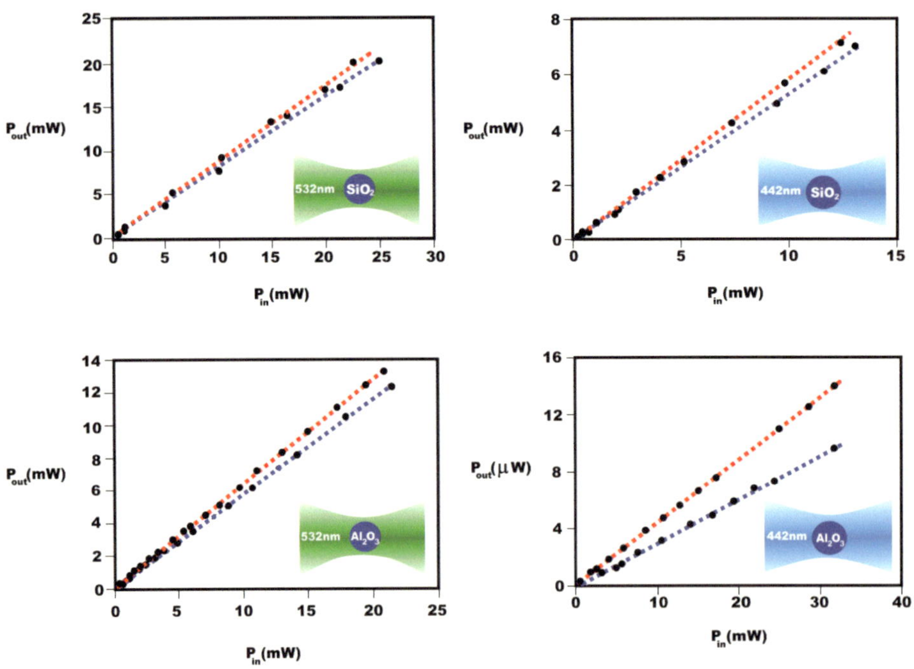

Figure 8. The integral output power P_{out} as a function of input power P_{in} of green and violet radiation propagating in the oil (red curve) and the HDN (blue curve).

http://dx.doi.org/10.5772/50212

The results of measurements of the change in Δn of the HDN based on Al_2O_3 nanoparticles obtained by z-scan are shown in Figure 9. Considering the absorption of radiation, we used z-scans with open aperture [9].The obtained results demonstrate a linear absorption of green radiation in the samples of pure oil and the HDN (normalized transmittance is equal to 1 for all z) within intensity range (0; 300) W/cm^2.However, the experiments revealed an absorption saturation of violet radiation. These experimental data have been used to adjust the normalized transmittance curves obtained with the use of the closed aperture technique, in accordance with the well-known method [47,48].

The results of the z-scan with closed aperture revealed the negative thermal change in the matrix refractive index and nonlinear negative change in the HDN refractive index. Since all physical processes in the matrix and nanoparticles, forming the change of the HDN refractive index under optical field, can be considered to a first approximation as independent processes $\Delta n_{HDN} = \Delta n_{matrix} + \Delta n_{nanopart}$, so we can assume that Δn (I,λ) for the HDN is a nonlinear function of input radiation intensity (Figure 9).

To study the effect of the matrix static permittivity on the electronic structure and optical properties of nanoparticles' array, we also investigated Al_2O_3nanoparticles suspended in isopropyl alcohol and distilled water according to the z-scan technique.The experiment showed a total absence of the nonlinear response in these media, being in agreement with our description in the introduction.

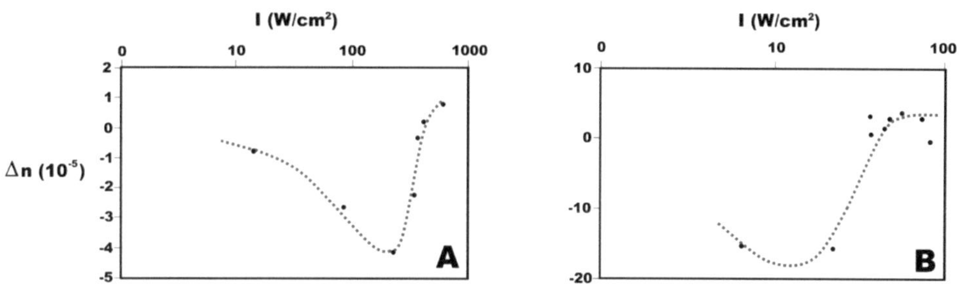

Figure 9. Change of refractive index Δn (I,λ) of the HDN based on Al_2O_3 nanoparticles under green (A) and violet (B) radiation. The curves were obtained by division of Δn for the HDN by Δn for the immersion oil.

The results of measurements of the refractive index and absorption coefficient of the HDN based on SiO_2 obtained by z-scan are shown in Figures 10 and 11. Since the transformer oil characteristics are not significantly changed within intensity range (0;500) W/cm^2, so the z-scan with close and open aperture informs only about changes in the HDN optical parameters caused by nanoparticles. The obtained results demonstrate the nonlinear refraction of laser radiation (Figure 10A,C and 11A,C) in the HDN within intensity range (0; 300)W/cm^2. In addition, the experiments revealed the nonlinear absorption of green and violet radiation in the HDN (Figure 10B,D and 11B,D) within the same intensity range. However, the HDN based on the nanoparticles of the narrow-band dielectrics (TiO_2 and ZnO) did not reveal any nonlinear changes in refraction and absorption under continuous low-intensity radiation.

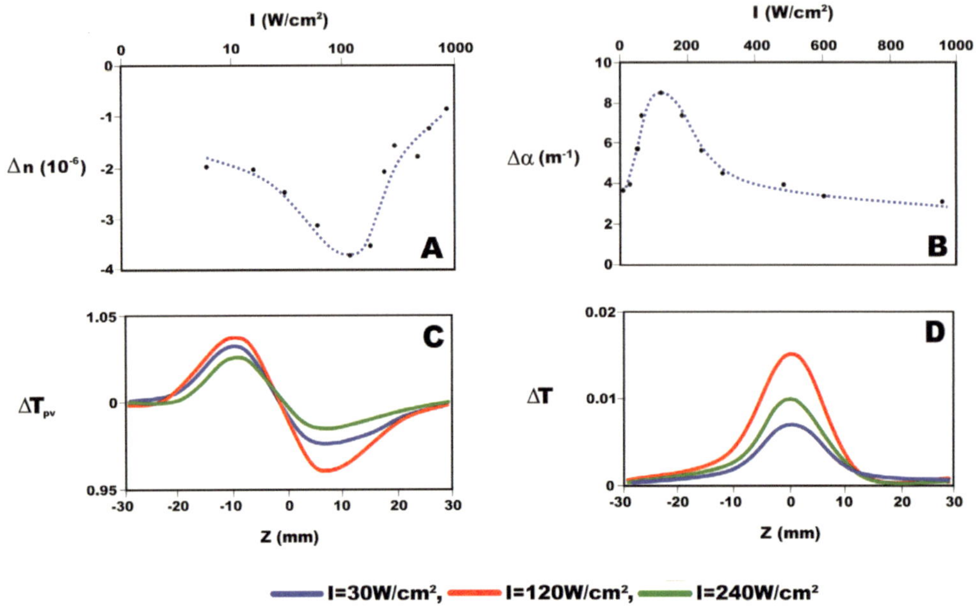

Figure 10. Change of refractive index Δn (A) and absorption coefficient Δα (B) of the HDN based on SiO_2 nanoparticles under green radiation; C and D are the approximation of z-scan results with close and open aperture, respectively.

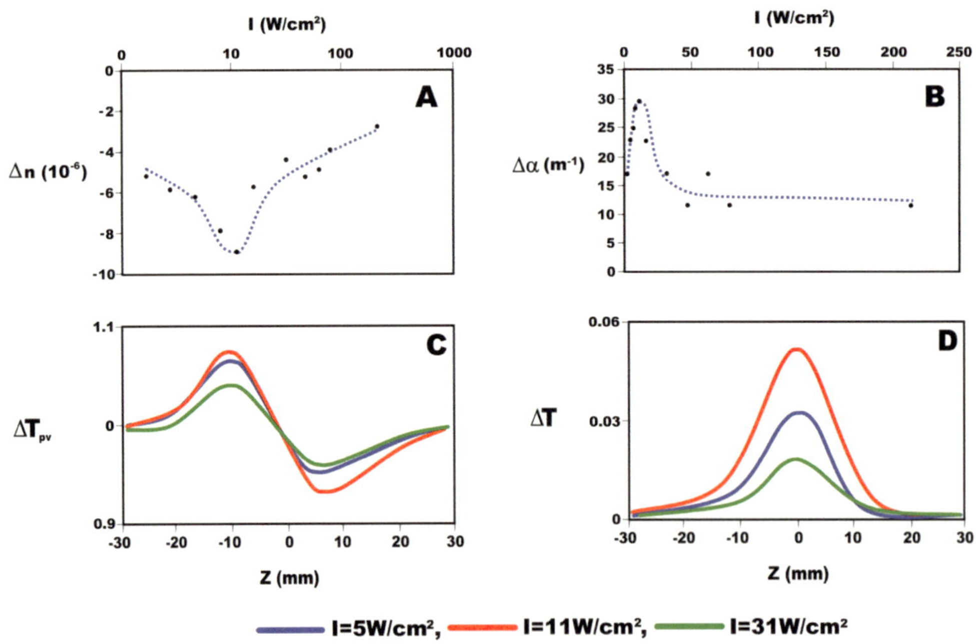

Figure 11. Change of refractive index Δn (A) and absorption coefficient Δα (B) of the HDN based on SiO_2 nanoparticles under violet radiation; C and D are the approximation of z-scan results with close and open aperture, respectively.

3. The Theory of The HDN Nonlinear Transmittance

3.1. Preface

The theoretical description of the physical and optical properties of heterogeneous nanocomposites containing nanoparticles is a complex problem. In fact, it seems impossible to correctly calculate the physical characteristics of an individual nanoparticle as a system consisting of a great number of particles obeying the quantum mechanics laws. Attempts to apply the well-known methods of solid state physics to describe the nanoparticles' properties run into problems, since it is not possible to disregard the effects caused by surface defects, as well as crystal lattice defects. It is known that the optical properties of a quantum mechanical system are associated with the features of the energy spectrum of charge carriers (electrons and holes).

At the present time, it is beyond all question that the optical and electric properties of nanoparticles have wide differences with that of the bulk samples due to the features of the energy spectra. These differences are caused by three effects. First, the band gap of nanoparticle charge carriers contain the allowed energies zone, herewith, the energy structure defined by the high density of surface structural defects and the irregular shape of nanoparticles. Second, the excitons and discrete energy spectra are formed below and into the conduction band due to the small nanoparticle size and size-quantization effect, respectively. In turn, the size quantization effect is caused by spatial confinement of the charge carriers' wave functions. Third, the electric dipole moments of electronic transitions in such quasi zero dimensional systems can be larger than that of the bulk sample. The formation of the above mentioned states is of threshold character, herewith, the threshold depends on the nanoparticle dimensions. Specifically, for a spherical nanoparticle (with the permittivity ε_2) dispersed in a medium (ε_1), such states can be formed if the nanoparticle radius $\alpha$is smaller than some critical radius α_c:

$$a \leq a_c = 6 \mid \beta \mid^{-1} a_{e,h} \tag{5}$$

where

$$\frac{\varepsilon_1 - \varepsilon_2}{\varepsilon_1 + \varepsilon_2} \tag{6}$$

Here $\alpha_{e,h}$is the Bohr radius of charge carriers in the nanoparticle material [49].

Some properties of the quantum states' spectrum can be clarified by studying the nanocomposites' transmittance spectra. As a rule, experimental studies are concerned with the transmittance spectra of nanoparticles' arrays embedded in a solid matrix or deposited on the transparent material surface. In this case, the electronic structure of nanoparticles is substantially influenced by the matrix material and the interaction between nanoparticles. Because of these effects, it is not possible to consider the transmittance spectra as the spectra of no interacting nanoparticles' arrays. Nanocomposites containing low concentrations of nanoparticles almost satisfy the condition for the lack of the above mentioned interactions, however, to study the optical properties of such composites, one cannot take into account the effects of the optical field on the distribution of the particles throughout degrees of freedom. In this case, given the low-intensity radiation, the optical field effect on the coordinates of a gravity centre of a nanoparticle can be disregarded, that cannot be said about the distribution of particles throughout the rotational degrees of freedom.

At the present time, there is no well-known theoretical approach taking into account not only the characteristic of nanoparticle dimensions, but also the orientation of nanoparticles in the external field of laser radiation, the dependences of the scattering and absorption cross sections on the propagating radiation intensity. In this context, it is necessary to develop a theoretical model of the scattering and absorption cross sections in dielectrics nanocomposites with the above mentioned features of such systems.

In this study, we suggest a semiphenomenological model of the optical transmittance of the array of noninteracting small sized ($\alpha<\alpha_c$) dielectric nanoparticles embedded in the dielectric matrix. We show that the basic mechanisms of low-threshold effects of nonlinear scattering and absorption of laser radiation in the HDN are: the photo induction of electric dipole moments of nanoparticles in the external optical field; the orientation of nanoparticles along the polarization direction of this field. In addition, we will discuss the behaviour of the HDN transmittance in the central frequency vicinity of the absorption band and the dependence of the band depth on the radiation intensity.

3.2. The Theoretical Approach

We consider the HDN consisting of the low concentration (the number of nanoparticles N per unit volume) of dielectric nanoparticles embedded in an isotropic transparent dielectric matrix with a small coefficient of viscosity and linear optical properties within the visible spectral range. In our case, the multiple scattering of radiation by nanoparticles and the nanoparticles' interaction with each other can be neglected. Let us introduce two coordinate systems with the same origin (Figure 12).

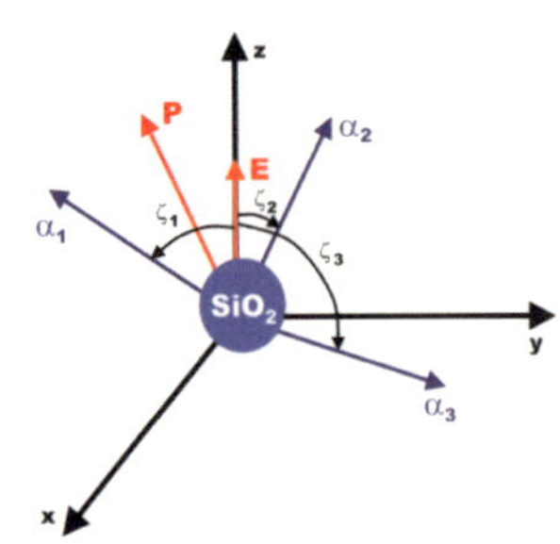

Figure 12. The coordinate system used in the theoretical study.

One of the systems $\{\alpha_1, \alpha_2, \alpha_3\}$ corresponds to the coordinates coinciding with the principle axes of the particle polarization tensor with the unit vectors (n_1, n_2, n_3). The other system is the Cartesian laboratory coordinate system {x, y, z} with the unit vectors (n_x, n_y, n_z). We suggest that the electromagnetic wave polarized along the zaxis E ={0,0,E} is incident on the composite. We chose the xaxis to be directed collinearly with a wave vector.

The optical transmittance of the HDN depends on the extinction coefficient, the path of the light beam in the material and the optical reflectance from the HDN boundary. For normal incidence of the light beam onto the boundary of the planar nanocomposite layer arranged normally to the xaxis, the transmittance expression can be written as [50]

$$T(\omega, N) = \frac{(1 - R^2)^2 \quad e^{\beta L}}{1 - R^2 \quad e^{-2\beta L}} \tag{7}$$

Here, β is the extinction coefficient, R is the optical reflectance of the boundary (in experiments R is much smaller than unit) and L is the interaction length of light beam with the HDN.

In the case of single scattering approximation, the extinction coefficient can be expressed in terms of the scattering $\sigma^s(\omega,a)$ and absorption $\sigma^a(\omega,a)$ cross sections of the HDN unit volume as

$$\beta(\omega, a) = \sigma^a(\omega, a) + \sigma^s(\omega, a) + a^m(\omega) \tag{8}$$

where $\alpha^m(\omega)$ is the extinction coefficient of the matrix material and a is the characteristic of nanoparticle dimension. For the above indicated orientation of the nanocomposite layer, the scattering and absorption cross sections in the laboratory coordinate system can be expressed in terms of the polarizability component of the HDN unit volume $\chi_{zz}(\omega,a)$ by the relations [50,51]

$$\begin{aligned} \sigma^a(\omega, a) &= \frac{4\pi\omega}{c} \mathrm{Im}\chi_{zz}(\omega, a) \\ d\sigma^s(\omega, a) &= \frac{\omega^4}{c^4} \mid \chi_{zz}(\omega, a) \mid {}^2 \sin^2\theta \quad d\Omega \end{aligned} \tag{9}$$

Here θ is the angle of the vector directed along the scattering direction and cis the light speed in vacuum.

We introduce the effective polarizability tensor for nanoparticle in the matrix $\alpha = \{\alpha_{ij}\}$ in such a way that the components of the nanoparticle electric dipole moment Pinduced by the external plane polarized monochromatic electromagnetic field E with the frequency ω are determined directly in terms of the external field rather than the local field $P_i=\alpha_{ij}E_j$. In the coordinate system $\{\alpha_1, \alpha_2, \alpha_3\}$, the polarization vector of the nanoparticle is

$$P = \sum_{j}^{3} a_{ij}(n_j E)n_j \tag{10}$$

If the vector E is directed along the z axis, the zcomponent of the polarization vector is

$$P_z = \sum_{j}^{3} a_{ij} E(n_j n_z)^2 \sum_{j}^{3} a_{ij} E \cos^2\theta_j \tag{11}$$

Here θ_j is the angle between vector E and the α_j axis; this angle specifies the nanoparticle orientation in the external electromagnetic field in the laboratory coordinate system. Since the nanoparticles are randomly oriented, we assume that the polarizability tensor of the medium $\chi=\{\chi_{ij}\}$ is diagonal and the polarization vector of the HDN unit volume in the laboratory coordinate system is $P_z=\chi_{zz}E$. Comparing this expression with (11), we obtain

$$\chi_{zz} = N\left(a_{11}\cos^2\theta_1 + a_{22}\cos^2\theta_2 + a_{33}\cos^2\theta_3\right) \tag{12}$$

After simple transformation, taking into account that $\cos^2\theta_1 + \cos^2\theta_2 + \cos^2\theta_3 = 1$, we can obtain an expression that relates the component of χ_{zz} in the laboratory coordinate system with the diagonal components of the nanoparticle polarizability tensor in the principle axes system:

$$\chi_{zz} = N\left(a_0 + \Delta a_1 Q_1 + \Delta a_2 Q_2\right) \tag{13}$$

where

$$\begin{aligned} a_0 &= \frac{a_{11} + a_{22} + a_{33}}{3} \\ \Delta a_1 &= a_{11} - a_{33} \\ \Delta a_2 &= a_{22} - a_{33} \end{aligned} \tag{14}$$

The values averaged over all possible orientations

$$\begin{aligned} Q_1 &= \left\langle \cos^2\theta_1 - \frac{1}{3} \right\rangle \\ Q_2 &= \left\langle \cos^2\theta_2 - \frac{1}{3} \right\rangle \end{aligned} \tag{15}$$

are the orientation order parameters of the nanoparticles ensemble in the external field. The angle distribution function of nanoparticles and, hence, the order parameters Q_1 and Q_2, depend on the laser radiation intensity and, via the components α_{ij}, on the radiation frequency. The quantities Q_1 and Q_2 as functions of the intensity exhibit the saturation at I>Ip irrespective of the matrix material.

The low-threshold nonlinear optical response takes place if the transmittance spectrum of the nanoparticles' array exhibits the broad absorption bands lacking in the bulk sample spectrum [31-33]. The polarizability tensor components $\alpha_{ij}$of the nanoparticle are to reach their maxima corresponding to dipole transitions of charge carriers from the state <n|to the state |g> within this frequency region. In addition, it is known that the diagonal tensor components in the coordinate system of the principal axes within this frequency region can be expressed as [52]

$$a_{jj}(\omega)=\sum_{n,g}\frac{|\langle n|er_j|g\rangle|^2}{\Box(\omega-\omega_{ng}+i\Gamma_{ng})}\Delta\rho_{ng} \tag{16}$$

The summation is performed over all allowed optical transitions of charge carriers of the nanoparticles with the frequency transition $\omega_{ng}$from the states <n|to the states |g>, being the component of the electric dipole moment of the transition $p^j{}_{ng}$=<n|er_j|g> and the transitions width Γ_{ng}. We can write the expression for only one nonzero polarizability tensor component in the laboratory coordinate system related to the individual nanoparticle using expressions (13) and (16), and introducing the definition $\Delta\omega_{ng}=\omega-\omega_{ng}$:

$$\chi(\omega, Q_1, Q_2)=N\sum_{n,g}\left[\frac{A_{ng}\Delta\omega_{ng}}{\Box\left(\Delta\omega_{ng}^2+\Gamma_{ng}^2\right)}-i\frac{A_{ng}\Gamma_{ng}}{\Box\left(\Delta\omega_{ng}^2+\Gamma_{ng}^2\right)}\right]\Delta\rho_{ng} \tag{17}$$

The next definition is included into the expression (17)

$$A_{ng}(Q_1, Q_2)=\frac{1}{3}|p_{ng}|^2+Q_1\left(|p_1^{ng}|^2-|p_2^{ng}|^2\right)+Q_2\left(|p_3^{ng}|^2-|p_2^{ng}|^2\right) \tag{18}$$

The quantity A_{ng} is proportional to the squared magnitude of the dipole moment of transitions from the state <n|to the state |g> provided certain optical radiation intensities, frequencies and specified parameters of the nanocomposite matrix. The population difference induced by radiation between the states <n| and |g> is a function of the incident radiation intensity. Using a two-level system approximation [52], this difference is

$$\Delta\rho_{ng}(I)=\left(1-\frac{I/I_S}{\Delta\omega_{ng}^2+\Gamma_{ng}^2\left(1+I/I_S\right)}\Gamma_{ng}^2\right)\Delta\rho_{ng}^0 \tag{19}$$

where $\Delta\varrho^0{}_{ng}$ is the thermal-equilibrium difference and I_S is the intensity of saturation, when the $\Delta\varrho/2$ carriers are in the upper energy level. Separating the real and imaginary parts of the polarizability tensor component (17) and taking into account the expression (19) we introduce the definitions

$$P = \int \sin^2\theta d\Omega$$

$$B_{ng}(\omega, T) = \frac{I/I_S}{\Delta\omega_{ng}^2 + \Gamma_{ng}^2\left(1 + I/I_S\right)}\Gamma_{ng}^2 \quad (20)$$

we obtain the integrated scattering and absorption cross sections of the united volume of the HDN in a single scattering approximation:

$$\sigma_a(\omega, a, I) = \frac{4\pi\omega N}{c\square}\sum_{n,g}\frac{A_{ng}\Gamma_{ng}\Delta\rho_{ng}^0}{\Delta\omega_{ng}^0 + \Gamma_{ng}^2}(1 - B_{ng}) \quad (21)$$

$$\sigma_S(\omega, a, I) = \frac{\omega^4 P N^2}{c^4\square^2}\sum_{n,g}\sum_{k,l}\left\{A_{ng}A_{kl}\frac{(\Delta\omega_{ng}\Delta\omega_{kl} + \Gamma_{ng}\Gamma_{kl})\Delta\rho_{ng}^0\Delta\rho_{kl}^0}{\left(\Delta\omega_{ng}^2 + \Gamma_{ng}^2\right)\left(\Delta\omega_{kl}^2 + \Gamma_{kl}^2\right)}(1 - B_{ng})(1 - B_{kl})\right\} \quad (22)$$

The dependence of the cross sections on the nanoparticle dimensions can be found by knowing the function of the relation between A_{ng} and the nanoparticle dimension. Given $\alpha<\alpha_c$ the dipole moment of the nanoparticle is proportional to its dimension. Therefore, as follows from expression (18), we can separate out the dependence of A_{ng}on the nanoparticle dimension as $A_{ng}=S_{ng}(I)a^2$. Here, $S_{ng}(I)$ is a function of the radiation intensity and depends on the nanoparticle shape.

Let us estimate the ratio between the scattering and absorption cross sections. We assume that transitions occur from only one level <n| and the width of the excited level Γ_g has a low-dependence on g. Taking into account that the frequencies ω and ω_g are of one magnitude order and the thermal equilibrium difference between the states is close to unity, and following the expressions (21) and (22), we obtain the next

$$\frac{\sigma_S(\omega, a)}{\sigma_a(\omega, a)} \approx \frac{NP\omega^3 a^2}{4\pi c^3\square\Gamma}\sum_{n,g}\{S_{ng}(I)\quad(1 - B_{ng}(I))\} \quad (23)$$

The quantity of σ_S/σ_a does not exceed $N*10^{-9}$ in any intensity region provided the nanoparticle dimensions α=(10;100)nm in the frequency range $(10^{13};10^{16})$Hz and $\Gamma=10^9$Hz. Given $N*10^{-9}<1$, the scattering cross section can be omitted from the expression for the extinction coefficient.

We can define $S_{ng}(I)=c_{ng}I$ and follow the expression (21) within radiation intensities region $I/I_S<<1$, we obtain the next

$$\sigma_a(\omega, a, I)\approx \frac{4\pi\omega N}{c\square}a^2 I\sum_{n,g} c_{ng}\frac{\Gamma_{ng}}{\left(\Delta\omega_{ng}^2+\Gamma_{ng}^2\right)^2}\Delta\rho_{ng}^0 \tag{24}$$

The absorption cross section reaches a maximum at some intensity $I=I_p$ under increasing radiation intensity, that follows from equation (21). It corresponds to the complete nanoparticle's orientation along the electric vector of the optical field and to the maximum value of $A_{ng}(I)$. This effect is responsible for a sharp enhancement of radiation absorption by the HDN unit volume. A further intensity increasing yields a noticeable increase of $B_{ng}(\omega,I)$ and decrease of the absorption cross section at a constant value of $A_{ng}(I)$. Given $I>>I_S$ the value of $B_{ng}(\omega,I)$ becomes approximately equal to unity resulting in an increase of the HDN transmittance. In this case, the absorption cross section can be written as

$$\sigma_a(\omega, a, I)\approx \frac{4\pi\omega N}{c\square}a^2\sum_{n,g} S_{ng}\frac{I_S}{I\quad \Gamma_{ng}}\Delta\rho_{ng}^0 \tag{25}$$

i.e., it is inverse proportional to the radiation intensity.

The broad optical absorption bands are manifested in the electronic structure of dielectric nanoparticles and absent in the corresponding bulk sample. In addition, the allowed electron energy sub-band (excitons, impurities, etc.) with the width $\Delta\omega_1$ lying in the band gap and adjoining the conduction band bottom, as well as the size-quantization levels (minibands) with the width $\Delta\omega_2$ in the conduction band, are typical for the electronic structure of HDN electrons.Taking into account the electronic structure of the nanoparticle, we substitute the summation over |g> states with integration from $(\omega_n - \Delta\omega_1)$ to $(\omega_n + \Delta\omega_2)$ with state densities of exciton g_1 and quantum-size g_2 levels, respectively. Let us choose one of the absorption bands as an example. Changing the summation by the integration over the frequency in (21) and introducing the definitions

$$\Delta\omega_n=\omega-\omega_n$$
$$F(I)=\sqrt{\frac{I_S}{I+I_S}} \tag{26}$$

we obtain the expression for the absorption cross section of light within the absorption band:

$$\sigma_a = \frac{4\pi\omega N a^2}{c\square} F(I) \left[g_1 S_1 \arctan\left(\frac{\Delta\omega_1 \Gamma_n F(I)}{\Gamma_n^2 + F^2(I)\Delta\omega_n(\Delta\omega_n + \Delta\omega_1)} \right) + g_2 S_2 \arctan\left(\frac{\Delta\omega_2 \Gamma_n F(I)}{\Gamma_n^2 + F^2(I)\Delta\omega_n(\Delta\omega_n - \Delta\omega_2)} \right) \right] \quad (27)$$

The quantities S_1 and S_2 are defined as the average form factors of nanoparticles $S_{ng}(I)$ for transitions to the upper and lower energy bands, respectively.
We may obtain the expression for the HDN optical transmittance in the absorption band with the central frequency ω_n from expression (7) and (27):

$$T(\omega, N, I) \approx \exp\left\{ -L \frac{4\pi\omega N}{c\square} D F(I) \right\} \quad (28)$$

where

$$D = a^2 g_1 S_1 \arctan\left(\frac{\Delta\omega_1 \Gamma_n F(I)}{\Gamma_n^2 + F^2(I)\Delta\omega_n(\Delta\omega_n + \Delta\omega_1)} \right) + a^2 g_2 S_2 \arctan\left(\frac{\Delta\omega_2 \Gamma_n F(I)}{\Gamma_n^2 + F^2(I)\Delta\omega_n(\Delta\omega_n - \Delta\omega_2)} \right) \quad (29)$$

3.3. The Theoretical Outputs and Discussion

It follows from expression (28) that the optical transmittance of the HDN essentially depends on the laser radiation intensity I (Figure 13). This dependence exhibits a minimum I_p corresponding to the lowest light transmittance of the HDN. As intensity is changed near I_p, we can see the effect of limitation of low-intensity radiation. The insert in Figure 13 is the experimental result obtained from Figure 10B. Since $I_{out}=I_{in}e^{-\alpha L}$ and $T=I_{out}/I_{in}$ provided low reflection and absorption, we suppose $T=e^{-\alpha L}$ and use data of α from Figure 10B. The theoretical and experimental results are in good agreement.

Curves from Figure 14 point out the basic features of the dependences of the HDN transmittance on the radiation wavelength. In the general case, the transmittance spectrum is asymmetric, since there is the difference between $\Delta\omega_1$ and $\Delta\omega_2$. The insert in this figure is the experimental spectrum of the HDN based on SiO_2 nanoparticles (Figure 4B). The behaviour of the experimental curve reflects the features of the theoretical one.

The largest dipole moment is induced at the central frequency ω_n in the absorption band. The expression for D at the central frequency of the absorption band ($\omega=\omega_n$, $\Delta\omega_n=0$) is given by

$$D = a^2 g S \left\{ \arctan\left(\frac{\Delta\omega_1 F(I)}{\Gamma_n} \right) + \arctan\left(\frac{\Delta\omega_2 F(I)}{\Gamma_n} \right) \right\} \quad (30)$$

http://dx.doi.org/10.5772/50212

Therefore, the HDN transmittance near the central frequency can be written as

$$T(I)=\exp\left\{-L\,\frac{4\pi\omega_n N}{c\square}a^2gS\left\{\arctan\left(\frac{\Delta\omega_1 F(I)}{\Gamma_n}\right)+\arctan\left(\frac{\Delta\omega_2 F(I)}{\Gamma_n}\right)\right\}F(I)\right\} \tag{31}$$

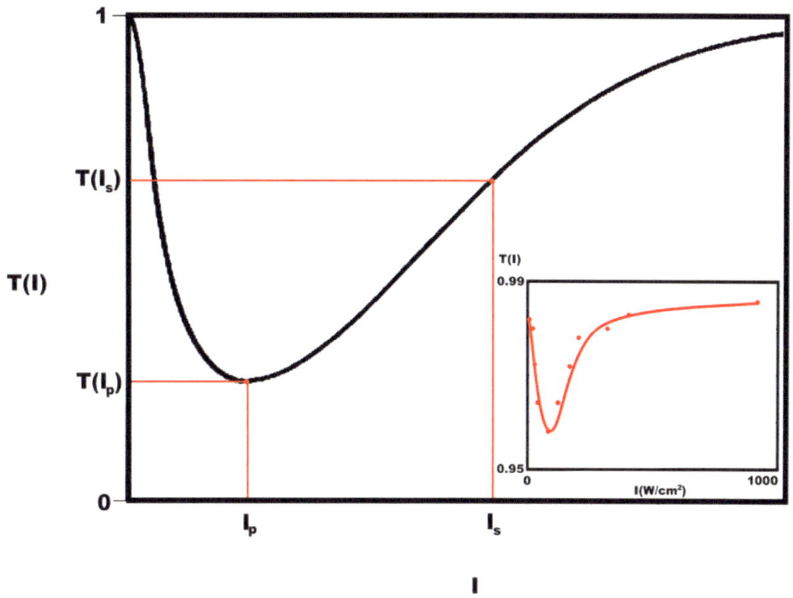

Figure 13. The theoretical dependence of the HDN transmittance on the intensity of input radiation. The insert is the experimental dependence of the HDN transmittance according with Figure 10B.

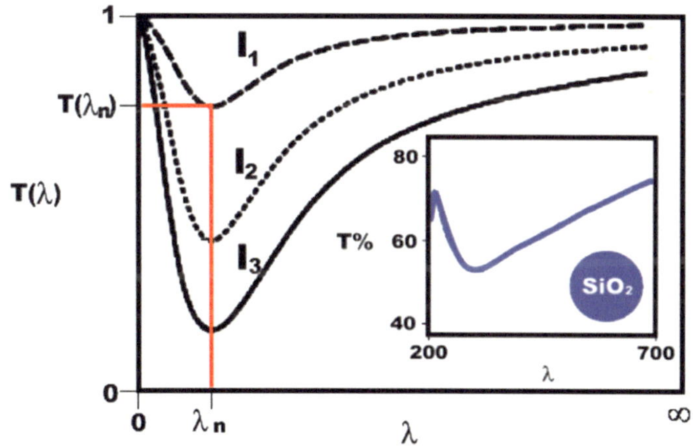

Figure 14. The theoretical dependence of the HDN transmittance on the wavelength of input radiation ($I_1>I_2>I_3$). The insert is the experimental spectrum of the HDN based on SiO_2 nanoparticles (Figure 4B).

As one can see from expression (31), the depth of the absorption band in the transmittance spectrum depends on the radiation intensity and the nanoparticle dimension. The orientation of nanoparticles along the vector E requires high radiation intensities provided a solid HDN. In the case of a liquid matrix, this situation corresponds to the range of intensities $I>I_p$. Here we may assume that all particles are oriented along the direction of the vector E of the external optical field, so the order parameters are constant and S_n are independent of the intensity. This indicates that the behaviour of the transmittance of solid and liquid matrices is similar. Therefore, the transmittance at the central frequency is

$$T(I)=\exp\left\{-L\,\frac{4\pi\omega_n N}{c\square\Gamma_n}a^2(gS_n)_{g=n}(\Delta\omega_1+\Delta\omega_2)F^2(I)\right\} \tag{32}$$

Expression (32) exponentially approaches to unity rapidly when the radiation intensity is increased and nanoparticle dimension, the summation ($\Delta\omega_1$ + $\Delta\omega_2$) and multiplication of gS, become larger.

Apart from transmittance, scattering and spectral properties of the HDN, the theory can describe the behaviour of light refraction in the HDN (Figures 15,16). Using expressions (16) and (19), we can obtain the theoretical dependence of the refractive index on the intensity. Since the value of $\Delta n(I)$ is negligible, the medium refractive index can be written as follows:

$$n(I,\lambda)\approx n_0+\frac{2\pi\chi_{zz}(I,\lambda)}{n_0} \tag{33}$$

where n_0 is the HDN refractive index in the absence of radiation and χ_{zz} is defined by expression (13).

Since χ_{zz} is determined by α_{jj}, so we may simplify equation (16) to carry out the integration over frequency, herewith, to assume $\Gamma_{ng}=\Gamma_n$ and the state density g_1 (g_2) and the values of Q_1 (Q_2) are independent of the frequency ω. Picking out the refraction real part from the resulting expression, we obtain [9,45]

$$n(I,\omega)=n_0+\frac{\square}{2}\sum_n\left\{A_{ng}(Q_1,Q_2)\Delta\rho^0\left[g_1\ln\frac{(\omega-(\omega_n-\Delta\omega_1))^2+\Gamma_n^2\left(1+I/I_S\right)}{(\omega-\omega_n)^2+\Gamma_n^2\left(1+I/I_S\right)}+g_2\ln\frac{(\omega-\omega_n)^2+\Gamma_n^2\left(1+I/I_S\right)}{(\omega-(\omega_n+\Delta\omega_2))^2+\Gamma_n^2\left(1+I/I_S\right)}\right]\right\} \tag{34}$$

where $A_{ng}(Q_1,Q_2)$ is determined by expression (18).

Equation (34) indicates that the term A_{ng} does not vanish in the case of propagation of unpolarized light through the medium (Q=0) and the nonlinear response of the dielectric nanocomposite is not equal to zero even in case of a solid matrix.

The important conclusion from equations (18) and (34) is that the modulus of photo-induced electric dipole moments $|p_{ng}|$ mainly defines the magnitude of nonlinear optical response under continuous low-intensity radiation. In general, the orientation parametersand dipole moment modulus reach their maxima with the increase of input power, however, this increase diminishes the population difference; hence the change of Δn tends to zero. It is these two competing processes that define the nonlinear features of the HDN refractive index.

The numerical simulation of the change in the HDN refractive index was carried out using equation (34). Since the photon energy is less than nanoparticle band gap, so, the dipole transition of electrons to the exciton state is most probable (g_2=0). In the case of the low-intensity continuous radiation and low concentration of nanoparticles, equation (34)can be rewritten as follows [9]:

$$\Delta n(I,\omega)\approx A(I)\ln\frac{(\omega-(\omega_n-\Delta\omega_1))^2+\Gamma_n^2\left(1+{}^{I}\!/_{I_S}\right)}{(\omega-\omega_n)^2+\Gamma_n^2\left(1+{}^{I}\!/_{I_S}\right)} \tag{35}$$

where the factor A(I) defines the dependence of A_{ng} on the radiation intensity as follows:

$$A(I)=A_0\quad\Delta\rho^0\left(1-e^{-aI}\right) \tag{36}$$

This dependence takes into account the magnitude of A_{ng}, which varies from zero to its maximum value with the increase of the external radiation intensity. The theoretical curves (Figures 15,16, solid lines) of the dependence Δn on the radiation intensity have been constructed by means of equation (35) for the HDN based on Al_2O_3 and SiO_2 nanoparticles, irradiating by green and violet radiation. The parameters for good approximation were calculated according to the next considerations: Γ was taken from T spectrum (Figure 4); I_S, α and $A_0\Delta\varrho^0$ were calculated by means of the three-equation system (35) with known parameters I and Δn (Figures 9 A-B, 10A, 11A, dotted curves).

The lack of nonlinear refraction and absorption of low-intensity continuous visible laser radiation in the HDN based on the nanoparticles of narrow-band dielectrics is caused by the absence of absorption band in the used frequency range (200;700)nm. So, in order for the nonlinear optical properties of such HDN to be observable we must use high-energy pulsed radiation or change the input radiation frequency. On the one hand, if we use the pulsed radiation we may get the typical nonlinearity of the nanocomposites, caused by nonlinear behaviourof excitons near the edge of fundamental absorption (Figure 6). The high energy is required, since there is the small dipole moment of electron transition to exciton states. On the other hand, if we change the input radiation frequency it is possible to find the defect energy levels in the HDN infrared spectrum. So, the nontypical nonlinearity can take place under low-intensity infrared radiation.

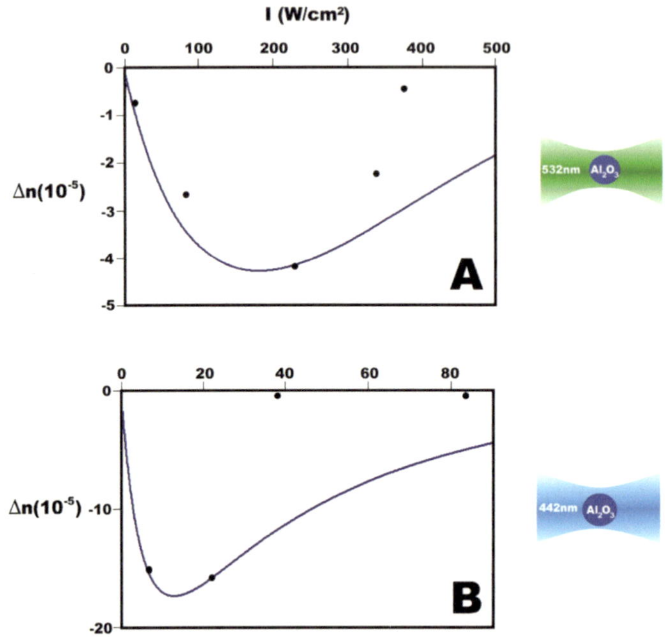

Figure 15. Theoretical curves of dependence of refraction index of green (A) and violet (B) radiation on its intensity in the HDN based on Al_2O_3 nanoparticles (dotted curves are the experimental results from Figure 9).

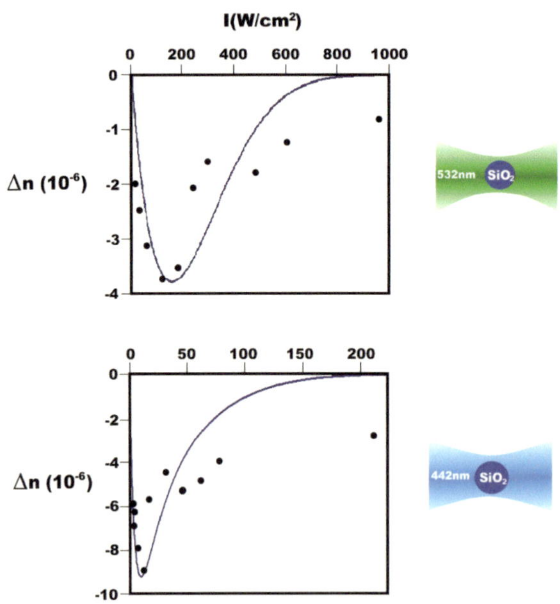

Figure 16. Theoretical curves of dependence of refraction index of green (A) and violet (B) radiation on its intensity in the HDN based on SiO_2nanoparticles (dotted curves are the experimental results from Figures 9A and 10A).

http://dx.doi.org/10.5772/50212

4. Conclusion

The experimental study of changes of optical characteristics of the dielectric nanostructures based on Al_2O_3, SiO_2, TiO_2, ZnO nanoparticles and theoretical description of these characteristics allows estimating the conditions of observing the low-threshold optical nonlinearity under low-intensity optical fields.

The ability to observe this nonlinearity is directly connected with the peculiarities of the energy spectrum of nanoparticle charge carriers. Because of the wide band gap of the bulk dielectric material, it is not possible to excite electron transitions to the conduction band by a visible light. The energy spectrum of nanoparticle electrons is of a different structure: the band gap has defect levels containing a lot of electrons due to a high density of crystal defects on the nanoparticle's surface; the small size and shape of nanoparticle leads to strong broadening of the band of high-density exciton states from the bottom of the conduction band up to defect levels. The existence of an absorption band in visible light spectrum is observed only for nanoparticles of broad-band dielectrics (Al_2O_3, SiO_2). The absorption band in the energy spectrum of electrons of narrow-band dielectric (TiO_2, ZnO) nanoparticles is not manifested in a visible light spectrum, however, it can be manifested within the infrared region and adjoins the bottom of the conduction band.

Comparing the experimental and theoretical results we conclude that the low-threshold nonlinearity of the HDN optical parameters (Δn, $\Delta\alpha$ and the scattering cross section) is caused, mainly, by transitions of electrons from defect levels to exciton states and, hence, photo excitation of electric dipole moments. However, experiments have shown that the nonlinear behaviour of the HDN optical parameters takes place when the matrix permittivity ε_{stat} is less than that of the nanoparticles (e.g., oil permittivity). Otherwise, the positive polarization charges, concentrated along the nanoparticle's inner surface, destroy the defect states. This explains the absence of nonlinear optical properties in the HDN based on water and alcohol matrices.

In view of the effect of giant oscillator strength, the magnitude of the photo excited dipole moment is enormous. It is the great value of the oscillator's strength for electron transition to the exciton states that is responsible for the low-threshold of the nonlinearity. As it follows from the theory, the dipole moment orientation along the external E field makes a minor contribution to the nonlinearity, therefore, this response can also be observed under pulsed and unpolarized laser radiation in solid matrices.

In addition, a qualitative agreement between experimental and theoretical results was also obtained and the proposed theory model of optical nonlinearity can be applied to explain the number of phenomena in physics of nanoscale dielectrics, e.g., proteins and blood bodies [53].

Acknowledgements

This work was supported by RFBR Grant No. 11-02-98514 r_vostok_a and FEB RAS Grant Nos. 12-I- OFN-05, 12-I- OFN-04, 12-I-P24-05, 12-II-UO-02-002.

Author details

Vladimir Dzyuba[1,2*], Yurii Kulchin[1,2] and Valentin Milichko[1,2]

*Address all correspondence to: vdzyuba@iacp.dvo.ru

1 Institute of Automation and Control Processes of Russian Academy of Science, Vladivostok, Russia

2 Far Eastern Federal University, Vladivostok, Russia

References

[1] Alivisatos, A. P., Harris, A. L., Levinos, N. J., Steigerwald, M. L., & Brus, L. E. (1988). Electronic states of semiconductor clusters: Homogeneous and inhomogeneous broadening of the optical spectrum. *The Journal of Chemical Physics*, 89(7), 4001-4011.

[2] Bawendi, M. G., Carroll, P. J., Wilson, William L., & Brus, L. E. (1992). Luminescence properties of CdSe quantum crystallites: Resonance between interior and surface localized states. *The Journal of Chemical Physics*, 96(2), 946-954.

[3] Bang, Jin Ho., & Kamat, Prashant V. (2009). Quantum Dot Sensitized Solar Cells. A Tale of Two Semiconductor Nanocrystals: CdSe and CdTe. *ACS Nano*, 3(6), 1467-1476.

[4] McGuire, John A., Sykora, Milan., Robel, Istva´n., Padilha, Lazaro A., Joo, Jin., Pietryga, Jeffrey. M., & Klimov, Victor. I. (2010). Spectroscopic Signatures of Photocharging due to Hot-Carrier Transfer in Solutions of Semiconductor Nanocrystals under Low-Intensity Ultraviolet Excitation. *ACS Nano*, 4(10), 6087-6097.

[5] Ivanov, S. A., & Achermann, M. (2010). Spectral and Dynamic Properties of Excitons and Biexcitons in Type-II Semiconductor Nanocrystals. *ACS Nano*, 4(10), 5994-6000.

[6] Hashimoto, Tadanori., Yamamoto, Tsuyoshi., Kato, Tomohiro., Nasu, Hiroyuki., & Kamiya, Kanichi. (2001). Z-scan analyses for PbO-containing glass with large optical nonlinearity. *Journal of Applied Physics*, 90(2), 533-537.

[7] Anderson, Mark S. (2003). Enhanced infrared absorption with dielectric nanoparticles. *Applied Physics Letters*, 83(14), 2964-2966.

[8] Miheev, O. P., & Sidorov, A. I. (2004). Optical nonlinearity of nanoparticles of wide-gap semiconductors and insulators in visible and near infrared spectral region. *Technical Physics*, 74(6), 77-82.

[9] Dzyuba, Vladimir., Milichko, Valentin., & Kulchin, Yurii. (2011). Nontypical photoinduced optical nonlinearity of dielectric nanostructures. *Journal of Nanophotonics*, 5, 053528, 1-13.

[10] Yu, Baolong, Zhu, Congshan, & Gan, Fuxi. (1997). Optical nonlinearity of Bi2O3 nanoparticles studied by Z-scan technique. *Journal of Applied Physics*, 82(9), 4532-4537.

[11] Major, A., Yoshino, F., Aitchison, J. S., & Smith, P. W. E. (2004). Ultrafast nonresonant third-order optical nonlinearities in ZnSe for photonic switching at telecom wavelengths. *Applied Physics Letters*, 85(20), 4606-4608.

[12] He, J., Ji, W., Ma, G. H., Tang, S. H., Elim, H. I., Sun, W. X., Zhang, Z. H., & Chin, W. S. (2004). Excitonic nonlinear absorption in CdS nanocrystals studied using Z-scan technique. *Journal of Applied Physics*, 95(11), 6381-6386.

[13] Larciprete, M. C., Ostuni, R., Belardini, A., Alonzo, M., Leahu, G., Fazio, E., Sibilia, C., & Bertolotti, M. (2007). Nonlinear optical absorption of zinc-phthalocyanines in polymeric matrix. *Photonics and Nanostructures - Fundamentals and Applications*, 5, 73-78.

[14] Ganeev, R. A., Suzuki, M., Baba, M., Ichihara, M., & Kuroda, H. (2008). Low- and high-order nonlinear optical properties of BaTiO3 and SrTiO3 nanoparticles. *Journal of Optical Society of America B*, 25(3), 325-333.

[15] Ganeev, R. A., Zakirov, A. S., Boltaev, G. S., Tugushev, R. I., Usmanov, T., Khabibullaev, P. K., Kang, T. W., & Saidov, A. A. (2003). Structural, optical, and nonlinear optical absorption/refraction studies of the manganese nanoparticles prepared by laser ablation in ethanol. *Optical Quantum Electronics*, 35, 419-423.

[16] Ganeev, R. A., Ryasnyansky, A. I., Tugushev, R. I., & Usmanov, T. (2003). Investigation of nonlinear refraction and nonlinear absorption of semiconductor nanoparticle solutions prepared by laser ablation. *Journal of Optics A*, 5, 409-417.

[17] Ganeev, R. A., & Usmanov, T. (2007). Nonlinear-optical parameters of various media. *IOP Quantum Electron*, 37(7), 605-622.

[18] Kulchin, Yu. N., Shcherbakov, A. V., Dzyuba, V. P., & Voznesenskiy, S. S. (2009). Interaction of collinear light beams with different wavelengths in a heterogeneous liquid-phase nanocomposite. *Technical Physics Letters*, 35(7), 640-642.

[19] Chen, Fang-Chung., Chu, Chih-Wei., He, Jun., Yang, Yang., & Lin, Jen-Lien. (2004). Organic thin-film transistors with nanocomposite dielectric gate insulator. *Applied Physics Letters*, 85(15), 3295-3297.

[20] Schrier, Joshua., Demchenko, Denis O., Wang, Lin-Wang., & Alivisatos, A. Paul. (2007). Optical Properties of ZnO/ZnS and ZnO/ZnTe Heterostructures for Photovoltaic Applications. *Nano Letters*, 7(8), 2377-2382.

[21] Dutta, Kousik., & De, S. K. (2007). Electrical conductivity and dielectric properties of SiO2 nanoparticles dispersed in conducting polymer matrix. *Journal of Nanoparticle Research*, 9, 631-638.

[22] Chu, Daobao, Yuan, Ximei, Qin, Guoxu, Xu, Mai, Zheng, Peng, Lu, Jia, & Zha, Longwu. (2008). Efficient carbon-doped nanostructured TiO2 (anatase) film for photoelectrochemical solar cells. *Journal of Nanoparticle Research*, 10, 357-363.

[23] Liu, Li., Wang, Ning., Cao, Xia., & Guo, Lin. (2010). Direct Electrochemistry of Cytochrome c at a Hierarchically Nanostructured TiO2 Quantum Electrode. *Nano Research*, 3-369.

[24] Hensel, Jennifer., Wang, Gongming., Li, Yat., & Zhang, Jin. Z. (2010). Synergistic Effect of CdSe Quantum Dot Sensitization and Nitrogen Doping of TiO2 Nanostructures for Photoelectrochemical Solar Hydrogen Generation. *Nano Letters*, 10, 478-483.

[25] Pellegrini, Giovanni., Mattei, Giovanni., & Mazzoldi, Paolo. (2009). Light Extraction with Dielectric Nanoantenna Arrays. *ACS Nano*, 3(9), 2715-2721.

[26] Redel, Engelbert., Mirtchev, Peter., Huai, Chen., Petrov, Srebri., & Ozin, Geoffrey. A. (2011). Nanoparticle Films and Photonic Crystal Multilayers from Colloidally Stable, Size-Controllable Zinc and Iron Oxide Nanoparticles. *ACS Nano*, 5(4), 2861-2869.

[27] Chen, F. C., Chuang, C. S., Lin, Y. S., Kung, L. J., Chen, T. H., & Shien, D. H. P. (2006). Low-voltage organic thin-film transistors with polymeric nanocomposite dielectrics. *Organic Electronics*, 7, 435-439.

[28] Park, Young-Shin., Cook, Andrew K., & Wang, Hailin. (2006). Cavity QED with Diamond Nanocrystals and Silica Microspheres. *Nano Letters*, 6(9), 2075-2079.

[29] Han, Jiaguang., Woo, Boon Kuan., Chen, Wei., Sang, Mei., Lu, Xinchao., & Zhang, Weili. (2008). Terahertz dielectric properties of MgO nanocristals. *The Journal of Physical Chemistry C*, 112, 17512-17516.

[30] Balasubramanian, Balamurugan., Kraemer, Kristin L., Reding, Nicholas A., Skomski, Ralph., Ducharme, Stephen., & Sellmyer, David J. (2010). Synthesis of Monodisperse TiO2-Paraffin Core-Shell Nanoparticles for Improved Dielectric Properties. *ACS Nano*, 4(4), 1893-1900.

[31] Landes, C., Braun, M., Burda, C., & El -Sayed, M. A. (2001). Observation of Large Changes in the Band Gap Absorption Energy of Small CdSe Nanoparticles Induced by the Adsorption of a Strong Hole Acceptor. *Nano Letters*, 1(11), 667-670.

[32] Kulchin, Yu. N., Shcherbakov, A. V., Dzyuba, V. P., Voznesenskii, S. S., & Mikaelyan, G. T. (2008). Nonlinear-optical properties of heterogeneous liquid nanophase composites based on high-energy-gap Al2O3 nanoparticles. *IOP Quantum Electronics*, 38(2), 154-158.

[33] Ho, Ching-Hwa., Chan, Ching-Hsiang., Tien, Li-Chia., & Huang, Ying-Sheng. (2011). Direct Optical Observation of Band-Edge Excitons, Band Gap, and Fermi Level in Degenerate Semiconducting Oxide Nanowires In2O3. *Journal of Physical Chemistry C*, 115, 25088-25096.

[34] Mu, R., Tung, Y. S., Ueda, A., & Henderson, D. O. (1996). Chemical and Size Characterization of Layered Lead Iodide Quantum Dots via Optical Spectroscopy and Atomic Force Microscopy. *Journal of Physical Chemistry*, 100, 19927-19932.

[35] Li, Jingbo., & Wang, Lin-Wang. (2003). Shape Effects on Electronic States of Nanocrystals. *Nano Letters*, 3(10), 1357-1363.

[36] Al-Hilli, S. M., & Willander, M. (2006). Optical properties of zinc oxide nano-particles embedded in dielectric medium for UV region: Numerical simulation. *Journal of Nanoparticle Research*, 8, 79-97.

[37] Stroyuk, Oleksandr. L., Dzhagan, Volodymyr. M., Shvalagin, Vitaliy. V., & Kuchmiy, Stepan Ya. (2010). Size-Dependent Optical Properties of Colloidal ZnO Nanoparticles Charged by Photoexcitation. *The Journal of Physical Chemistry C*, 114, 220-225.

[38] Azpiroz, Jon M., Mosconi, Edoardo., & De Angelis, Filippo. (2011). Modeling ZnS and ZnO Nanostructures: Structural, Electronic, and Optical Properties. *The Journal of Physical Chemistry C*, 115, 25219-25226.

[39] Pokutnyi, S. I. (2006). Optical absorption and scattering at one-particle states of charge carriers in semiconductor quantum dots. *Semiconductor*, 40(2), 217-223.

[40] Pokutnyi, S. I. (2007). Excition states in semiconductor quantum dots in the modified effective mass approximation. *Semiconductor*, 41(11), 1323-1328.

[41] Sinha, Sucharita., Ray, Alok., & Dasgupta, K. (2000). Solvent dependent nonlinear refraction in organic dye solution. *Journal of Applied Physics*, 87(7), 3222-3226.

[42] Mosconi, Edoardo., Selloni, Annabella., & De Angelis, Filippo. (2012). Solvent Effects on the Adsorption Geometry and Electronic Structure of Dye-Sensitized TiO2: A First-Principles Investigation. *Journal of Physical Chemistry C*, 116, 5932-5940.

[43] Dzyuba, V. P., Krasnok, A. E., Kulchin, Yu. N., & Dzyuba, I. V. (2011). A model of nonlinear optical transmittance for insulator nanocomposites. *Semiconductor*, 45(3), 295-301.

[44] Dzyuba, V. P., Krasnok, A. E., & Kulchin, Yu. N. (2010). Nonlinear refractive index of dielectric nanocomposites in weak optical fields. *Technical Physics Letters*, 36(11), 973-977.

[45] Kulchin, Yu. N., Dzyuba, V. P., & Voznesenskiy, S. S. (2011). Threshold optical nonlinearity of dielectric nanocomposite. In: Boreddy Reddy (ed.), *Advances in Diverse Industrial Applications of Nanocomposites*, Rijeka, InTech, 261-288.

[46] Wakaki, M., Kudo, K., & Shibuya, T. (2007). Physical Properties and Data of Optical Materials. CRC Press Taylor and Francis Group. New York. 10.1201/9781420015508

[47] Said, A. A., Sheik-Bahae, M., Hagan, D. J., Wei, T. H., Wang, J., Young, J., & Van Stryland, E. W. (1991). Determination of bound-electronic and free-carrier nonlinearities in ZnSe, GaAs, CdTe and ZnTe. *Journal of Optical Society of America B*, 9(3), 405-414.

[48] Van Stryland, E. W., & Sheik-Bahae, M. (1998). Z-scan measurement of optical nonlinearities. In: Kuzyk M. G., DirkC. W. (eds), *Characterization Techniques and Tabulation for Organic Nonlinear Materials*, Marcel Dekker Inc., New York, 655-692.

[49] Pokitnyi, S. I., & Efremov, N. A. (1991). Theory of Macroscopic Local Single-Particle Charge States in Quasi-Zero-Dimensional Structures Surface Local States. *Phys. Status Solidi B*, 165, 109-118.

[50] Bohren, C., & Huffman, D. (1983). *Absorbtion and Scattering of Light by Small Particles*, Wiley, New York.

[51] Landau, L. D., & Lifshitz, E. M. (1984). Course of Theoretical Physics, Volume 8: Electrodynamics of Continuous Media. Pergamon, New York.

[52] Shen, Y. R. (1984). The Principles of Nonlinear Optic. Wiley, Hoboken, NJ.

[53] Kulchin, Yu. N., Dzyuba, V. P., & Milichko, V. A. (2010). Optical Nonlinearity of a Biological Liquid Nanocomposite. *Pacific Science Review*, 12(1), 4-7.

Chapter 16

Effect of Nano-TiN on Mechanical Behavior of Si_3N_4 Based Nanocomposites by Spark Plasma Sintering (SPS)

Jow-Lay Huang and Pramoda K. Nayak

Additional information is available at the end of the chapter

http://dx.doi.org/10.5772/50547

1. Introduction

Ceramic nanocomposites are often defined as a ceramic matrix reinforced with submicron/ nano sized particles of a secondary phase. The advantages of these nanocomposites include: improved mechanical properties, surface properties, high thermal stability and superior thermal conductivity. It is very fascinating/interesting for the researchers to synthesize these composites as the incorporation of few percent nanosized particles changes the materials property substantially. Niihara et al.,[35], [36] have reported that the mechanical properties of ceramics can be improved significantly by dispersing nanometer-sized ceramic particles into ceramic matrix grains or grain boundaries. According to their observation, 5 vol% of silicon carbide nanoparticles into alumina matrix increases the room temperature strength from 350 MPa to approximately 1 GPa. Other strength improvements through similar approaches have been observed in alumina-silicon nitride, magnesia-silicon carbide, and silicon nitride-silicon carbide composite systems.

Apart from the basic mechanical properties such as such as micro hardness, facture strength, and facture toughness [9; 23; 44], nanocomposites also exhibit electro conductive, wear resistance, creep resistance and high temperature performance [10; 24; 37; 38, 39] However, the degree of improvement in these properties is dependent on the type of composite system involved.

1.1. Novel Synthesis of Ceramic nanocomposite

Chemical Vapor Deposition (CVD) is a very preferable method to disperse the nano-sized second phases into the matrix grains or at the grain boundaries [33]. However, the CVD process is not applicable to fabricate the large and complex shaped component for the mass produc-

INTECH
open science | open minds

© 2012 Huang and Nayak; licensee InTech. This is a paper distributed under the terms of the Creative Commons Attribution License (http://creativecommons.org/licenses/by/3.0), which permits unrestricted use, distribution, and reproduction in any medium, provided the original work is properly cited.

tion and also it is very expensive. Processing route is another technique to prepare ceramic nanocomposites. Following the initial work of [34], several research groups have tried to synthesize the nanocomposites using different processing route such as conventional powder processing [6; 8], hot press sintering [25; 47] sol-gel processing [30; 50] and polymer processing [5; 16]. The ceramic nanocomposites can be synthesized using microwave plasma [48; 49]. The main advantage of this technique is that the reaction product does not form hard agglomerates because of the specific conditions during synthesis.

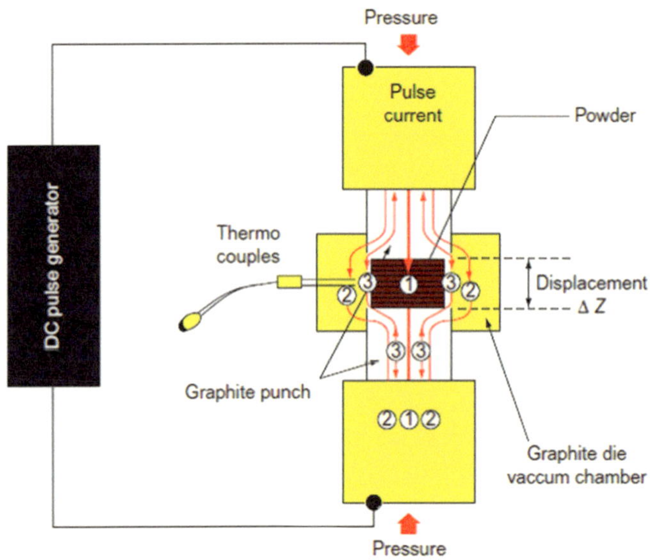

Figure 1. Schematics of Spark Plasma Synthesis (SPS) process

Recently developed Spark plasma sintering (SPS) is a novel sintering technique that uses the idea of pressure driven powder consolidation under pulsed direct electric current passing through a sample compressed in a graphite matrix. It is also known as the field assisted sintering technique or pulse electric current sintering. This newly developed sintering technique is regarded as an energy-saving technology due to the short process time and fewer processing steps. This technique was first described by Raichenko, 1987 and the key characteristics of this SPS are given as follows:

i. The generation of local electric discharge plasma and its effect on the material.

ii. The combined effect of external fields, such as a force field and electric field, on the densification and phase formation in a particulate system.

iii. The influence of electric current in the near surface layers of conductors and semiconductors (the so-called "skin-effect").

iv. The rapid and nonuniform heating/cooling throughout the sample, causing large temperature gradients.

The schematic of SPS process is shown in Fig.1 [41].

1.2. Advantages of SPS over other synthesis method

The most impressive advantage of SPS is its applicability to sinter materials of various types of chemical bonding and electric conductivity. Novel materials have been prepared from powders of ceramic dielectrics, conductors, semiconductors, amorphous alloys, and, sometimes, polymers. The traditional driving force involved in commonly used consolidation techniques such as solid state sintering and hot press sintering are: surface tension, external pressure, chemical potential due to the gradient of surface curvature, concentration gradient in multicomponent system etc. In SPS technique, the additional driving forces include electromechanical stress, high local temperature gradients creating thermal stresses intensifying thermal diffusion, and dislocation creep. These additional driving forces are responsible for much faster transport mechanism, that accelerates rapid sintering, which is observed in SPS.

1.3. TiN/Si_3N_4 Ceramic Nanocomposites

Si_3N_4 ceramics are regarded as one of the important high temperature structural materials. These ceramics have attracted much attention due to their good mechanical and chemical properties and due to their reliability at room and elevated temperatures [14; 19]. High strength and high-toughness Si_3N_4 matrix composites such as whisker-reinforced of particulate-reinforced ceramics, have been developed to improve the mechanical reliability of Si_3N_4 ceramics [2; 13]. However, these composites are extremely hard and machining using conventionals tools is difficult, which limits the widespread application of these materials in many fields. If, sintered Si_3N_4 bodies can be made electro conductive, electrical discharge machining (EDM) technique can be applied to manufacture complex components [32]. It has been reported that introduction of electro conductive second phase can improve the mechanical properties and electroconductivity of Si_3N_4 ceramics [14; 18; 43].

TiN exhibits a number of desirable properties, including high hardness, good chemical durability, high electrical conductivity and is a popular second phase additive due to its good compatibility with Si_3N_4. It often incorporated into the β-Si_3N_4 matrix as cutting-tool materials [4; 14; 17; 28]. There are two advantages to the Si_3N_4 based composites. First of all, the good physical properties of TiN, such as high melting point, hardness, strength and chemical stability, as well as its good erosion and corrosion resistance, enable it to be an excellent toughening material [7; 28; 51]. Secondly, the electrical resistance of Si_3N_4 can be substantially decreased, which consequently makes electric-discharge machining possible [17; 21].

2. Spark Plasma Sintering of TiN/Si_3N_4 nanocomposite

As described in the previous section, SPS is a newly developed sintering technique and it is beneficial to consolidate Si_3N_4 based nanocomposites in a short time. Some researchers have already reported that TiN/Si_3N_4 based nanocomposites with excellent mechanical properties and conductivity can be processed through a chemical route and sintered by SPS [1; 22].

However, due to the complexity of these processing techniques, they are not suitable for large scale production. The planetary milling process has been introduced in their study. Moreover, the details of microstructural development of Si_3N_4 and TiN have not described, especially in the presence of a pulse direct current through the sintering compact during a sintering cycle.

In the present study, we have prepared TiN/Si_3N_4 nanocomposite using SPS from Si_3N_4 and TiN nano powders. The Si_3N_4 and TiN nano powders were applied because they are sensitive to the microstructural changes during the sintering process. The relationship between microstructure and performance, like mechanical properties and electrical conductivity, of these TiN/Si_3N_4 nanocomposites are discussed. Finally, the effect of nano-TiN on the mechanical behavior of Si_3N_4 based nanocomposite has been investigated in sufficient details.

2.1 Experimental Details

2.1.1. Preparation of TiN/Si_3N_4 nanocomposite powder

Commercially available Si_3N_4 nano powder (SM131, Fraunhofer-Institut fur Keramische Technologien and Sinterwerkstoffe, Dresden, Germany) doped with sintering additives of 6 wt% Y_2O_3 and 8 wt% Al_2O_3 was taken as raw material for the matrix phase. It contains 90 wt% β-phase and 10 wt% α-phase, with a manufacturer-determined average particle size of 70nm by the Rietveld method. Nanosized TiN with size of ~30nm (Hefei Kiln Nanometer Technology Development, Hefei, China) was used as secondary phase and it was mixed with Si_3N_4 nano powders. The composition was chosen to yield a TiN content of 5, 10, 15, 20, and 30 wt % in the final product. The specimen designations and corresponding TiN/Si_3N_4 ratios in volume percentage (vol.%) for each composite are shown in Table 1. The mixing powders were ultrasonically dispersed in ethanol for 15 min, and then mixed by planetary milling at a rotation speed of 300rpm for 6 h using a 375 ml nylon bottle with Si_3N_4 balls. The powder mixture was dried in a rotary evaporator, iso-statically cold-pressed into round ingots at a pressure of 200 MPa, crushed and then passed through a #200 sieve for granulation.

TiN/Si_3N_4 content ratio (wt%)	Designation	TiN/Si_3N_4 content ratio (vol%)
5	5TN	3.31
10	10TN	6.75
15	15TN	10.06
20	20TN	14.00
30	30TN	21.81

Table 1. Specimen designation of composites for different TiN content.

2.1.2. Preparation of sintered bodies by SPS

The granulated powders were loaded into a graphite mold with a length of 50 mm and inner and outer diameters of 20 and 50 mm, respectively. A graphite sheet was inserted into the small gap between the punches and mold to improve the temperature uniformity effectively. The graphite mold was also covered with carbon heat insulation to avoid heat dissipation from the external surface of the die. After the chamber was evacuated to a pressure of 10 Pa, the sample was heated to 1600°C under a uniaxial pressure of 30 MPa by SPS (Dr. Sinter 1050, Sumitomo Coal Mining, Kawasaki, Japan). All the SPS measurements were carried out with a heating rate of 200°C/min and holding time of 3 min. A 12 ms-on and 2 ms-off pulse sequence was used. The heating process was controlled using a monochromatic optical pyrometer that was focused on the surface of the graphite mold.

2.1.3. Characterization of sintered bodies

The effective densities of the sintered composites were measured by the Archimedes principle. Phase identification was performed by an X-ray diffractometer (XRD; Model D-MAX/IIB, Rigaku, Tokyo, Japan). Cell dimensions were determined from XRD peak data using UNITCELL with a Si standard. A semiconductor parameter analyzer (HEWLETT PACKARD 4140B, USA) was used to determine the electrical resistivity of the samples. The upper surfaces of the sintered samples were polished down to 1μm. Hardness was measured with a Vickers hardness tester (AKASHI AVK-A, Japan) and by applying a micro-hardness indent at 196N for 15 s. Fracture toughness was measured by the Vickers surface indentation technique [12]. The polished and plasma etched surfaces were used for microstructural characterization by field emission scanning electron microscope (FESEM, XL-40FEG, Philips, The Netherlands). A thin specimen was prepared with a focused ion beam system (FIB, SEIKO, SMI3050, Japan). Transmission electron microscopy (FEGTEM, Tecnai G2 F20, Philips, Eindhoven, Netherlands) was used to characterize the TiN grain of the sintered sample.

2.2. Results and Discussion

2.2.1. Phase Identification of nanocomposite powders by XRD

Fig.2 shows the typical X-ray diffraction patterns of sintered TiN/Si_3N_4 composites with varying TiN content. These composites consist of the β-Si_3N_4 phase as a major phase along with coexistence with secondary TiN phase. The intensity of TiN peaks continues to increase with increasing TiN content. The value of the lattice constant for TiN is 4.25Å, approaching that of pure TiN. On the other hand, the values for a_0 and c_0 for the β-Si_3N_4 phase are 7.61 and 2.91Å, respectively, which are somewhat deviated from those for pure β-Si_3N_4 (+0.01Å). This result suggests that a tiny amount of Si–N may be replaced by Al–O in the particle dissolution and coarsening stages of liquid phase sintering and form the β-SiAlON phase [45].

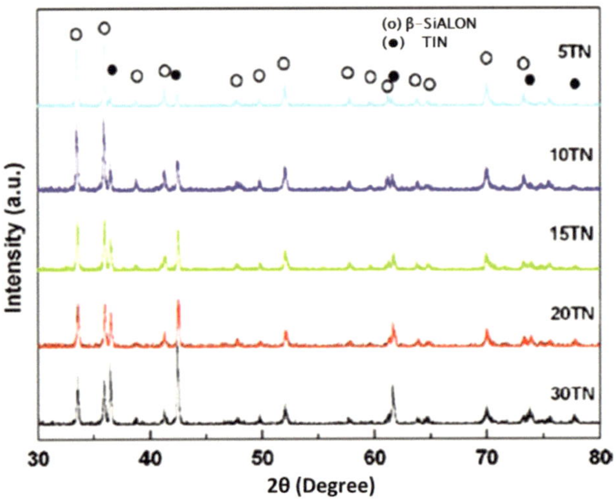

Figure 2. X-ray diffraction patterns of composites with different TiN content.

2.2.2. Densification Behavior

The apparent density of samples containing up to 30 wt% TiN is presented in Fig.3. The apparent density is found to increase with the increase of TiN content. The increase in density is as predicted, because the theoretical density of TiN (5.39 g/cm^3) is substantially greater than that of monolithic Si_3N_4 (3.19 g/cm^3) (Lide, 2002). No obvious pores are also observed on the polished surfaces of the samples (Fig.4), which suggest the TiN/Si_3N_4 composites are near full densification.

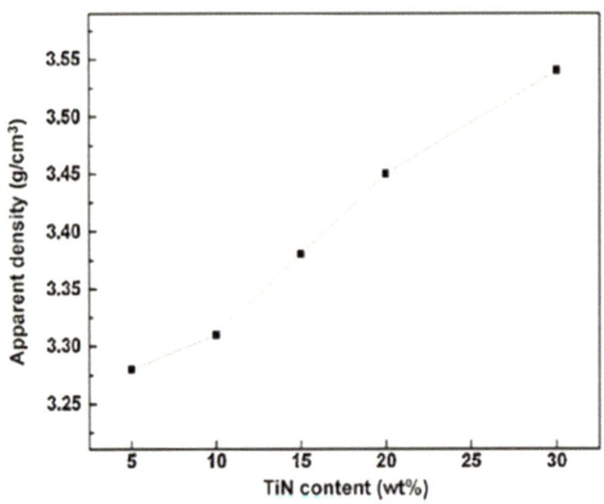

Figure 3. Variation of apparent density of the composites with TiN content.

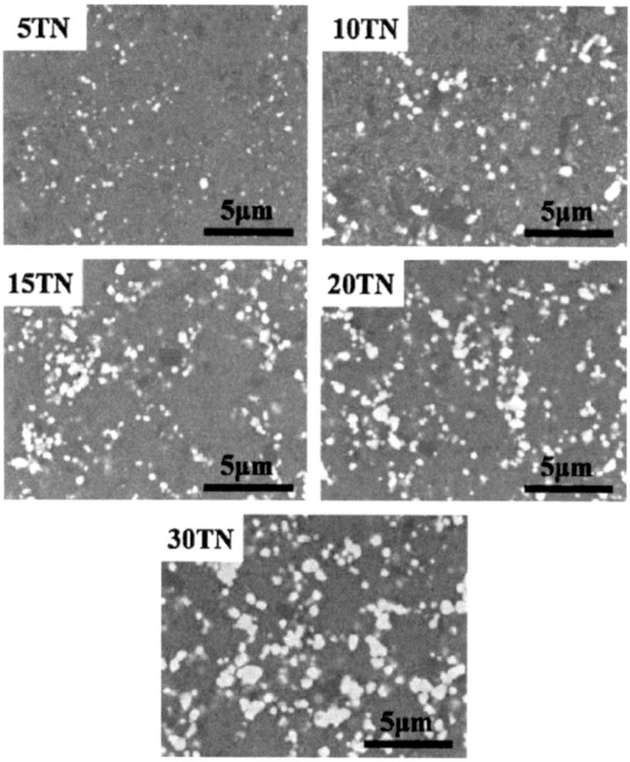

Figure 4. Backscattered SEM images of polished TiN/Si_3N_4 composites with varying TiN content. The brighter phase is TiN phase and the darker phase is β-SiAlON matrix.

2.2.3. Microstructure Observation of Nanocomposites

The backscattered electron images in SEM of the polished surface of composites with varying TiN content are shown in Fig. 4. As stated earlier, the samples were sintered at 1600°C for 3min with a heating rate of 200°C/min in a vacuum. The lighter and heavier atoms in the backscattered images show up as the gray and white regions corresponding to the β-SiAlON matrix (including glassy phase) and TiN particles. Therefore, the TiN particles are distributed homogeneously in the β-SiAlON matrix. However, most of the TiN appear as submicrosized grains, which are much larger than the size of the starting nano powders.

The typical bright field and dark field images of the TiN grain for the as-sintered composite containing 10 wt% TiN content are shown in Fig. 5(a) and (b), respectively. Fig. 5(c) presents the [0 1 1] selected area diffraction pattern (SAD) for the submicrosized TiN grain, and it shows the existence of a twin structure. The results suggest that grain growth and coalescence of TiN occurs in the composite during the spark plasma sintering process in a short time.

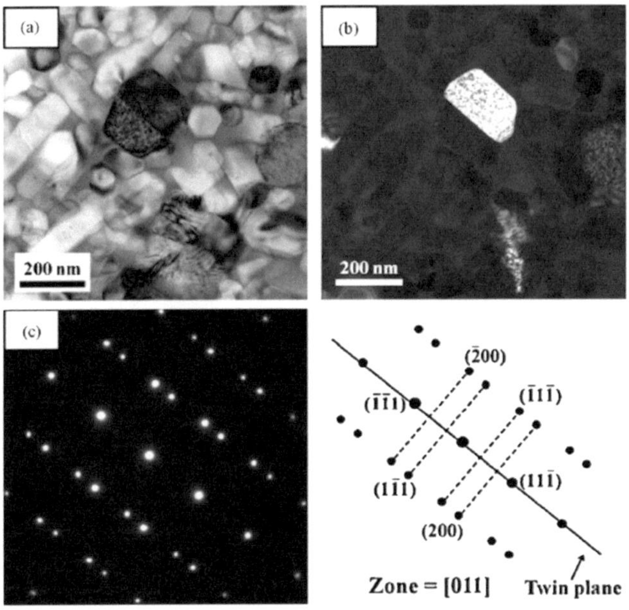

Figure 5. a) Bright field and (b) dark field micrographs, and (c) [0 1 1] selected area diffraction patterns of TiN in spark plasma sintered TiN/Si_3N_4 composite containing 10 wt% TiN.

The typical micrographs of β-SiAlON grains with different TiN content are presented in Fig. 6. In general, the TiN in TiN/Si_3N_4 based composite inhibits grain boundary diffusion and reduces the grain size of the Si_3N_4 matrix [20]. However, for the special case of 10TN among all these composites, the large, elongated grains can be obtained. The conductive phase of TiN might play an important role in the microstructural development of TiN/Si_3N_4 based composites. The electrical resistivity of TiN (3.34×10–7Ω.m) [14] is in the range of metallic materials. Although a tiny current appears as measured by a semiconductor parameter analyzer, it is reasonable that a large current might be induced in the presence of pulsed electrical field during sintering. A leakage current might go through the sintering compact during a heating process, and a similar phenomena is also proposed in ferroelectric ceramics [30; 45], and TiCxOyNz/Si_3N_4 based nanocomposites [31]. Therefore, it is expected that a direct current might hop across conductive TiN grains embedded in the insulating β-SiAlON matrix when applying a pulsecurrent. A temporary high temperature might occur in the specimen, and consequently accelerate the grain coarsening behavior of β-SiAlON during a sintering cycle.

Except for the case of the 10TN composite, most of the β-SiAlON grains for the composites have an equiaxial shape with a grain size of less than 200nm (as shown in Fig. 6), whereas a tiny amount of elongated grains with a grain width of 100nm were observed. For the composite of 5 TN, the percolation concentration is too low (i.e. the interparticle distance of TiN is large) to allow a pulse current to pass through the sintering body [52]. For the samples of 15 TN, 20 TN, and 30 TN, the TiN phase significantly inhibits the grain growth of the β-SiA-

http://dx.doi.org/10.5772/50547

lON matrix, even though it is possible for a pulse current to pass through the samples during sintering.

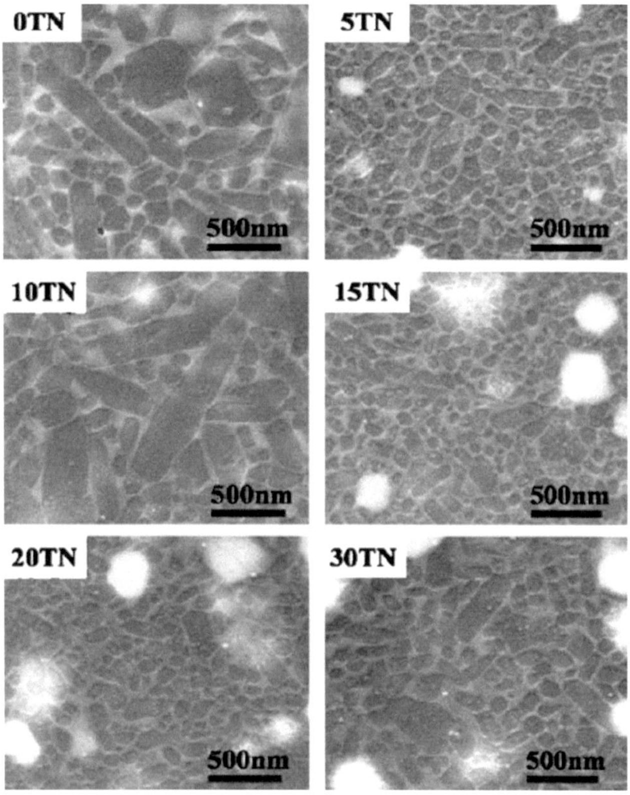

Figure 6. SEM micrographs showing the etching surface of TiN/Si_3N_4 composites with varying TiN content.

2.2.4. Electrical Properties

The change in electrical resistance for the above composites with varying TiN content is shown in Fig. 7. The electrical resistance substantially decreases from 2.43×10^{10}(5 TN) to 1.93×10^{8}(20 TN) (Ω.m) with the increase in TiN content, whereas it suddenly increases to a value of 4.19×10^{10} (Ω.m) for composite 30 TN. It has been reported that if the fraction of conductive TiN phase in the composite is under the degree of percolation threshold, the insulating property of the composites is maintained as the electrical resistance of non-conductive Si_3N_4 matrix [52]. This suggests the TiN phase does not form a connective network. As it is evidenced from Fig. 4, the submicrosized TiN grains are nearly isolated from each other. Moreover, the change of electrical resistance possibly depends on the grain size of the conductive phase [11]. Compared with the special cases for 20TN and 30 TN, the larger particle size of TiN for 30TN increases the interparticle distance, leading to a higher magnitude of electrical resistance over composite 20 TN.

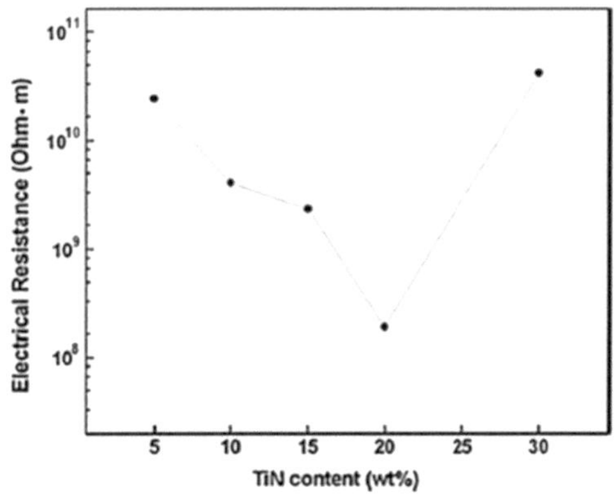

Figure 7. Change of electrical resistance of TiN/Si_3N_4 composites with varying TiN Content.

2.2.5. Mechanical Properties

The mechanical performance of the above nanocomposites has been studied from hardness and toughness measurements. Indentation hardness is a measure of resistance of a sample to permanent plastic deformation due to a constant compression load from a sharp object. This test works on the basic premise of measuring the critical dimensions of an indentation left by a specifically dimensioned and loaded indenter. Vickers hardness is a common indentation hardness scale, which was developed by Smith & Sandland [42]. Vickers test is often easier to use than other hardness tests since the required calculations are independent of the size of the indenter, and the indenter can be used for all materials irrespective of hardness. The influence of the TiN content of TiN/Si_3N_4 based nanocomposites on the Vickers hardness is shown in Fig. 8. The hardness value for these composites decreases with an increasing amount of TiN phase in the β-SiAlON matrix, and a similar trend was also observed by Lee et al., [31].

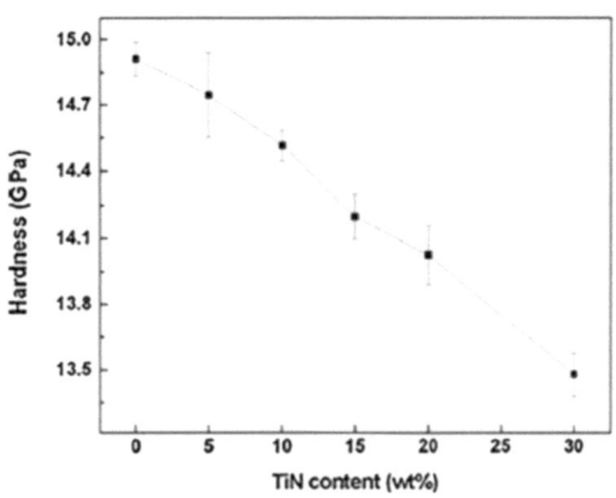

Figure 8. Vickers hardness of TiN/Si_3N_4 composites with various TiN content.

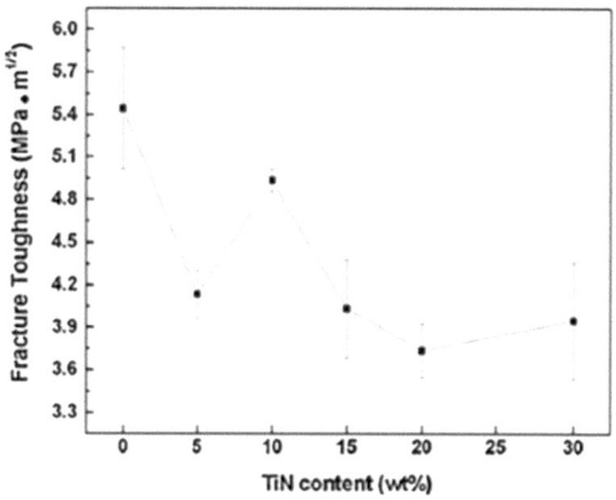

Figure 9. Fracture toughness of TiN/Si_3N_4 composites measured by Vickers surface indentation technique for various TiN content.

Fracture toughness describes the ability of a material containing a crack to resist fracture, and is one of the most important properties of any material for virtually all design applications. It is also a quantitative way of expressing a material's resistance to brittle fracture when a crack is present. The fracture toughness of the composites containing various compositions of TiN measured by the indentation technique for composites is shown in Fig. 9. The monolithic Si_3N_4 ceramics have the highest value of 5.4 MPa.m$^{1/2}$.Among the TiN/Si_3N_4 composites, the toughness reaches a maximum value of 4.9 MPa.m$^{1/2}$ for the composite containing 10 wt% TiN, whereas the other composites have values lower than 4.2 MPa.m$^{1/2}$. An equation derived by Buljan et al., 1988, which expresses the increase in toughness as function of grain size under the assumption that the grain shapes are the same, is given by

$$dK_c = CK_c^0\left(\frac{dD}{D_0} - 1\right) \tag{1}$$

Where C is a coefficient dependent on the mode of fracture; $K_c{}^0$ and and D_0 are the initial toughness and grain size; dK_c and dD are the respective changes in toughness and diameter. Hence, the changes in grain size and shape are directly related to toughness. An increase in Si_3N_4 grain size results in increasing fracture toughness. Although the addition of TiN does not improve the mechanical properties of Si_3N_4 based composites, the special sintering behavior produced by pulse direct current (grain coarsening effect for Si_3N_4 based grain) may occur in a SPS process.

3. Conclusions

i. By utilizing Si_3N_4 and TiN nano powders as starting materials, a series of near-fully dense TiN/Si_3N_4 based nanocomposites containing varying TiN contents (5–30 wt %) have been fabricated successfully by a spark plasma sintering technique.

ii. A grain coalescence of the TiN phase has been demonstrated by TEM. The conductive TiN grains in the insulating Si_3N_4 matrix are observed to be isolated from each other. From the microstructural observations, the composites appear to be insulating materials.

iii. For the nanocomposite of 5 TN, 15 TN, 20 TN, and 30 TN, the TiN phase inhibits the grain growth of Si_3N_4 based grains during sintering. Hence, the nanosized Si_3N_4 based crystallites are maintained in size as the raw material.

iv. The spark plasma sintered TiN/Si_3N_4 based composite containing 10 wt% TiN achieves the largest grain size and the highest toughness of 4.9 MPa. $m^{1/2}$ compared to the other composites. A possible pulse current sintering mechanism might occur, which causes a temporary high temperature in the sintering compact, and then accelerates the grain coarsening of Si_3N_4 based grains.

Acknowledgements

Authors are thankful to National Science Council of Taiwan for its financial support under the contract No: NSC 99-2923-E-006-002-MY3 to carry out the present work.

Author details

Jow-Lay Huang* and Pramoda K. Nayak

*Address all correspondence to: JLH888@mail.ncku.edu.tw

National Cheng Kung University, Taiwan

References

[1] Ayas, E., Kara, A., & Kara, F. (2008, July). A novel approach for preparing electrically conductive α/β SiAlON-TiN composites by spark plasma sintering. *Journal of the Ceramic Society of Japan*, 116(1355), 2008, 812-814, 0914-5400.

[2] Buljan, S. T., Baldoni, J. G., & Huckabee, M. L. (1987). *Si_3N_4-SiC composite, American Ceramic Society Bulletin*, 66(2), 347-352, 0002-7812.

[3] Buljan, S. T., Baldoni, J. G., Neil, J., & Zilberstein, G. (1988). Dispersoid-toughened silicon nitride composites,. *ORNL/Sub/8522011/1, GTE.*

[4] Balakrishnan, S., Burnellgray, J. S., & Datta, P. K. (1995). in: S. Hampshire, M. Buggy, B. Meenan, N. Brown (eds.) Preliminary studies of TiN particulate-reinforced

Si(3)N(4) matrix composite (SYALON 501) following exposure in oxiding and oxy-chloridising environments. Key Engineering Matererials, 1013-9826 , 99(1), 279-290.

[5] Borsa, C. E., & Brook, R. J. (1995). Fabrication of Al_2O_3/SiC nanocomposites using a polymeric precursor for SiC. In Ceramic Transactions Ceramic. Processing and Science, The American Ceramic Society, H. Hausner, G. L. Messing and S.-I. Hirano, (Ed.), Westerville, OH , 51, 653-657.

[6] Borsa, C. E., Ferreira, H. S., & Kiminami, R. H. G. A. (1999, May). Liquid Phase Sintering of Al_2O_3/SiC Nanocomposites. *Journal of the European Ceramic Society*, 19(5), 1999, 615-621, 0955-2219.

[7] Blugan, G., Hadad, M., Janczak-Rusch, J., Kuebler, J., & Graule, T. (2005, April). Fractography, mechanical properties, and microstructure of commercial silicon nitride-titanium nitride composites. *Journal of the American Ceramic Society*, 88(4), 2005, 926-933, 1551-2916.

[8] Carroll, L., Sternitzke, M., & Derby, B. (1996, November). Silicon carbide particle size effects in alumina based nanocomposites. *Acta Meterialia,*, 44(11), 1996, 4543-4552, 1359-6454.

[9] Dusza, J., Šajgalík, P., & Steen, M. (1999). Fracture Toughness of Silicon Nitride/Silicon Carbide Nanocomposite at 1350°C. *Journal of the American Ceramic Society*, 82(12), 3613-3615, 1551-2916.

[10] Dusza, J., Kovalčík, J., Hvizdoš, P., Šajgalík, P., Hnatko, M., & Reece, M. (2004). Creep Behavior of a Carbon-Derived Si_3N_4-SiC Nanocomposite. *Journal of the European Ceramic Society*, 24(12), 3307-3315, 0955-2219.

[11] Deepa, K. S., Kumari, Nisha. S., Parameswaran, P., Sebastian, M. T., & James, J. (2009). Effect of conductivity of filler on the percolation threshold of composites. Applied Physics Letters, 0003-6951 , 94(14), 142902.

[12] Evans, A. G., & Charles, E. A. (1976, July). Fracture Toughness Determinations by Indentation. *Journal of the American Ceramic Society*, 59(7-8), 1976, 371-372, 1551-2916.

[13] Greil, P., Petzow, G., & Tanaka, H. (1987). Sintering and hipping of silicon nitride-silicon carbide composite materials, Ceramics International (September 1986) 0272-8842 , 13(1), 19-25.

[14] Gogotsi, Y. G. (1994). Review: particulate silicon nitride based composite. Journal of Materials Science, January (1994). 1573-4803 , 29(10), 2541-2556.

[15] Gao, L., Li, J. G., Kusunose, T., & Niihara, K. (2004, February). Preparation and properties of TiN-Si_3N_4 composites,. *Journal of the European Ceramic Society*, 24(2), 2004, 381-386, 0955-2219.

[16] Galusek, D., Sedláček, J., Švančárek, P., Riedel, R., Satet, R., & Hoffmann, M. (2007). The influence of post-sintering HIP on the microstructure, hardness, and indentation

fracture toughness of polymer-derived Al_2O_3-SiC nanocomposites. *Journal of the European Ceramic Society*, 27(2-3), 1237-1245, 0955-2219.

[17] Guo, Z., Blugan, G., Kirchner, R., Reece, M., Graule, T., & Kuebler, J. (2007, September). Microstructure and electrical properties of Si_3N_4-TiN composites sintered by hot pressing and spark plasma sintering,. *Ceramics International*, 33(7), 2007, 1223-1229, 0272-8842.

[18] Herrmann, M., Balzer, B., Schubert, C., & Hermel, W. (1993). Densification, microstructure and properties of Si_3N_4-Ti(C,N) composites. *Journal of the European Ceramic Society*, 12(4), 287-296, 0955-2219.

[19] Huang, J. L., Chen, S. Y., & Lee, M. T. (1994). Microstructure, chemical aspects and mechanical properties of TiB_2/Si_3N_4 and TiN/Si_3N_4 composites. Journal of Materials Research, 0884-2914 , 9(9), 2349-2354.

[20] Huang, J. L., Lee, M. T., Lu, H. H., & Lii, D. F. (1996). Microstructure, fracture behavior and mechanical properties of TiN/Si3N4 composites, *Materials Chemistry and Physics*, 45(3), 203-210, 0254-0584.

[21] Kawano, S., Takahashi, J., & Shimada, S. (2002). Highly electroconductive TiN/Si_3N_4 composite ceramics fabricated by spark plasma sintering of Si_3N_4 particles with a nano-sized TiN coating. *Journal of Materials Chemistry*, 12(2), 361-365, 1364-5501.

[22] Kawano, S., Takahashi, J., & Shimada, S. (2003, April). Fabrication of TiN/Si_3N_4 ceramics by spark plasma sintering of Si_3N_4 particles coated with nanosized TiN prepared by controlled hydrolysis of $Ti(O\text{-}i\text{-}C_3H_7)_4$. *Journal of the American Ceramic Society*, 86(4), 2003, 701-705, 1551-2916.

[23] Kašiarová, M., Dusza, J., Hnatko, M., Lenčéš, Z., & Šajgalík, P. (2002). Mechanical Properties of Recently Developed Si3N4 + SiC Nanocomposites. *Key Engineering Materials*, 223, 233-236, 1013-9826.

[24] Kašiarová, M., Rudnayová, E., Kovalčík, J., Dusza, J., Hnatko, M., Šajgalík, P., & Merstallinger, A. (2003). Wear and creep characteristics of a carbon-derived Si_3N_4/SiC micro/nanocomposite. Materialwissenschaft und Werkstofftechnik, April (2003). 1521-4052 , 34(4), 338-342.

[25] Kašiarová, M., Dusza, Ján., Hnatko, M., Šajgalík, P., & Reece, M. J. (2006). Fractographic Montage for a Si_3N_4-SiC Nanocomposite. Journal of the American Ceramic Society May (2006). , 1551-2916 , 89(5), 1752-1755.

[26] Lee, B. T., Yoon, Y. J., & Lee, K. H. (2001, January). Microstructural characterization of electroconductive Si3N4-TiN composites,. *Materials Letters*, 47(1-2), 2001, 71-76, 0016-7577X.

[27] Lide, D. R. (2002). Experimental Data: Evaluation and Quality Control. CRC Handbook of Chemistry and Physics 83rd ed. CRC Press Boca Raton April (2002).

[28] Liu, C. C., & Huang, J. L. (2003). Effect of the electrical discharge machining on strength and reliability of TiN/Si_3N_4 composites. *Ceramics International*, 29(6), 679-687, 0272-8842.

[29] Liu, X. K., Zhu, D. M., & Hou, W. C. (2006). Microwave permittivity of SiC-$A1_2O_3$ composite powder prepared by sol-gel and carbothermal reduction. *Transactions of Nonferrous Metals Society of China*, 16, s494-s497, 1003-6326.

[30] Liu, J., Shen, Z. J., Nygren, M., Kan, Y. M., & Wang, P. L. (2006). SPS processing of bismuth-layer structured ferroelectric ceramics yielding highly textured microstructures. *Journal of the European Ceramic Society*, 26(15), 3233-3239, 0955-2219.

[31] Lee, C. H., Lu, H. H., Wang, C. A., Nayak, P. K., & Huang, J. L. (2011, March). Influence of Conductive Nano-TiC on Microstructural Evolution of Si3N4-Based Nanocomposites in Spark Plasma Sintering. *Journal of the American Ceramic Society*, 94(3), 2011, 959-967, 1551-2916.

[32] Martin, C., Mathieu, P., & Cales, B. (1989). Electrical discharge machinable ceramic composite. *Materials Science and Engineering: A*, 109, 351-356, 0921-5093.

[33] Niihara, K., & Hirai, T. (1986). Super-Fine Microstructure and Toughness of Ceramics. *Bull. Cerum. Soc. Jpn.*, 21(7), 598-604.

[34] Niihara, K., & Nakahira, A. (1988). Strengthening of oxide ceramics by SiC and Si_3N4 dispersions. In: Proceedings of the Third International Symposium on Ceramic Materials and Components for Engines, The American Ceramic Society, V. J. Tennery, (Ed.), Westerville, Ohio , 919-926.

[35] Niihara, K., & Nakahira, A. (1991). Strengthening and toughening mechanisms in nanocomposite ceramics. *Ann. Chim. (Paris)*, 16, 479-486, 0151-9107.

[36] Niihara, K. (1991). New design concept for structural ceramics-Ceamic nanocomposites. *The Centennial Memorial Issue of The Ceramic Society of Japan*, 99(10), 974-982, 0914-5400.

[37] Nakahira, A., Sekino, T., Suzuki, Y., & Niihara, K. (1993). High-temperature creep and deformation behavior of Al_2O_3/SiC nanocomposites. *Ann. Chim. (Paris),*, 18, 403-408, 0151-9107.

[38] Ohji, T., Nakahira, A., Hirano, T., & Niihara, K. (1994). Tensile creep behavior of Alumina/ Silicon carbide nanocomposite. *Journal of the American Ceramic Society*, 77(12), 3259-3262, 1551-2916.

[39] Ohji, T., Jeong-K, Y., Choa-H, Y., & Niihara, K. (1998, June). Strengthening and toughening mechanisms of ceramic nanocomposites. *Journal of the American Ceramic Society*, 81(6), 1998, 1453-1460, 1551-2916.

[40] Raichenko, A. I. (1985, January). Theory of metal powder sintering by an electric-pulse discharge. *Soviet Powder Metallurgy and Metal Ceramics*, 24(1), 1985, 26-30, 0038-5735.

[41] Ragulya, A. V. (2010). Fundamentals of Spark Plasma Sintering. Encyclopedia of Materials: Science and Technology , 978-0-08043-152-9 , 1-5.

[42] Smith, R. L., & Sandland, G. E. (1922). An Accurate Method of Determining the Hardness of Metals, with Particular Reference to Those of a High Degree of Hardness. Proceedings of the Institution of Mechanical Engineers , LCCN 0801 8925, I, 623-641.

[43] Sinha, S. N., & Tiegs, T. N. (1995). Fabrication and properties of Si_3N_4-TiN composite. Ceramic Engineering and Science Proceedings July (1995). , 0196-6219 , 16(4), 489-496.

[44] Šajgalík, P., Dusza, J., & Hoffmann, M. J. (1995). Relationship Between Microstructure, Toughening Mechanisms and Fracture Toughness of Reinforced β-Si_3N_4 Ceramics. *Journal of the American Ceramic Society,*, 78(10), 2619-2624, 1551-2916.

[45] Shen, Z., & Nygren, M. (2001). Kinetic aspects of superfast consolidation of silicon nitride based ceramics by spark plasma sintering,. *Journal of Materials Chemistry*, 11(1), 204-207, 0959-9428.

[46] Shen, Z. J., Peng, H., Liu, J., & Nygren, M. (2004). Conversion from nano- to micron-sized structures: Experimental observations. *Journal of the European Ceramic Society*, 24(12), 3447-3452, 0955-2219.

[47] Šajgalík, P., Hnatko, M., Lojanová, Š., Lenčéš, Z., Pálková, H., & Dusza, J. (2006). Microstructure, hardness, and fracture toughness evolution of hot-pressed SiC/Si_3N_4 nano/micro composite after high-temperature treatment. International Journal of Materials Research , 1862-5282 , 97(6), 772-777.

[48] Vollath, D., Szabó, D. V., & HauBelt, J. (1997). Synthesis and Properties of Ceramic Nanoparticles and Nanocomposites. *Journal of the European Ceramic Society,*, 17(11), 1317-1324, 0955-2219.

[49] Vollath, D., & Szabó, D. V. (2006). Microwave Plasma Synthesis of Ceramic Powders. In: Advances in Microwave and Radio Frequency Processing Part-IX, Springer , 619-626.

[50] Xu, Y., Nakahira, A., & Niihara, K. (1994). Characteristics of Al_2O_3-SiC nanocomposite prepared by sol-gel processing. *Journal of the Ceramic Society of Japan*, 102(3), 312-315, 0914-5400.

[51] Yoshimura, M., Komura, O., & Yamakawa, A. (2001, May). Microstructure and tribological properties of nano-sized Si 3N4. *Scripta Materialia*, 44(8-9), 2001, 1517-1521, 1359-6462.

[52] Zivkovic, L., Nikolic, Z., Boskovic, S., & Miljkovic, M. (2004, June). Microstructural characterization and computer simulation of conductivity in Si3N4-TiN composites,. *Journal of Alloys and Compounds*, 373(1-2), 2004, 231-236, 0925-8388.

Chapter 17

Synthesis and Characterization of Ti-Si-C-N Nanocomposite Coatings Prepared by a Filtered Vacuum Arc with Organosilane Precursors

Seunghun Lee, P. Vijai Bharathy, T. Elangovan,
Do-Geun Kim and Jong-Kuk Kim

Additional information is available at the end of the chapter

http://dx.doi.org/10.5772/51059

1. Introduction

Many deposition tools such as magnetron sputtering, plasma enhanced chemical vapor deposition (PECVD), arc ion plating (AIP), and filtered vacuum arc (FVA) have been introduced for synthesizing nanocomposite films. Table 1 summarizes previous works of nanocomposite coatings. Nanocomposites based on TiN have been investigated dominantly with the incorporation of silicon or carbon contents. The incorporation methods such as an alloy arc cathode, addition of reactive gas, and additional magnetron sputtering have been used to deposit ternary or quaternary composition nanocomposite films. The magnetron sputtering and PECVD have been firstly used to grow nanocomposite films due to the simplicity of controlling a composition ratio becuase precise control of additional components is important to make a nanocomposite structure. For example, Ti-Si-N nanocomposite films have showed the maximum hardness at Si content of 9±1 at.% [1]. After that, a vacuum arc discharge has been applied to the nanocomposite coatings becuase the vacuum arc process has many advantages against other CVD or PVD processes. A vacuum arc plasma exhibits high ionization ratio more than 90%. Also the ion energy in a vacuum arc is in the range of 10-100 eV. Hence, the effect of ion energy on the film structure appears significantly [2]. Nevertheless, the vacuum arc method cannot avoid the problem of macro particles. Macro particles generated from arc spots are the main drawback of the vacuum arc process. The macro particles form micro cracks or pin holes, resulting in a bad corrosion resistance when coatings are exposed in some corrosive environments.

INTECH
open science | open minds

© 2012 Lee et al.; licensee InTech. This is a paper distributed under the terms of the Creative Commons Attribution License (http://creativecommons.org/licenses/by/3.0), which permits unrestricted use, distribution, and reproduction in any medium, provided the original work is properly cited.

In FVA, magnetic filters have been introduced to transport plasma except the macro particles. The filters transport charged particles selectively using electromagnetic fields. However, neutral macro particles collide with a wall by an inertia drift. Main issues are the efficient removal of the macro particles and the minimization of ion loss through a magnetic filter wall. The effective way to reduce the macro particles is based on the spatial separation of the trajectories of macro particles and ions [3]. If the magnetic field is curved such as the field inside a curved solenoid, electrons follow the curvature. The electrons are said to be magnetized. In contrast, ions are usually not magnetized because the gyration radius of ion is much larger than that of electron. Nevertheless the ions are forced to follow the magnetic field lines due to the electric fields between electrons and ions. Therefore ions and electrons are transported along magnetic field lines [4]. Various FVA methods have been widely applied to the nanocomposite deposition without any macro particle problems. Magnetic filtering technology removes efficiently the macro particles and result in a smooth film surface [5,6].

Method	Material	Hardness	Ref.
Arc	Ti-Si-N	45 GPa	[1]
Arc	Ti-Al-N/Cr-N	37 GPa	[7]
Arc	Ti-Al-Si-N	34 GPa, 42.4 GPa	[8,9]
Arc	Ti-Al-N	35.5 GPa	[10]
Arc, magnetron sputter	TiN-Cu, CrN-Cu, MoN-Cu	27-42 GPa	[11]
Arc, magnetron sputter	Ti-Si-N	45-55 GPa	[12,13]
FVA, magnetron sputter	Ti-Si-N	45 GPa	[14]
FVA, magnetron sputter, E-beam evaporation	Ti-Cr-N,Ti-B-C	43.2 GPa	[15,16]
FVA	Ti-Si-N	40.1 GPa,	[17]
PECVD	Ti-Si-N	3500 HK(kg/mm^2), 40 GPa	[18,19]
PECVD	Ti-Si-C-N	48 GPa	[20]
PECVD	Ti-Si-C-N	52 GPa	[21]
Magnetron sputter	Ti-Si-N	38GPa, 45 GPa	[22,23]

Table 1. Nanocomposite coatings by various methods.

Several studies found the maximum efficiency and the optimum condition for the curved magnetic filters. An analysis of plasma motion along the toroidal magnetic field has been shown that plasma that is transported by the magnetic field in a guiding duct should satisfy the following relation [24],

$$B > M_i V_3 / Zea \tag{1}$$

where M_i is the ion mass, V_o is the translational velocity, Z is the charge multiplicity of the ion, e is the electron charge, and a is the minor radius of the plasma guiding duct. The transport of heavy metal ions having an energy of even a few tens of eV requires strong magnetic field above 1 Tesla to fulfill the inequality in Eq. (1). However, it is practically impossible to provide a stable burning of the direct current arc discharge in the strong magnetic field. Therefore, it is reasonable to consider heavy-element plasma flow transport in a curvilinear system with crossed electric and magnetic fields using the principles of plasma optics as a guide [25,26]. In this case, the required magnetic field is determined by the following condition, $\rho_e < a < \rho_i$, where ρ_e and ρ_i are the electron and ion Larmor radius, respectively. The required field is significantly lower than the fields defined by the expression in Eq. (1). Electron Larmor radius is

$$r_e = (m_e k T_e)^{1/2} / eB \tag{2}$$

where m_e is the electron mass, k is Boltzmann constant, and T_e is the electron temperature.

Note that electrons are only magnetized, while the ions are not. The electrons move along the magnetic field lines. Due to the highly conductive plasma, the magnetic field lines are equi-potentials. Considering a plasma diffusion in vacuum, electrons have higher mobility than ions due to smaller mass except at a sheath boundary. However, electrons expand with the same velocity as ions because electrostatic forces keep the electrons and ions together. And a cross field diffusion is given by the Bohm formula, $D_B = kT_e/16B$, though the cross field diffusion coefficient, D, is proportional to B^{-2} in the classical theory[27].

Anders mentioned about the criterion of system efficient, K_s, which is generally considered as the ratio of the total ion flow at the exit of the system, I_i, to the arc discharge current, I_a, as follows,

$$K_s = I_i / I_a \tag{3}$$

The system coefficient is typically 1% [6]. There is a general agreement that the transport efficiency is maximized by focusing the plasma into the duct and biasing the duct to a positive potential of 20 V. Predictions of the available maximum transmission vary between 11 and 25% depending on the ion energy [28]. In practice, the transport of plasma produced by pulsed high current arcs (HCA) was showed that the system coefficient was 7% [6]. For linear FVA, the maximum value of the system efficiency reached 8% when arc current Ia was adjusted in the range of 100–110 A and the magnetic filter field was ~ 20 mT [29].

To supplement the accuracy of system coefficient, a particle system coefficient, K_p, is proposed to eliminate the influence of the various ion charge states by considering the mean ion charge state, Z_{av}, of the used metal as follows.

$$K_p = I_i / Z_{av} I_a \tag{4}$$

Because the average ion charge state is taken into account, the particle system coefficient is more closely related to the deposition rate [30]. However this can be particularly insufficient for filter optimization when the system used a graphite cathode that generates solid rebounding macro particles.

The problem has been solved by the numerical calculation of the particle trajectories using a two-dimensional approximation [31]. In the calculation, it assumes that the macro particles are solid spheres, the inner surfaces of the plasma guide, and the intercepting fins are smooth, the repulsion of particles from a guiding duct wall is partially elastic, the particles are emitted froma cathode spot with equi-probability in any direction. The computing results make it possible to estimate the ratio of the pass of macro particle flow,N_{ex}, to the flow, N_{ent}, generated by the cathode spot. The ratio N_{ex}/N_{ent} characterizes the likelihood of an macro particle passing through the system. The results of simulations indicate that the absence of a direct line-of-sight between the cathode and the substrate is not always sufficient to provide the required degree of macro particle removal from the plasma. The results of computations performed for various magnetic filters are presented in Table 2 [27].

Filter type	Knee (45°)	Torus (45°)	Rectang.	Dome	Torus (90°)	Retil.	Radial	Wide apert.
N_{ex}/N_{ent} [%] (predicted)	1.7	25.0	17.0	1.7	0	4.4	0	0
Transport [%] (measured)	3.0	2.5	2.5	2.5	1.5	1.8	8.4	~6.0

Table 2. Fitlering (N_{ex}/N_{ent}) and transporting properties of magnetic plasma filters [27].

There are various types of filters as shown in Fig. 1[32]. Most types are used magnetic fields to transport plasma without macro particles. Several types only use the collisional reduction of macro particles. In most cases a plasma is transported from the cathode to the substrate, and the droplets are eliminated by a plasma transportation wall, guiding duct. Many review papers of the filtered arc system and technology have been reported [6,33-38]. A typical filtered arc system with its different electromagnetic plasma transportation duct or droplet filter configurations is shown in Fig. 1(a)–(h). Electromagnetic coils transporting plasma in the out of line of sight direction can be positioned in the chamber, instead of placing them outside of the filter duct. The off-plane double bend filter is nicknamed FCVA and is now commercially available [39,40]. Most FAD units have electromagnetic coils outside of the plasma duct and have baffles inside the duct wall. However, some types have freestanding coils inside the plasma duct or the chamber. Other interesting filters have been developed. Examples are shown in Fig. 1(i)–(l). In the Venetian-blind filter, the plasma passes between the vane lamellae, and the droplets are caught or reflected by the lamellae [41,42]. A coaxial filter is operated with a large current pulse, and the plasma is driven by a self magnetic field [43]. An electrostatic filter can be used with a pulsed arc having a laser trigger [44]. However, recently only the laser triggered arc is used without the electrostatic filter. Mechanical filters can be used in a pulse arc [45], which may also be used in pulsed laser deposition [46].

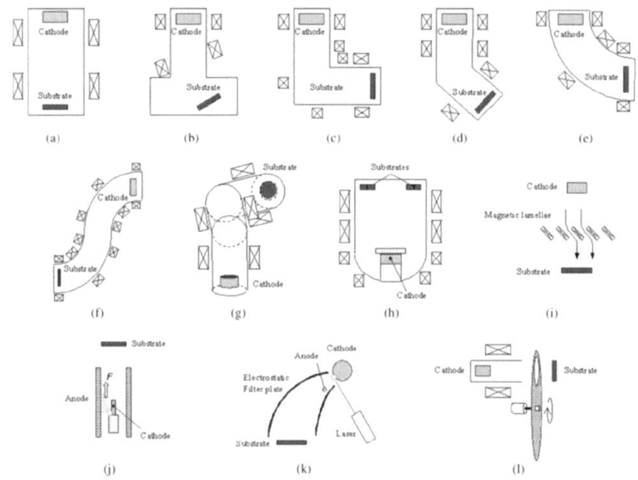

Figure 1. Various types of filter systems [32]. (a) Rectilinear. (b) Bent. (c) Rectangular. (d) Knee. (e) Torus. (f) S-shape. (g) Off-plane double bend. (h) Dome. (i) Venetian blind. (j) Co-axial (pulse). (k) Electrostatic filter with laser trigger (pulse). (l) Mechanical pulse.

2. Nanocomposite Films Prepared by a FVA With an Organosilane Precursor

2.1. Basic Configuration of Deposition System for an Organosilane Incorporated FVA

Ti-Si-C-N quaternary nanocomposite coatings were prepared by using a filtered vacuum arc deposition system. Figure 2 represents the schematic diagram of the FVA coating system. The deposition system consists of a water-cooled cathode and anode, plasma guiding duct, and magnet coils. The diameter and height of the chamber is 800 and 650 mm, respectively. The vacuum arc discharge from cathode emits high energy (~ 60 eV) ions and generates dense plasma above 10^{13} cm^{-3}. An arc spot also makes neutral macro particles, which cause several problems such as rough surface, pinhole, and micro cracks in coatings. To remove the macro particles the plasma guiding duct and magnet coil were used. Coaxial magnetic fields about 15 mT were generated by six magnetic coils. The pumping system consists of rotary pump (900 l/min) and oil diffusion pump (1500 l/s). The ultimate pressure of deposition chamber was 2×10^{-6} mTorr by using rotary and diffusion pumps. MKS mass flow controllers were used to regulate the flow rate of tetramethylsilane (TMS) (99.99%), argon (99.999%) and nitrogen (99.99%) gases.

The samples are mounted on the rotational substrate holder with a rotational speed of 3 rpm for the deposition of the coatings. The deposition process consisted of Ar ion bombardment cleaning, Ti(100 nm)/TiN(200 nm) interlayer deposition to improve an adhesion strength. The thickness of the Ti-Si-C-N nanocomposite coating was about 0.4 to 0.6 μm. The experimetal details are shown in Table 3.

Structural characterization of as-obtained samples was done by X-ray diffraction (Shimadzu XRD-6000, Cu Kα radiation λ = 1.5406 Å, scanning rate 1° min−1) and transmission electron microscopy (JEOL JEM-3100 FEF-UHR, 300 kV). The crystallite size was determined by Scherrer formula and lattice parameter of coating was calculated by Bragg law for the cubic system.

Elemental analysis and chemical nature of coatings were performed with X-ray photoelectron spectroscopy (XPS), using a VG Scientific ESCALAB 250 spectrometer with a Mg Kα X-ray source. Hardness are assessed by means of a nanoindentation system (MTS, nano indenter XP) using a Berkovich diamond indenter. Coating hardness was determined from the loading and unloading curves employing depth-sensing hardness testers. The applied load was gradually increased to 3000 μN at a loading rate of 150 μN/s, and was held at this maximum value for 10 s. Testing was done using the constant-displacement-rate mode until a depth of 200 nm was reached and the values from fifteen indents were averaged for each condition. Adhesion strength was measured by a scratch tester (J&L, Scratch Tester). The applied load on the diamond tip (tip radius 200 μm, conical angle 120°) was continuously increased at a rate of 0.25 N/s, while the tip advanced at a constant speed of 0.05 mm/s.

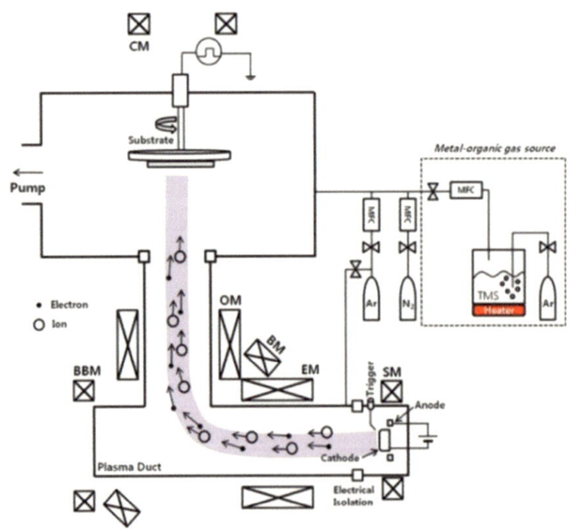

Figure 2. Experimental schematics of filtered vacuum arc with organosilane vaporizations.

Arc voltage	24 V
Arc current	60 A
Duct voltage	25 V
Duct current	43A
Substrate voltage	0 ~ -400 V
Base pressure	2.3×10^{-6} Torr
Total pressure	8.2×10^{-4}Torr
Deposition rate	9-38 nm/min
N2 flow rate	50 sccm
TMS flow rate	10 sccm
Substrate Temperature	< 150°C

Table 3. Experiment details.

2.2. Ti-Si-C-N Nanocomposite Film Prepared by a FVA With TMS Gas

2.2.1. Composition and Chemical Analysis

XPS analysis is used to identify a chemical composition on the surface of the nanocomposite coating. Surface pre-sputtering eliminated the surface oxide at air-exposed samples. The elemental compositions for Ti, Si, C and N as a function of TMS flow rate from 5 to 20 sccm are given in Fig. 3. A strong increase of the Si (2.1±1 to 12.2±1 at. %) content is observed up to the TMS flow rate of 20 sccm. Ti and N content showed decreasing trend from 18.8±1 to 9.83±1 at.% and 24.8±2 to 19.14±2 at. %, respectively. Whereas the carbon content remained constant in the range of 31± 2 at. %.

The XPS spectra of of Ti 2p, Si 2p, and N 1s are shown in Fig. 4. Fig. 4 (a) depicts the XPS spectrum of Ti 2p. The characteristic doublet of Ti $2p_{3/2}$ and $2p_{1/2}$ are clearly observed. The Ti $2p_{3/2}$ peak position shows positive shift from 460.85 eV, which reveals a good agreement value of 460.5 eV for TiN. This peak position shifts to higher value at silicon content of 3 at. % and then it moves to lower binding energy (BE) side at silicon content at 8 at. %. The second peak position Ti $2p_{1/2}$ is showing at 455.15 eV matching with Ti- nature bonding position of TiN. It is well known that TiN_x has flexible chemical states governed by the compactness of other small non-metal atoms filled in the octahedral voids of titanium [47]. Fig 4 (b) shows the high resolution spectrum of N 1s region. The three peaks correspond to 398.5, 396.8 and 400.6 eV, are in agreeing with the binding energies of Si_3N_4, TiN and TiC, respectively. With increasing silicon content the N1s peak position showeda negative shift from 396.8 to 396.6 eV. Fig 4 (c) shows Si 2p spectra peaks, which are observed at 101.8, 101.1 and 102.8 eV, are attributed to the Si_3N_4, the Si (2p)–C, and Si (2p)–O bonds, respectively. The incorporation of TMS at low Si content (2 at.%) results in Si_3N_4 formation, whereas SiC formation is dominant at the high Si content (12 at.%). The reason is that CH_3 in TMS is not dissociated perfectly in arc plasma so the carbon content are incorporated and react with Si contents. From the XPS results, it was concluded that Si in Ti–Si–C–N coatings existed mainly as amorphous silicon nitride with some silicon carbide.

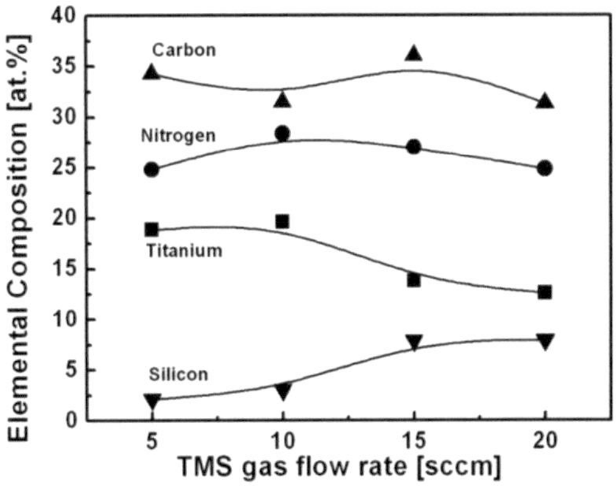

Figure 3. Chemical composition of Ti-Si-C-N nanocomposite films as a function of TMS flow rate.

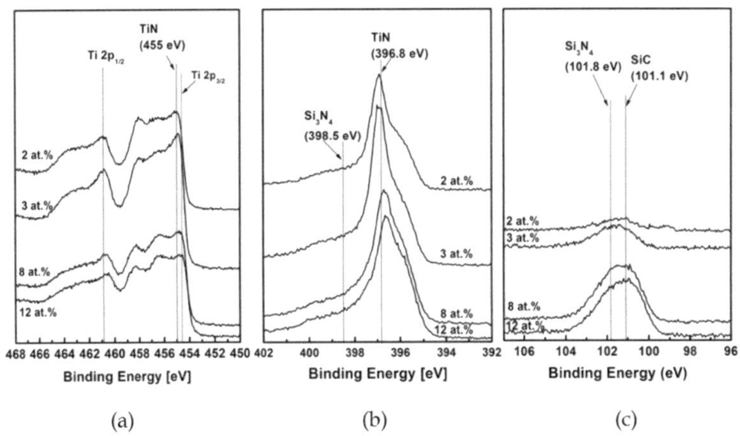

Figure 4. XPS spectra of Ti-Si-C-N nanocomposite films. (a) Ti (2p), (b) Si (2p),(c) N (1s).

2.2.2. Microstructure Analysis

X-ray diffraction (XRD) is used to investigate the crystalline phases of the Ti-Si-C-N coating. Figure 5 shows the XRD pattern of Ti-Si-C-N coating with different silicon contents. The Ti-Si-C-N nanocomposite coatings with silicon content range of 2.1 to 16.2 at. %, exhibit the diffraction peaks at the angles of $2\theta = 36.01°$, 43.63° and 50.79° which are corresponding to (111), (200) and 220) TiCN reflections respectively [14,48-52]. These positions of the three peaks are coinciding with the values obtained in the JCPDS card [53]. Also, peak is observed at 38° that are attributed to the diffraction of stainless steel, which is the substrate. No signals from the phase formation of Si_3N_4 or from titanium silicide can be observed [54,55]. Note that the amorphous phase Si_3N_4, deduced from the XPS results analysis, has been confirmed with the XRD measurement.

http://dx.doi.org/10.5772/51059

The crystalline size of the Ti-Si-C-N coating is calculated from the TiN (111) diffraction peak. The TiN (111) peak is fitted by using a Gaussian function to calculate the crystallite size from FWHM by using Scherrer equation. Adding Si, the crystallite sizes decreases from 3 to 2 nm, which shows that nanocrystalline phases are formed. It representsthat the addition of silicon could reduce the grain coarsening in the Ti-S-C-N nanocomposite coatings.

Figure 6 shows HRTEM picture acquired on the Ti-Si-C-N coating contain 3 at. % of Si. The lattice spacing of these crystallites is 0.212 nm. The amorphous phase has an irregular shape and a boundary surrounding the TiN nanocrystallites. The growth direction of (111) plane is clearly identified in this image. The image shows the fine grain structure and reveals that these fine grains are largely oriented in the direction of growth. An arrow indicated the growth direction. TiN nanocrystals with an average grain size of about 10 nm were separated by less than 1 nm thick brighter Si_3N_4 tissues. Fig. 6 (b) shows a STEM image of Ti-Si-C-N coating and the intensities from Ti, Si, C, and N obtained from an EDX line scan acquired from the STEM image. In the matrix, the Ti and the Ti signals are high and small for silicon small signals are observed from the XRD, XPS and TEM analysis results, it could be confirmed that the Ti–Si-C–N coating obtained in this experiment consisted of nanosized TiN crystallites surrounded by thin amorphous phase of Si_3N_4 [56].

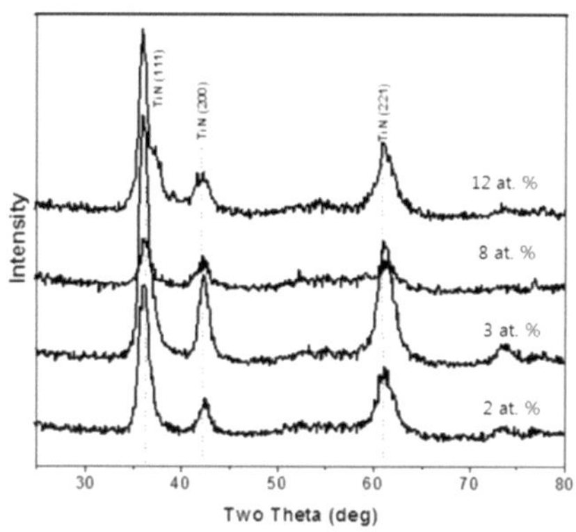

Figure 5. X-ray diffraction patterns at various Si contents from 2 to 12 at. %.

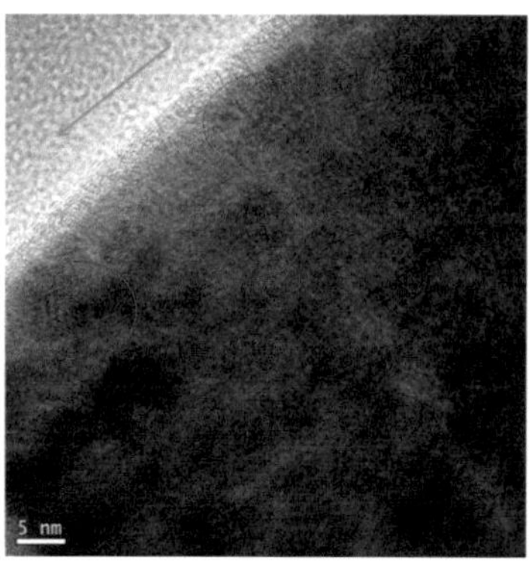

Figure 6. HRTEM image of Ti-Si-C-N at 3 at.% of Si content.

2.2.3. *Mechanical Properties*

The nano-indentation technique permits to extract the surface mechanical properties from depths of nanometers. Hardness can be calculated using Oliver and Pharr method [57]. The hardness decreases with increasing the depth of indentation at extremely small depths. This indentation sizes effect is expected for soft metal films and has been related to strain gradient plasticity. In the present study, nano-hardness of the Ti-Si-C-N coatings was obtained as a function of depth up to a maximum depth of 180 nm. The steel substrate hardness is around 2 GPa and Young's modulus is 200 GPa as obtained by nano-indentation experiments. Load versus indentation depth curves from multiple experiments by using the same maximum load and from 25 different sample locations were averaged and standard deviations were calculated and reported. Figure 7 represents the hardness of Ti-Si-C-N coatings as a function of displacement into surface. The hardness values of Ti-Si-C-N coatings are 18, 35, 32 and 27 GPa at a contact depth of 40 nm at 2.1, 3, 7 and 8 at. % of Si, respectively. The maximum hardness of ~35 GPa is obtained for the silicon content of 3 at. %. The hardness value of 48 GPa was achieved adding 10 at.% Si and 30 at.% C by Dayan Ma et.al [58]. Also he was explained it with carbon contents. Suddeep mabiraham et. al [49] also achieved the hardness of the Ti-Si-C-N films at the Si content of 9.2. at % with maxiumum hardness around 55 GPa achieved with Si content of 8.9 at.%. by using PVD method in [59]. So far many people have been reported the hardness of the Ti-Si-C-N coating at above 5 at. % of the silicon content. Present results show that the maximum increases in hardness for Ti-Si-C-N coating within 3 at. % of silicon, which proves such hardening effects, is known to occur in transition metal nitride systems with a few at. % of silicon content and this increased hardness effect due to hindrance effect of the segregated Si_3N_4 on TiN grain boundary sliding which is the predominant deformation mechanism in nanocrystalline materials [60]. On the other hand, the hardness reduction with a further increase in Si content is observed [58]. The decrease in hardness is due to increasing contributions from the soft substrate.

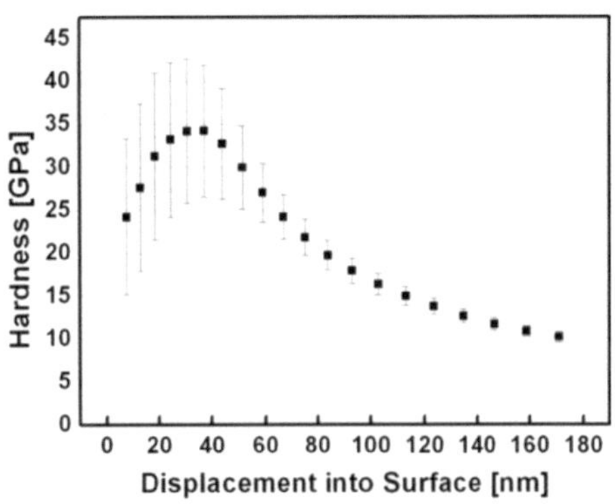

Figure 7. Nano indentation of Ti-Si-C-N films with 3 at. % of Si.

2.2.4. Adhesion Properties

Scratch tests with ramping loads up to 100 N were conducted on the films deposited on stainless steel substrates. The critical load required to cause the first delamination at the edge of the scratch track (adhesive failure) characterizes the adhesion properties of coatings. Optical microscopy of the films subjected to the scratch tests (Not shown here). The critical loads obtained for the above four coatings were 37, 44, 33 and 30 N, respectively. Strong acoustic signals were observed for the critical load exceeding 30 N. The lowest value of COF of 0.16 is obtained for the hardest film with Si content of 2.1 at.% and it is 0.17- 0.22 for the rest of nanocomposite films and agrees with the literature [59, 61]. For FVA deposition, the intrinsic energy of ions emitted from a cathode source is considerably higher as compared to the evaporated or sputtered atoms. This higher ion energy up to 60 eV condenses deposited film and enhances adhesion. Beside that the friction coefficient and penetration depth versus force show several inflections for the Ti-Si-C-N with 8 at. % Si.

2.3. Substrate Bias Effects on Ti-Si-C-N Nanocomposite Films

2.3.1. Structural Properties

The deposition rate of Ti-Si-C-N nanocomposite coating was found to decrease gradually from 18 to 8 nm/min, when the bias voltage increased from 0 to −400 V. The decreasing trend of deposition rate has been explained on the basis of removal of impurities and densification of films due to the energetic ion bombardment when increased substrate bias voltage [62].The XRD diffraction of Ti-Si-C-N nanocomposite coating for different substrate bias voltage is as shown in Fig. 8. It matches well with the ICDD PDF 42-1489 revealing the NaCl type crystal structure. However, the diffraction peaks are shifted towards the higher angle side indicating the presence of compressive stress in the coatings [63]. At zero bias voltage, coating shows predominant presence of (111) and (200) orientations. With increase in substrate bias, (220) was found to dominate the spectra especially at substrate biases of − 300 V and − 400 V. This behavior has been observed in binary, ternary and quaternary coatings and is attributed to the decrease in sputtering rate at higher substrate bias. There is no further detail of the Ti-Si-C-N coating with effect of bias voltage by using FVA system. These result show that substrate bias voltage has a strong influence on the structural properties of the deposited films and will be correlated with changes in hardness of the coatings. Fig. 9 shows the calculated lattice parameters from the XRD pattern for TiN peak position using equation (1). The lattice constant of TiN changes from 4.31599 to 4.24762 Å with the different substrate bias voltages. For −100 V bias voltage, TiN films are over stoichiometric corresponding to the highest lattice parameter (ICDD PDF 42-1489) a =4.29 Å). The shift could be attributed to higher residual stress in the coatings and changes in the composition of the coating. The calculated lattice parameter was shown 4.24762 Å at −400 V, it is slightly lower than that of the value reported standard lattice parameter. Usually these kinds of stress with increasing bias voltage are commonly observed in thin films grown by physical vapour deposition methods. With increasing bias voltage, the ion bombardment encouraged mobility of atoms was major effect which implies that the decreasing trend of the stress because of enhanced annihilation of defects [64].

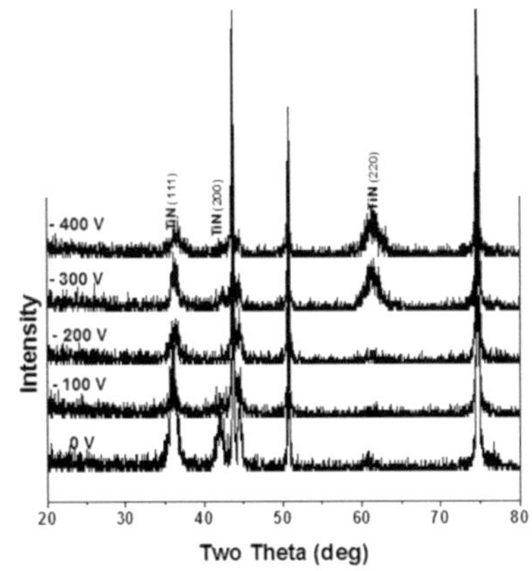

Figure 8. X-ray diffraction patterns of Ti-Si-C-N nanocomposite coatings various substrate bias voltage.

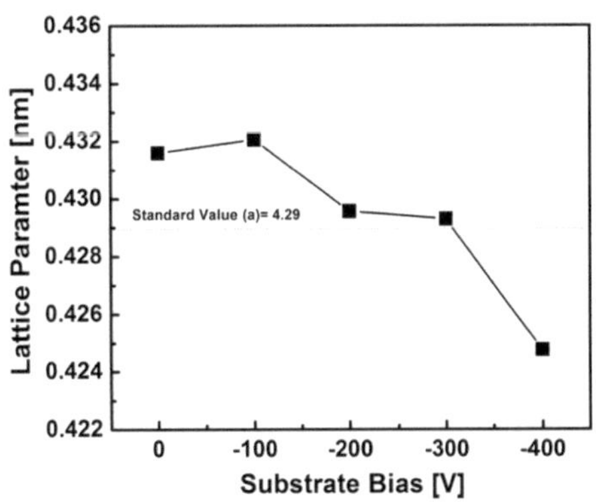

Figure 9. Lattice parameter of the Ti-Si-C-N nanocomposite coatings various substrate bias voltage.

2.3.2. Texture Orientation

The changes in the preferred orientation of Ti-Si-C-N nanocomposite coating as a function of the substrate bias are qualitatively estimated in terms of texture coefficients (TC). The TCs, determined by TC = $I_m(hkl)/I_0(hkl)/(1/n)\{I_m(hkl)/I_0(hkl)\}$ for (111) and (220) reflection [65], as a function of substrate bias are shown in Fig 10. Where $I_m(hkl)$ is the observed intensity of the (hkl) plane, $I_0(hkl)$ is the standard data (JCPDS) of the (hkl) plane, and N is the total number of diffraction peaks. When the TC value is larger than 1, a preferred orientation exists in the sample. The texture coefficient of the (111) orientation is significantly higher than that of the (220) orientation in the substrate bias range 0 to −100 V. When deposited lower film thickness it has been shown that the influence of surface prevails over strain energy and (100) orientation expected, but it is higher film thickness have (111) orientation expected with vice versa [66]. However we observed the contrary result compared to other researchers for nitride coatings with substrate bias. This may be due to the ion input energy or channeling effect is not the major role for the change of orientation between (111) and (220) plane. At −300 V, the trend of texture has completely changed from (111) to (220) and texture coefficient value shows extensively higher value than that of the (111) and it was maintained at the higher bias voltage of -400 V. It attributed to the mutual effects of both Ti and Si elements for preferred growth orientation of the Ti-Si-C-N nanocomposite coating at higher substrate bias voltage.

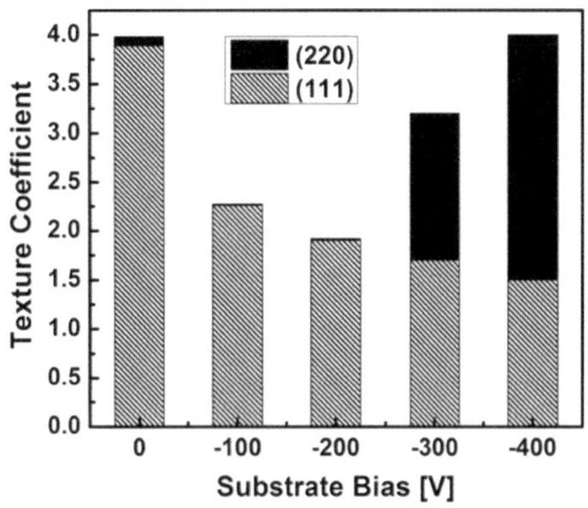

Figure 10. Texture Coefficient of the Ti-Si-C-N nanocomposite coatings various substrate bias voltage.

2.3.3. Chemical Composition and Analysis

To determine the composition of the Ti-Si-C-N nanocomposite coating, XPS analysis has been performed. The samples are sputtered with argon ions to remove the oxide top layer. Even though we found uniform oxygen concentration with respect to substrate bias voltage, could be due to the TMS gas which is used in the present studies. With increasing the negative bias voltage from 0 to −300 V, the silicon concentration increased from 5.17 to 8.14 at. %. Silicon has lower electro negativity than carbon; amorphous carbon atom bonded to a silicon atom attracts electrons from the silicon atom, which condense the Si-H bonds. The number of silicon atoms attached to Si-Hx is predicted to increase as silicon content increases in the Ti-Si-C-N nanocomposite coating [67]. In contrast, decreasing trend was observed for carbon content in the range from 40 to 24 at. %. It may be due to carbon atom is lighter than the Si atom and also the energy of the Ar ion increases with cause to build strong bombarding to the substrate with increasing bias voltage. Further with increasing substrate bias voltage about −400V, the silicon concentration shows decreasing trend in the range of 4 at.%, it is also lower value than substrate bias voltage range of silicon content from 0 to -300 V. The N and Ti content are seen to be almost independent respect to the substrate bias.

The chemical state of the Ti-Si-C-N nanocomposite coating was analyzed by the XPS measurements and is as depicted in Fig. 11. Fig 11 (a) depicts the C 1s peak spectra shows five component's namely, C1, C2, C3, C4 and C5 were observed in the spectrum zone. The two peaks of C1 (at 285.7 eV) and C2 (at 284.5 eV) corresponded with the position of C-N and C=C phase formation. Another three peak were peaks observed C3 (at 282.5 eV), C4 (at 281.2 eV) C5 (at

288.4 eV) corresponded with the position of Ti-C phase formation. As compared Ti-C peak position, the C=C peak position accounts for only a small fraction of the total C1 spectra, indicating that a small fraction of C atoms are bonded to Ti atoms, and most of the C atoms exist as amorphous carbon [68,69]. Fig 11(b) depicts the Si 2p peak spectra shows three components namely S1, S2 and S3 were observed in the spectrum zone. These two peaks of S1 (at 100.8 eV) and S2 (at 102.3 eV) corresponded with the position of SiC and Si_3N_4 phase formation. Another one of weak component was observed at about S3 (at 102.8 eV) corresponds to Si-O bonds. The peak intensity gradually increased with an increase of bias voltage, which implies that increasing trend of silicon concentration. Fig. 11(c) depicts the N1s 2p peak spectra shows three components namely N1, N2 and N3 were observed in the spectrum zone. These two peaks of N1 (at 398.28 eV) and N2 (at 396.64 eV) corresponded with the position of Si_3N_4 and TiN phase formation. Another one of the weak component at about N3 (at 400.8 eV) corresponds to C-N bonds. The formation of Si_3N_4 phases was confirmed by XPS analyses for nanocomposite coatings. Figure 12 shows the HRTEM image of Ti-Si-C-N nanocomposite coatings deposited at a bias voltage of −100 V. These coatings are nanocomposite coatings of TiN nano-crystalline (black area) embedded in an amorphous matrix, which are clearly distinguished from the particles by high-resolution TEM image. The SAED patterns in the TEM analysis did not reveal any crystalline silicon nitride. Finally from the XRD, TEM and XPS result together confirmed that the Ti-Si-C-N coatings had nanocomposites structure of nanosized TiN crystallites embedded in an amorphous Si_3N_4 matrix.

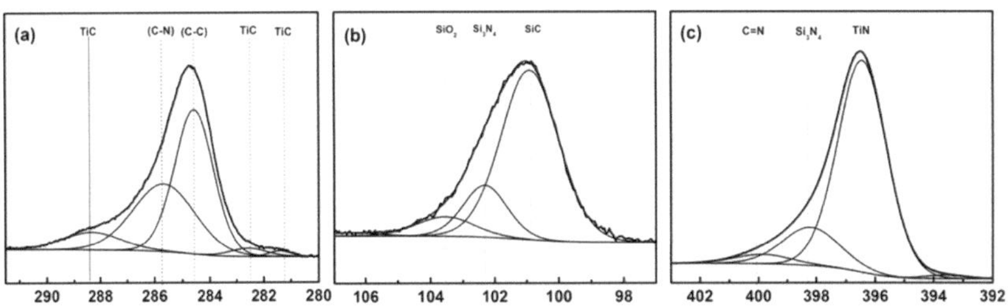

Figure 11. XPS spectra of Ti-Si-C-N nanocomposite coatings with various substrate bias voltage (a) C (b) Si and (c) N.

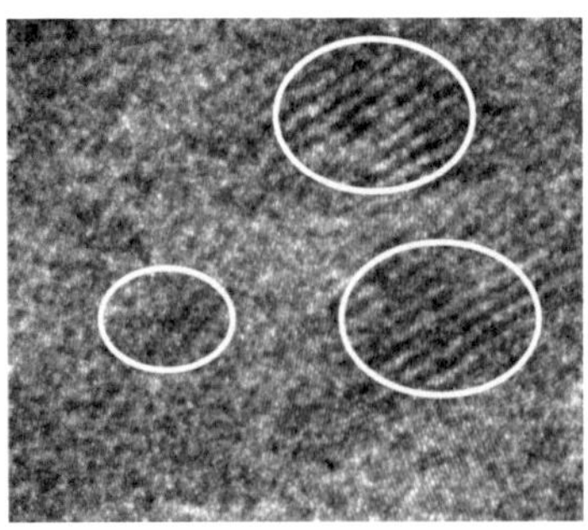

Figure 12. HRTEM image of Ti-Si-C-N coating at -100 V.

2.3.4. Nano Mechanical Properties

Figure 13 shows the effect of substrate biasing on the hardness of Ti-Si-C-N nanocomposite coating. It was clearly seen that the hardness increases initially with the increase in the indentation depth upto 30 nm. After that the hardness starts to decrease with increase in the penetration depth and finally attains saturation. This saturation in hardness value is observed in all the coatings. Hence this clearly indicates that at higher penetration depths, the obtained hardness may be due to the substrate effect. Hence, as per the Oliver-Pharr [70], the extracted hardness values are only from the 10 % of the film thickness. Hence, for the un-biased Ti-Si-C-N coating,at the depth of about 30 nm, the peak hardness was found to be 49 GPa. A maximum value of 49 GPa has been obtained for the coatings deposited at a substrate bias of − 100 V. This peak in hardness value is well corroborated with XRD, XPS and TEM studies. Furthermore increasing trend of hardness with bias voltage have been also studied many researchers by using different PVD synthesis methods. Usually enhancement of the packing density in plane of (111) were improved with increasing the substrate bias voltage and also from in our XRD pattern already confirms about orientation (111) has been formed at this bias range of coating. It is well known resultant microstructure of the film depends on the ion bombardment [71], and this in turn affects the hardness of the film. Also from the TEM analysis we have confirmed nanocrystalline phase formation with amorphous phase formation in this range of bias (−100 V) coating, it may be another reason for peaking hardness at this range of nanocomposite coating. Finally with increase the substrate bias the energy of the bombarding ions increases, causing structural modification, which is responsible for changes in the level of the hardness. However, a drastic drop of hardness observed from 49 to 20 GPa over the range −100 and −200 V was observed. The film hardness between −200 to −300 V substrate biases remained constant at a value of about 18.5 GPa. We already pointed out in our XRD result shows the preferred orientation of (220) and stress formation with more defects this causes decreasing hardness. Therefore we can conclude that the hardness of nanocomposite coating depends directly on the substrate bias of nanocomposite coating in this present studies.

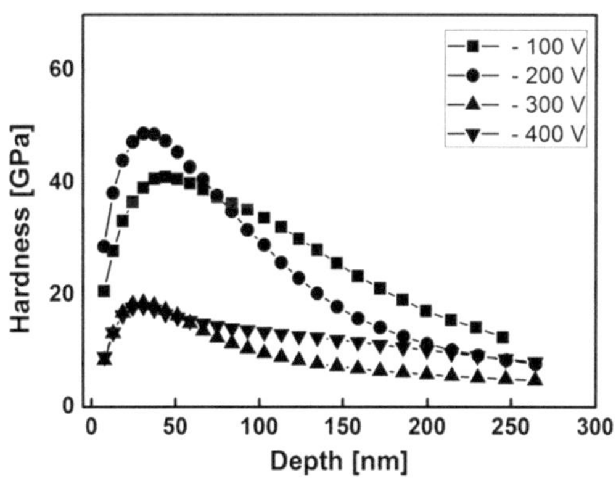

Figure 13. Nanohardness as a function of displacement into surface at various substrate voltages.

3. Summary

Ti-Si-C-N nanocomposite thin films on stainless steel were prepared by using FVA technique at constant gas mixture of argon and nitrogen flow rate with room temperature. The nanocomposite films with silicon content in the range of 2.1 to 16.2 at. % was prepared on stainless steel substrate with different TMS gas using FVA technique. From the XRD pattern, we have confirmed nancomposite structure like nc-TiCN/a-Si_3N_4 formation. The nanocrystallite size of the samples decreases with the silicon content. Nanohardness measurement indicated a peak hardness of ~49 GPa and Young's modulus of ~245 GPa for the films with Si content of 3 at.%. All these results show that Ti-Si-C-N nanocomposite coatings are suitable for surface coatings applications requiring low roughness, moderate hardness and low friction coefficient.

In the variation of substrate bias, we found (111) orientation at lower bias voltage range from 0 to − 100 V, whereas the (220) orientation has confirmed at higher bias voltage range from −200 to −400V. XPS result of Ti-Si-C-N coatings has confirmed the formation of nc-Ti(C)N/a-Si_3N_4 phase with respect bias voltage. The highest hardness around at 49 GPa has achieved phase at the bias voltage of -100 V. Further increasing voltage the hardness was decreased due to stress and orientation behavior on this coating. By changing the bias voltage to change microstructure and chemical natural of the films and to improve tribological applications such as hardness and adhesion properties, it shows promising future in industrial fields.

Author details

Seunghun Lee, P. Vijai Bharathy, T. Elangovan, Do-Geun Kim* and Jong-Kuk Kim

*Address all correspondence to: dogeunkim@kims.re.kr

Korea Institute of Materials Science, Changwon, Republic of Korea

References

[1] Yang, Sheng-Min, Chang, Yin-Yu, Wang, Da-Yung, Lin, Dong-Yih, & Wu, WeiTe. (2007). *Journal of Alloys and Compounds*, 440, 375-379.

[2] Miller, H. C. (1981). *J. Appl. Phys.*, 52, 4523.

[3] Aksenov, I. I., Strel'nitskij, V. E., & Vasilyev, V. V. (2003). *D.Yu. Zaleskij. Surf. and Coat. Technol.*, 163-164, 118-127.

[4] Sanders, D. M., & Anders, A. (2000). *Surf. Coat. Technol.*, 133/134, 78-90.

[5] Aksenov, I. I., Vakula, S. I., Padalka, V. G., & Khoroshikh, V. M. (1980). Sov. 350 mA when the arc current was 60 A and the extrac-. *Phys. Tech. Phys.*, 25, 1164.

[6] Kim, Jong-Kuk, Lee, Kwang-Ryeol, Eun, Kwang Yong, & Chung, Kie-Hyung. (2000). *Surf. Coat. Technol.*, 124, 135-141.

[7] Chang, Yin-Yu, Wang, Da-Yung, & Hung, Chi-Yung. (2005). *Surf. Coat. Technol.*, 200, 1702-1708.

[8] Tanaka, Y., Ichimiya, N., Onishi, Y., & Yamada, Y. (2001). *Surf. Coat. Technol.*, 146-147, 215-221.

[9] Li, Chen., Yong, Du., Wang, Ai J., Wang, She. Q., & Zhou, Shu. Z. (2009). *Int. Journal of Refractory Metals & Hard Materials*, 27, 718-721.

[10] Bujak, J., Walkowicz, J., & Kusinski, J. (2004). *Surf. Coat. Technol.*, 180-181, 150-157.

[11] Ozturk, A., Ezirmik, K. V., Kazmanl, K., Urgen, M., Eryılmaz, O. L., & Erdemir, A. (2008). *Tribology International*, 41, 49-59.

[12] Choi, Sung Ryong, Park, In-Wook, Kim, Sang Ho, & Kim, KwangHo. (2004). *Thin Solid Films*, 447-448, 371-376.

[13] Kim, KwangHo, Choi, Sung-ryong, & Yoon, Soon-young. (2002). *Surf. Coat. Technol*, 298, 243-248.

[14] Kim, Do-Geun, Svadkovski, Igor, Lee, Seunghun, Choi, Jong-Won, & Kim, Jong-Kuk. (2009). *Current Applied Physics*, 9, S179-S181.

[15] Gorokhovsky, V. I., Bowman, C., Gannon, P. E., Van Vorous, D., Voevodin, A. A., Muratore, C., Kang, Y. S., & Hu, J. J. (2008). *Wear*, 265, 741-755.

[16] Gorokhovsky, V., Bowman, C., Gannon, P., Van Vorous, D., Voevodin, A. A., Rutkowski, A., Muratore, C., Smith, R. J., Kayani, A., Gelles, D., Shutthanandan, V., & Trusov, B. G. (2006). *Surf. Coat. Technol.*, 201, 3732-3747.

[17] Chang, Chi-Lung, Chen, Jun-Han, Tsai, Pi-Chuen, Ho, Wei-Yu, & Wang, Da-Yung. (2008). *Surf. Coat. Technol.*, 203, 619-623.

[18] Park, In-Wook, & Kim, KwangHo. (2002). *Journal of Materials Processing Technology*, 130-131, 254-259.

[19] Lee, Eung-Ahn, & Kim, KwangHo. (2002). *Thin Solid Films*, 420-421, 371-376.

[20] Ma, S.L., Ma, D.Y., Guo, Y., Xu, B., Wu, G.Z., Xu, K.W., & Chu, Paul K. (2007). *Acta-Materialia*, 55, 6350-6355.

[21] Guo, Yan, Ma, Shengli, Xuand, Kewei, & Bell, Tom. (2008). *Nanotechnology*, 19, 215603.

[22] Kim, Soo Hyun, Kim, Jong Kuk, & Kim, KwangHo. (2002). *Thin Solid Films*, 420-421, 360-365.

[23] Rebouta, L., Tavares, C. J., Aimo, R., Wang, Z., Pischow, K., Alves, E., Rojas, T. C., & Odriozola, J. A. (2000). *Surf. Coat. Technol.*, 133-134, 234-239.

[24] Khizhnyak, N. A. (1965). *Sov. Phys. Technol. Phys.*, 35, 847.

[25] Aksenov, I. I., Belous, V. A., & Padalka, V. G. (1978). USSR Authors Certificate No. 605425, (Rus.).

[26] Aksenov, I. I., et al. (1978). *Prib. Tekhn. Ehksp.*, 5, 236, (Rus.).

[27] Anders, A., Anders, S., & Brown, I. G. (1995). *Plasma Sources Sci. Technol.*, 4, 1-12.

[28] Martin, P. J., & Bendavid, A. (2001). *Thin Solid Films*, 394, 1-15.

[29] Aksenov, I. I., Vasilyev, V. V., Druz, B., Luchaninov, A. A., Omarov, A. O., & Strel'nitskij, V. E. (2007). *Surf. Coat. Technol.*, 201, 6084-6089.

[30] Byon, E., Kim, J.-K., Kwon, S.-C., & Anders, A. (2004). *IEEE Trans. Plasma Sci.*, 23, 433-439.

[31] Aksenov, I. I., Zaleskij, D.Yu., & Strel'nitskij, V. E. (2000). 1st International Congress on Radiation Physics, High Current Electronics and Modification of Materials, September, Tomsk, Russia. Proceedings 3, 130.

[32] Takikawa, Hirofumi, & Tanoue, Hideto. (2007). *IEEE Trans. Plasma Sci.*, 35(4), 992-999, Aug. 2007.

[33] Karpov, D. A. (1997). *Surf. Coat. Technol.*, 96(1), 22-33, Nov. 1997.

[34] Martin, P. J., Bendavid, A., & Takikawa, H. (1999). *J. Vac. Sci. Technol. A, Vac. Surf. Films*, 17(4), 2351-2359, Jul. 1999.

[35] Anders, A. (1999). *Surf. Coat. Technol.*, 120/121, 319-330.

[36] Martin, P. J., & Bendavid, A. (2001). *Surf. Coat. Technol.*, 142-144, 7-10.

[37] Anders, A. (2002). *Vacuum*, 67(3/4), 673-686, Sep. 2002.

[38] Aksenov, I. I., Strel'nitskij, V. E., Vasilyev, V. V., & Zaleslij, D. Y. (2003). *Surf. Coat. Technol.*, 163/164, 118-127.

[39] Shi, X., Tay, B. K., Tan, H. S., Liu, E., Shi, J., Cheah, L. K., & Jin, X. (1999). *Thin Solid Films*, 345(1), 1-6, May 1999.

[40] Tay, B. K., Zhao, Z. W., & Chua, D. H. C. (2006). *Mater. Sci. Eng. R*, 52(1-3), 1-48, May 2006.

[41] Ryabchikov, A. I., & Stepanov, I. B. (1998). Rev. Sci. Instrum., 69(2), 810-812, Feb. 1998.

[42] Zimmer, O. (2005). *Surf. Coat. Technol.*, 200(1-4), 440-443, Oct. 2005.

[43] Chun, S. Y., Chayaraha, A., Kinomura, A., Tsubouchi, N., Heck, C., Horino, Y., & Fukui, H. (1999). Jpn. J. Appl. Phys., 38(4B), L467-L469, Apr. 1999.

[44] Meyer, C. F. H., & Scheibe, J. (1999). Presented at the Int. Conf. Metallurgical Coatings Thin Films (ICMCTF), San Diego, CA, Paper B4-9.

[45] Taki, Y., Kitagawa, T., & Takaki, O. (1997). *J. Mater. Sci. Lett.*, 16(7), 553-556, Apr. 1997.

[46] Yoshitake, T., Shiraishi, G., & Nagayama, K. (2002). Appl. Surf. Sci., 197/198, 379-383.

[47] Ning Jiang, Y. G., Shen, H. J., Zhang, S. N., & Bao, X. Y. Hou. (2006). *Materials Science and Engineering: B*, 135, 1-9.

[48] Guo, Y., Shengli, M., & Xu, K. (2007). *Surf. Coat. Technol.*, 201, 5240-5243.

[49] Abraham, S., Choi, E. Y., Kang, N., & Kim, K. H. (2007). *Surf. Coat. Technol.*, 202, 915-919.

[50] Guo, Y., Ma, S. L., Xu, K. W., Bell, T., Li, X. Y., & Dong, H. (2008). *Key Engineering Materials*, 373, 188-191.

[51] Qin, C. P., Zheng, Y. G., & Wei, R. (2010). *Surf. Coat. Technol.*, 204, 3530-3538.

[52] Shtansky, D. V., Levashov, E. A., Sheveiko, A. N., & Moore, J. J. (1999). *Metallurgical And Materials Transactions A*, 30, 2439-2447, JCPDS File No: 42-1489.

[53] Diserens, M., Patscheider, J., & Lévy, F. (1998). *Surf. Coat. Technol.*, 108, 241-246.

[54] Phinichka, N., Chandra, R., & Barber, Z. H. (2004). J. Vac. Sci. Technol. A., 22, 477-482.

[55] Johnson, L. J. S., Rogström, L., Johansson, M. P., Odén, M., & Hultman, L. (2010). Thin Solid Films, 519, 1397-1403.

[56] Oliver, W.C., & Pharr, G.M. (1992). *J. Mater. Res.*, 7, 1564-1583.

[57] Dayan, M., Shengli, M., & Xu, K. (2005). *Surf. Coat. Technol.*, 200, 382-386.

[58] Jeon, J. H., Choi, S. R., Chung, W. S., & Kim, K. H. (2004). *Surf. Coat. Technol.*, 188-189, 415-419.

[59] Veprek, S., & Reiprich, S. (1995). *Thin Solid Films*, 268, 64-71.

[60] Xu, H., Nie, X., & Wei, R. (2006). *Surf. Coat. Technol.*, 201, 4236-4241.

[61] Sundgren, J.-E., Johansson, B.-O., Hentzell, H. T. G., & Karlsson, S.-E. (1983). *Thin Solid Films*, 105, 385.

[62] Zerkout, S., Achour, S., & Tabet, N. (2007). *J. Phys. D: Appl. Phys.*, 40, 7508.

[63] Pfeiler, M., Kutschej, K., Penoy, M., Michotte, C., & Mitterer, Kathrein M. (2007). *Surf. Coat. Technol.*, 202, 1050.

[64] Lee, D. N. (1989). *Journal of Material Science*, 24, 4375.

[65] Zhao, J. P., Wang, X., Chen, Z. Y., Yang, S. Q., Shi, T. S., & Liu, X. H. (1997). *J. Phys. D: Appl. Phys*, 30, 5.

[66] Zhang, X., Weber, W. H., Vassell, W. C., Potter, T. J., & Tamor, M. A. (1998). *J. Appl. Phys.*, 83, 2820.

[67] Jiang, N., Shen, Y. G., Zhang, H. J., Bao, S. N., & Hou, X. Y. (2006). *Materials Science and Engineering*, B 135, 1.

[68] Fallon, P. J., Veerasamy, V. S., Davis, C. A., Robertson, J., Amaratunga, G. A., Milne, W. I., & Koskinen, J. (1993). *Phys. Rev. B*, 48, 4777.

[69] Oliver, W. C., & Pharr, G. M. (2004). *Journal of Material Research*, 19, 3.

[70] Chun, Sung-Yong. (2010). *Journal of the Korean Physical Society*, 56, 1134.

Chapter 18

Study of Multifunctional Nanocomposites Formed by Cobalt Ferrite Dispersed in a Silica Matrix Prepared by Sol-Gel Process

Nelcy Della Santina Mohallem, Juliana Batista Silva,
Gabriel L. Tacchi Nascimento and
Victor L. Guimarães

Additional information is available at the end of the chapter

http://dx.doi.org/10.5772/51154

1. Introduction

Surface science has a long history, involving the development of colloids, particulate material, thin films and porous materials. These materials have been known and used for centuries, without a profound knowledge of their real physical–chemistry characteristics. The detailed study of their properties was only possible with the emergence of more sophisticated spectroscopic techniques, and high-resolution eletron microscopes [1].

With the development of nanoscience and nanotechnology, new materials began to be studied like nanoparticles, porous materials that are formed by a network of nanoparticles, and nanocomposites. There are infinite possibilities of production of nanocomposites and one of them is the formation of nanoparticles inside porous matrices that can have their texture and morphology tailored by thermal treatment or templates [2].

For many applications, the textural properties, such as porosity and specific surface area, are as important as the chemical composition. Thus, the growing demand for porous products in the industry of nanotechnology, especially for magnetic nanocomposites, has led to the increase in the studies related to these properties [1, 2].

Magnetic materials have been used by man for centuries, since ancient people discovered the natural magnets called lodestones. The term "magnet" is used for magnetic materials that produce their own magnetic field. Other magnetic materials have magnetic proprieties only in response to an applied magnetic field. There are several types of magnetic materials

INTECH
open science | open minds

© 2012 Mohallem et al.; licensee InTech. This is a paper distributed under the terms of the Creative Commons Attribution License (http://creativecommons.org/licenses/by/3.0), which permits unrestricted use, distribution, and reproduction in any medium, provided the original work is properly cited.

that have been used in diverse devices and systems for industrial products. Some traditional applications of these materials are in cores for motors, generators and transformers, microwave devices, magnetic media used in computers, recording devices, and magnetic cards, among others [3].

There are various metallic elements (Fe, Ni, etc) that have magnetic properties due to their crystalline atomic structure whose spins align spontaneously. Some alloys formed by metallic elements and others including the earth rare elements also have excellent magnetic properties (alnico, samarium-cobalt and neodymium-iron-boron magnets). Finally, the ferrites are a known class of magnetic materials formed by metallic oxides.

With the advancement of the material sciences and the emergence of the nanoscience and nanothecnology, new kinds of magnetic materials have been developed and studied in the last years, such as the magnetic nanoparticles, ferrofluids and magnetic nanocomposites. With these materials, new applications could be tested in areas such like electronic, catalysis and biomedicine, among others [4].

1.1. Ferrites

Ferrites are chemical compounds obtained as powder or ceramic body with ferrimagnetic properties formed by iron oxides as their main component, Fe_2O_3 and FeO, which can be partly changed by others transition metals oxide. The ferrites can be classified according their crystalline structure: hexagonal ($MeFe_{12}O_{19}$), garnet ($Me_3Fe_5O_{12}$) and spinel ($MeFe_2O_4$), where Me represents one or more bivalent transition metals (Mn, Fe, Co, Ni, Cu, and Zn). The ferrites are classified as "soft" or "hard" magnets, according to their magnetic properties, which refers to their low or high magnetic coercivity, respectively. Hard magnets are not easily demagnetized (curve a), due to their high coercivity and soft magnets are easily magnetized and demagnetized (curve b) with application of a magnetic field, due to their low coercivity. The characteristic magnetic hysteresis curves of these type of magnets are shown in Figure 1 [3,5,6].

The intermediate magnets, generally used in magnetic media, must have coercivity sufficiently high for withholding the information, but sufficiently low to allow for the information to be deleted (curve c) [5,6].

These magnetic ceramics [6] are important in the production of electronic components, since they reduce energy losses caused by induced currents and they act as electric insulators. They can be used in simple function devices such as small permanent magnets, until as sophisticated devices for the electro-electronic industry.

Recently, these materials have been discovered as good catalysts [7,8,9] and biomaterials [10,11].

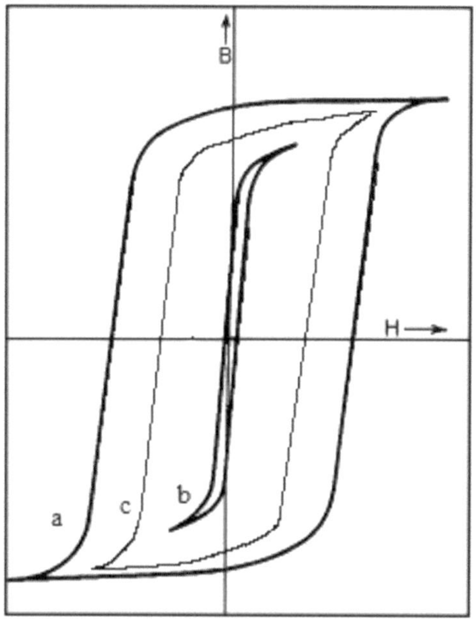

Figure 1. Magnetic hysteresis curves of (a) hard, (b) soft and (c) intermediate magnets.

1.1.1. Cobalt ferrites

Cobalt ferrite [3, 6], an intermediate magnet, is an important multifunctional magnetic material not only for its magnetic properties but also for its biomedical and catalytic applications, which depend on their textural and morphological characteristics. Cobalt ferrite, that has great physical and chemical stability, has been used in the production of permanent magnets, magnetic recording such as audio and videotape and high-density digital recording disks, magnetic fluids and catalysts. This ferrite has spinel inverse structure and exhibits a large coercivity, differently from the rest of the spinel ferrites. The magnetization of the $CoFe_2O_4$ crystal has anisotropic character because depends on its orientation. The strong magnetic flux promoted by the superexchange interaction is directioned along of the magnetization direction, and generally may be coinciding with the crystallographic axes. The magneto-crystalline anisotropy is related with the spin-orbit coupling. In polycrystalline materials, the magnetization measured corresponds to a mean value.

Recently these metal-oxide nanoparticles have been the subject of much interest because of their unusual optical, electronic and magnetic properties, which often differ from the bulk. These nanoparticles should have single domain, of pure phase, having high coercivity and intermediary magnetization [12].

The properties of the cobalt ferrite are changed according to the form of obtainment of the material, as bulk, particles or nanoparticles. The nanocrystalline particles have a high surface/volume ratio, and thus, they present different properties from those of bulk materials

Various authors [3,4,6,12,13] described the saturation magnetization and coercivity measured at room temperatures as a function of crystallite size and these values change from 30 to 80 emu g^{-1} for saturation magnetization and of 0.5 to 5.4 kOe for coercivity for crystallite size varying from 4 to 50 nm.

The effect of thermal vibrations is largest in very small particles, especially in materials with low anisotropy. The magnetic moments assume random orientations, at room temperature, for nanoparticles with size below the limit of 4-10 nm, resulting in superparamagnetic behavior [14-16].

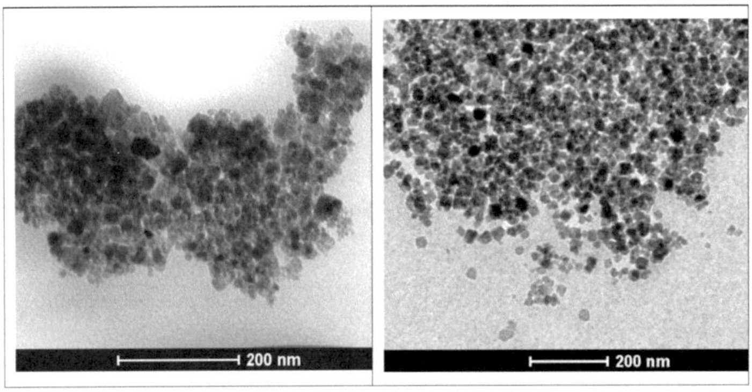

Figure 2. Cobalt ferrite nanoparticles obtained by (a) coprecipitation and (b) hydrothermal processes.

Superparamagnetic materials magnetize and demagnetize more easily than the other ones due to their dimension being equivalent to a magnetic domain. The magnetic domain of very small particles is different from that observed in bulk structures. There is a critical diameter below which the formation of a domain wall results in an increase of the total energy. The mono-domain size for $CoFe_2O_4$ nanoparticles has been estimated between 10 and 70 nm [17]. Crystallites with diameter smaller than 10 nm have superparamagnetic behavior, while with diameters larger than 70 nm (critical particle size/Dc) show multi-domain microstructure, with the consequent decrease in coercivity [17]. The existence of multiple domains separated by walls governs the magnetic behavior. The magnetization and demagnetization processes driven by an external field are characterized by the nonexistence of the hysteresis, characteristic of superparamagnetic materials.

Because of these interesting characteristics, nanocrystalline ferrites have been extensively studied with emphasis on the particle size variation and the influence of this variation in the mechanical, biomedical, magnetic and catalytic properties. In order to achieve desired properties, it is necessary to obtain high-density powders with a small and uniform grain size, and controlled stoichiometry. Hence, there is the need to develop fabrication processes relatively simple that induce the formation of controlled particle size materials. Some nanoparticles can be achieved more easily by using chemical methods [3,4,8,12,18,19], such as coprecipitation, hydrothermal synthesis and sol-gel process, among others, but generally the nanoparticles tend to agglomerate due to their high reactivity. Figure 2 shows some cobalt ferrite nanoparticles obtained by coprecipitation and hydrothermal processes.

Due to these problems with high reactivity, agglomeration and aggregation of the nanoparticles, and the possibility of the development of new materials with peculiar properties, it has been synthesized nanocomposite materials formed by metal or metallic oxide nanoparticles dispersed in ceramic or vitreous matrices, avoiding the agglomeration and improving the dispersion and the distribution of the nanoparticles inside the system [20-34].

1.2. Nanocomposites

A composite is considered as a multiphase material with significant proportion of the properties of the constituent phases, whose final product has its property improved. There is the possibility of combining various types of materials in a single composite, in order to optimize their properties according to the desired application. When one of the phases has nanometric dimension, the system is called nanocomposite. [25,34].

Nanocomposite materials formed by metal or metallic oxide nanoparticles dispersed in ceramic or vitreous matrices have important applications due to the possibility of developing more reactive materials with new properties. The interest in the preparation of magnetic nanocomposites has increased in the last years due to the properties presented by these materials, which depends on the particle size, concentration and distribution of the particles in the matrix. Nanosystems such as Fe/SiO_2, Ni/SiO_2, Fe_3O_4/SiO_2, $CoFe_2O_4/SiO_2$, $NiFe_2O_4/SiO_2$ have been intensively studied in the last years, revealing different behavior from those of bulk magnetic systems and serving as models for the study of small particles [25-34].

The texture of the matrix and the interaction between the magnetic nanoparticles and the host matrix can be used to control the magnetic properties and the stability of these materials.

The magnetic nanoparticles dispersed in a inert matrix act as isolated nanomagnets, eliminating energetic losses, and producing the coupling between neighboring nanoparticles, which improve their magnetic properties. These nanocomposites can have high chemical and structural stability, high catalytic activity and high mechanical resistence.

The crystallite size control inside the matrix is justified by the existence of an average diameter range of single domain crystallites, between 10 nm $< d <$ 80 nm, depending on the desired optimal magnetic properties [36,37]. When the ferrite concentration is low (< 30%), the crystallites are isolated, having single domains and showing superparamagnetism. Concentrations above 50% of ferrite provoke the agglomeration of the crystallites, which results in multi-domains [20]. Another important characteristic of nanocomposites in general is the texture of the matrix, as pore distribution, pore size and specific surface area, which has large influence in their final characteristics, as the transport and interaction of fluids within their connected network formed by micro, meso and macropores [38-41]. Important materials to be used as porous matrices are xerogels and aerogels, material obtained by sol-gel process.

In the last years, the sol-gel process has been used to produce magnetic nanocomposites by incorporation of ultra-fine magnetic nanoparticles with a high surface/volume ratio in different matrices. The nanocomposites formed have different properties from the magnetic bulk.

1.3. Sol-gel process

The sol-gel chemistry is based on mechanisms of hydrolysis and condensation of precursors containing metal (s) of interest, called "sol", resulting in an M-O-M oxide network that form a wet gel. There are two types of precursor: an aqueous solution of an inorganic salt or a metal alcoxide compound. The gel may be formed by polymerisation (gel polymer) or aggregation of colloidal particles subject to the physic-chemical conditions of the medium (colloidal gel). In either case, a three-dimensional solid network of the gel retains a liquid phase in its pores [42-49].

In practice, the network structure and the morphology of the final product depend on the relative contributions of the reactions of hydrolysis and condensation. These contributions may be controlled by varying the experimental conditions: the type of metal, type of organic binder, the molecular structure of the precursor, water/alkoxide ratio, type of catalyst and solvent, temperature and concentration of the alkoxide.

After the gelification, the wet gel is subjected to aging, to occur the polymerization processes, syneresis, and neck formation between the particles, leading to increase in connectivity and strength of the gel structure. The gel obtained is formed by a solid structure impregnated with the solvent. After the aging, various drying processes can be used to convert the wet gel in a porous material, denominated xerogel or aerogel.

The sol-gel process allows the preparation of materials in various forms such as powders, thin films, and monoliths, with desirable properties such as hardness, chemical durability, thermal and mechanical resistance and with different textures. The final product (xerogel or aerogel) can be tailored by different temperatures of thermal treatment leading to materials with different specific surface areas and porosities.

1.4. Xerogels and aerogels

The drying is one of the more important steps in sol-gel process because it is possible to obtain different materials by changing the drying routes. During the drying, the solvent adsorbed inside the porous gel is removed. During this process the gel network can collapse.

There are several types of drying processes; among them we can mention the controlled drying and the supercritical drying. In the controlled process, the solvent is evaporated slowly at room temperature and pressure, generating a contraction on the material, provoking the decreasing in the pore size due to the surface tension. The dry gels obtained by this process are called xerogels and have high porosity and specific surface area [27,50,51].

In the supercritical drying, the wet gels are put in a reactor at high temperature and pressure, above the critical point of the system, where there is no discontinuity between the liquid and gaseous phase, avoiding capillary forces. The dry gels obtained are called aerogels and have higher porosity than the xerogel. [27,52,53].

In this work we studied the characteristics of nanocomposites formed by cobalt ferrites dispersed in silica matrix (xerogel and aerogel) obtained at different thermal treatment temperatures. Techniques such as X-ray diffraction (XDR), spectroscopy in the infra-red region, force atomic microscopy, transmission electron microscopy, scanning electronic microscopy equipped with energy dispersive X-ray (EDS) and wavelength dispersive (WDS) probes and gas adsorption were used to study the morphological and structural changes of the materials as a function of the thermal treatment temperature. The results were used to evaluate the mechanism of formation of the nanocomposites and relate their characteristics with magnetic and catalytic properties.

2. Experimental

2.1. Silica Matrix and Cobalt Ferrite Nanocomposite

The inert matrices formed by porous pure silica were obtained by mixing tetraethylorthosilicate, ethyl alcohol, water (1/3/10) and nitric acid, used as a catalyst. The nanocomposite precursor solution was obtained by mixing cobalt and iron nitrates, ($Co(NO_3)_2.6H_2O$ and $Fe(NO_3)_2.9H_2O$) with the matrix precursor, to form nanocomposites with 30 wt% of ferrite. The solutions were stirred for 1 h for homogenisation and left to rest for gelation, which takes place due to the hydrolysis and polycondensation of the metallic alkoxides. The wet gelswere aged at 60 °C for 24 h and dried at 110 °C for 12 h, leading to the formation of xerogels. Aerogels were formed by supercritical drying of the wet gels in an autoclave under 180 atm of N_2 and raising temperature up to 300 °C at 5 °C min^{-1}, temperature and pressure adequate to exceed the critical point of the mixture ethyl alcohol/water. The system was kept in this condition for 2 h. All xerogel and aerogel were heated between 300 and 1,100 °C for 2 h.

2.2. Caracterization Techniques

The structural evolution of the samples were analyzed in an X-ray diffractometer (Rigaku, Geigerflex 3034) with CuKa radiation, 40 kV and 30 mA, time constant of 0.5 s and crystal graphite monochromator. Crystallite sizes were determined by Scherrer equation (D = 0.9k/b cos h, where D is the crystallite diameter, k is the radiation wavelength and h the incidence angle). The value of b was determined from the experimental integral peak width using silicon as a standard. The values were corrected for instrumental broadening.

Spectra in the infrared region were obtained in an ABB Bomem equipment, model MB 102.

The composite compositions were evaluated by an electron microprobe (Jeol JXA, model 8900RL) with energy dispersive (EDS) and wavelength dispersive (WDS) spectrometers.

The morphology was obtained by scanning electron microscopy (high resolution SEM - Quanta 200 - FEG - FEI), by transmission electron microscopy (high resolution TEM - FEI) and by atomic force microscopy (Dimension 3000, Digital Instruments Nanoscope III - LNLS).

Sample textural characteristics were determined by N_2 gas adsorption (Quantachrome, model Nova 1200) at liquid nitrogen temperature. The samples dried at 110 °C were outgassed at 100 °C for 3 h. The others ones were outgassed at 200 °C for 3 h before each experiment. Specific surface areas and total pore volumes were obtained by the Brunauer-Emmett-Teller (BET) equation and the Barrett-Joyner-Halenda (BJH) method. These measurements were used to evaluate the total porosity, by the equation $P = 1 - \varrho_{ap}/\varrho_{th}$, where ϱ_{ap} is the apparent density and ϱ_{th} is the theoretical density. True densities were obtained in a helium picnometer (Quantachrome) and apparent densities were obtained by mercury picnometry.

The magnetic measurements were made in a Lake Shore vibrating sample magnetometry (VSM) at 300 K with a maximum applied magnetic field of 1 Tesla.

The nanocomposites were tested as catalysts in the oxidation of chlorobenzene in air. The catalytic reactions were carried out in a fixed bed reactor with 25 mg of catalyst. Chlorobenzene at 0.1% was introduced in the air stream (30 mL min-1) by a saturator at 0°C. The reaction products were analyzed by gas chromatography (Shimadzu/GC 17A) with a flame ionization detector (FID) and an Alltech Econo-Cap SE capillary column (30 mm 90. mm 9 0.25 lm).

3. Results and Discussion

The xerogels and aerogels silica matrices and nanocomposites were obtained in the monolithic form, without cracks.

3.1. Structural characterization

SiO_2 xerogel and aerogel treated up to 900°C exhibit amorphous behavior. A narrowing of the XRD peak accompanied by an increase in its intensity with increasing in the temperature of the preparation indicate an increase in the structural organization of the samples (Figure 3a and 3b). Characteristic reflections of crystobalite and tridimite appear at 1100°C for both the samples. The intensity of the xerogel peaks are larger than the aerogel ones.

The xerogel nanocomposites exhibit amorphous behavior up to 300 °C (according the X-ray difractometer resolution). $CoFe_2O_4$ crystalline particles with cubic spinel structure are detected by XRD inside the amorphous silica matrix above this temperature (Figure 4a). During the formation of the $CoFe_2O_4$ nanocrystals, no traces of intermediate products are found even at temperatures as high as 1100°C, indicating that the ferrite particles were formed without binding to the matrix. The magnetic nanoparticles also avoided the formation of either crystobalite or tridimite phases.

The aerogel nanocomposites exhibit amorphous behavior up to 700 °C (Figure 4b). The $CoFe_2O_4$ phase is detected by XRD only above this temperature.

http://dx.doi.org/10.5772/51154

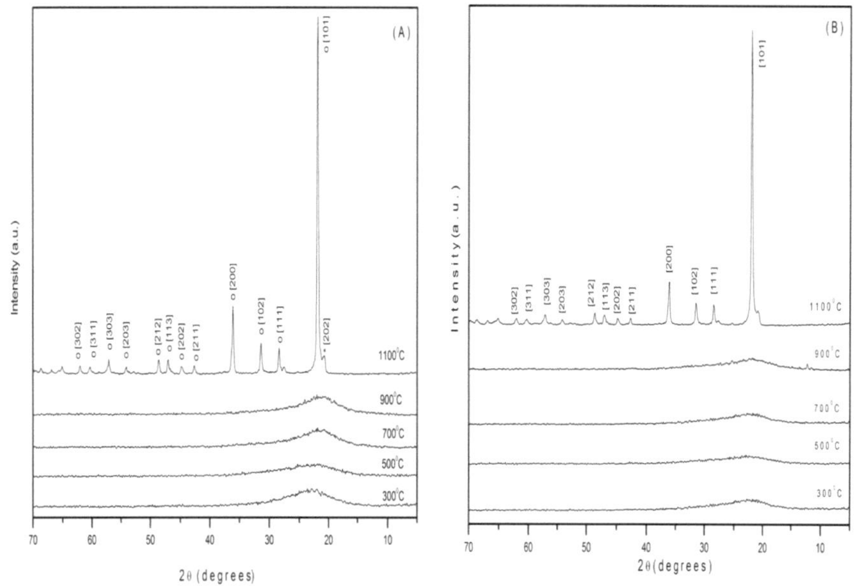

Figure 3. X-ray diffraction patterns of SiO_2 (a) xerogel and (b) aerogel, thermally treated in air for 2 hours at various temperatures. Crystobalite (o) and tridimite (*)

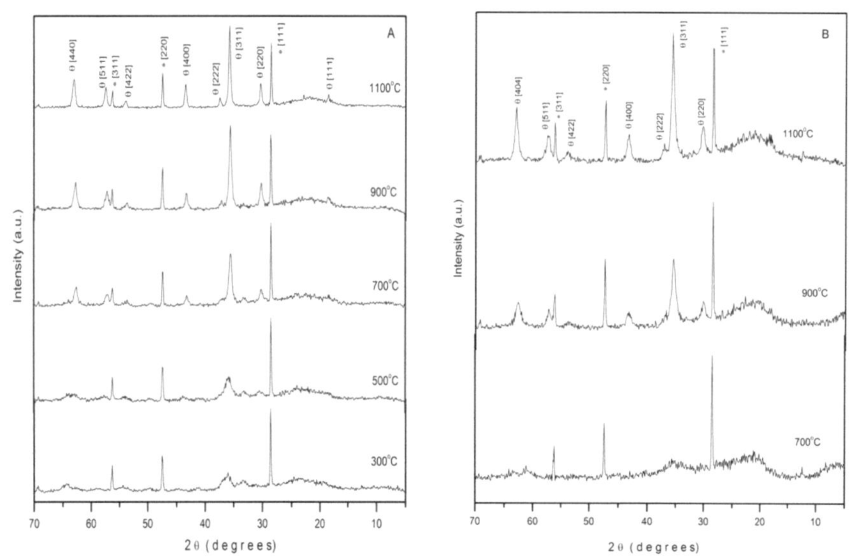

Figure 4. X-ray diffraction patterns of $CoFe_2O_4/SiO_2$ (a) xerogel and (b) aerogel, thermally treated in air for 2 hours at various temperatures. (Si) silicon, (Co) $CoFe_2O_4$.

Figure 5a shows the IR spectra of the xerogel samples obtained after heat-treating of the dried gel at various temperatures for 2 hours. The IR spectrum of the sample dried at 300°C has absorptions characteristic of the silica network at 1080, 810, and 460 cm^{-1}. The 1086 cm^{-1} band with the shoulder at 1160 cm^{-1} is due to the asymmetric stretching bonds Si-O-Si of the SiO_4 tetrahedron associated with the motion of oxygen in the Si-O-Si anti-symmetrical stretching. The 810 cm^{-1} band is associated with the Si-O-Si symmetric stretch and the band at 461 cm^{-1} with either Si-O-Si or O-Si-O bending. The weak band in the 950cm^{-1} is due to stretching of the Si-OH. With increasing in temperature this band disappears due to the condensation reactions which change the Si-OH groups on Si-O-Si. The sample heated at 900^{0}C shows a decrease in intensity of the bands characteristic of the silica matrix, suggesting the rearrangement process of silica network, according XRD results. The aerogels have similar spectra.

Figure 5b shows the IR spectra of the xerogel nanocomposites. The IR spectrum of the sample dried at 300°C also has absorptions characteristic of the silica network and at 968 cm^{-1} we observed the band composed of the contributions from Si-O-H and Si-O-Fe vibrations. The band at 584 cm^{-1} is related to Fe-O stretching. The 968 cm^{-1} band disappears with the increase in temperature, showing that the weak bond between Si and Fe is broken [54]. We can observe a slight shift of the 584 cm^{-1} band to the left. Co-O stretching vibration characteristic band also appear at 461 cm^{-1}. The weak band at 675 cm^{-1} can be due to the cobalt ion in tetrahedral centers in the matrix pores.

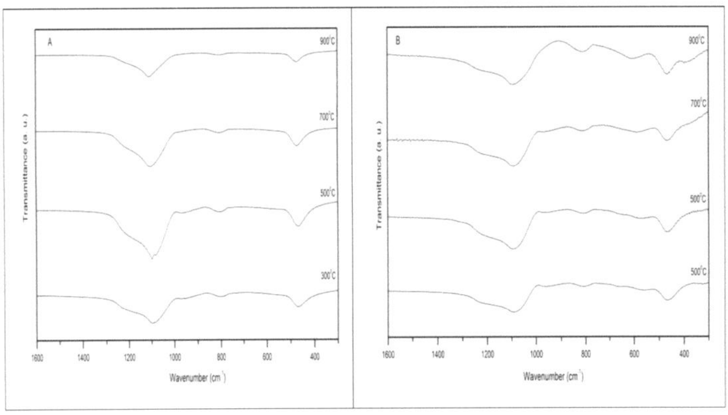

Figure 5. a) Infrared spectra of SiO_2 xerogel and (b) infrared spectra of $CoFe_2O_4/SiO_2$ xerogel treated at different temperatures.

IR spectra show that there was no formation of by-product, confirming X-ray diffraction results. All results suggest that the iron ions had interaction with the silica matrix when the nanocomposite was heated at low temperatures. These interactions disappeared with increasing in heating temperature, showing that they are weak bonds. These results suggest that the cobalt ions have been diffused by the porous matrix and have been bonded to iron to form the ferrite within the pores, without any binding or interaction with the silica network.

3.2. Textural Characteristics

3.2.1. Silica xerogels and nanocomposite xerogels

http://dx.doi.org/10.5772/51154

The xerogel is a typical porous material formed by a silica network with micro, meso and macro pores interconnected for all the bulk. Micro pores are pores smaller than 2 nm in diameter, meso pores are the pores with diameters between 2 and 20 nm, and macro pores are larger than 20 nm.

Monolithic porous matrices (Figure 6), without defects after drying, changed in size after thermal treatments at high temperatures. The shape of the samples were defined by the template. The silica xerogels are optically transparent in all temperatures of preparation.

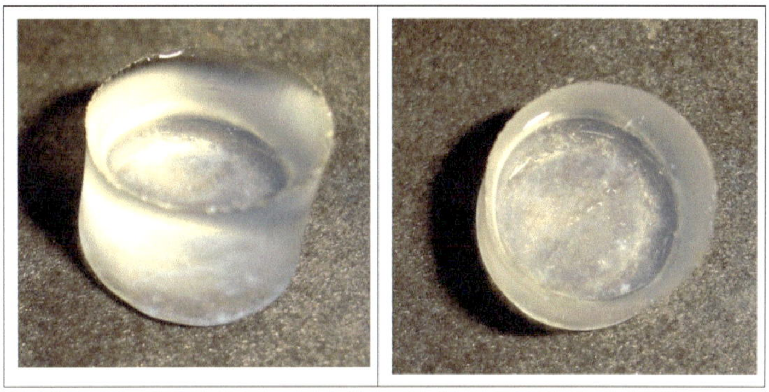

Figure 6. Silica xerogel.

AFM and TEM images (Figure 7) show the microstructure of a typical xerogel.

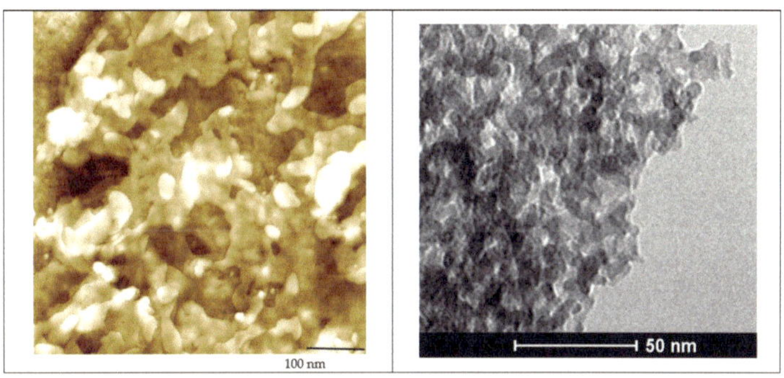

Figure 7. a) atomic force microscopy and (b) transmission electron microscopy images of a typical silica xerogel prepared at 500 °C.

The textural characteristics of the silica matrix xerogel changed substantially with thermal treatment. The specific surface area (Figure 8) decreased gradually for samples prepared above 300°C, and between 500 and 700°C decreased rapidly due to the densification process of the material. The silica xerogel porosity (Figure 9) remained constant for samples prepared up to 700 °C and decreased sharply until 900 °C, due to the collapse of the pores. The sample heated at 1100°C became a material with a continuous silica network without pores.

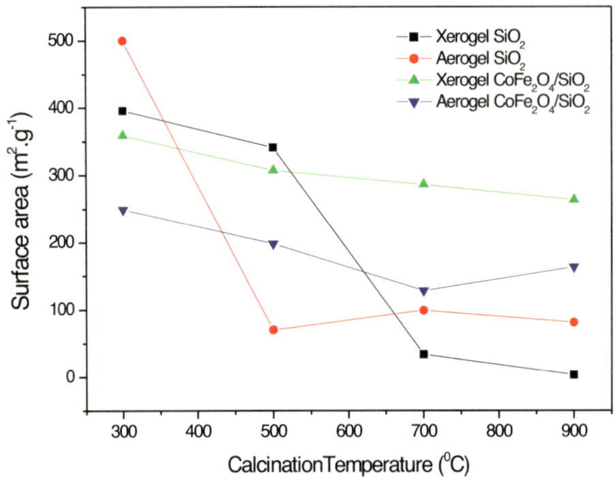

Figure 8. Variation of surface area as a function of temperature for composite $CoFe_2O_4$ in SiO_2 matrix and SiO_2. Error: 5%

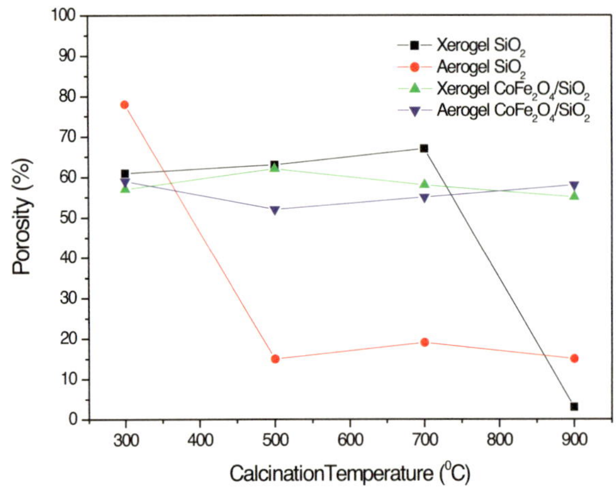

Figure 9. Variation of the porosity of silica matrices and cobalt ferrite nanocomposites as a function of temperature. Error: 5%

The formation of the ferrite nanoparticles inside the pores of the xerogel matrix reinforced the silica structure, keeping stable the pore network at high temperatures. In this case, the specific surface area decreases about 14% (Figure 8) and the total porosity remain almost constant between 300^{0}C and 900^{0}C (Figure 9). The shrinkage of the material structure occured only in samples heated above 900 ^{0}C.

http://dx.doi.org/10.5772/51154

Figure 10 shows the microstructure of the ferrite nanocomposite. With the increase in temperature, the ferrite grows inside the silica matrix.

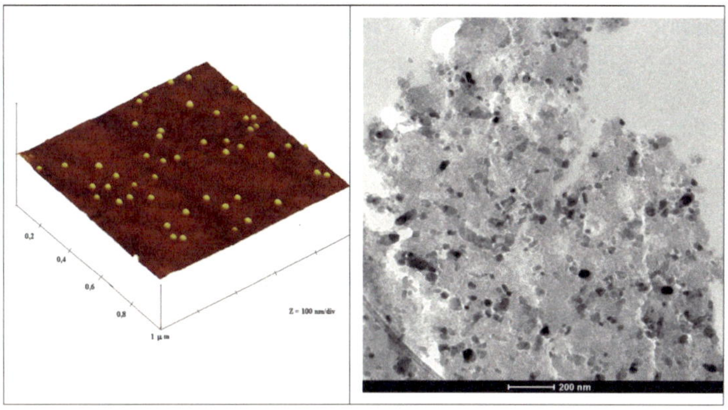

Figure 10. Atomic force microscopy and transmission electron microscopy images of cobalt ferrite nanocomposite prepared at 900 ^{0}C.

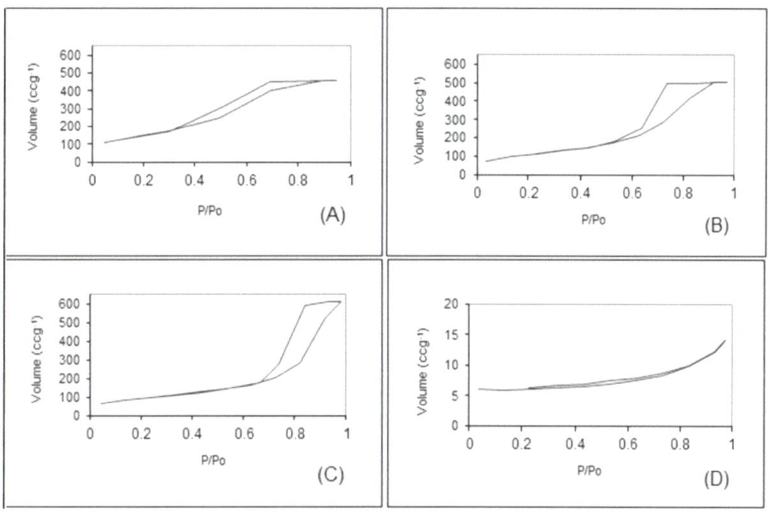

Figure 11. Adsorption-desorption isotherms of silica xerogels heated at (a) 300^{0}C, (b) 500^{0}C, (c) 700^{0}C and (d) 900^{0}C.

Figure 11 shows the adsorption-desorption isotherms of the SiO_2 xerogel at different thermal treatment temperatures. The sample prepared at 300^{0}C adsorbed about 450 $cm^3.g^{-1}$ of N_2. The xerogel adsoved more gas with increasing in the heating temperature due to the liberation of the organic compounds of their pores. The xerogel prepared at 700^{0}C adsorbed about 600 $cm^3.g^{-1}$. The isotherm of the sample heated at 300 ^{0}C presented characteristic intermediary of meso and microporous materials. The isotherms of the samples treated at 500^{0}C and

700^{0}C presented mesoporosity characteristics (isotherm type IV according the BDDT classification [39,41]). The samples heated at 900^{0}C presented isotherm type III, without hysterese, characteristic of non-porous material.

With the formation of ferrite nanoparticles inside the xerogel the material prepared at all temperatures became mesoporous corroborating the reinforcement in the xerogel microstructure.

All nanocomposite xerogel isotherms shown in Figure 12 are characteristic of mesoporous materials. The nanocomposite without thermal treatment adsorbed about 200 cm^3/g of N_2,

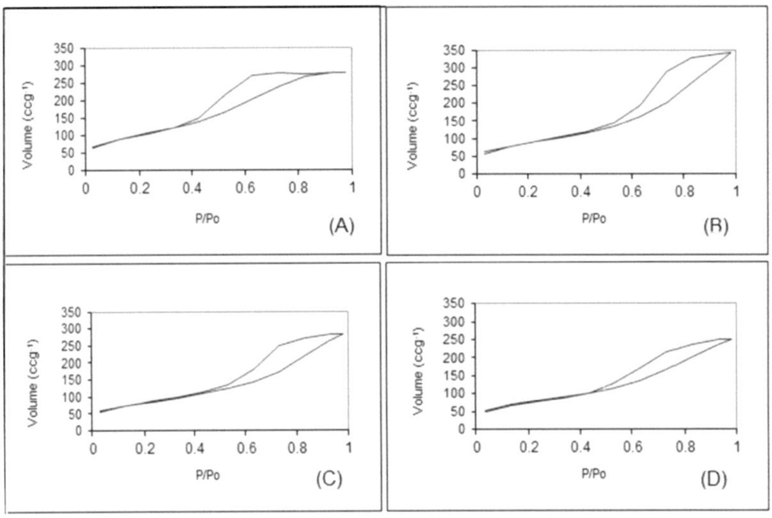

Figure 12. Adsorption/desorption isotherms of $CoFe_2O_4$ / SiO_2 xerogel, heated at (a) 300^{0}C, (b) 500^{0}C, (c) 700^{0}C and (d) 900^{0}C.

while nanocomposites prepared at higher temperatures adsorbed about 300 cm^3/g of N_2, due to the elimination of solvents. All isotherms (type IV [39,41]) presented characteristics of mesoporous materials. The hysteresis curves are type H2, characteristics of pores with indefinite form and size, according to the AFM and TEM images (Figure 10).

3.2.2. Silica aerogels and nanocomposite aerogels

The aerogel is an extremely porous material also formed by a silica network with micro, meso and macro pores interconnected for all the bulk. This material is much more porous than the xerogel, mainly as prepared. Monolithic aerogel matrices (Figure 13), without defects after drying, changed sharply in size after thermal treatments at high temperatures. The silica aerogels are slightly opaque due to the macroporosity, whose pores with the same size than the light wavelength interfere with the optical transparence.

http://dx.doi.org/10.5772/51154

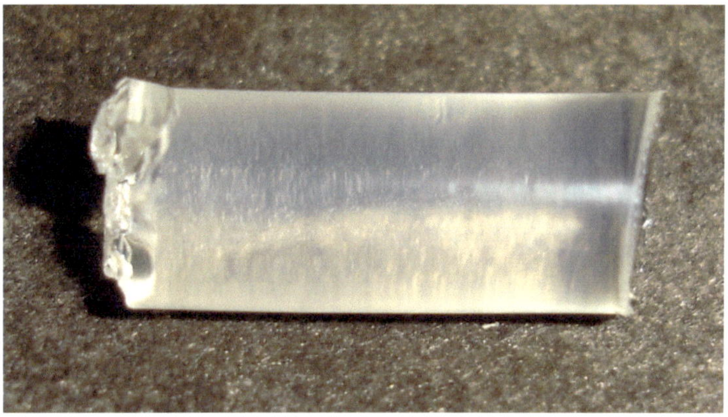

Figure 13. Silica aerogel.

Figure 14a shows a like-smoke structure of the aerogel. It is very difficult to obtain SEM and AFM images of this kind of material, due to their high porosity. Almost any preparation can destroy the network formed by the interconnection of the nanoparticles. Figure 14b shows the porous network structure of the aerogel, evidencing the necks formed by silica nanoparticles that led to the formation of micro, meso and macropores.

The as-prepared silica aerogels presented higher specific surface area and porosity than the xerogels obtained from the same precursor (Figure 8 and 9). Nevertheless, contrary to what happens with the xerogel, the porous network collapsed at low temperature of preparation

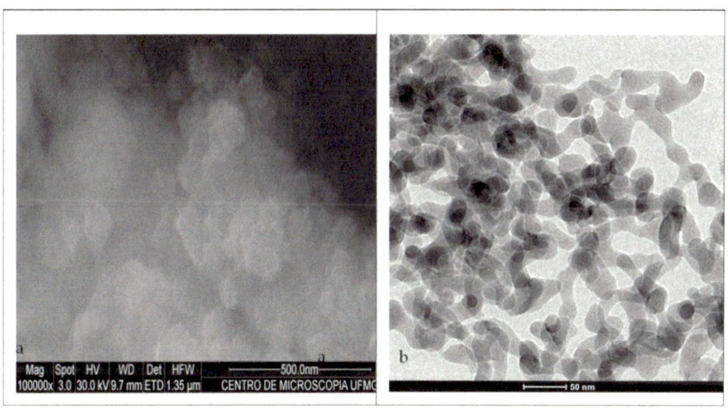

Figure 14. Scanning (a) and transmission (b) electron microscopies images.

of about 500 ^{0}C. The specific surface area and porosity changed from 500 m^2/g and 80 % for aerogels prepared at 300 ^{0}C to 50 m^2/g and 15 % for aerogels prepared at 500 ^{0}C. These values have been kept constant until temperatures of about 900 ^{0}C, when occurred the total collapse of the porous.The formation of the ferrite nanoparticles inside the pores of the aerogel matrix also reinforced their silica structure as happened with the xerogel matrices, but the values of specific surface area were lower than the xerogel ones, as seen in Figure 8. In this case, the specific surface area Decreased of 250 to 150 m^2/g. The aerogel porosity values increased in comparison to the silica aerogel matrix and remained almost constant between 300ºC and 900ºC, with similar values to the xerogel ones. The shrinkage of the material network also occured above 900 ºC.

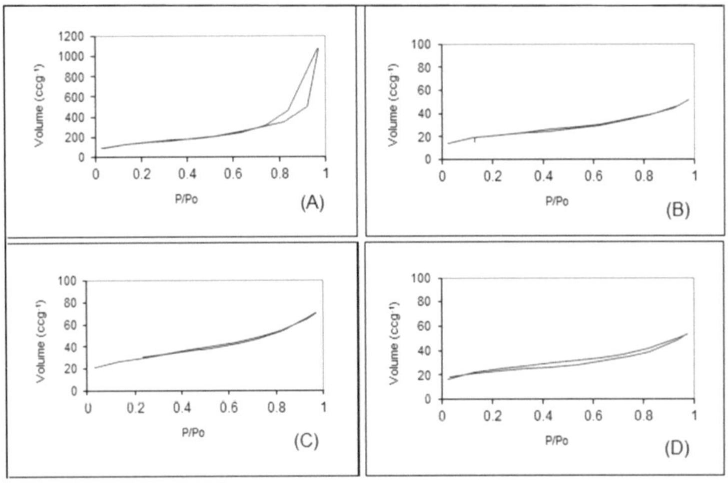

Figure 15. Adsorption-desorption isotherms of silica aerogels heated at (a) 300ºC, (b) 500ºC, (c) 700ºC and (d) 900ºC.

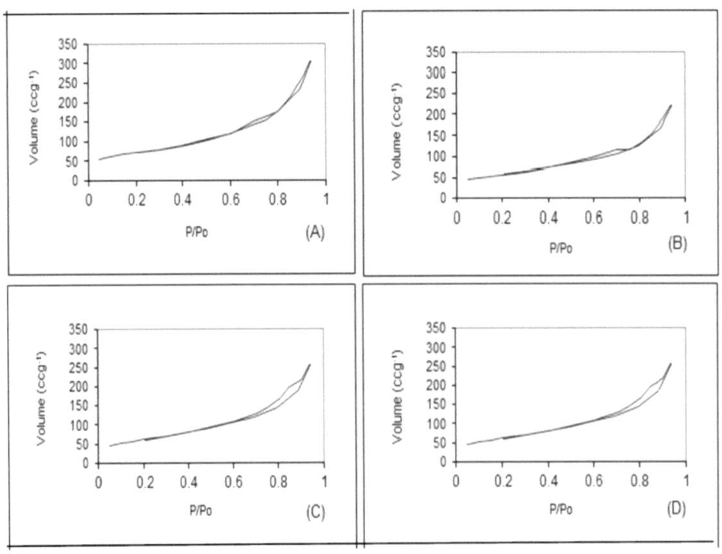

Figure 16. Adsorption/desorption isotherms of $CoFe_2O_4$ /SiO_2 aerogel obtained for sol-gel method and heated at (a) 300ºC, (b) 500ºC, (c) 700ºC and (d) 900ºC.

Figure 15 shows the silica aerogel isotherms. The isotherm of the sample prepared at 300⁰C presented characteristics of mesoporous material (39,41), with large N_2 adsorption, of about 1200 $cm^3.g^{-1}$. This material is very fragile and when submitted at high heating temperatures its structure was annihilated, adsorbing only 60$cm^3.g^{-1}$ when prepared between 500 and 900 °C. Samples prepared between these temperatures did not present hysteresis and the adsorption-desorption curves are characteristics of macroporous materials [39,41].

All nanocomposite aerogel isotherms shown in Figure 16 are characteristic of macroporous materials. The as-prepared aerogel nanocomposite adsorbed ~300 $cm^3.g^{-1}$ of N_2, value much lower than that presented by the silica aerogel. The nanocomposites prepared at higher temperatures adsorbed about 250 $cm^3.g^{-1}$ of N_2, values larger than the silica aerogels prepared in similar temperatures.

3.2. Mechanism of formation

Backscattered electron images of the cobalt ferrite nanocomposites prepared between 300 and 700 °C showed defined white regions distributed throughout the sample, whose EDS analyses detected mostly the presence of cobalt clusters. In the gray region, Si, Fe, O and traces of Co were detected (Figure 17). The clusters disappeared with the increase in the preparation Temperature, suggesting that the cobalt ions diffused into the composite, binding to iron to form the ferrite. At temperatures above 900°C, EDS analyses detected a homogeneous distribution of Co and Fe in the composite. These results corroborate XRD and IR results.

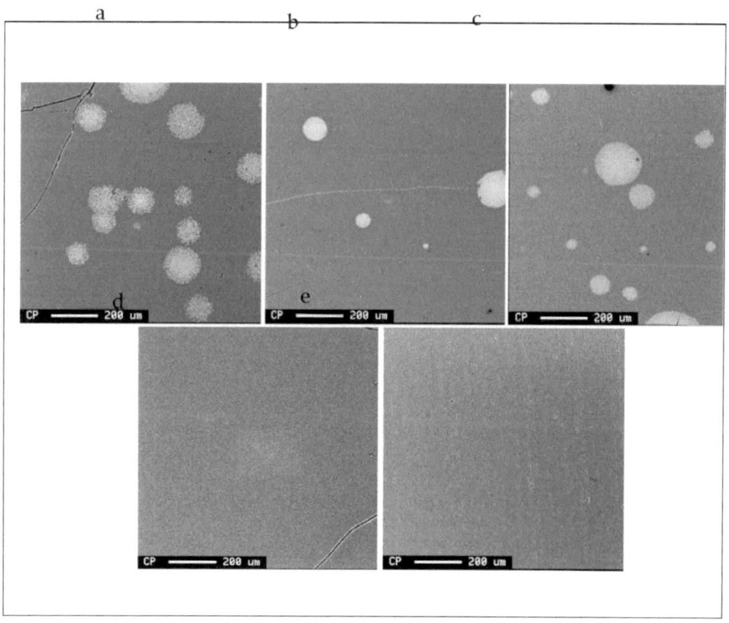

Figure 17. SEM images (backscattered electrons) of xerogels prepared at (a) 300⁰C, (b) 500⁰C, (c) 700 ⁰C, (d) 900 ⁰C and (e) 1100 ⁰C.

Figure 18 shows the WDS mapping, used to confirm the mechanism of formation of nanoparticles inside the porous silica matrix. The different concentration of each metallic ion is shown by the color evolution. By this mapping, it is possible to observe with more acuity the diffusion process of the ions as a function of the temperature of preparation.

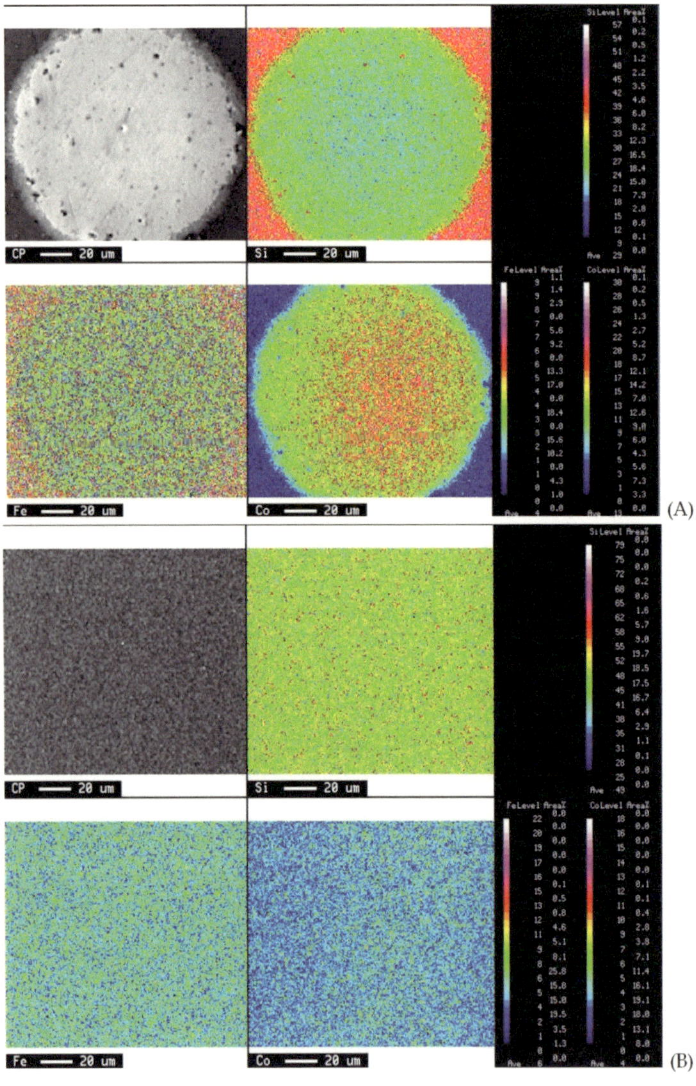

Figure 18. WDS mapping of the nanocomposite prepared at (a)500°C and (b)1100°C

Figure 19 shows the proposed model of the formation mechanism of the cobalt ferrite nanocomposites. The precursor solution is prepared by the mixing of Si alkoxide, alcohol, water and Fe and Co nitrates, and some catalysts (Figure 19 a). The wet gels are formed by hydrolysis and polycondensation of the sol constituents, maintaining the same ions distribution.

After drying, the elimination of water and organic residues occurs, and the xerogel (or aerogel) is formed by a silica network with iron ions distributed by the network and weakly bonded to the silicon (Figure 19 b). The Co ions are agglomerated in definite regions forming the clusters. With increasing temperature of preparation, the cobalt ions diffuse by the silica network, forming a chemical bond with the iron, which has its weak bond with the silicon broken (Figure 19 c). The pores diminish in amount and in size, and the magnetic nanoparticles grow inside these pores with the increase in temperature, leading to the encapsulation of the nanoparticles by the silica matrix (Figure d and e).

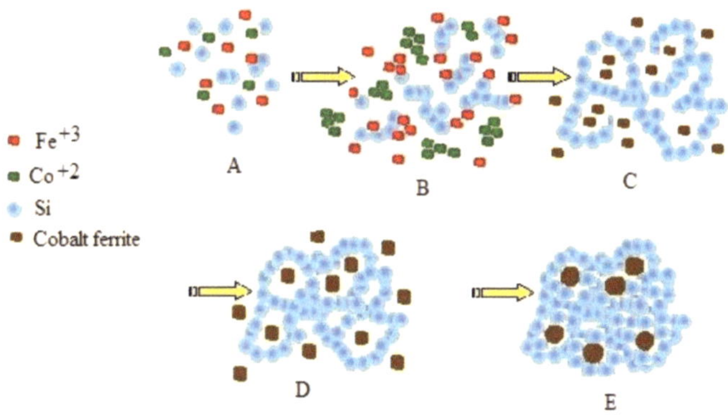

Figure 19. A proposed model of the formation mechanism of cobalt ferrite nanocomposites.

3.3. Properties and applications

Magnetic xerogels and aerogels can be considered as multifunctional materials due to the possibility of use their properties in multiple applications. The desired application can be obtained by tailoring the characteristics of the material.

Multifunctional materials are composites or systems capable of performing multiple functions simultaneously, depending of the involved phases, their structural, morphological and textural characteristics, improving system performances and reducing the redundancy between the composite components and their individual functions.

For example, porous xerogel and aerogel nanocomposites have interesting catalytic properties when tested in the oxidation of chlorobenzene in air. Figure 20 shows the performance of the xerogel nanocomposites prepared at various temperatures, compared to SiO_2. It is clear that the more porous nanocomposite had the best performance. Figure 21 shows the best performance of the xerogel and the aerogel prepared at 500 °C compared with ferrite powders obtained by coprecipitation.

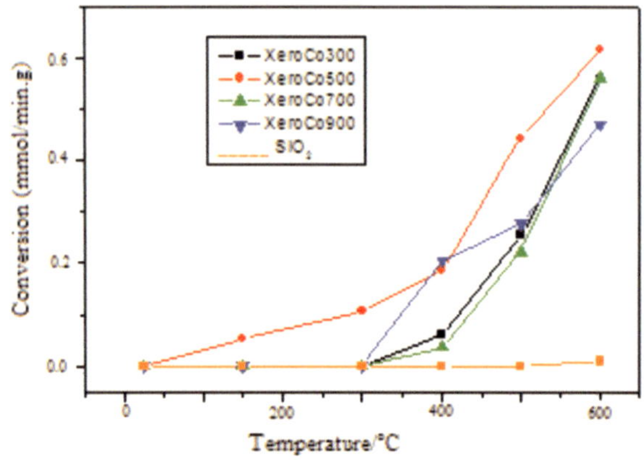

Figure 20. Catalytic oxidation of chlorobenzene in the presence of $CoFe_2O_4/SiO_2$ xerogels thermally treated at various temperatures.

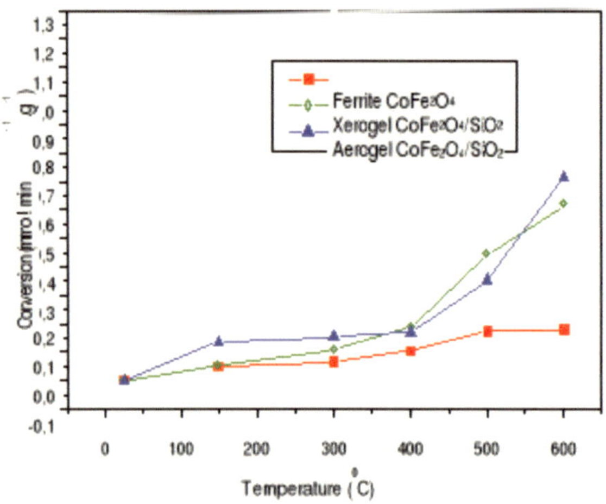

Figure 21. Catalytic oxidation of chlorobenzene in the presence of $CoFe_2O_4/SiO_2$ xerogels, aerogel and ferrite.

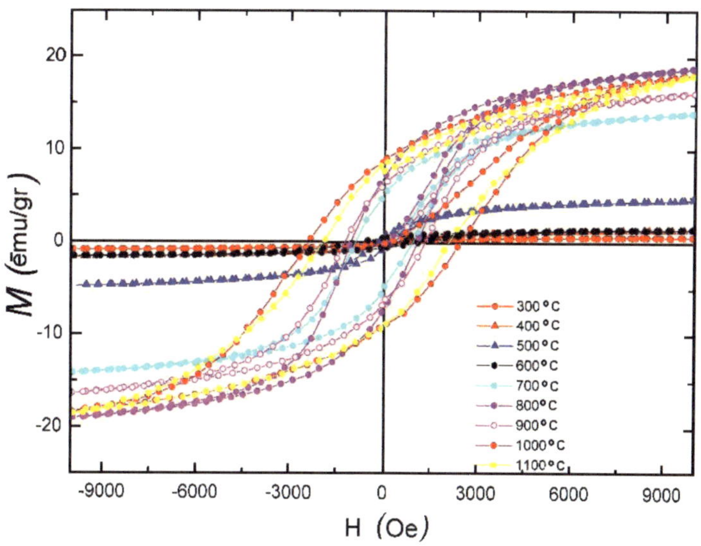

Figure 22. Hysteresis curves of the xerogel nanocomposites prepared at several temperatures.

The magnetic properties also are tailored by temperature. Figure 22 shows the hysteresis curve of the cobalt ferrite nanocomposites prepared at several temperatures, whos exerogels showed since superparamagnetic characteristics when prepared at low temperatures and intermediate magnetism at high temperatures. Due to these properties, these nanocomposites can be used as electronic devices or in biomedicine in cancer treatment by hyperthermia and drug release controlled by magnetic field.

4. Conclusion

Nanocomposites formed by cobalt ferrite nanoparticles dispersed in porous silica matrix ($CoFe_2O_4/SiO_2$) were prepared by the sol-gel process. The presence of the magnetic nanoparticles inside of the inert porous matrices of xerogels and aerogels reinforced their structure, avoiding large changes in specific surface area, porosity and in the microstructure of the matrix after preparation temperature, which varied between 300 and 900 °C. These characteristics influence the properties of the nanocomposites, such as their chemical reactivity, catalytic activity and magnetization. Due to the possibility of tailoring the textural and morphological characteristics of these types of multifunctional nanocomposites, they are promising candidates for many technological applications in electronic, catalysis and biomedical areas.

Acknowledgements

This work was supported by CNPq and Fapemig (Brazilian agencies). The authors acknowledge the use of the infrastructure of the Centre of Microscopy and Laboratory of Microanalyses/UFMG, and Laboratório Nacional de Luz Síncrotron – Brazil

Author details

Nelcy Della Santina Mohallem[1*], Juliana Batista Silva[2], Gabriel L. Tacchi Nascimento[3] and Victor L. Guimarães[1]

*Address all correspondence to: nelcy@ufmg.br

1 ICEx – Universidade Federal de Minas Gerais, Brazil

2 Centro de Desenvolvimento de Tecnologia Nuclear - CNEN/CDTN, Brazil

3 Nanum Nanotechnology Ltda, Brazil

References

[1] Moore, J. H., & Spencer, N. D. (2001). Ed. Encyclopedia of Chemical Physics and Physical Chemistry. Taylor & Francis

[2] Brushar, B. (2004). Handbook of Nanotechnology. New York, Springer.

[3] Valenzuela, R. (1994). Magnetic Ceramics. Cambridge, Cambridge University Press, 10.1017/CBO9780511600296.

[4] Coey, J. M. (2001). Magnetism in future. *Journal of Magnetism and Magnetic Materials.*, 226-230, 2107-2112.

[5] Snelling, E. C., & Giles, A. D. (1986). Ferrites for Inductors and Transformers. 2 a. ed. NY,

[6] Kingery, W. D., Bowen, H. K., & Uhlmann, D. R. (1976). Introduction to Ceramics 2. ed. New York, John Wiley & Sons.

[7] Rajaram, R. R., & Sermon, P. A. (1985). Adsorption and catalytic properties of $Co_xFe_{3-x}O_4$ spinels. *J. Chem. Soc., Faraday Trans.*, 81, 2577-2591.

[8] Ramankutty, C. G., & Sugunan, S. (2001). Surface properties and catalytic activity of ferrospinels of nickel, cobalt and copper, prepared by soft chemical methods. *Applied Catalysis A: General.*, 212, 39-51.

[9] Khalf-Alla, M., El -Salaam, A. B. D., Said, A. A., Hassan, E. A., & El -Wahab, abd. M. M. (1994). Structure and electronic effects of cobalt ferrites, CoxFe3-xo4 on catalytic decomposition of isopropyl alcohol. *Collect. Czech. Chem. Commun.*, 59, 1939-1949.

[10] Sharifi, H., & Shokrollahi, S. A. (2012). Ferrite-based magnetic nanofluids used in hyperthermia applications Ibrahim. *J. Magnet. Magnet. Mat.*, 324, 903-915.

[11] Bocanegra-Diaz, A., Mohallem, N. D. S., & Sinisterra, R. D. (2003). Preparation of a ferrofluid using cyclodextrin and magnetite. *J. Braz. Chem. Soc.*, 14, 936-941.

[12] Silva, J. B., Brito, W. ., & Mohallem, N. D. S. (2004). Influence of the heat treatment on cobalt ferrite nanocrystal powders. *Materials Science and Engineering. B, Solid State Materials for Advanced Technology*, 112, 182-187.

[13] Hench, L. L., & West, J. K. (1990). Principles of Electronic Ceramics. Wiley.

[14] Grigorova, M., Blythe, H. J., & Blaskov, V. (1998). Magnetic properties and Mössbauer spectra of nanosized $CoFe_2O_4$ powders. *J. Magnet. Magnet. Mat*, 183, 163-172.

[15] Qu, Y., et al. (2006). The effect of reaction temperature on the particle size, structure and magnetic properties of coprecipitated $CoFe_2O_4$ nanoparticles. *Mater. Lett*, 60, 3548-3552.

[16] Song, Q., & Zhang, Z. J. (2006). Correlation between spin-orbital coupling and the superparamagnetic properties in magnetite and cobalt ferrite spinel nanocrystals. *J. Phys. Chem. B. V. 110*, 23, 11205-11209.

[17] Chinnasamy, C. N., et al. (2003). Unusually high coercivity and critical single-domain size of nearly monodispersed $CoFe_2O_4$nanoparticles. *App. Phys. Lett.*, 83.

[18] Haneda, K., & Morrish, A. H. (1981). Magnetic structure of small NiFe2O4 particles. *J. Appl. Phys.*, 52.

[19] Haneda, K., & Morrish, A. H. (1988). Noncollinear magnetic structure of CoFe2O4 small particles. *J. Appl. Phys.*, 63.

[20] Silva, J. B. ., & Mohallem, N. D. S. (2001). Preparation of composites of ferrites dispersed in a silica matrix. *J.Magnet. Magnet. Mat*, 226, 1393-1396.

[21] Casula, M. F., Corrias, A., & Paschina, G. (2000). Nickel oxide-silica and nickel-silica aerogel and xerogel nanocomposite materials. *Journal Materials Reserch.*, 15, 2187-2194.

[22] Casula, M. F., Corrias, A., & Paschina, G. (2001). Iron oxide-silica aerogel and xerogel nanocomposite materials. *J Non- Crystall. Solids.*, 293-295, 25-31.

[23] Estounes, C., Lutz, T., & Happich, J. (1997). Nickel nanoparticles in silica gel: preparation and magnetic properties. *J. Magnet. Magnet. Mat.*, 173, 83-92.

[24] Huang, X. H., & Chen, Z. H. (2004). Sol-gel preparation and characterization of CoFe2O4-SiO_2 nanocomposites. *Solid State Commun.*, 132, 845-850.

[25] Kasemann, R., & Schmitd, H. K. (1994). A new type of a sol-gel derived inorganic-organic nanocomposite. *Materials Research Society.*, 346, 915-921.

[26] Silva, J. B. ., Diniz, C. F. ., Ardisson, J. D. ., Persiano, A., & Mohallem, N. D. S. (2004). Cobalt ferrite dispersed in a silica matrix prepared by sol-gel process. *Journal of Magnetism and Magnetic Materials*, 272-76, 1851-1853.

[27] Silva, J. B., Mohallem, N. D. S., Alburquerque, A. S., Ardisson, J. D. ., Novak, M. A., & Sinnecker, E. (2009). Magnetic studies of cofe2o4/sio2 aerogel and xerogel nanocomposites. *Journal of Nanoscience and Nanotechnology*, 9, 1-8.

[28] Souza, K. C. ., Mohallem, N. D. S. ., & Sousa, E. M. B. (2010). Mesoporous silica-magnetite nanocomposite: facile synthesis route for application in hyperthermia. *Journal of Sol-Gel Science and Technology*, 53, 418-427.

[29] Xianghui, H., & Zhenhua, C. (2006). Preparation and characterization of $CoFe_2O_4$/ SiO_2 nanocomposites. *Chin. Sci. Bull.*, 51(20), 2529-2534.

[30] Xiao, S. H., Luob, K., & Zhanga, L. (2010). The structural and magnetic properties of cobalt ferrite nanoparticles formed in situ in silica matrix. *Materials Chemistry and Physics*, 123-385.

[31] Julian, C., Alcazar, G. A., & Montero, M. I. (1999). Mössbauer analysis od the phase distribution present in nanoparticulate Fe/SiO_2 samples. *Journal of Magnetism and Magnetic Materials.*, 203, 175-177.

[32] Rohilla, S., Kumar, S., Aghamkar, P., Sunder, S., & Agarwal, A. (2011). Investigations on structural and magnetic properties of cobalt ferrite/silica nanocomposites prepared by the coprecipitation method. *Journal of Magnetism and Magnetic Materials*, 323-897.

[33] Gharagozlou, M. (2010). Study on the influence of annealing temperature and ferrite content on the structural and magnetic properties of $x(NiFe_2O_4)/(100-x)SiO_2$ nanocomposites. *Journal of Alloys and Compounds*, 495-217.

[34] Gan, Y. X. (2012). Structural assessment of nanocomposites. *Micron*, 43-782.

[35] Sivakumar, N., Narayanasamy, A., Chinnasamy, C. N., & Jeyadevan, B. (2007). Electrical and magnetic properties of chemically derived nanocrystalline cobalt ferrite. *J. App. Phys.*, 102.

[36] Rajendran, M., Pullar, R. C., et al. (2001). Magnetic properties of nonocystaline $CoFe_2O_4$ powders prepared at room temperature: variation with crystallite size. *Journal of Magnetism and Magnetic Materials.*, 232, 71-83.

[37] George, M., et al. (2007). Finite size effects on the electrical properties of sol-gel synthesized $CoFe_2O_4$ powders: deviation from Maxwell-Wagner theory and evidence of surface polarization effects. *J. Phys. D: App. Phys.*, 40, 1593-1602.

[38] Meyer, K., & Lorenz, P. (1994). Porous solids and their characterization. *Cryst. Res. Technol.*, 29, 903-930.

[39] Lowell, S., & Shields, J. E. (1984). Power surface area and porosity. 2 a ed.,John Wiley N. Y.

[40] Gelb, L., & Gubbins, K. E. (1998). Characterization of porous glasses. *Langmuir*, 14, 2097-2111.

[41] Gregg, S. J., & Sing, K. S. W. (1982). Adsorption, surface area and porosity. 2 a ed. NY,

[42] Aegerter, M. A. (1989). Sol-Gel Science and Technology. .London World Scientific Publishing

[43] Brinker, C. J., & Scherer, G. W. (1990). Sol-gel science: The physics and chemistry of sol-gel processing. Academic Press Inc.

[44] Livage, J., Henry, M., & Sanchez, C. (1988). Sol-gel chemistry of transition metal oxides. *Prog. Solid State Chem.*, 18.

[45] Klein, E. . (1994). Sol-Gel Optics: processing and applications. Kluwer Academic Publishers.

[46] Mac, Kenzie. J. D. (2003). Sol-gel research- achievements since 1981 and prospects for the future. *J. Sol-Gel Sci. Techn.*, 23-27.

[47] Panjonk, G. M., Venkateswara, A., & Sawant, B. M. (1997). Dependence of monolithicity and physical properties of TMOS silica aerogels on gel aging and drying conditions. *Journal Non-Crystalline Solids.*, 209.

[48] Mohallem, N. D. S., Santos, D. I., & Aegerter, M. A. (1986). Fabricação de sílica vítrea pelo método sol-gel. Cerâmica, , 32(197), 109-116.

[49] Burneau, A., Gallas, J. P., & Lavalley, J. C. (1990). Comparative study of surface hidroxyl groups of fumed and precipitated silicas. *Langmuir*, 6, 1364-1372.

[50] Casu, M., & Casula, M. F. (2003). Textural characterization of high temperature silica xerogels. *Journal of Non-Crystalline Solids.*, 315, 97-106.

[51] Rao, A. V., Haranath, P. B. W., & Risbud, P. P. (1999). Influence of temperature on the physical properties of TEOS silica xerogels. *Ceramics International.*, 25, 505-509.

[52] Stolarski, M., & Pniak, M. S. B. (1999). Synthesis and characteristic of silica aerogels. *Applied Catalysis A: General.*, 177, 139-148.

[53] Schneider, M., & Baiker, A. (1995). Aerogels in catalysis. *Catal. Ver. Sci. Eng.*, 37, 515-556.

[54] Waldron, R. D. (1955). Infrared Spectra of Ferrites. *Physical Review.*, 99(6), 1725-1732.

Chapter 19

Interfacial Electron Scattering in Nanocomposite Materials: Electrical Measurements to Reveal The Nc-MeN/a-SiN_x Nanostructure in Order to Tune Macroscopic Properties

R. Sanjinés and C. S. Sandu

Additional information is available at the end of the chapter

http://dx.doi.org/10.5772/51123

1. Introduction

The impressive number of outstanding physical, chemical and mechanical properties of transition metal nitrides MeN (Me stands for transition metal Ti, V, Cr, Zr, Mo, Nb, Ta,..) makes then very attractive materials for many industrial applications as protective and decorative coatings [1,2], superconducting nanostructured thin films for single photon detectors [3,4], diffusion barriers in microelectronic devices [5,6], catalytic films [7,8], and also as materials for biomedical applications [9,10]. Depending on the oxidation states of the transition metal, the Me-N system can exhibit a rich variety of stable or metastable crystallographic phases. Thus, the tetragonal Me_2N and the cubic fcc structures are preferred for IVB-VA compounds (TiN, VN, ZrN) while for VB and VIB-VA compounds the stable phase is the hexagonal one (NbN, MoN, TaN and WN). In particular, as thin films MeN can be easily integrated in microelectronic devices and commonly used as diffusion barriers in magneto-resistive random access memory, resistors, excellent barrier diffusion against Cu, or as preferred barrier absorber material for EUV mask [4-8].

To further improve the performances and efficiency of MeN functional properties, nanocrystalline or amorphous ternary systems, such as Me-Al-N, Me-Si-N, and other Me-X-N forming highly stable compounds have been also investigated [11-30]. By addition of Al or Si to binary MeN, hardness, thermal stability and chemical inertness of the films can be improved [11-16]. In particular TiSiN, TaSiN, NbSiN and WSiN thin films have been mainly investigated as diffusion barriers and electrodes for phase change random access memory (PRAM) devices. [21-30]. The addition of Si leads to the formation of a nanocomposite

INTECH
open science | open minds

© 2012 Sanjinés and Sandu; licensee InTech. This is a paper distributed under the terms of the Creative Commons Attribution License (http://creativecommons.org/licenses/by/3.0), which permits unrestricted use, distribution, and reproduction in any medium, provided the original work is properly cited.

(nanocrystallites of MeN + amourphous SiN_x) or a solid solution single phase $Me_{1-x}Si_xN$ material [13-16]. In nanocomposite thin films (nc-MeN/a-SiN_x), crystallite sizes are of the order of few nanometers. The density of point defects (vacancies, interstitials, antisites), the grain size, the grain surfaces, and boundary regions play an increased role on physical properties. The arrangement and the chemical composition of the so-called "amorphous" minority phase (SiN_x) are crucial for electrical and mechanical properties [17-20]. The location, composition and the thickness of the amorphous phase must therefore be known precisely.

Usually these films are deposited by CVD [12, 15] or PVD [11, 14, 16] techniques; among the PVD techniques, magnetron reactive sputtering is often used as a low-temperature film growth technique. The macroscopic properties of these films such as mechanical, optical or electrical strongly depend on chemical composition and nanostructure of the resulting films which are influenced by the deposition parameters such as the substrate temperature, the flux and kinetic energy of impinging atomic and ionic species on the surface of the growing film, and the condensation rate.

The aim of this paper is to give a general overview on the relationship between the electrical and structural properties of binary MeN and nanocomposite nc-MeN/a-SiN_x thin films deposited by reactive magnetron sputtering. In particular we will focus on the possibility to use electrical measurements and electron scattering models to obtain pertinent information concerning the chemical composition, thickness and continuity of the insulating layer covering conducting nanocrystallites in nanocomposite films. It is not the purpose of this paper to develop further the models describing film nanostructure. This has already been extensively covered in much of the cited literature. The limitations of characterization techniques, such as HRTEM, XRD and XPS, in revealing such composite nanostructures, as described by various physical models, motivate us to employ unconventional investigation techniques such as electrical measurements in order to evidence, for example, the continuity of the insulating SiN_x-layer on conducting MeN-crystallites. The case of a special type of nanocomposite materials: nanocrystallites of Phase1 surrounded by a very thin interfacial layer of Phase2, obtained as a result of self-segregation is one of the most difficult to investigate. Instead, the goal of this paper is to discuss the ability of electrical measurements to support such models.

2. Film morphology and nanostructure

Depending on the deposition conditions binary transition metal nitride MeN thin films deposited by reactive magnetron sputtering usually crystallize with strong (111) or (200) preferential orientation and exhibit elongated crystallites in the grow direction [16-20,31] as one can notice from XRD or SEM measurements (Fig. 1). In MeXN (X=Si,Ge,B,Cu..), the addition of X leads to important modification of the films morphology. Thus, as a function of increasing X content (C_X), the average crystallite size, *d*, in many systems such as Ti-Si-N, Ti-Ge-N, Ti-Sn-N, Nb-Si-N, Zr-Si-N, Ta-Si-N, decreases from tens on nm to about 2 nm [16-20, 32-35]. Whether a ternary single-phase or composite multiphased system is formed depends on the chemical reactivity of the involved atoms and on the deposition conditions. In many case X atoms can substitute metal atoms in the fcc MeN lattice up to a critical concentration (limit of solubility, α_X). The segregation of X atoms on the MeN crystallite surface is mainly responsible for the limitation of their growth. It results in the formation of a nanocomposite material composed of a thin amorphous phase on the MeN crystallite surfaces. Frequently a relationship$d \propto 1/C_X$ is observed in MeXN films (see Fig. 2) suggesting that in this regime the increase in the X content determines a simultaneous increase in the surface-to-volume ratio of the MeN crystallites, which is realized by a subsequent decrease in the average crystallite sizes.

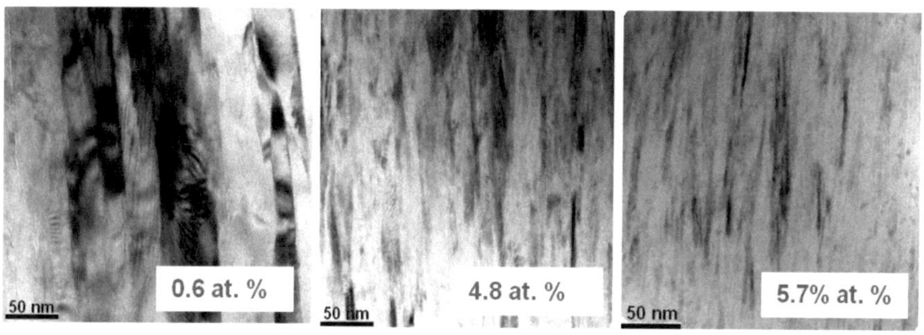

Figure 1. TEM images showing the evolution of the Zr-Si-N films morphology with increasing Si content.

3. Model for the Me-X-N film formation

The sketch given in Fig. 3 illustrates the growth model for the formation of Me-X-N ternary system. As a function of the X content, three X concentration regions can be identified. In the case of PVD deposition techniques such as magnetron sputtering, the film growth is frequently made out of thermodynamical equilibrium. Consequently the addition of X atoms in small quantities into the MeN lattice presents the introduction of structural points defects (substitutions, interstitials, vacancies), which might perturb the crystallite growth. This region 1 is called *Region* 1 or the region of pseudo-solubility of X atoms in MeN. The limit of the pseudo-solubility α_X of X depends on deposition conditions (substrate temperature and bias). Once the X content exceeds α_X the additional X atoms increasingly segregate and accumulate at the grain boundary regions. This concentration region is denoted as *Region* 2, in this region the surface of each X crystallite is progressively coated by a growing XN_y tissue layer up to a certain limit, referred to as the so-called X coverage level, X_{cov}. When X_{cov} =1 (full coverage), further increase of the X content leads to the formation of ultrathin XN_y layer surrounding completely the surface of the MeN crystallite and hindering the crystallite growth. Thus, in the *Region* 3 the, microstructure is strongly altered as a consequence of X segregation.

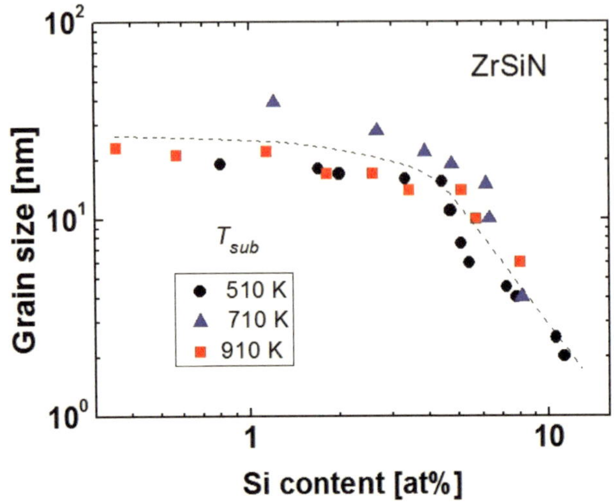

(a)

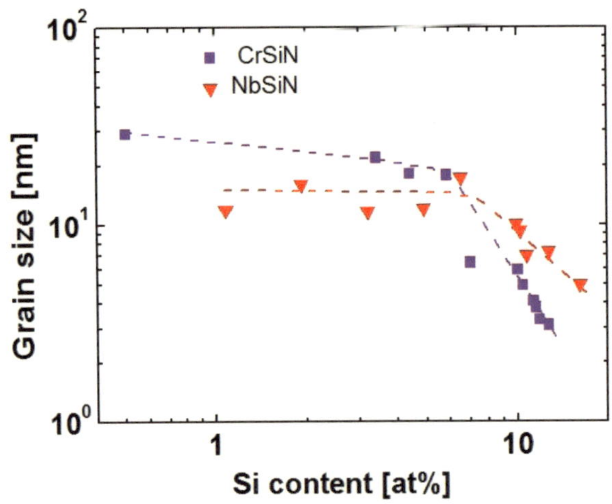

(b)

Figure 2. (a)Grain size vs. Si content in Zr-Si-N films. (b) Grain size vs. Si content in Cr-Si-N and Nb-Si-N thin films.

The degree of X surface-coverage of a crystallite of a typical size d can be determined in terms of C_X and C_{Me} concentrations considering a simple model. In a cubic shaped crystallite of volume $V_C = d^3$. For a fcc-NaCl-type structure each unit cell of volume a^3 contain 4 atoms, then the density of Me atoms in V_C is given by $N_{Me/Vc} = (4/a^3)d^3$ while its surface density is $N_{Me/Surf} = (2/a^2)(6d^2)$ (a is the lattice constant). The relation between the number of Me surface atoms and that of the volume is $N_{Me/Surf}/N_{Me/Vc} = (3\ a/d)$. Under the assumption that the segregated X atoms occupy the surface Me sites, the degree of the X coverage (X_{cov}) in terms of C_X and C_{Me} atomic per cent is then given by

http://dx.doi.org/10.5772/51123

$$X_{cov} = \left(\frac{N_{X/Surf}}{N_{Me/Surf}} \right) = \frac{N_{X/Surf}}{N_{Me/Vc} \left(3\frac{a}{d} \right)} = \frac{(C_X - \alpha_L)}{(C_{Me} + \alpha_L)\left(3\frac{a}{d} \right)} \tag{1}$$

In equation (1) the quantity α_L is the limit X solubility and takes into account of the amount of X atoms that are incorporated in the MeN:Si crystal lattice. As state above, in Me-X-N system the films generally exhibit a pronounced needle-like, columnar structure, with elongated crystallites where the length to width ratio higher than L/d=10 is observed. For such a situation, the relation (1) can be easily modified by introducing the vertical grain extension *L* as an integer multiple of the in plane crystallite dimension *L*=n*d*, note that in case of cubic-shaped crystallites n=1. Therefore the Si coverage for elongated crystallite is

$$X_{cov} = \frac{(C_X - \alpha_L)}{(C_{Me} + \alpha_L)\left(2 + \frac{1}{n} \right)\left(\frac{a}{d} \right)} \tag{2}$$

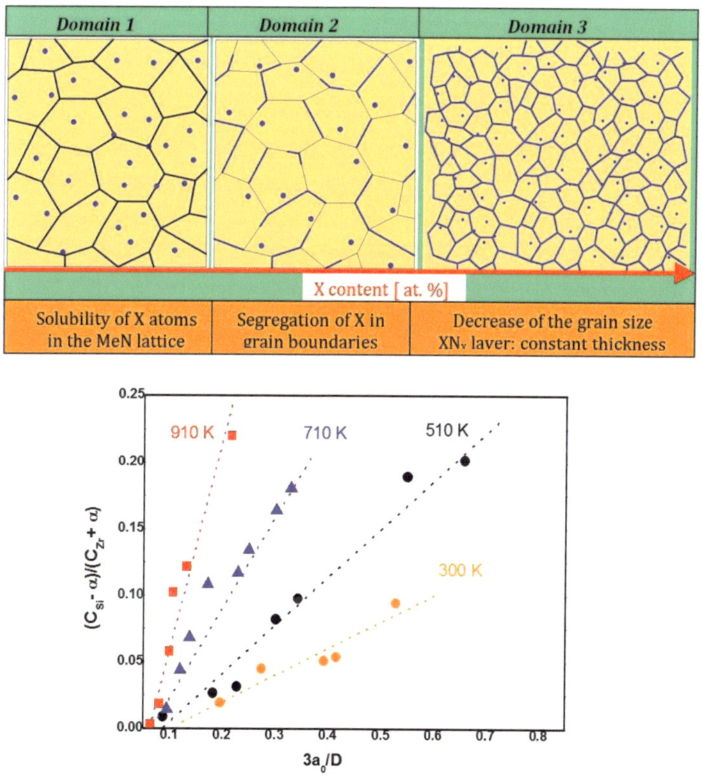

Figure 3. (a) Physical model describing the evolution of nanostructure with increasing X element content. (b) Correlation between secondary phase segregation at the grain boundaries and nanostructure in Zr-Si-N films deposited at various temperatures.

Interestingly, the relation (1,2) predicts that if X_{cov} remains constant, the average crystallite size d and the X content follow a linear relationship $C_X \approx cte \times \frac{1}{d}$, which is observed in many Me-X-N systems. Fig. 3b for example illustrates that the dependence of $\frac{(C_X - \alpha_L)}{(C_{Me} + \alpha_L)}$ on $\frac{3a}{d}$ for the ZrSiN films is linear and that X_{cov} can be evaluated from the slope of the curve.

4. Electrical properties

The electrical resistivity is strongly dependent on the film nanostructure. Not only the type of polycrystalline major phase and grain boundary phase (metal-like conductor, semiconductor or insulator), but also the grain size of crystalline phase, the thickness of grain boundary phase and the global film density are the main parameters that influence the resistivity of nanocomposite thin films. The thicknesses of the minority grain boundary phase (such as SiN_x, a-C, BN, or $TiGe_y$) can be calculated using the model for the film formation described in the section 3. Due to the fact that the charge carrier scattering is very sensitive to grain size and nature of the grain boundary regions, it should more convenient to plot the d.c. electrical resistivity values as a function of the grain size rather than to consider the atomic concentration C_X of the minority phase.

In the results presented in this section, the reported grain size values were obtained from XRD measurements, most of which were acquired in grazing incidence configuration. This value represents the mean value of the crystallite size in an oblique direction at about 15°-30° with respect to the film normal, the grain size values obtained from grazing incidence XRD are much closer to the lateral size of the crystallites. So, to a first approximation, these values could be considered as more suitable for calculating electrical parameters, due to the fact that the electrical resistivity is measured in the plane of the film. Obviously, some adjustment could be made in order to take into account the real lateral size of the crystallites, which can be obtained from TEM in cross-section.

4.1. Nanostructure and RT d.c. electrical resistivity

Depending on the atomic concentration of the minority phase and on the chemical composition of the main crystalline phase, the room temperature (RT) resistivity of MeXN nanocomposites can change over two or more orders of magnitude. It is worth noting that rather to plot the RT resistivity as a function of the atomic concentration of the minority phase it is more instructive to represent the RT resistivity as a function of the grain size in order to extricate the contribution of the structural film modification on the carriers transport properties. In this section, we will consider the nature of composites, how they can be classified from their dc electrical resistivity behavior, how these reflect the electrical properties of the constituent materials, and, in the next section, to what extent they can be modelled. Depending on the electrical nature of the polycrystalline mayor phase (metal-like conductor or semiconductor) and grain boundary tissue phase (conductor or isolator) three types of nanocomposites will be discussed: metal-like conductor/insulator (M-I), metal-like conductor/conductor (M-M), and semiconductor/insulator (S-I).

4.1.1. Metal-like conductor/Insulator (M-I) interfaces

The room temperature electrical resistivity of Zr-Si-N films, deposited at various temperatures and bias voltages are shown in Fig. 4a [19]. The influence of the crystallite size on the resistivity is clearly observed in the case of films deposited without bias at 510, 710 and 910 K. These films present a nanocomposite structure nc-ZrN/a-SiN_x. The formation of an amorphous SiN_x insulating (a-SiN_x) layer on the ZrN nanocrystallite (nc-ZN) surface is responsible for significant increases in resistivity only for the films with silicon coverage Si_{cov} greater than 0.5 ML. It should be mentioned that 0.5 ML coverage layer corresponds to 1 ML of SiN_x between two adjacent ZrN crystallites. Wherever such SiN_x layers (thicker than 1.0 ML) are formed, it is observed a significant gap between the resistivity values at the same crystallite size value but at different values for SiN_x thickness. The effect of grain boundary scattering on film resistivity is enhanced as grain size is decreased. This corresponds to the increase of the gap between the resistivity values of films deposited at 510 K, 710 K and 910 K with crystallite size reduction. Thus, the grain boundary scattering is enhanced in the case of the films showing higher SiN_x surface coverage. The increase in resistivity with increasing Si content related to the formation of nanocomposite material showing an insulating and continuum layer between conducting nanocrystallites, has been reported in Zr-Si-N (nc-ZrN/SiN_x) [14], Nb-Si-N (nc-NbN/SiN_x) [18], Ta-Si-N (nc-TaSiN/SiN_x) [36] and Ti-B-N films (nc-TiB/BN) [37].

4.1.2. Conductor/Conductor (M-M) interfaces

In the case of M-M nanocomposites, the presence of a different conducting phase at the grain boundaries of conducting crystallites, does not strongly affect the resistivity behavior. Small changes in the densification, chemical composition of the films and high density of point defects at the grain boundary regions could induce the observed variations. In nitrogen-deficient or nitrogen-rich binary $MeN_{1\pm x}$ thin films, the resistivity can strongly depend on the chemical composition. Thus, in $ZrN_{1\pm x}$ and $TaN_{1\pm x}$ large variations of the resistivity (one to two orders of magnitude) as a function of the N content are observed. The N-deficiency also affects the electrical properties of Me-Si-N nanocomposites. For example, in the case of N-deficient $(ZrSi)_yN_x$ (with $x \leq 0.5$) films deposited at RT without bias, or at 300 K and 510 K with -150 V bias, it is observed that the resistivity does not change significantly with decreasing grain size (increasing Si content) as shown in Fig. 4a. The Si compositional independent behavior of the resistivity is supposed to be originated from direct percolation of the conducting ZrN_{1-x} crystallites and/or ZrN_{1-x} crystallites separated by low degree of nitridation of the SiN_x grain boundary phase. In fact, the Si_{cov} surface coverage was found to be too small (about 0.3 ML) to completely encapsulate the ZrN crystallites. We have also obtained similar results on N-deficient $(TaSi)_yN_x$ nanocomposite films [36]. Resistivity Si compositional independent behavior was also reported for nanocomposite TiN/SiN_x films by Jedrzeovski [38].

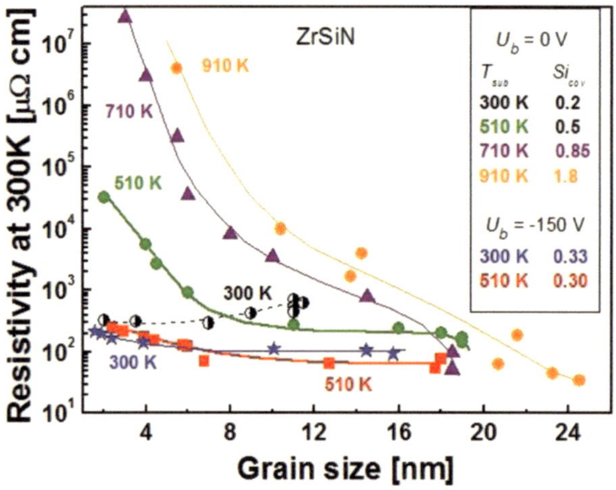

(a)

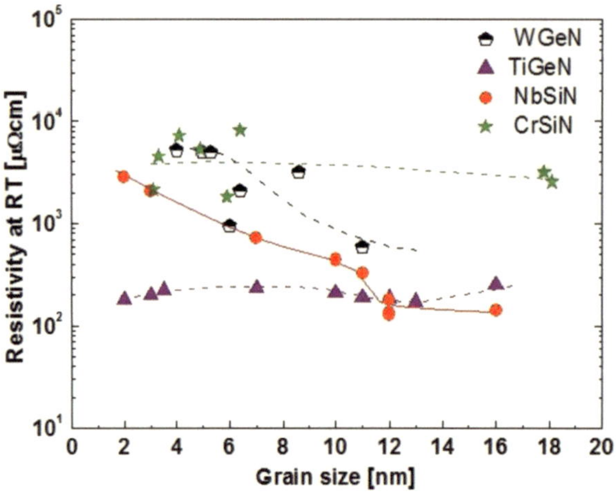

(b)

Figure 4. a) Resistivity vs. grain size for Zr-Si-N films deposited at various temperatures and biases. (b) Resistivity vs. grain size for various films.

Furthermore, by comparing the evolution of the resistivity with decreasing grain size for Ti-Ge-N and W-Ge-N composite films, a different behavior is observed (Fig. 4b). This difference gives us information about the electrical nature of the grain boundary phase: conducting $TiGe_x$ phase in the case of Ti-Ge-N films [32] and insulating GeN_x phase in the case of W-Ge-N films [33], which is similar to the insulating SiN_x phase in Nb-Si-N films. In the case of the WC_x-C films, changes in the phase composition from nc-W_2C/nc-WC to nc-WC/a-C are responsible for resistivity variation correlated to the variation of the crystallite size and the presence of high density of point defects [34]. The situation is similar for TiBC films though the presence of three phases, nc-TiB, nc-TiC and a-C, and the large solubility of B in TiC make it difficult the interpretation of results [35]. In WC-C and TiBC nanocomposites, the grain boundary regions composed of a-C do not play a significant role. The main free path of the electrons is mainly limited by the high density of point defects in the amorphous samples whilst lattice defects and grain size predominate in presence of nanocrystalline binary or ternary phases [34,35].

4.1.3. Semiconductor/Insulator (S-I) interfaces

Some MeN such as ScN and CrN are semiconductors. As far as we know, the electrical properties of ScN/SiN_x composites have not been published. In the case of CrN/a-SiN_x system, variation of resistivity with the grain size was also observed [39]. But in the case of a semiconductor material, small variation in the chemical composition of CrN_x crystallites strongly influences the electrical resistivity of the film as shown in Fig. 4b. This could explain the dispersion of the points for the same value of the grain size. This case is the most difficult to model unambiguously. The temperature dependence of the intrinsic resistivity in semiconductor materials masks the temperature dependence of the grain boundary scattering.

4.2. Temperature dependence of d.c. resistivity.

Measuring the electrical resistivity as a function of the temperature gives further information on the main mechanisms responsible of the charge carriers scattering linked to structural changes due to the addition of the second constituent. Fig. 5 shows the temperature dependent d.c. electrical resistivity $\rho(T)$ curves of NbSiN films deposited at 510 K as a function of the Si content [18]. The $\rho(T)$ curves progressively change from metallic-like to non-metallic-like behavior as the Si content in the films increases. These characteristic trends are often observed in (M-I) type of nanocomposites as a function of the concentration of the insulating minority phase. Fig. 6 a and 6b shows few $\rho(T)$ curves of selected nanocomposite films such as ZrSiN, TiGeN, WC-C, and TiBC for specific grain size. In Fig. 6c are presented $\rho(T)$ curves of $Cr_{0.92}Si_{0.08}N_{1.02}$ and CrN_y for $0.93 \leq y \leq 1.15$. Detailed results concerning temperature dependent electrical resistivity can be found in [18] for NbSiN, in [34, 35] for WC-C and TiBC, and in [19] for ZrSiN.

In the case of (M-I) nanocomposites (Fig. 6a), the temperature dependence of resistivity can easily be correlated with film nanostructure (grain size and thickness of the insulating phase). The effect of the electron scattering at grain boundaries is enhanced by the presence of a thin insulating barrier. Thus, the resistivity $\rho(T)$ of Zr-Si-N films with large grain size exhibits metallic behavior (see [19]) while those having small grains exhibit a negative temperature coefficient of resistivity (TCR $=\frac{1}{\rho}\frac{\partial \rho}{\partial T}$). Similar behavior was reported by Piloud in the case of TiBN films [40]. In all these works the authors correlates the negative TCR with the diminution of the crystallite size and the presence of an insulating phase between conducting crystallites.

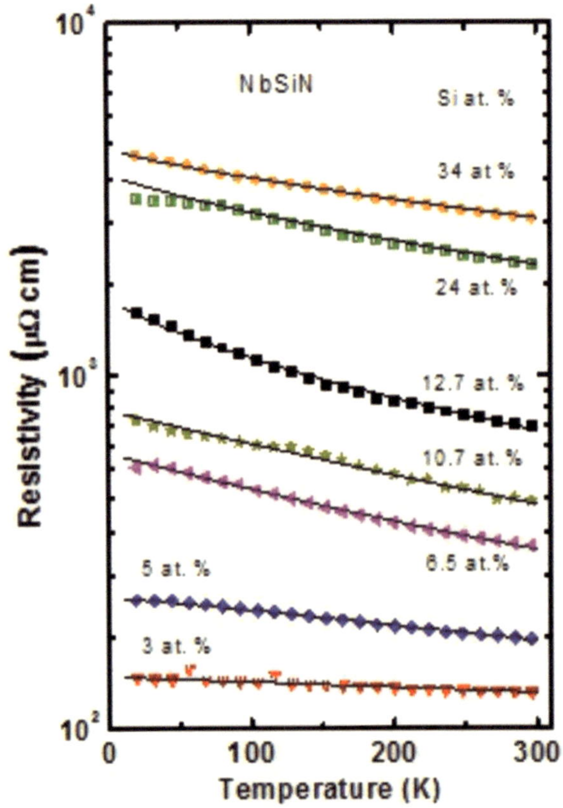

Figure 5. Resistivity vs. Temperature variation for Nb-Si-N films with various Si content.

In the case of (C-C) nanocomposite (TiGeN, WC-C and TiBC), the temperature dependence of resistivity is flat, so the TCR values are low (see Fig. 6b). The resistivity variations for 3 types of films having grain size of about 3nm are similar. The resistivity variation behavior cannot be correlated with the thickness of the phase present at grain boundaries, probably because of a high transmission probability G of charge carriers at the grain boundaries. The absence of the energy gap at GB should be responsible for this.

In the case of (S-I) nanocomposites (CrSiN) the temperature dependence of resistivity (Fig. 6c) cannot easily be correlated with film grain size and scattering probability, because the dependence of polycrystalline semiconducting materials on temperature masks the nanostructure related effects. The change in the N content in CrN_x crystallites significantly influences the resistivity behavior. The resistivity behavior of CrN_x changes from metallic to semiconducting with increasing N content. The formation of a nanocomposite CrN/SiN film with an insulating SiN_x phase between semiconducting CrN crystallites could explain the further increase in resistivity.

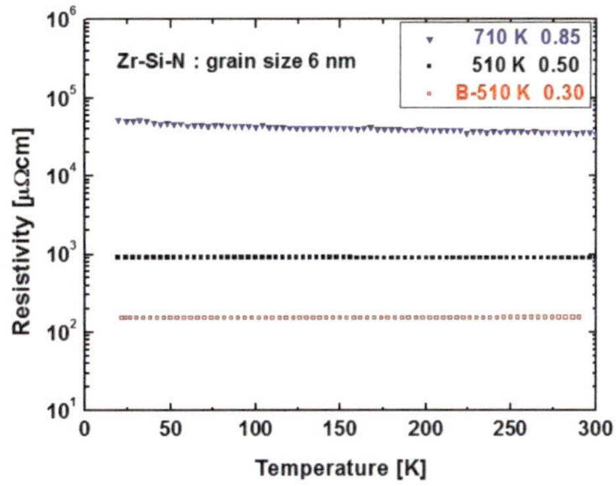

(a)

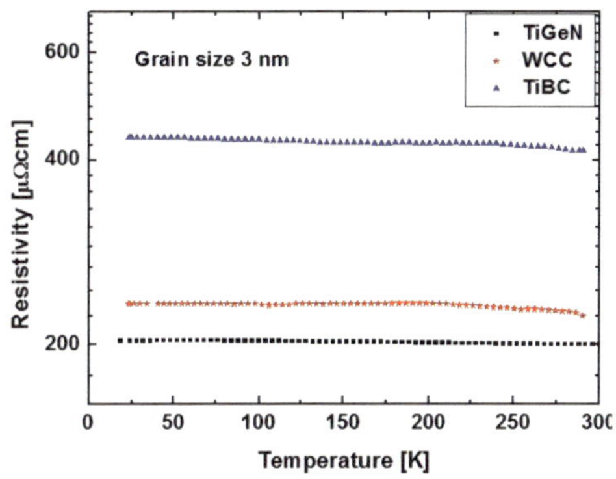

(b)

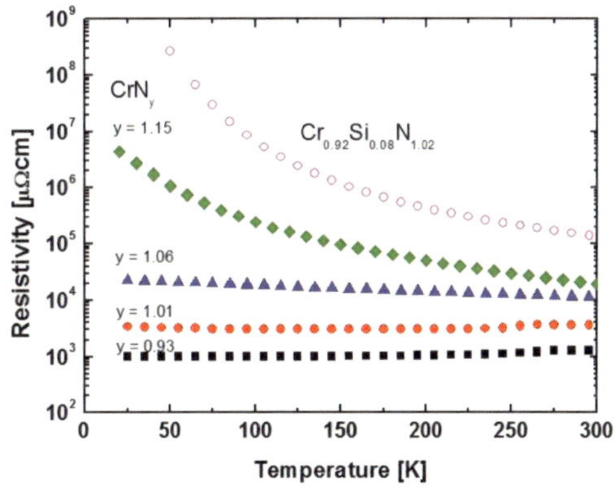

(c)

Figure 6. a) Resistivity vs. Temperature variation for Zr-Si-N films with a grain size of 6 nm. (b) Resistivity vs. Temperature variation for various films with a grain size of 3 nm. (c) Resistivity vs. Temperature variation for CrN films with various N/Cr atomic ratios.

5. Grain boundary scattering model

It is frequently observed that the electrical conductivity of thin polycrystalline films strongly deviates from that of the corresponding bulk single-crystalline material. The conductivity is reduced, which commonly is explained by a reduction of the mean free path of electrons (mfp), and often a negative coefficient of resistivity TCR is observed. In the case of quasi-amorphous or heavily distorted materials negative TCR values have been explained by the hopping mechanism or by a week localization of a two-dimensional electron system. However, these models cannot explain all negative TCR values. Based on many experimental results G. Reiss, H. Hoffman *et al.* [41] proposed the grain boundary scattering model for the d.c. resistivity of polycrystalline thin film materials. The authors state that all electrons reflected by the grain boundaries along one mfp do not contribute to the resulting current and the reduction of the conductivity depends exponentially on the number of grain boundaries per mfp. In this model, an effective mean free path $L_G = L\,G^{(L/D)}$ is introduced to describe the electron scattering at the grain boundaries including the grain size effect; the d.c. electrical conductivity is given by $\sigma = \sigma_B G^{(L/D)}$ where σ_B is the bulk conductivity, G is the probability for an electron to pass a single grain boundary and D is the mean grain size. Under the condition $L/D \ll 1$ the conductivity is reduced to the Drude conductivity without grain boundary effect. The model also predicts a change of the sign of TCR from positive to negative values when L and G fulfil the condition $(L/D)\ln(1/G) > 2$.

Thus, the dc electrical resistivity is then given by

$$\rho_g = \left(\frac{m_e^* v_F}{Ne^2}\right)\left(\frac{1}{L}\right)G^{-(L/D)} = \left(\frac{K}{L}\right)G^{-(L/D)} \tag{3}$$

where m_e^* is the effective masse of the charge carriers, v_F is the Fermi velocity, N is the density of the charge carriers, D is the grain size parameter, L is the inner-crystalline mean free path and G is the mean probability for electrons to pass a single grain boundary. The inner-crystalline mean free path L, describing the volume scattering of electrons, is limited by a temperature invariant elastic scattering at lattice defects and acoustic phonons, namely l_e, and by the temperature dependent inelastic scattering, l_{in}, $L^{-1} = l_e^{-1} + l_{in}^{-1}$. The inelastic mean free path is approximated by $l_{in} \approx \alpha T^{-p}$ where α and p are material specific constants.

In nanocomposite materials composed of a main polycrystalline phase (TiN, ZrN, NbN, TaN, CrN, WC, etc) and amorphous minority or tissue phase (SiN_x, GeN, TiB, a-C, etc.) the grain size of the main material, and the thickness and nature of the grain boundary regions can be easily tailored by the volume concentration of the minority phase. The equation (3) gives the possibility, by a simple fitting procedure of $\rho(T)$ curves, to obtain pertinent information on the main scattering parameters such as G, D and l_e. For the theoretical modeling via the relation (3), at the first approximation grain sizes obtained from XRD or HRTEM measurements can be used. Regarding the factor $K = \left(\frac{m_e^* v_F}{N e^2} \right)$, the Fermi velocity and the electron density N are typically in the range of $v_F \approx 1.0\ 10^8$ cm s^{-1} and N= (4-10) 10^{22} cm^{-3}. More precise N values can be obtained from Hall effect or from optical measurements for stoichiometric or defective MeN$_x$ and MeC$_x$. During the fitting procedure, these values can be adjusted to obtain the best fits.

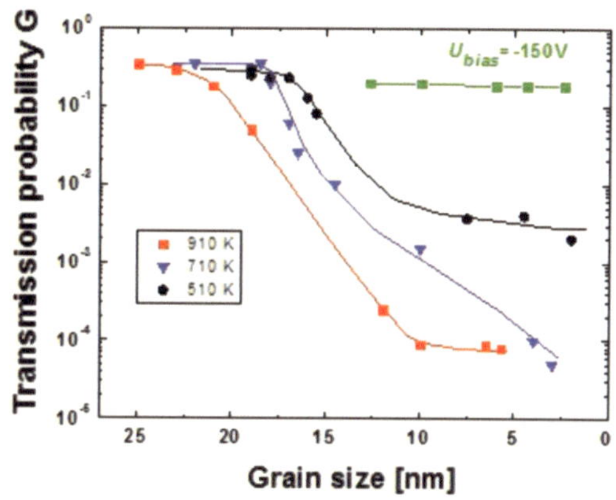

Figure 7. Mean probability G vs. grain size for ZrSiN films deposited at 510, 710 910 K and under bias (lines are added to aid the eye).

The transport mechanisms in nc-NbN/a-SiN$_x$, nc-ZrN/a-SiN$_x$, and nc-TaN/a-SiN$_x$ have been satisfactory described by the grain boundary scattering model. In the case of the Zr-Si-N system, the electrical properties of Zr$_z$Si$_y$N$_x$ films deposited various temperatures, namely, at 300 K (without substrate heating), 510 K, 710 K and 910 K have been investigated in details. It is important to point out that by increasing the substrate temperature the solubility limit of Si, α_L , in the ZrN lattice decreases whereas the Si coverage, Si_{cov}, increases. Thus, the values of the pairs (α_L ,Si_{cov}) are (5 at. %, 0.2), (4 at. %, 0.5), (2 at. %, 0.8) and (1 at. %, 1.8) for the Zr$_z$Si$_y$N$_x$ films deposited at 300 K, 510 K, 710 K and 910 K, respectively. The main probability for electrons to pass the grain boundary G is related to the formation of the SiN$_x$ coverage layer as shown in Fig. 7. In films deposited at 710 K and 910 K, pure ZrN and Zr$_z$Si$_y$N$_x$ films with low Si content (<0.5 at. %) exhibit high G values, G=0.25-0.35. But, for Si content > 1 at%, where the solubility limit is low and the Si coverage important, G deeply decreases down to small values in good correlation with the high thickness values of the SiN$_x$ grain boundary layer (1.6 ML and 3.6 ML, respectively) in these films. For films deposited at 510 K, G decreases slowly and at higher Si content (> 3 at. %) in good agreement with the higher Si solubility and lower thicknesses of the SiN$_x$ layer observed in these films. The electron transmission probability coefficient, G gives us information concerning the continuity and thickness of the insulating phase between conducting grains. In the case of nanocomposites with SiN_x covering layers thinner than 1.0 ML (300 K ZrSiN and 510K ZrSiN with -150 V bias), *G* is larger than 0.05. So, a small scattering probability at grain boundaries implies a small barrier at grain boundaries or the percolation of ZrN crystallites. The effect of the nitrogen content on the electrical nature of the SiN_x grain boundary layer has been investigated in ZrSiN (deposited at 510 K and 710 K with -150 V bias) and in TaSiN films (deposited at 653 K). For N-deficient $(ZrSi)_yN_x$ and $(TaSi)_yN_x$ nanocomposites, the transmission probability *G* remains in the range of 0.1-0.2 over the full investigated Si compositional range(0 – 12 at. %). These results clearly indicate that Si segregation in N-deficient MeSiN films does not lead to the formation of an effective electrically insulating SiN_x layer.

6. SiN_x thickness and resistivity

6.1. Tunneling effect

When two metallic electrodes are separated by an insulating layer (M-I-M structure) the action of the insulating layer is to introduce a potential barrier Φ between the electrodes inhibiting the flow of electrons. However, if the insulating layer is sufficiently thin the current can flow through the insulating region by tunnel effect [42,43]. In the case of electron tunnelling experiments the tunnelling probability is found to be exponentially dependent on the potential barrier width, the tunnelling current is $I_T \propto e^{-\sqrt{\Phi}d} \approx e^{-2.4d}$ and the tunnelling conductance can change by about one order of magnitude for the change $\Delta d \approx 0.1$ nm. Fig. 8 was constructed by considering the thickness of the SiN_x covering layer, as calculated by using the 3-step model for the film formation in the case of ZrSiN films, and the measured resistivity values taken in the region where we have a nanocomposite ZrN/SiN_x structure, as far for the grain size of 4, 6, 8 and 10 nm. The resistivity tends to increases exponentially with the thickness of the SiN_x layer in the range 1.0-3.6 ML (corresponding to a separation distance of 0.2-0.8 nm between metallic crystallites) suggesting that the transport of the electrons across the thin barrier layer seems to occur by tunnelling.

For a M-I-M structure with an insulating layer of thickness *d*, the tunnelling probability T_p for a with a rectangular barrier with an effective barrier height $e\phi_B$ is given by:

$$T_P = exp\left(-2\left[\frac{2m_e^* e\phi_B}{\hbar^2}\right]^{1/2} d\right) \approx exp\left(-\alpha_T\sqrt{\phi_B}d\right) \tag{4}$$

If the effective masse in the insulator is$m_e^* \approx m_e$, the ϕ_B en volts and *d* in Å then α_T=1. The tunnelling conductivity σ_T is given by

$$\sigma_T = \varepsilon_0 \omega_D^2 \tau_T = \left(\frac{Ne^2}{m_e^* v_F}\right) l_e T_P \tag{5}$$

where τ_T is the tunneling relaxation time $\tau_T = \frac{l_e T_P}{v_F}$ and l_e the effective main free path. The tunneling conductivity decreases exponetially with increasing the thickness of the insulating layer. Fig. 9 shows the relationship of the tunneling resistivity and the thickness of the insulating layer with the tunneling probability as calculated from Eq's (4) and (5) for the ZrSiN system. T_P and σ_T have been calculated for two different electron densities N= (1.8-3.6) 10^{22} cm^{-3} for the ZrN, $v_F = 10^8$ cm s^{-1}, l_e=(5-10) nm and for two different effective barrier height values, ϕ_B = 0.6 V and ϕ_B= 1 V. Considering that 1 ML of SiN_x corresponds to about 0.22 nm, the tunneling model predicts that for ϕ_B= 1 V the tunneling probability decreases from 10^{-1} to 10^{-4} and the resistivity increases from about 10^2 μΩ cm up to 10^5 μΩ cm when the thickness of the SiN_x layer change from 1 ML to 4 ML. For lower ϕ_B values, equivalents insulating layers lead to low resistivity values. In Fig. 7 it is shown the transmission probability G, obtained by fitting the ρ(T) experimental curves using the grain boundary scattering model, as a function of the crystallite size (deduced from XRD) for the ZrSiN films deposited at various temperatures. It is worth noting that for films exhibiting comparable crystallite sizes, for instace 12 nm, but with different Si coverages, G values are in the rage of 10^{-1}, 10^{-2} and 10^{-3} corresponding to SiN_x thicknesses of 1ML, 1,6 ML and 3.6 ML, respectively. Though the tunneling conductivity in nanopolycrystallite materials is undoubtedly complexe, the correlation between T_P and G is remarkable. These trends suggest that tunneling conduction should be envolved as one of the the conduction mechanisms responsible for electrons to cross de grain boundary layer between two adjacent crystallites; in particular in the case of elongated crystallites where the length to width ratio higher than 10 have been reported from HRTEM investigations for MeN/SiNx nanocomposites [31,32].

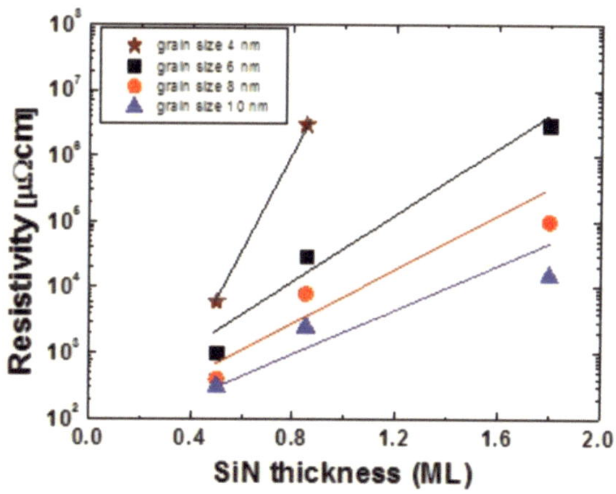

Figure 8. Resistivity vs. thickness of SiN_x interfacial layer: ZrSiN films with 4, 6, 8, and 10 nm crystallite size at 510 K (0.5 ML), 710 K (0.85 ML) and 910 K (1.8 ML). (lines are added to aid the eye).

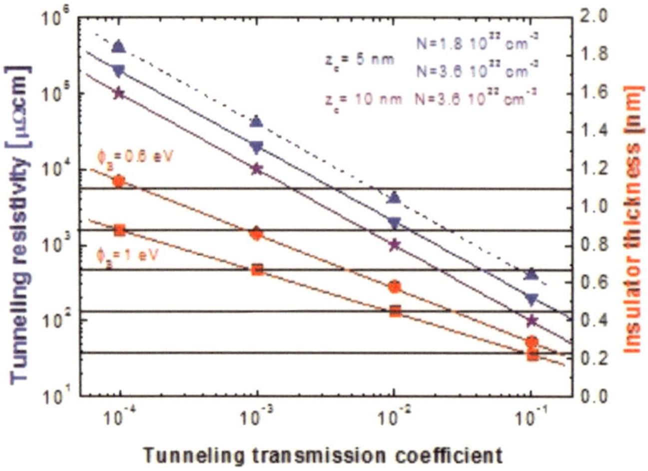

Figure 9. Tunneling resistivity and interfacial insulating thickness vs. tunneling coefficient.

It will be useful to estimate the barrier height $e\phi_B$ in such M-I-M structures. Knowing this value, the transmission probability across metallic-insulating-metallic structures can be calculated as a function of SiN_x thickness. By determining G from fitting the resistivity dependence on temperature, we can extract the SiN_x thickness from electrical measurements. We will not speculate that the calculated G values are sufficiently precise to then extract the energy gap at the grain boundaries. Rather, we would just like to highlight the good correlation between structural and electrical properties.

6.2. I-V characterisation

To investigate if the observed conductivity in SiN_x thin film results from tunnelling of electrons through the SiN_x thin film current-voltage (I-V) measurements should be performed on Me-SiNx-Me structures. For this purpose, SiN_x films sandwiched between ZrN or TaN have been prepared by magnetron sputtering. These structures have been deposited at 740 K and with bias voltage of -150 V in order to obtain smooth surfaces leading to relatively sharp interfaces. The I-V characteristics for $ZrN/SiN_x/ZrN$ structures with different SiN_x thicknesses are show in Fig. 10. The effect of the SiN_x thickness is clearly noticed by comparing the I-V curves with that of the structure without SiN_x layer. For SiN_x thicknesses small than 5 nm (namely the ultrathin regime) the I-V curves show ohmic behaviour, while for thicknesses higher or equal to 5 nm the I-V curves exhibit a symmetric non-linear behaviour. Similar curves have been also observed in the case of TaN/SiNx/TaN structures. The linear behaviour observed in the ultrathin regime can be interpreted in terms of electron tunneling process. A Poole-Frenkel type resistance describes the S-shaped curves, often observed in thin films. In the case of ideal symmetric M-I-M structure the tunnelling current for V< ϕ_B is given by [42]

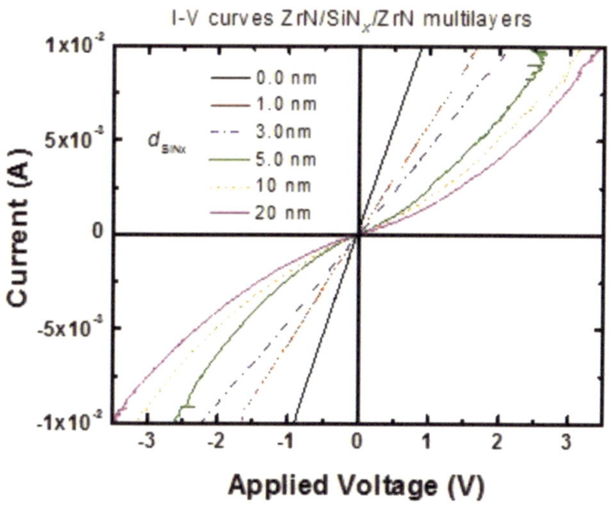

Figure 10. I-V curves in ZrN/SiN/ZrN multilayer films with various insulating SiN layer thicknesses.

$$I = I_0\left[(\phi_B - V/2)exp(-A\sqrt{(\phi_B - V/2)}) - (\phi_B + V/2)exp(-A\sqrt{(\phi_B + V/2)})\right] \tag{6}$$

and for low V range

$$I = \frac{(2m\phi_B)^{1/2}e^2}{2h^2 d} Vexp\left[-2d\sqrt{\frac{2me\phi_B}{\hbar^2}}\right] \tag{7}$$

Earlier studies of the current transport mechanisms in silicon nitride thin films, performed on structures such as Au/Si_3N_4/Mo and Au/Si_3N_4/Si, have shown that the current transport is essentially independent of the substrate material, the film thickness and the polarity of the electrodes [44]. In these studies the Si_3N_4 thickness was in the range of 30 to 300 nm. Depending on the ambient temperature and the electric field three different conduction mechanisms have been identified: Ohmic-type, Poole-Frenkel emission and Fowler-Nordheim tunneling. The Poole-Frenkel mechanism is mainly due to field-assisted excitation from traps and is often observed on defective materials while the Fowler-Nordhein conduction depends on free carriers tunnelling through high quality Si_3N_4 at high electric fields. Ohmic conduction was attributed to the hopping of thermally excited electrons from one isolated state to another.

$$\text{Poole-Frenkel} \qquad I_{PF} = C_{PF} V exp(-e\phi_B + aV^{1/2} / kT) \tag{8}$$

$$\text{Fowler-Nordhein} \qquad I_{FN} = C_{FN} V^2 exp(-b / V) \tag{9}$$

$$\text{Ohmic-type} \qquad I_{Om} = C_{Om} V exp(-e\phi_O / kT) \tag{10}$$

Tao et al [45] have been investigated the effect of N vacancies (Si-Si bonds) and O substitutions (Si-O bonds) on the current-transport properties of $SiN_{1.06}$, $SiN_{1.33}$ and $SiO_{1.67}N_{0.22}$ thin films. The thickness of the Si nitride and of the Si oxynitride layers in Al/SiNO/Si/In structures was typically 15 nm. The results of these studies have been well correlated with the nature of the insulating layer. Thus, all the films exhibit an Ohmic regimen at low electrical fields. The ohmic resistivity depends on the nature of the film; Si-rich films exhibit lower resistivity values while oxynitrides films show the highest values, as the carriers are generated by thermal excitation from traps it was concluded that the density of traps is higher in Si-rich films than in oxynitrides. At intermediate and high electrical fields, Poole-Frenkel emission is the dominant conduction mechanism in Si-rich SiN_x films whereas Fowler-Nordhein tunnelling is mainly involved in oxynitrides films but absent in Si-rich films. Both Poole-Frenkel (at intermediate electrical fields) and Fowler-Nordhein (at high fields) mechanisms are present in nearly stoichiometric Si_3N_4 films.

Based on all these studies we can conclude that the tunnelling current-transport in ultrathin SiN_x layers is very sensitive to N vacancies and to the presence of oxygen atoms. Therefore, Nc-MeN/a-SiN_x nanocomposite thin films containing silicon nitride layers with similar thicknesses but with different chemical composition (sub-stoichiometric or nearly stoichiometric Si_3N_4, oxynitride) can exhibit different electrical properties. Thus, the effects of N-deficiency on the electrical properties of ZrN/SiN_x and TaN/SiN_x nanocomposites as discussed in the section 4 can be interpreted in terms of the presence of high density of free carriers at the grain boundaries thereby leading to high tunnel currents. In addition, the difficulty with real interfaces in thin films is that even if the chemical composition were well controlled surface roughness would increase the local electrical field rising up unexpectedly the tunnel currents.

7. Conclusion

Nanocomposite materials present a high degree of complexity due to small grain size, high curvature radius of nanocrystallites and, in general, a very thin minority phase layer situated at the grain boundaries. Correlating electrical resistivity measurements with film nanostructure provides information concerning the thickness and continuity of the interfacial layer covering conducting nanocrystallites in conducting-insulating nanocomposite films. Aside from some constraints, the possibility to measure experimentally, albeit indirectly, such small interfacial layer thicknesses constitutes an important breakthrough in precise characterization of such nanostructures.

http://dx.doi.org/10.5772/51123

Acknowledgements

The authors wish to thank the Swiss National Science Foundation and the EPFL for financial support.

Author details

R. Sanjinés* and C. S. Sandu

*Address all correspondence to: rosendo.sanjines@epfl.ch

EPFL-SB-ICPM-LPMC, Ecole Polytechnique Fédérale de Lausanne, CH-1015 Lausanne, Switzerland

References

[1] Toth, . E. (1971). *Transition Metal Carbides and Nitrides*, Academic, New York.

[2] Holleck, H., & , J. (1986). *Vac. Sci. Technol. A*, 4, 2661.

[3] Kaloyeros, A. E., & Eisenbraun, E. (2000). *Annu. Rev. Mater. Sci.*, 30, 363.

[4] Riekkinen, T., Molarius, J., Laurila, T., Nurmela, A., Suni, I., & Kivilauhti, J. K. (2002). *Microelectron. Eng.*, 64, 289.

[5] Rossnagel, S. M. (2002). *J. Vac. Sci. Technol.*, B20, 2328.

[6] Wittmer, M. (1980). *Applied Physics Letters*, 36, 456.

[7] Daughton, J. M. (1992). *Thin Solid Films,*, 216, 162.

[8] Sun, X., Kolawa, E., Chen, J., Reid, J. S., & Nicolet, M. A. (1993). *Thin Solid Films*, 236, 347.

[9] Leng, Y. X., et al. (2001). Thin Solid Films, , 398-399.

[10] Wallrapp, F., & Fromherz, P. (2006). *J. Appl. Phys.*, 99, 114103.

[11] Diserens, M., Patscheider, J., & Lévy, F. (1998). *Surf. Coat. Technol.*, 108-109.

[12] Veprek, S., et al. (1999). *J. Vac. Sci. Technol.*, A 17, 2401.

[13] Musil, J. (2000). *Surf. Coat. Technol.*, 125, 322.

[14] Pilloud, D., Pierson, J. F., Marques, A. P., & Cavaleiro, A. (2004). *Surf. Coat. Technol.*, 180-181.

[15] Veprek, S., Maritza, J. G., & Veprek-Heijman, . (2008). *Surf. Coat. Technol.*, 202, 5063.

[16] Sandu, C. S., Sanjinés, R., Benkahoul, M., Medjani, F., & Lévy, F. (2006). *Surf. Coat. Technol.*, 201, 4083.

[17] Martinez, E., Sanjinés, R., Banakh, O., & Lévy, F. (2004). *Thin Solid Films*, 447-448.

[18] Sanjinés, R., Benkahoul, M., Sandu, C. S., Schmid, P. E., & Lévy, F. (2005). *J. Appl. Phys.*, 98, 123511.

[19] Sandu, C. S., Medjani, F., & Sanjinés, R. (2007). *Rev. Adv. Mater. Sci.*, 15-173.

[20] Sandu, C. S., Harada, S., Sanjinés, R., & Cavaleiro, A. (2010). *Surf. Coat. Technol.*, 204, 1907.

[21] Reid, J. S., Kolawa, E., Ruiz, R. P., & Nicolet, M. A. (1993). *Thin Solid Films*, 236, 319.

[22] Kim, D. J., Kim, Y. T., & Park, J. W. (1997). *J. Appl. Phys.,*, 82, 4847.

[23] Lee, Y. J., Suh, B. S., Kwom, M. S., & Park, C. O. (1999). *J. Appl. Phys.*, 85, 1927.

[24] Suh, Y. S., Heuss, G. P., & Misra, V. (2002). *Appl. Phys. Lett.*, 80, 1403.

[25] Letendu, F., Hugon, M. C., Agius, B., Vickridge, I., Berthier, C., & Lameille, J. M. (2006). *Thin Solid Films*, 513, 118.

[26] Olowolafe, J. O., Rau, I., Mr, K., Unruh, C. P., Swann, Z. S., Jawad, T., & Alford, . (2000). *Thin Solid Films*, 365, 19.

[27] Hübner, R., Hecker, M., Mattern, N., Hoffmann, V., Wetzig, K., Heuer, H., Wenzel , Ch , Engelmann, H. J., Gehre, D., & Zschech, E. (2006). *Thin Solid Films*, 500, 259.

[28] Cabral, C., Jr, Saenger, K. L., Kotecki, D. E., & Harper, J. M. E. (2000). *J. Mater. Res.*, 15, 194.

[29] Alén, P., Aaltonen, T., Ritala, M., Leskelä, M., Sajavaara, T., Keinonen, J., Hooker, J. C., & Maes, J. W. (2004). *J. Electrochem. Soc.,*, 151, G523.

[30] Jung, K. M., Jung, M. S., Kim, Y. B., & Choi, K. D. (2009). *Thin Solid Films,*, 517, 3837.

[31] Sandu, C. S., Sanjinés, R., & Medjani, F. (2008). *Surf Coat. Technol*, 202, 2278.

[32] Sandu, C. S., Sanjinés, R., Benkahoul, M., Parlinska-Wojtan, M., Karimi, A., & Lévy, F. (2006). Thin Solid Films ., 496, 336.

[33] Piedade, A. P., Gomes, M. J., Pierson, J. F., & Cavaleiro, A. (2006). Surf. Coat. Technol ., 200, 6303.

[34] Abad, M. D., Sánchez-López, J. C., Cusnir, N., & Sanjinés, R. (2009). *Journal of Applied Physics*, 105, 033510.

[35] Abad, M. D., Sanjinés, R., Endrino, J. L., Gago, R., Andersson, J., & Sánchez-López, J. C. (2011). *Plasma Process and Polymers*, 8, 579.

[36] Oezer, D., Ramirez, G., Rodil, S. E., & Sanjinés, R. submitted to J.A.P

[37] Pierson, J. F., Bertran, F., Bauer, J. P., Jolly, J., & Surf, Ž. (2001). *Coat.Technol.*, 142-144.

[38] Jedrzejowski, P., Baloukas, B., Klemberg-Sapieha, J. E., & Martinu, L. (2004). *J. Vac. Sci. Technol. A*, 22, 725.

[39] Martinez, E., Sanjinés, R., Banakh, O., & Lévy, F. (2004). *Thin Solid Films*, 447-448.

[40] Pilloud, D., Pierson, J. F., & Pichon, L. (2006). *Materials Science and Engineering B*, 131, 36.

[41] Reiss, G., Vancea, J., & Hoffman, H. (1986). *PhysRev.Lett*, 56, 2100.

[42] Simmons, J. G. (1963). *J. Appl. Phys.,*, 34, 1793.

[43] Fisher, J. C., & Giaever, I. (1961). *J. Appl. Phys.*, 32172.

[44] Sze, S. M. (1967). *J. Appl. Phys.*, 38, 2951.

[45] Tao, M., Park, D., Mohammad, S. N., Li, D., Botchkerav, A. E., & Morkoç, H. (1996). *Phil. Mag. B,*, 73, 723.